Combating and Controlling Nagana and Tick-Borne Diseases in Livestock

Caleb Oburu Orenge
Department of Veterinary Anatomy and Physiology, Faculty of Veterinary Medicine and Surgery, Egerton University, Kenya

A volume in the Advances in Environmental Engineering and Green Technologies (AEEGT) Book Series

IGI Global
PUBLISHER of TIMELY KNOWLEDGE

Published in the United States of America by
IGI Global
Engineering Science Reference (an imprint of IGI Global)
701 E. Chocolate Avenue
Hershey PA, USA 17033
Tel: 717-533-8845
Fax: 717-533-8661
E-mail: cust@igi-global.com
Web site: http://www.igi-global.com

Library of Congress Cataloging-in-Publication Data

Names: Orenge, Caleb Oburu, 1964- editor.
Title: Combating and controlling nagana and tick-borne diseases in livestock / Caleb Oburu Orenge.
Description: Hershey, PA : Business Science Reference, [2021] | Includes bibliographical references and index. | Summary: "The purpose of the book will provide an update on African Animal Trypanosomiasis and tick-borne diseases and provide the latest empirical research findings in the area of African Animal Trypanosomiasis (Nagana) and tick-borne diseases affecting Livestock"-- Provided by publisher.
Identifiers: LCCN 2020020899 (print) | LCCN 2020020900 (ebook) | ISBN 9781799864332 (hardcover) | ISBN 9781799864349 (paperback) | ISBN 9781799864356 (ebook)
Subjects: LCSH: Trypanosomiasis in animals--Africa. | Trypanosomiasis in cattle--Africa. | Tick-borne diseases in animals--Africa. | Livestock--Trypanotolerance--Africa.
Classification: LCC SF807 .C763 2020 (print) | LCC SF807 (ebook) | DDC 636.089/4533--dc23
LC record available at https://lccn.loc.gov/2020020899
LC ebook record available at https://lccn.loc.gov/2020020900

This book is published in the IGI Global book series Advances in Environmental Engineering and Green Technologies (AEEGT) (ISSN: 2326-9162; eISSN: 2326-9170)

British Cataloguing in Publication Data
A Cataloguing in Publication record for this book is available from the British Library.

All work contributed to this book is new, previously-unpublished material. The views expressed in this book are those of the authors, but not necessarily of the publisher.

For electronic access to this publication, please contact: eresources@igi-global.com.

Advances in Environmental Engineering and Green Technologies (AEEGT) Book Series

Sang-Bing Tsai
Zhongshan Institute, University of Electronic Science and Technology of China, China & Wuyi University, China
Ming-Lang Tseng
Lunghwa University of Science and Technology, Taiwan
Yuchi Wang
University of Electronic Science and Technology of China Zhongshan Institute, China

ISSN:2326-9162
EISSN:2326-9170

Mission

Growing awareness and an increased focus on environmental issues such as climate change, energy use, and loss of non-renewable resources have brought about a greater need for research that provides potential solutions to these problems. Research in environmental science and engineering continues to play a vital role in uncovering new opportunities for a "green" future.

The **Advances in Environmental Engineering and Green Technologies (AEEGT)** book series is a mouthpiece for research in all aspects of environmental science, earth science, and green initiatives. This series supports the ongoing research in this field through publishing books that discuss topics within environmental engineering or that deal with the interdisciplinary field of green technologies.

Coverage

- Water Supply and Treatment
- Sustainable Communities
- Biofilters and Biofiltration
- Policies Involving Green Technologies and Environmental Engineering
- Electric Vehicles
- Cleantech
- Alternative Power Sources
- Waste Management
- Green Transportation
- Pollution Management

IGI Global is currently accepting manuscripts for publication within this series. To submit a proposal for a volume in this series, please contact our Acquisition Editors at Acquisitions@igi-global.com or visit: http://www.igi-global.com/publish/.

The Advances in Environmental Engineering and Green Technologies (AEEGT) Book Series (ISSN 2326-9162) is published by IGI Global, 701 E. Chocolate Avenue, Hershey, PA 17033-1240, USA, www.igi-global.com. This series is composed of titles available for purchase individually; each title is edited to be contextually exclusive from any other title within the series. For pricing and ordering information please visit http://www.igi-global.com/book-series/advances-environmental-engineering-green-technologies/73679. Postmaster: Send all address changes to above address.

Titles in this Series

For a list of additional titles in this series, please visit: www.igi-global.com/book-series

Artificial Intelligence and IoT-Based Technologies for Sustainable Farming and Smart Agriculture
Pradeep Tomar (Gautam Buddha University, India) and Gurjit Kaur (Delhi Technological University, India)
Engineering Science Reference • ©2021 • 400pp • H/C (ISBN: 9781799817222) • US $215.00

Handbook of Research on Waste Diversion and Minimization Technologies for the Industrial Sector
Ashok K. Rathoure (M/s Akone Services, Lucknow, India)
Engineering Science Reference • ©2021 • 455pp • H/C (ISBN: 9781799849216) • US $295.00

Smart Farming Techniques and Knowledge-Based Agriculture Emerging Research and Opportunities
Petraq Papajorgji (European University of Tirana, Albania) and Francois Pinet (Irstea Centre de Clermont-Ferrand, France)
Engineering Science Reference • ©2021 • 230pp • H/C (ISBN: 9781799845409) • US $195.00

Solar Concentrating Modules With Louvered Heliostats Emerging Research and Opportunities
Dmitry Strebkov (Federal State Budget Scientific Institution, Russia & Federal Scientific Agroengineering Center VIM, Russia) Natalya Filippchenkova (Federal State Budget Scientific Institution, Russia & Federal Scientific Agroengineering Center VIM, Russia) and Anatoly Irodionov (Federal State Budget Scientific Institution, Russia & Federal Scientific Agroengineering Center VIM, Russia)
Engineering Science Reference • ©2021 • 267pp • H/C (ISBN: 9781799842767) • US $190.00

Smart Agricultural Services Using Deep Learning, Big Data, and IoT
Amit Kumar Gupta (Amity University Jaipur, India) Dinesh Goyal (Poornima Institute of Engineering and Technology, India) Vijander Singh (Manipal University Jaipur, India) and Harish Sharma (Rajasthan Technical University, India)
Engineering Science Reference • ©2021 • 280pp • H/C (ISBN: 9781799850038) • US $195.00

Eco-Friendly Energy Processes and Technologies for Achieving Sustainable Development
Mir Sayed Shah Danish (University of the Ryukyus, Japan) and Tomonobu Shah Senjyu (University of the Ryukyus, Japan)
Engineering Science Reference • ©2021 • 340pp • H/C (ISBN: 9781799849155) • US $195.00

Precision Agriculture Technologies for Food Security and Sustainability
Sherine M. Abd El-Kader (Electronics Research Institute, Cairo, Egypt) and Basma M. Mohammad El-Basioni (Electronics Research Institute, Cairo, Egypt)
Engineering Science Reference • ©2021 • 437pp • H/C (ISBN: 9781799850007) • US $195.00

IGI Global
PUBLISHER of TIMELY KNOWLEDGE

701 East Chocolate Avenue, Hershey, PA 17033, USA
Tel: 717-533-8845 x100 • Fax: 717-533-8661
E-Mail: cust@igi-global.com • www.igi-global.com

Editorial Advisory Board

Table of Contents

Detailed Table of Contents

African animal trypanosomiasis (AAT) is a major constraint to livestock productivity, particularly in cattle and in camels. This chapter covers some general aspects of the arthropod vectors of animal trypanosomiasis, the tsetse flies Glossina spp., and to a lesser extent the biting flies. This chapter covers the classification, morphology, basic biology, and the eco-distribution of tsetse flies. The role of tsetse flies in disease epidemiology has also been reviewed. The elementary biology of these vectors is quite well known and elucidated. However, with advances in molecular and other biological techniques, new insights related to tsetse biology have been obtained. This chapter will revisit these basics and include some updated information emanating from research done in the recent past. The final part of the chapter is devoted to a brief discussion on biting flies, the vectors of T. evansi, which causes camel trypanosomiasis.

The role of the Kenya Tsetse and Trypanosomiasis Eradication Council (KENTTEC) in the control of animal trypanosomiasis is premised on the fact that a large proportion of animal trypanosomiasis in Kenya is tsetse transmitted. Tsetse distribution in Kenya is characterized by eight discontinuous belts defined by topographical, environmental and land-use. KENTTEC's strategy for control of African animal trypanosomiasis is based on use of community-based organizations for spraying of livestock, control of the vector using various devices such as targets and traps, and development of strategies and policies for use of land after the intervention. The council has developed linkage with research institutions for adaptive and operational research. The council has initiated the development of national atlas by mapping tsetse and animal trypanosomiasis distribution in collaboration with stakeholders at the national, regional, and international levels.

This chapter describes the meaning of trypanotolerance and its implication as used with trypanotolerant livestock that are known and studied for many years in West Africa. This trait is observed in many West African breeds of cattle that survive in areas of tsetse fly challenge where the humped zebu cannot. Small ruminants and wildlife also exhibit increased resistance to infection with trypanosomes. Trypanotolerance is viewed as a more sustainable and environmentally friendly approach to control of trypanosomiasis given the limitations of the present methods of control and lack of vaccine. Little work has been conducted with East African cattle breeds. The Kenyan Orma Boran (OB) has been studied since 1983 to enhance its conservation and utilization in East African region. Nucleus breeding herd of superior genotypes was multiplied in Galana Agricultural Development Corporation (ADC) and disseminated to trypanosomosis endemic areas of Kenya. On evaluation, performance of the crosses was comparable to that of other Borans. Challenges of adoption in pastoral area are presented.

The non-tsetse-transmitted animal trypanosomoses are infections caused by three main mammalian trypanosome species, namely Trypanosoma evansi, T. equiprdum, and T. vivax. Their global distribution is much wider than tsetse-transmitted trypanosomoses and includes Africa and Latin America, Asia, and Euro-Asia. These trypanosomoses affect a very wide range of domestic animals and game and negatively impact on the development of the animal industry, thus directly affecting national economies and people's livelihoods.

The tropical and sub-tropical climate in Africa favours multiplication and maintenance of tick vectors and transmission of various pathogens to cattle. Key challenges including acaricide resistance, policy issues, transboundary animal movements, and inadequate veterinary services compromise effective control of tick-borne diseases (TBDs). This chapter discusses important host, pathogen, climatic, and management factors that impact the control of TBDs among cattle in Africa, and which affect the productivity and overall contribution to economic development. The economic losses in cattle production attributed to tick infestation and TBDs in Africa are also reviewed. The use of a sustainable integrated control approach, including vaccination, strategic tick control, surveillance for acaricide resistance, and multi-stakeholder involvement is also evaluated.

Ticks are important ectoparasites of livestock in tropical and sub-tropical countries. They are responsible for the enormous economic losses both through the direct effects of blood sucking and indirectly as vectors of disease pathogens and toxins. Morphologically, ticks are classified into two families known as Ixodidae (hard ticks) and Argasidae (soft ticks). In this chapter, the authors provide a description of the soft and hard ticks. Emphasis is laid on the genera morphology, species, mammalian hosts, and distribution and the diseases transmitted by the ticks species. Various methods for controlling ticks and tick-borne diseases using chemicals are described. A broad range of chemical acaricides used for the control of ticks on livestock are described. Conventional methods and unconventional methods for applying acaricides to protect livestock against ticks are described. Alternative methods for tick control are described. Only the most important hard and soft ticks and associated diseases are discussed within the limits of this review.

East coast fever, a disease of cattle caused by the protozoan parasite Theileria parva and transmitted by the three-host tick Rhipicephalus appendiculatus (the brown ear tick), is a major constraint to cattle production in Eastern, Central, and Southern Africa. In Kenya it is the most important tick-borne disease and a major constraint in cattle productivity. This is due to the high morbidity and mortality it causes in susceptible herds, the cost of control of the vector ticks, and the cost of treatment of clinical cases. Animals that recover from the disease also suffer from reduced productivity which can be long term. The limited distribution of the tick and the disease to only East, Central and Southern Africa also means that the market for therapeutic drugs and acaricides is small. Therefore, drug companies are not keen on funding research and development of new drug and acaricide molecules when resistance occurs.

Bovine anaplasmosis is one of the most important tick borne diseases of ruminants worldwide causing significant economic losses in the livestock industries due to the high morbidity and mortality in susceptible cattle herds. Bovine anaplasmosis, caused by Anaplasma marginale, is an infectious but non-contagious disease. The mode of transmission of bovine anaplasmosis includes mechanical (blood contaminated fomites (needles, ear tagging, dehorning and castration equipment), biological (tick bites) and transplacental (mother to fetus). Bovine Anaplasmosis occurs in tropical and subtropical regions worldwide. Cattle of all ages are susceptible to infection with A. marginale, but the severity of disease increases with age. The common clinical sign of bovine anaplsmosis includes; fever, anorexia, rapid loss of body condition, severe decrease in milk production, pale and icteric mucous membranes, increased heart and respiratory rates, muscle weakness and depression. Diagnosis of bovine anaplasmosis can be made by demonstration of A. marginale on stained blood smears from clinically infected animals during the acute phase of the disease, but it is not reliable for detecting infection in pre-symptomatic or carrier animals. Instead, serological demonstration of antibodies and confirmation of antigen with molecular detection tools are used for diagnosis. Anaplasmosis can be treated by administration of oxytetracyclines, however oxytetracyclines do not clear the parasite from carrier animals. Control measures for bovine anaplasmosis vary with geographical location and include maintenance of Anaplasma free herds, vector control, administration of antibiotics and vaccination. Intensive acaridae application to control ticks has a number of limitations, therefore, immunization together with strategic tick control is recommended for exotic and crossbred cattle. Further studies on epidemiology of bovine anaplasmosis is needed

Heartwater (HW) is an acute, febrile, tick-borne disease of cattle, sheep, goats, and wild ruminants characterized by nervous signs and high mortality. The disease is caused by a rickettsia agent, Erlichia ruminantium, formally classified as Cowdria ruminantium. The disease is transmitted by several ixodid ticks of the genus Ambylomma. Chemoprophylaxis has been used as a method to facilitate the movement of heartwater susceptible stock into heartwater endemic areas while allowing them to acquire immunity by limited tick exposure.

Bovine babesiosis is considered one of the most economically important tick-borne diseases in tropical and subtropical areas. Thus, it is important to understand the disease in terms causative agent, vectors, hosts, and distribution. Babesia is the causative agent of the disease and has a life cycle both in the invertebrate and vertebrate hosts. The major bovine species are B. bovis, B. bigemina, found in the tropics and sub tropics while B. divergens and B. major are found in the temperate regions. Bovine babesiosis is transmitted by the one-host cattle ticks, formerly in the genus Boophilus and now Rhipicephalus. In endemic areas are tick management, immunization, and anti-babesia drugs or by a combination. Chemotherapy is important for controlling the disease either to treat field cases or to control artificially induced infections. Conventional drugs for treating bovine babesiosis are Imidocarb dipropionate and Diminazene aceturate. Ticks are controlled by application of acaricides, but tick resistance and environmental pollution are drawbacks.

Canine ehrlichiosis is an infection of canids causing a clinical disease in domestic dogs and asymptomatic infections in wild canids. Currently, the disease assumes a cosmopolitan distribution. This chapter summarises published information on the disease from across Africa. Some studies were able to demonstrate experimental infections in these canids. Different diagnostic methods, cell culture, direct microscopy, serology, hematology, and molecular methods were employed in different studies for detection of ehrlichiosis. Treatment of the disease mainly involved use of oxytetracycline, doxycycline, imidocarb disproportionate, and levamisole. In severe cases, management has involved administration of supportive therapy such as blood transfusion. Generally, though available, the information on different aspects of the disease in the Africa is scant and fragmented. There is still a need to generate more information on the epidemiology, diagnosis, clinical aspects, and treatment of the disease.

Nairobi sheep disease virus (NSDV) is a one of the viruses transmitted by tick vectors and causes Nairobi sheep disease (NSD) – a disease of small ruminants. The disease was first identified as a problem in sheep and goats in Kenya in 1910. The virus belongs to the genus Nairovirus and family Bunyaviridae. An equally pathogenic variant of NSDV is Ganjam virus, which is found in India and Sri Lanka. Both viruses are closely related to the human pathogen Crimean-Congo haemorrhagic fever virus (CCHFV), a fever transmitted through infected tick bites or through contact with infected animal blood or tissues during and immediately after slaughter. Other members of the genus include Dugbe virus, Hazara virus, and Kupe virus, isolated from cattle ticks in East Africa. The virus is spread by hard (Ixodid) ticks and appears to be dependent on the tick vector for dissemination, with no direct transmission between animals. The transmission of tick-borne viruses (TBV) occurs primarily during tick feeding and is a complex process, known to be promoted by tick saliva constituents.

Ticks are distributed worldwide and have an enormous medical and veterinary importance owing to the direct damage they cause and indirectly as vectors of a large variety of human and animal pathogens. The family Argasidae (soft ticks) comprises five genera and with about 193 species. Among all the argasid ticks, only four Argas and two Ornithodoros species are competent to transmit diseases. This chapter describes the various ticks of the argasidae family, diseases they transmit, and strategies for their control. A description of the two important genera, Ornithodorus and Argas, that belong to the family argasidae are provided. Emphasis is on the mammalian hosts affected, tick species involved, morphological features (with relevant pictorials), geographic distribution, life cycle, and economic importance. A detailed description of the two most important diseases transmitted by argasidae ticks namely African swine fever (ASF) and avian spirochaetosis is given. Emphasis is laid on the historical background, epidemiology, clinical signs, and strategies for their control.

Preface

AN OVERVIEW OF THE SUBJECT MATTER

Livestock production is undoubtedly the mainstay of many sub-Saharan African farmer livelihoods. On the other hand, livestock production potential in these region is limited by endemic livestock diseases which include Nagana (African Animal Trypanosomiasis-AAT) and tick-borne diseases (TBDs). Control of these diseases in the region goes a long way in reducing the livestock constraints in addition to freeing substantial junks of land rendered unusable due to occupation of the vectors and disease. AAT and TBDs causes various disease conditions in different livestock species. In this book these various disease conditions are reviewed, highlighting the specific definition of the disease and its etiology, the risk factors that predisposes the animals to infection and epidemiology of the disease, the disease's clinical manifestation and signs, diagnostic approaches employed, treatment available and more emphasis on the available methods for their control.in different animals.

In AAT main control measures zero in in the use of treatment chemotheraphy and chemo prophyaxis and use of genetically trypanotolerant animals and Orma Boran is used as an example. However, these methods of control are not 100% successful due to a number of challenges like prohibitive cost of the drugs used, as most of the farmers are resource poor and rarely have any other source of income other than the animals. The other challenge is the gradual development of drug resistance by the parasites transmitting the diseases mainly due to improper use (usually application of substandard amounts) by the farmers themselves or unqualified animal health service providers. Drugs to treat AAT are available, but efficacy is limited, particularly as there is concern over increasing disease resistance to the drugs, which were developed 40 to 50 years ago.

Furthermore, the dream of having new drug molecules in the market in the near future is gloomy, mainly because the development of new drug molecules is very expensive and not very many commercial drug companies may be willing to venture onto this business unless assured of adequate immediate returns. Another key concern is drug safety, in terms of side-effects for animal health. The development of new drugs is required to effectively protect and treat cattle against the disease.

For the section of TBDs, the tick vectors transmitting the diseases and the specific disease conditions affecting different livestock species are also reviewed. The tick vectors include the hard (Ixididae) and argasidae (soft ticks). Examples of hard ticks are *Ambylomma*, *Boophilus*, *Hyalomma*, *Rhipecephalus*, *Haemophysalis*, etc. The diseases include ECF, Anaplasmosis, Babesiosis and Nairobi sheep disease / Ganjam disease. Examples of disease conditions transmitted by soft ticks are African Swine fever (ASF) and spirochaetosis in chicken. African swine fever, caused by a large DNA virus of the *Asfarviridae* family, which also infects soft ticks of the genus *Ornithodoros*. The ASF-virus causes a hemorrhagic fever with high mortality rates in domestic pigs; some isolates can cause death of animals as quickly as a week after infection.

WHERE THE TOPIC FITS IN THE WORLD TODAY

The title of the book 'Combating and controlling African Animal trypanosomosis and Tick borne disease in Livestock' is relevant and fits in the world today in addressing the problem of food security and poverty eradication. This is more so because, the animals affected by these diseases belong to poor resource farmers whose entire livelihoods depends on these animals. Furthermore, the land available for these animals to be reared is the same area occupied by the vectors transmitting the diseases. With the increasing animal and animal products (beef, dairy, skins and hides) and animal draught power due to the ever increasing human population, there is equally an increasing need to have enough animals in good health and production status and enough high quality animal products, to meet the demands. Efforts towards controlling these endemic livestock diseases will have multiple effects that include ensuring some reasonable level of food security/food safety to the ever increasing human population and also sustaining the livelihoods of rural poor resource livestock farmers and currently, these are the main concerns of most third world governments.

TARGET AUDIENCE

The target audience for this book include Veterinarians (practicing clinicians, consultants and lecturers, taking care of the animal health aspect of the animals and contributing to body of knowledge on the subject matter); Animal productionists/scientists (improving productivity); zoologists, genecists (using genetics and genomic approaches to unravel genetics of disease resistance e.g. in the identification of chromosomal regions controlling disease resistance traits for example in enhancing trypanotolerance as a control strategy for Nagana); Animal breeders (breeding of trypanotolerant breeds of animals as a sustainable approach of control of Nagana, breeding of animals resistant to East coast fever and other tick borne diseases); statisticians and epidemiologists (for disease dynamics); pharmaceutical industry (for drug trials, new molecule/drug developments, vaccine developments especially for Tick borne diseases, development of chemicals and technologies to control the vectors-tsetse flies and ticks). It also targets researchers (both socio-economists and life scientists), undergraduate and post graduate students in Biological and Life sciences (Biology, Veterinary Medicine, Biochemistry, Animal Science, Animal production systems, animal nutrition and production. The book also targets livestock farm managers and livestock policy makers

DESCRIPTION OF EACH CHAPTER

Although chapter numbering is continuous, the first six chapters deal with Nagana (Section 1) while the next nine chapters are devoted to tick borne diseases in Section 2.

Chapter 1 gives an introduction of African Animal Trypanosomosis (AAT) or Nagana. It deals with the definition of the disease and the causative agents, the trypanosomes. It also gives a brief description of the vectors transmitting the disease, the tsetse fly. It describes the distribution of the vector (tsetse fly) and cattle in Africa, the effects of the disease in livestock production, the species host range affected, the pathogenesis of the disease, the general clinical manifestation of the disease in the animals, current laboratory diagnostic methods used, general treatment methods and general control methods employed for the disease.

Chapter 2 deals with the epidemiology and economic importance of Nagana in livestock. It starts with a brief introduction and a distribution description of the main vectors for trypanosomes, the tsetse fly in Africa, the area occupied and the transmission cycle. The chapter also gives a summary of major characteristic features of syndromes caused by the various trypanosome parasites. The parasite life cycle and transmission is dealt with here. This chapter also gives a summary of *Glossina,* tsetse fly species and the host animals they affect; tsetse fly species life cycle and reproduction; ecology of Tsetse fly (dispersal, population dynamics and seasonality); the reservoir and the host (animals in the cycle of animal trypanosomosis). Finally the chapter underscores the risk factors of Animal trypanosomosis (anthropogenic changes and the epidemiology of African Animal Trypanosomosis, African Animal trypanosomosis and the changing environment) and the economic importance of the disease (direct and indirect economic losses) in various livestock species, notably cattle, sheep and goats and camels.

Chapter 3 of the book gives an overview of trypanosome vectors and their role in the transmission of the disease, African Animal Trypanosomosis (AAT) or Nagana. The main vector of veterinary and medical importance is here identified as the *Glossina* species of tsetse fly. The chapter starts with giving a summary of the currently described tsetse fly species by sub-generic division. It then gives salient identifying morphological features and geographical distribution of the Tsetse fly. The chapter gives a detailed description of how Tsetse fly locates its hosts, the feeding mechanisms it uses, and Tsetse microbiota (first, second and third endosymbiont). This is the chapter where the role of Tsetse fly in disease transmission is explained. There is a detailed description of transmission cycle of trypanosomes and finally the chapter gives a snapshot of the role of biting flies in the African Animal Trypanosomosis epidemiology

Chapter 4 is devoted to the role of a special state agency in the control of African Animal Trypanosomosis; a case of Kenya Tsetse and Trypanosomiasis Eradication Council (KENTTEC), a successor to Pan African Tsetse and Trypanosomiasis Eradication Campaign (PATTEC) program initiated by African Heads of States in Lomé Togo in 2000 to eliminate Tsetse and Trypanosomosis in Africa. The agency has over the years employed some of the following technologies in the control of Tsetse fly: Targets and traps, Push-pull technology (a bait technology based on attractive olfactory and visual cues identified from tsetse hosts), insecticide treated cattle (ITC), adoption of trypanotolerant livestock breeds. Other roles played by the agency in the control of tse tse and trypanosomosis are having collaborative linkages with stakeholders (KALRO, ICIPE, ILRI, FAO, OIE, universities), Farmers sensitization, training, and capacity building and active and passive monitoring of animal trypanosomosis (to establish disease prevalence and incidence in known tsetse hotspots and suspected areas), use of GIS to map status of AAT in the country and create a central data management. The agency also embracing progressive control pathways (PCP) for the planning of eradication of Tsetse and trypanosomosis in the country, contribution to policy development in related sectors. Finally the chapter gives case studies of Tsetse control in Lake Bogoria and Mwea National Reserve and its environs.

Chapter 5 identifies use of 'trypanolerance' as a sustainable control strategy for AAT. It starts by defining what trypanotolerance is and the use of trypanotolerant breeds in the control of AAT. It goes on to identify the main types of trypanolerant breeds like the West African cattle types (Muturu, Baoule,

Laguna, Samba, Dahomey and the N'dama), the East African cattle types (the humpless *Bos taurus* and the humped *B. indicus).* This chapter puts emphasis on one type of East African cattle type, the Orma Boran for the control of AAT. It gives a brief history of the breed, how phenotypic selection was done to have a more tolerant breed, and its multiplication for subsequent dissemination to tsetse infested areas and a follow up on their performance in those areas as case studies. Finally the chapter gives a long term Orma Boran breeding programme for future utilization of the genetic resource

Chapter 6 of the book is the last chapter of section one that deals with AAT. This chapter specifically deals with non- tsetse transmitted trypanosomosis (NTTT), caused by three main trypanosome species, namely: *Trypanosoma evansi*, *T. equiperdum* and *T. vivax*. Their global distribution is much wider than tsetse-transmitted trypanosomoses and includes Africa and Latin America, Asia and Euro-Asia. The chapter deals with each one of these trypanosomes (*Trypanosoma Evansi*, *T. equiperderm* and *T. Vivax*) and under each of them, there is a description of transmission, hosts involved, prevalence, geographical distribution, pathogenesis and clinical signs, diagnosis, postmortem findings, the disease impact, prevention and control and conclusion.

Chapter 7 is the first chapter of section two that deals with control of tick borne diseases in livestock. The chapter introduces the section by underscoring the epidemiology and economic importance of tick borne diseases as it shows that ticks and tick-borne diseases (TBDs) cause significant economic and health losses to livestock and humans, especially in the tropical and sub-tropical regions. The chapter gives a summary of important tick borne diseases of cattle in Africa, among them Anaplasmosis, Heartwater (Cowdriosis), Babesiosis and Ttheileriosis. Transmission of the various diseases and the corresponding vectors are discussed here. Next, the chapter discusses epidemiology of the TBDs i.e. factors related to the management systems/practices, socio-economic status, climate, vertebrate host and pathogen factors that influence the distribution and prevalence of tick-borne pathogens. Finally, economic losses due to ticks (vectors) and tick borne diseases and the control of TBDs among cattle herds, in Africa are underscored in this chapter.

Chapter 8 is devoted to a review of vectors (both soft and hard ticks) transmitting tick borne disease (TBDs) in livestock. In this chapter, these vectors are introduced as obligate blood sucking ectoparasites of vertebrates (particularly mammals, birds and terrestrial reptiles), falling into two well-defined families, namely, the *Ixodidae* or hard ticks and the *Argasidae* or soft ticks. In this chapter, the hard ticks are dealt with first in respect to classification, life cycle, morphological features, their mammalian hosts and distribution, diseases they transmit in livestock and the various methods used in their control e.g. chemical control through acaricides- arsenical, chlorinated hydrocarbons, organophosphates, carbamates and synthetic pyrethroids; conventional chemical application of acaricides including dipping, spray races, spot treatment and hand dressing and unconventional chemical application which include use of ear tags, neck bands, tail bands and pour-on. The chapter spells out other alternative control methods for these *Ixodidae* ticks as biological, ecological and genetical and the use of anti-tick vaccines. Equally for *argasidae* family of ticks, the chapter highlights their general morphology and life cycle, and also identifies various species of veterinary importance and the diseases they cause.

Starting from Chapter 9onwards, individual tick borne diseases are discussed and on this series, East coast fever (ECF) is the first one. A detailed introduction of the disease and the life cycle of T*heileria Parva* in the vector, *Rhipicephalus appendiculatus* tick and mammalian host, is given in this chapter. The distribution of *Rhipicephalus appendiculatus* in Africa, factors affecting its distribution, factors affecting survival of *T. parva* in the tick, factors affecting transmission of *T. parva* from ticks to cattle, factors affecting transmission of *T. parva* from cattle to ticks, and the target cell types in *T. parva* infection, are discussed in details. Next, the clinical manifestation of the disease (subacute, chronic and cerebral forms), pathogenesis and histopathology, gross pathology, hematology, immune responses against *T. parva* and immune responses against the tick and diagnosis are discussed. Finally, a detailed discussion of various control measures (application of acaricides, chemotherapy, infection and treatment immunization) is given.

Chapter 10 deals with Anaplasmosis in ruminants, which is introduced as an infectious, non-contagious haemotropic disease of cattle, sheep and goats caused by obligate intra-erythrocytic bacterium *Anaplasma* species. The etiology and the recent reclassification of the genus anaplasma organisms is given. Disease transmission by the vector is underscored including vertical transmission (transplacental) in cattle. Next is the description of tick vectors of Anaplasma-*Rhipecephalus decoloratus* (African blue tick) and *Rhipicephalus microplus.* Geographical distribution of Anaplasmosis worldwide, across the tropical and sub-tropical regions is given in this chapter and it follows the distribution of the vector. Epidemiology and risk factors, pathogenesis, clinical manifestation, diagnosis, treatment and control of Anaplasmosis is presented in this chapter. Finally the chapter gives a snapshot of economic importance of the disease, Anaplasmosis.

Chapter 11 deals with another tick borne disease, Heartwater (formerly called cowdriosis), an acute, febrile, tick-borne disease of cattle, sheep, goats and wild ruminants characterized by nervous signs and high mortality and caused by a rickettsia agent, *Erhlichia ruminantium.* The etiology of the disease and life cycle of the vector is given a detailed description here. A description of the disease transmission and epidemiology, risk factors and immune mechanisms is given. Protective and acquired immunity, innate resistance, antigenic and immunogenic variation are other aspects of immunity discussed in the chapter. Presentation of the clinical disease, economic importance and diagnosis (serology and molecular diagnosis) is given. Finally, control of the disease (treatment, tick control and immunization) is discussed in the chapter.

Chapter 12 discusses a very important tick borne disease, Babesiosis, with synonyms Red water, Texas fever, Tick fever, Piroplasmosis, Tristeza are infectious, non-contagious diseases, affecting most domestic and wild mammals, mainly but not exclusively in (sub)-tropical regions. The causative agents in the various host animals is presented in this chapter: *Babesia bigemina* in cattle and buffaloes, *B. Canis* and *B. gibsoni* in canine, *Babesia motasi* and *Babesia ovis in* sheep and goats, *Babesia caballi* in equine, *Babesia trautmanni* in swine and *Babesia cati* (large forms) and *Babesia felis* in feline. In this chapter, Bovine Babesiosis is discussed under transmission, life cycle in the tick and life cycle in cattle and pathogenesis. Diagnosis of bovine babesiosis is done through preparation of thin and thick blood smears, brain smear, indirect fluorescent Antibody test (IFAT), Enzyme linked Immunosorbent Assay (ELISA), Immunochromatography Test (ICT). The chapter further gives DNA probes, Polymerase Chain Reaction (PCR), Reverse Line Blot Hybridization (RLB), Real Time PCR (RT-PCR) and Loop Mediated Isothermal Amplification (LAMP) as molecular methods of diagnosis of Babesiosis. Finally the chapter outlines control methods used as vector control, immunization against Babesia, infection and treatment method, natural infection controlled by imidocarb/Treatment of clinical cases, live attenuated vaccines, vaccines based on exoantigens and immunization against ticks

Chapter 13 deals with Canine Ehrlichiosis and describes it as an infection of canids, causing a clinical disease in domestic dogs and asymptomatic infections in wild population of canids. This chapter summarises the available information on the disease in the African continent, as collated from published articles and indicates that the disease is prevalent in a number of countries, with infections detected in domestic and wild canids. The introductory part of the chapter gives a description of the disease through its definition, aetiology and epidemiology and further, a description of the vector-ecology and its life cycle. The chapter gives prevalence of the disease in Africa over a period of 30 years. Discussed in the chapter is the clinical picture and different diagnostic techniques for detecting infections with *Ehrlichia* species and these include; cell culture, direct microscopy, serology, hematology and molecular methods. Treatment of the disease as it mainly involves the use of oxytetracycline, doxycycline, imidocarb disproportionate and levamisole is discussed here. The chapter discusses several attempts of Ehrlichiosis disease management as involving vector control, treatment of infected dogs with appropriate drugs.

Chapters 14 and 15 discuss two viral diseases transmitted by tick vectors.

Chapter 14 highlights the first disease, Nairobi sheep disease or Ganjam, a disease of sheep and goats (small ruminants), first detected in Kenya in 1910, and is caused by Nairobi disease virus. The virus causing the disease is spread by hard (*Ixodidae*) ticks, and appears to be dependent on the tick vector for dissemination, with no direct transmission between animals. The chapter discusses the Nairobi sheep disease vectors and virus transmission, as it is done commonly by the brown ear tick, *Rhipicephalus appendiculatus* in Africa. The chapter further gives insights into tick pathogen interaction, the molecular biology of Nairobi sheep disease virus, its pathogenesis, its epidemiology and risk factors, clinical signs and diagnosis. Finally the chapter discusses treatment and control, and prevention of the disease.

Chapter 15 discusses the second viral disease caused by tick vectors and that's African swine fever in swine and spirochaetosis in chicken. The chapter gives further insight that the tick vectors transmitting these disease are generally called the soft ticks (the family Argasidae) comprising of about 193 known tick species. It discusses the multi host life cycle of the argasiade ticks. The chapter discusses the first disease caused by argasidae ticks-African Swine fever, a highly contagious haemorrhagic viral disease of domestic and wild pigs, which is responsible for serious economic and production losses. Historical background, epidemiology, clinical signs, prevention and control measures of ASF are discussed in this chapter. The second disease discussed in this chapter is Avian Spirochaetosis, caused by the bacteria *Borrelia anserine.* Pathogenesis, treatment and control through chemotherapy, biological control, appropriate housing, and enhancement of biosecurity measures and use of anti-tick vaccines are discussed in this chapter.

CONCLUSION

In conclusion, this book impacts the animal resource industry field positively. This book describes a myriad of livestock diseases transmitted by G*lossina* species of tsetse fly and tick vectors by giving a detailed insights on the general effects the diseases have on the livestock, the risk factors that predisposes the animals to infection and disease dynamics, their clinical manifestation, diagnostic approaches employed, treatment available and various methods available for their control. By availing this information in the most current and updated form, to the animal health and production practitioners, researchers and educationists in this area, relevant and appropriate policy makers and indeed its target audience, the book impacts positively in the agricultural field. The consumers of the information are made to be up to date as far as the information and technology or skill is concerned. Furthermore the information is available in a single handy repository and thus making it highly accessible. Consequently, proper and professional application of these information will result into gradual and cumulative improvement in animal health and animal production aspects as a result the diseases having been put under control. This will further impact positively on the farmer's livelihood and general economic improvement. Generally speaking, the book contributes to revolutionilization of the animal resource industry.

Further, this book has contributed immensely to the subject matter by generating, collecting and collating information from different sources and putting it together as a repository for ease of access. It has expanded the body of knowledge by including new and latest information, technology and skills both for the laboratory and field based scientists. In conclusion, this book impacts positively, the agricultural and animal industry field and also contributes immensely to the subject matter.

Section 1

Introduction of African Animal Trypanosomosis (AAT)

Chapter 1
Introduction of African Animal Trypanosomosis (AAT)/Nagana

Monicah W. Maichomo
Kenya Agricultural and Livestock Research Organization, Kenya

Caleb Oburu Orenge
Department of Veterinary Anatomy and Physiology, Faculty of Veterinary Medicine and Surgery, Egerton University, Kenya

Daniel Ochieng Gamba
Kenya Tsetse and Trypanosomiasis Eradication Council, Kenya

ABSTRACT

African animal trypanosomosis (AAT), a livestock disease, also known as Nagana, tsetse fly disease, or tsetse disease, is a widespread tsetse-borne disease complex caused by unicellular protozoan parasites belonging to the genus Trypanosoma. It is one of the major constraints to the expansion of livestock rearing and livestock-based industries in Africa. It also constrains mixed farming, human health, and livelihood in tropical Africa. The tsetse fly is the major vector of the disease. Acute disease is characterized by marked depression, intermittent fever, anorexia, anemia, blood-tinged diarrhea, and adenopathy, sometimes petechiae on mucosa, abortion, and death if not treated. Diagnosis is made by observing trypanosome parasites by direct microscopic examination of blood, lymph nodes, edema fluid, or tissues. Treatment is by chemotherapy and chemoprophylaxis. Control is achieved through vector control treatment using available molecules and use of a few available trypanotolerant breeds of animals.

INTRODUCTION

Trypanosomosis is a zoonotic disease affecting both humans and livestock, hence the name Human African trypanosomosis (HAT) and Animal African trypanosomosis (AAT) respectively. Different livestock species are affected and the geographic distribution follows that of the tsetse fly vector. Human and livestock hosts coming into contact with the vector are thus infected with a possibility of either developing acute to chronic forms of the disease depending on various factors. This chapter introduces

DOI: 10.4018/978-1-7998-6433-2.ch001

the reader to the dynamics of AAT, including its epidemiology, pathogenesis, symptoms, treatment and control. Further details are however contained in the following specific chapters.

AFRICAN ANIMAL TRYPANOSOMOSIS: AN INTRODUCTION

Epidemiology of African Animal Trypanosomosis

The history of African trypanosomosis dates back to antiquity with evidence suggesting that the disease was known in ancient Egypt (Griffith, 1898). However, it was only in 1903 that a Scottish pathologist and microbiologist, David Bruce, provided unequivocal evidence that the disease is transmitted through the bite of an infected tsetse fly. African Animal Trypanosomosis (AAT), the disease in livestock, also known as Nagana, Tsetse fly disease or Tsetse disease is a widespread tsetse-borne disease complex caused by unicellular protozoan parasites belonging to the genus *Trypanosoma*. It is one of the major constraints to the expansion of livestock rearing and livestock-based industries in Africa.

Trypanosomosis is of major economic importance in Sub Saharan Africa, many parts of Asia, and in some parts of tropical Latin America. Of importance to note is that humans are normally not susceptible to the trypanosomes that cause African animal trypanosomosis. The trypanolytic activity of apolipoprotein L-I (apoL-I), found in human blood, is thought to be the major innate defense mechanism against animal trypanosomes. Genetic defects in apoL-I have been detected in some people who were infected by trypanosomes not expected to be pathogenic for humans (e.g., *T. evansi*, which causes surra or Trypanosomosis in camels). Partial resistance of *T. vivax* and *T. congolense* to human serum has also been demonstrated in the laboratory under certain conditions (Spickler, 2018). Distribution of AAT follows that of biological vector, the tsetse fly (see figure 1). In Sub-Saharan Africa (SSA), it is in the area referred to as the tsetse fly belt which occurs approximately between latitude 15^0 N and $20\text{-}30^0$ S. A few species of tsetse flies have also been detected in parts of the southwestern Arabian Peninsula. Trypanosomes, particularly *T. vivax,* can spread beyond the tsetse fly belt by animal movements and transmission through mechanical vectors. *T. vivax* has become established in parts of South and Central America and the Caribbean, which are free of tsetse flies. Despite many and various approaches to its control over the past 100 years, the tsetse belt remains unchanged and the effect of the trypanosome species that it transmits continues to affect millions of poor people and their livestock (Kristjanson et al., 1999; Muhanguzi et al., 2017). However, sustained control efforts have reduced the number of new cases of human African Trypanosomosis (HAT) to 977 in 2018. This notwithstanding, the estimated human population at risk is 65 million people (WHO, 2020).

The epidemiology of the disease is complex and is greatly influenced by Agro Economic Zones (AEZ), livestock management system and farming practices (Uilenberg, 1998). Age, herd size, season and sex are significant risk factors for occurrence of AAT observed in Tanzania (Ngongolo K et al., 2019). Geographic location in view of presence of the tsetse vector and seasonal migration are positively associated with AAT prevalence (Odeniran & Ademola, 2018). The problem of AAT was recognized by African stockmen long, before the cause of the disease was known, and many pastoralists associated the disease with the presence of tsetse flies (Uilenberg, 1998). Infection rates in cattle in endemic areas vary considerably and could be over 50% (Maichomo et al., 2005; Radostits et al., 2000). Studies have shown that the prevalence in Nigeria varies between 16-47% (Majekodunmi et al., 2013), and 16.3% and 2.4%

Figure 1. Distribution of tsetse and cattle in Africa (source; Researchgate.net)

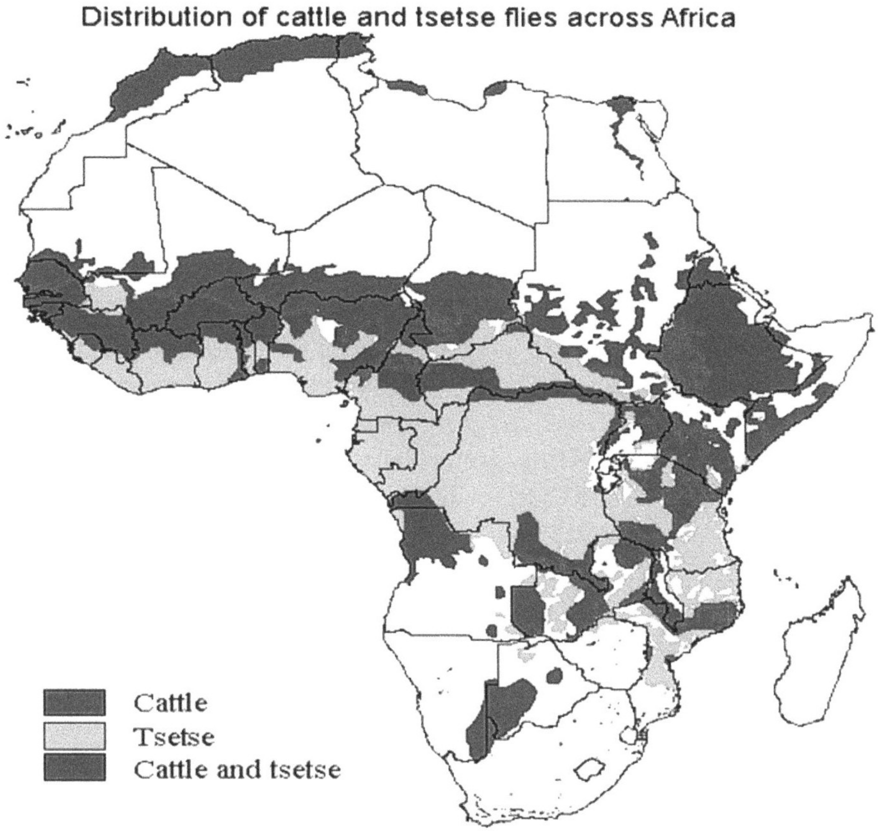

in Uganda (Muhanguzi et al., 2017) and South Africa (Alingu et al., 2014) respectively. Morbidity rates during outbreaks are variable and may reach 70% in cattle infected with *T.vivax* (Radostits et al., 2000).

Species of *Glossina* responsible for transmission also determines the epidemiology of tsetse-transmitted trypanosomosis whereby savannah species are better vectors of the pathogenic trypanosomes of livestock (Leak, 1998). In addition, daily activity patterns of the tsetse species involved and the grazing patterns of the herds greatly influence the epidemiology of trypanosomosis (Leak, 1998). During dry season, the southward migration of livestock in West Africa is greater than usual, and the owners deliberately choose the risk of trypanosomosis over starvation where huge losses may occur unless the animals are treated in time (Leak, 1998). The epidemiology of mechanically-transmitted trypanosomosis (*T. evansi*, *T. vivax*) is influenced by seasonal climatic differences, host preferences and daily behaviour patterns of biting flies (Radostits et al., 2000). The chronic disease sometimes becomes more clinically apparent during the dry season, when immunosuppressive factors such as poor nutritional state of the animal diminish its defenses, even when the initial infection occurred during the rains. Wild animals and chronically infected domestic animals are the reservoirs of infection hence the relative abundance of wildlife in East Africa compared to West Africa may partly explain the decreasing prevalence in the West (Radostits et al., 2000).

THE EFFECTS OF AAT IN LIVESTOCK PRODUCTION

Animal African trypanosomosis (AAT) is one of the biggest constraints to livestock production and a threat to food security in sub-Saharan Africa. It impairs the development of agriculture in endemic areas which constitute 41% of land in sub-Saharan Africa but which carry only 26% of the ruminant population (Kristjanson et al., 1999). Highly productive 'exotic' animals are excluded while local breeds loose production, particularly during high tsetse challenge (Radostits et al., 2000). It is estimated that 10 million km^2 of Africa is tsetse infested, exposing 60 million cattle to the risk of trypanosomosis. AAT costs about 4.5 billion USD a year to the rural economy attributed to loss of production, high mortalities and degraded human livelihood and health in tropical Africa (Yaro et al., 2016). In Africa, 80% of traction power is non-mechanized and availability of draught oxen to a family unit can result in a six-fold increase in agricultural output. Furthermore, the manure produced by livestock is essential for the production of food and cash crops, and is a potential source of energy in the form of biogas. Eighty-five percent (85%) of land in Kenya is arid and semi-arid (ASAL) and trypanosomosis is a constraint to improved livestock productivity in 60% of these ASAL's (KETRI, 1997).

Trypanosomosis results in poor growth, weight loss and low milk yield, reduced capacity for draught power, infertility and abortion (Murray et al., 1990). Growth retardation and productivity loss in trypanosomosis infected East African Zebu cattle has been demonstrated by Rowlands et al. (1994) in a longitudinal study in South-West Ethiopia, where he recorded 4-24% mortalities that were higher in parasitaemic than aparasitaemic calves. In the year 2011/2012, bovine trypanosomosis accounted for 64.6% of the total annual deaths, costing (US$200)/household in Ethiopia (Seyoum et al., 2013). In Kenya, up to 30% production losses associated with trypanosomosis prevalence rates ranging from 0-50% under natural tsetse challenge has been reported of (Maichomo et al., 2005; Mugunieri et al., 2003). In Nigeria, it has been estimated that AAT reduces the total number of livestock in an area by 25–50%, and predicted that with an elasticity of 0.2, AAT can reduce the agricultural gross domestic product (GDP) by 5–10% (Isaac et al., 2017). Production losses up to 20% across a range of parameters such as mortality, calving rate, draft power, meat and milk production have also been reported (Swallow, 2000). Some livestock breeds are recognized as being trypanotolerant exhibiting spontaneous self-cure where recovery may occur and partial immunity may be acquired (Murray et al., 1990). However, a study undertaken by ITC showed that even among the trypanotolerant N'dama cattle, emaciation, general weakness and abortion are encountered during the rainy season, probably when trypanosomosis challenge is high. Similar observations were equally noted with the Orma boran breed (Maichomo et al., 2005).

CAUSATIVE ORGANISMS OF AFRICAN ANIMAL TRYPANOSOMOSIS, SPECIES AFFECTED AND TRANSMISSION

Animal African trypanosomosis (AAT) of livestock that is caused by species of organisms transmitted by tsetse flies, the most important of which are *T. congolense*, *T. vivax* and *T. brucei* subsp. *brucei*. These three causative organisms belong to subgenus Nannomonas, Duttonella and Trypanozoon subgenus, respectively. Through deoxyribonucleic acids (DNA) probing, four genotypes/subtypes of *T. congolense,* causing pathology including anemia, high parasitaemia and leucopenia, have been identified in different ecological zones: *T. congolense* "forest type", *T. congolense* "savannah type", *T. congolense* "Kilifi type" and *T. congolense* "Tsavo type" (Bengaly et al., 2002; Clausen et al., 1998; Majiwa et al., 1993).

Table 1. Occurrence of African trypanosomes in domestic animals, reservoir hosts and in laboratory animals. (Source: www.fao.org).

Trypanosome species	Domestic animals affected	Reservoir hosts	Laboratory animals
T. congolense	Cattle, camels, horses, dogs, sheep, goats, pigs	Several groups of wild mammals	Rats, mice, guinea pigs, rabbits
T. simiae	Pigs	Warthog, bush pig	Rabbits, monkeys
T. godfreyi	Pigs	Warthog	None susceptible
T. vivax	Cattle, sheep, goats, domestic buffalo, horses	Several groups of wild mammals	Usually none susceptible
T. uniforme	Cattle, sheep, goats	Various wild ruminants	None susceptible
T. brucei brucei	Horses, camels, dogs, sheep, goats, cattle, pigs	Several groups of wild mammals	Rats, mice, guinea pigs, rabbits
T. brucei gambiense, *T. brucei rhodesiense*	Human sleeping sickness; affect domestic animals as *T. brucei bruce*	Several groups of wild mammals (particularly *T. brucei rhodesiense*)	As for *T. brucei brucei* (after initial adaptation where *T. brucei gambiense* is concerned)
T. evansi	Camels, horses, dogs, domestic buffalo, cattle	Several wild mammals in Latin America	As for *T. brucei brucei*
T. equiperdum	Horses, donkeys, mules	None known	As for *T. brucei brucei* (after initial adaptation)
T. theileri and *T. ingens* (subgenus *Megatrypanum*)	Cattle, domestic buffalo (not pathogenic)	Various wild ruminants	None

Trypanosoma. simiae, *T. suis*, *T. godfreyi* are known to affect pigs while *T. Suis* is suspected to be carried in African wild suids (Spickler, 2018). In Cattle, sheep and goats, the most important disease agents are, *T. vivax, T. congolense* and *T. brucei brucei*, while in pigs and camels, it is *T. simiae* and *T. evansi*, respectively (Table 1). Domestic animals and wildlife act as reservoirs for the human disease (sleeping sickness) caused by *T. brucei gambiense, T. brucei rhodesiense* (Msangi et al., 1999).

The unicellular organisms causing AAT have been found in many species of mammals, including all domesticated animals and some wildlife. Some wild ruminants such as the South American white-tailed deer, duikers, antelope and African buffalo, wild equids, felids, warthogs, capybaras, elephants, nonhuman primates and various rodents are susceptible to infection (Claxton et al., 1991). Although reptiles and birds are known to carry their own trypanosome species, *T. vivax* DNA has been detected by PCR in crocodiles and monitor lizards in Africa (Njagu et al., 1999). However, to be determined yet, is whether this organism can become established in reptiles or it is only transiently inoculated by insects.

Trypanosoma equiperdum that causes Dourine in horses is not transmitted by an invertebrate vector and it is primarily a tissue parasite that rarely invades the blood (Filip et al., 2005)._

Nagana is transmitted cyclically by the tsetse vector where the trypanosome parasite undergoes a cycle of development. Dissections of wild tsetse flies has revealed different developmental stages of trypanosomes in Glossina species. *Trypanosoma congolense* has been isolated specifically from *G. palpalis. G. swynnertoni, G. palidipes, G. morsitans, G. austeni, G. vanhoofi* and others. Tsetse flies become infected with bloodstream trypomastigotes when they take a blood meal from an infected host. The blood passes through the pharynx, the esophagous and the proventriculus before entering the crop. The proventriculus extrudes a chitinous tube, the peritrophic membrane. This membrane extends from the anterior part of the proventriculus to the hind-gut. The muscular proventriculus controls the release of blood from the crop to the mid-gut. The cycle of trypanosome development starts in the endoperitrophic space of the mid gut and takes about 2-3 weeks. The transformed parasites multiply as elongated promastigotes, and then migrate within the endoperitrophic space from the mid-gut to the labrum where they transform into epimastigotes. When they reach the hypopharynx, they transform further into metatrypanosomes, which are infective to the mammalian host. In the mammalian host the parasites multiply in the lymphatic tissues as metatrypanosomes. These are later ingested by other flies to complete the cycle

On the other hand, *T. vivax*, with the shortest cycle, can also be mechanically transmitted by biting flies such as *Haematopota*, *Liperosia*, *Stomoxys* and *Tabanus* species (Chartier et al., 2000). The latter two species are thought to be the most important vectors in South America, where there are no tsetse flies. Mechanical transmission of trypanosomes by biting flies depends on several factors: the feeding of the vectors, the ability of the trypanosome parasites to survive in the fly and their adaptability to the new host. Laboratory studies have also demonstrated the mechanical transmission of *T. congolense* (OIE, 1997). In mechanical transmission, a biting insect passes the blood forms from an infected animal to another in the course of interrupted feeding. The time between the two feeds is crucial for effective transmission because the trypanosomes die when the blood dries. Large biting insects such as tabanids carry more blood and are more likely to act as mechanical vectors than mosquitoes. Tsetse flies themselves can of course also act as mechanical vectors (Roberts et al., 1989). This mode of transmission has proved sufficient to maintain *Trypanosoma vivax* and *Trypanosoma evansi* in South and Central America, and the latter species in North Africa and Asia where no tsetse flies occur outside tropical Africa, apart from small pockets in the southwest of the Arabian Peninsula (Abebe & Jobre, 1996). The biting flies involved in mechanical transmission are highly mobile and abundant, and they have mouthparts that are adapted for transmission (Foil, 1989).

Trypanosome parasites can also be transmitted by iatrogenic means, which may occur when using the same needle or surgical instrument on more than one animal, at sufficiently short intervals that the blood on the needle or instrument does not dry. It can occur when animals are vaccinated or treated by injection, or when blood is collected from several animals in a row, without changing or disinfecting needles (Davilla & Silva, 2000). It may also occur when several animals are subjected at short intervals to a surgical intervention (dehorning, castration, etc.) without properly disinfecting the instruments. Trans-placental transmission has also been reported (Okech et al., 1996). The possibility of venereal transmission was suggested by the detection of *T. vivax* nucleic acids in the semen of experimentally infected goats and the demonstration that *T. cruzi,* the cause of Chagas disease) can be transmitted venereally (Rottcher et al., 1987). Dourine (trypanosomosis in horses) is transmitted during breeding and mortality in acute cases is about 50% especially in stallions (OIE, 2012). Trypanosomes do not survive for long periods outside the host. *T. brucei* may remain alive for a few hours in blood, and for a few days if it is refrigerated in blood with adequate oxygen. Animals infected with trypanosomes can become chronic carriers, and inapparent infections can be reactivated if the animal is stressed.

THE BIOLOGICAL VECTOR OF AFRICAN ANIMAL TRYPANOSOMOSIS, THE TSETSE FLY

The Genus Glossina is divided into three distinct taxonomic groups using morphological characteristics, the *fusca, palpalis* and *morsitans* group as described by (MJ Vreysen et al., 2013). *Fusca* group inhabits

Figure 2. The Tsetse Ecological Distribution (TED) model percentage probability map overlaid on a physiographic map of Kenya (Source: Research gate.net)

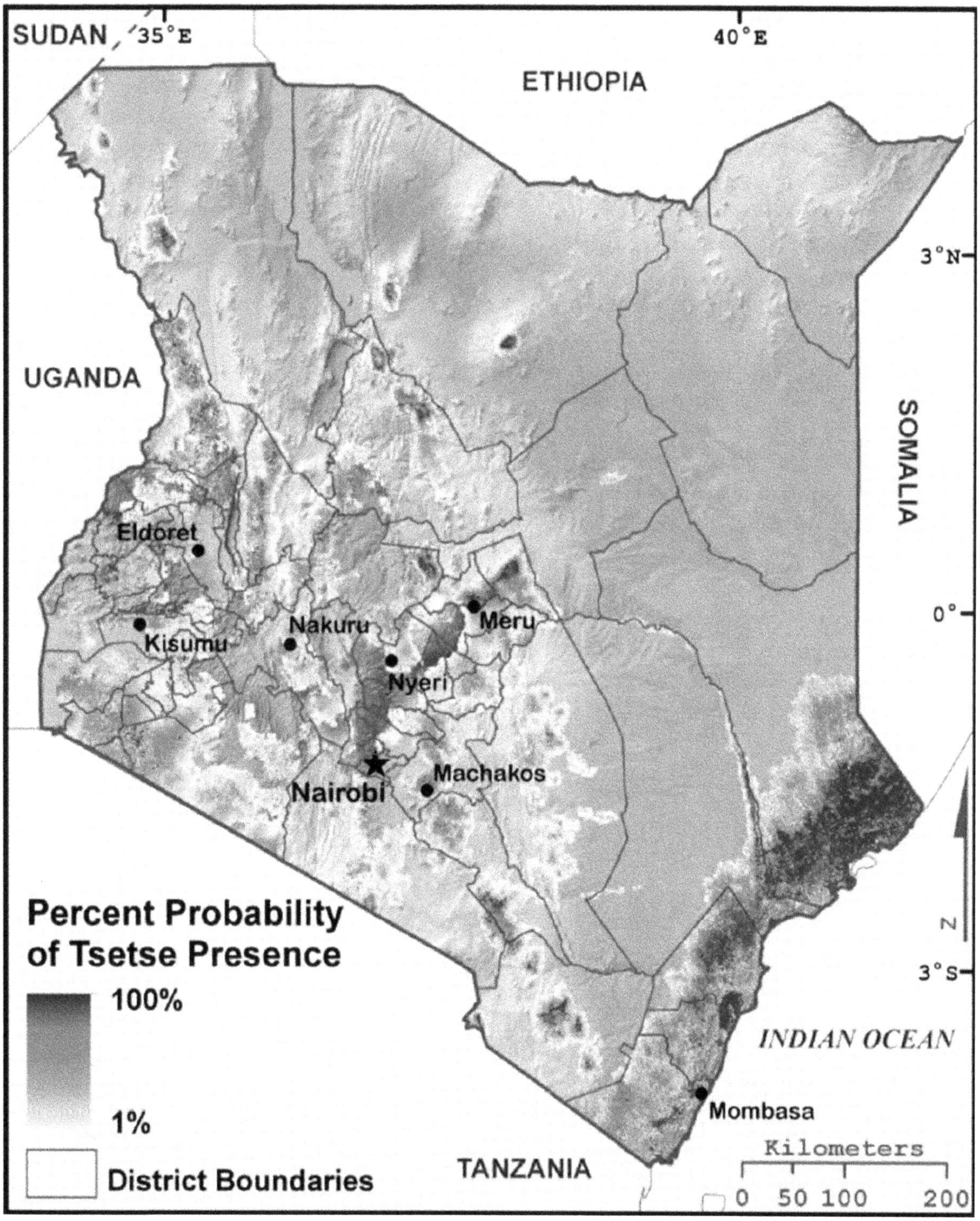

the great main forest in West and West-central Africa and is associated with streams and rivers in the savannah areas surrounding the great forests. *G. tabaniformis, G. haningtoni, G. nashi, G. nigrofusca, G. fusca, G. medicorum, G. fuscipleuris, G. schwetzi, G. severin, G. vanhoofi, G. brevipalpis, G. fuscipleuris* and *G. longipennis* belong to the *fusca* group (Leak, 1998). The *Palpalis* group is associated with rivers and lakes and is therefore similar to *fusca* group. They are important as vectors of bovine and porcine trypanosomosis, as well as of Gambian sleeping sickness due to *T. brucei gambiense* (Radostits et al., 2000). *Glossina palpalis, G. gambiense, G. fuscipes* and *G. tachinoides* of the *palpalis* group are found in western and central Africa (Hendrickx & Napala, 1997). The *morsitans* group occupies vast areas of the savannah grasslands in West and East Africa, and is probably the most economically important species in the transmission of trypanosomosis (Leak, 1998). Seven species belong to this group; *G. m. morsitans, G. m. orientalis, G. m. submorsitans, G. pallidipes, G. longipalpis, G. swynertoni* and *G. austeni* (Paling et al., 1987).

Tsetse flies infests an area of approximately 10 million km^2 in 38 countries of sub-Saharan Africa (MJ Vreysen et al., 2013). Ecological and climatic features, particularly temperature, rainfall, soil and vegetation type, and other climate variables influence the distribution of the tsetse flies (Leta et al., 2015). The infestation in each country is dependent on the geographic variation in tsetse species distribution and eco-epidemiology of the disease it transmits, as well as disparities in resource distribution, infrastructure and political stability within and between sub-Saharan Africa countries (Meyer et al., 2016).

In Kenya, tsetse - G. swynnertoni Austen (morsitans group) in the Maasai mara, G. brevipalpis New-stead, G. longipennis corti, G. fuscipleuris Austen (fusca group) and G. f. fuscipes Newstead (palpalis group) in the shores of Lake victoria (KETRI, 1997; Ngari et al., 2020; Okoth, 2007). It is estimated that 70 percent of Uganda, an area equivalent to 14,000km^2 is infested with 11 species of tsetse flies. Each species occupies different ecological niche with Glossina fuscipes fuscipes, G. pallidipes, *and G.* morsitans submorsitans being the most prevalent (Egeru et al., 2020). A total of 33% Tanzania land mass is infested by 10 species and sub species of Glossina. The species include both the Savannah (Morsitans): *G. morsitans morsitans, G. m. centralis, G. swynnertoni, G. pallidipes* and *G. austeni;* Forest species (Fusca): G. *longipennis, G. brevipalpis* and *G. fuscipleuris*, and to a lesser extent the palpalis (Nermohina): *G. fuscipes fuscipes, G. fuscipes martini*, which are restricted towards Lakes Victoria and Tanganyika respectively (Malele et al., 2011).

PATHOGENESIS OF THE DISEASE

A phase of local inflammation and a swelling (chancre) develops once tsetse fly injects infective metacyclic trypanosomes into the skin of the host. It is in this chancre that the trypanosomes divide and multiply before giving rise to the typical blood forms which invade the lymph nodes and lymphatics and then the bloodstream. Nagana, like other infectious diseases, starts with an increase in body temperature, as a result of the contact between the trypanosomes multiplying in the host and the defense system of the host. The surface proteins of the trypanosomes provoke the host in making specific antibodies against these proteins, and after a few days almost all of the trypanosomes in the blood are destroyed by these antibodies and the body temperature drops (Onyilagha & Uzonna, 2019). However, a few parasites survive as they have been able to replace their surface proteins by different ones, against which the antibodies cannot act. These surviving trypanosomes can multiply, and cause a new peak of parasitaemia and hyperthermia, until the organism of the host makes specific antibodies against the new surface proteins. This see-sawing process continues for a long time, as the trypanosome is able to make an almost unlimited number of antigenic variants, and the host responds to each of them, until either the antigenic repertoire of the trypanosome is finally exhausted, in which case self-cure of the host follows, or the ability of the host to react to all of the antigenic variants is overwhelmed, and the host dies.

The temperature curve (see figure 3) is a typical aspect of the disease where there are peaks every few days, especially in the beginning before the chronic phase sets in. As the disease progresses, the intervals between temperature peaks become longer and the peaks are shorter and antigenic variation becomes slower (Uilenberg, 1998).

Anaemia, one of the main symptoms of trypanosomosis is thought to be partially caused by phagocytosis of erythrocytes by the host's leucocytes. Auto-immunity phenomenon is initiated whereby the erythrocytes apparently become coated with material from lysed trypanosomes and this tricks the leucocytes into mistaking them for foreign invaders and removing them. This anaemia caused by phagocytosis (leucocytes) is increased by toxic substances from the trypanosomes that destroy erythrocytes directly by lysis (haemolysis). The hemopoietic system, tries to compensate for the loss of erythrocytes by increasing its activity but, later, in the chronic stages of disease, other toxins from the parasites have a depressing effect on the hemopoietic system, and the host is unable to produce as many erythrocytes as are removed (erythrocytes have a limited life span of 120 days, and now they are being removed even faster because of auto-immunity and haemolysis) (Anderson et al., 2018).

Figure 3. Temperature curve in an animal suffering from T.Congolense (source; FAO)

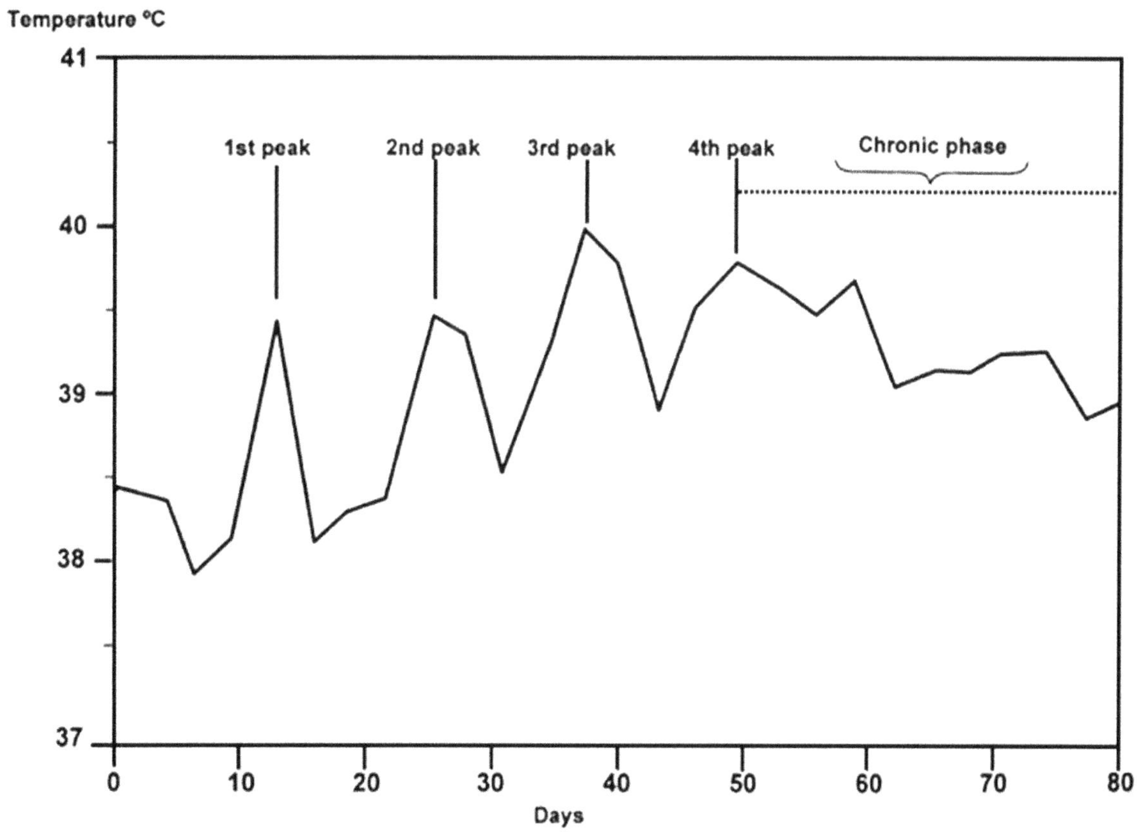

Trypanosomosis is also associated with *immunosupression*, i.e. the host's immune system becomes less efficient to deal with infections. This may be explained in part by the supression of the haemopoietic system, which not only provides erythrocytes but also leucocytes which are involved in the immune response. Animals affected by AAT often develop a lower antibody titre after vaccination against other diseases, and secondary infections which the host would normally control may also crop up during the disease. For example, it is common to find considerable numbers of *Babesia*, *Theileria* and/or *Anaplasma* parasites in the blood smears of animals suffering from AAT, in situations where normal animals are healthy carriers of these tick-borne infections.

Depending on the species involved, various organs are also affected by AAT, probably by trypanosome parasites themselves, toxins produced by the trypanosome parasites as well as anemia. *Trypanosoma congolense* is mainly confined to the blood, while *T. vivax* and *T. brucei* also invade the tissues. *Trypanosoma vivax* is found in the lymph and even in the chamber of the eye and *T. brucei* is well known to invade the central nervous system causing the human sleeping sickness (*T. brucei gambiense* and *T. brucei rhodesiense*). T. brucei also invades tissues in animals such as horses, goats and dogs, where it has been found to invade the nervous system in the later stages of dourine (*T. equiperduni*) (Uilenberg, 1998).

The heart is often inflamed (myocarditis), and heart failure is often the direct cause of death. Extensive myocarditis with the presence of trypanosomes in the heart muscle has been found in highly susceptible European cattle infected experimentally in Europe with pathogenic West African *T. vivax*, kept at rest in a stable and on a good level of nutrition, and which had shown no obvious signs of distress before they were slaughtered at the end of the experiment. Such infections could easily have caused fatal heart failure in African zebu cattle subjected to the stress of poor nutrition and walking long distances to watering points (Uilenberg, 1998).

Oedema is often present in AAT infections, particularly in horses and dogs, due to increased permeability of blood capillaries, and therefore leakage of blood plasma leading to the swellings.In chronic trypanosomosis the animal loses condition, and there is pronounced wasting. This is due to consumption of the fat reserves during the recurrent bouts of fever, and severe degenerative changes of the muscle cells and other tissue cells. There is also an increased breakdown of protein in muscles and elsewhere, leading to atrophic degeneration. The decreased supply of oxygen because of the anaemia is also an important factor (Anderson et al., 2018; Uilenberg, 1998). Studies involving animal models and humans suggest that high levels of pro-inflammatory cytokines might be associated with moderate to severe neuropathy, including elevated levels of counter-inflammatory cytokines in late stage disease (Kennedy, 2009). Counter-inflammatory cytokines (IL-10 and IL-6) have been associated with a reduction in the severity of neuropathology, suggesting a possible protective role (Kennedy, 2009).

CLINICAL SIGNS

Trypanosomosis is a collective term for a group of diseases brought about by one or more of the pathogenic trypanosome species (Radostits et al., 2000). Clinical signs therefore vary according to the trypanosome species and the virulence of the particular isolate. After an incubation period of 8-20 days, acute episodes last for a few days from which the animal dies or lapses into a sub- acute or chronic stage. Acute disease is characterized by marked immunosupression, intermittent fever, anorexia, anemia, blood-tinged diarrhea, adenopathy, sometimes petechiae on mucosa, abortion, and, if not treated, possibly death within a few weeks (Batista et al., 2006; Mwongela et al., 2005). Mortality rates of over 50 percent have been associated with *T. vivax*. *Trypanosoma congolense* causes acute disease lasting 4-6 weeks while *T. vivax* causes severe disease in West Africa as well as in dairy cattle in East Africa. Within a week of infection with trypanosomes, there is usually a pronounced decrease in packed cell volume (PCV), hemoglobin, red and white blood cell levels. When the nervous system is involved, animals showing rabies-like symptoms can die within 2-12 hours (Ramanatha et al., 1990).

Chronic cases of trypanosomosis may run a steady course, may be interrupted by severe illness or undergo spontaneous recovery. While *T. vivax* causes acute and often fatal disease in cattle, recovered animals develop a chronic wasting disease in East Africa which often ends in the death of the animal if not treated (Mwongela et al., 2005). *Trypansooma brucei brucei* and *T. congolense* species cause chronic disease whose signs appear as a relapsing fever (Naessens, 2006). Major symptoms in this stage are anemia, intermittent fever, edema, ocular discharge, pallor mucous membranes, weight loss, 'staring coat', tight skin over the ribs and pelvis, arched back, photophobia and swelling of the superficial lymph nodes often associated with *T. vivax* infections. In areas where reinfection is frequent, death will commonly occur within 2 to 4 months or longer, unless the animal is treated. Animals that survive are often infertile and of low productivity (Batista et al., 2006; Okech et al., 1996). In some instances, infected animals show no overt signs of disease but can succumb if stressed, for example, by work, pregnancy, poor nutrition, concurrent diseases, milking or adverse environmental conditions.

LABORATORY DIAGNOSTIC TESTS

Microscopical Diagnosis

A tentative diagnosis can be made if trypanosome parasites are observed by direct microscopic examination of blood, lymph nodes aspirates, edema fluid or tissues biopsy collected at necropsy. The parasites are most likely to be found in the blood during the initial stages of the infection and less likely to be detected in chronically ill animals, and are almost never seen in healthy carriers. Samples for microscopy can include unstained wet blood films and thick or thin smears stained with Giemsa or other agents. Wet films are examined by light microscopy (condenser aperture, phase-contrast or interference contrast), using 40x power to detect the motile trypanosomes. Trypanosomes are seen either directly, moving between the blood cells, or indirectly, as they cause the blood cells to move. This test is simple and inexpensive and if_trypanosomes are found, the disease is diagnosed on the spot. However, disadvantages of wet film method include; limited sensitivity (the detection limit is usually around 10^4 trypanosomes per ml of blood), the species of trypanosome cannot be identified and a field microscope has to be taken to the herd as the parasites lose their mobility after a limited time. *Trypanosoma vivax* can often be strongly suspected if the parasites move quickly forward through the microscopical field.

Amongst the direct examination techniques, stained thin blood films are generally regarded as more specific but less sensitive than the thick films. The actual specificity and sensitivity of these techniques is directly dependent on the volume of blood actually examined and the skill and experience of the examiner. Examination of Giemsa-stained blood smear (GSBS) remains the classic and most certain reference diagnostic test for trypanosome infection (OIE, 2018). Thick films are more likely to detect trypanosomes when their numbers are low, but parasite morphology is easier to observe in thin films (Paris et al., 1982). Usually, both a thin and thick smear are made from the same sample. Thick smears contain more blood than thin smears and, hence, have a higher diagnostic sensitivity. Thin smears on the other hand allow Trypanosoma subgenus identification.

Parasite Concentration Techniques

Examination of fresh or stained blood films has been modified to improve diagnostic sensitivity by concentrating the blood through centrifugation in a haematocrit tube, namely the haematocrit centrifuge technique (HCT) (Desquesnes, 2017). Other improvements include mini anion– exchange chromatography, hematocrit centrifugation, the quantitative buffy coat method and the dark-ground/phase-contrast buffy coat technique, BCT (Woo, 1970). The advantage of these techniques is that they increase the probability of detecting the organisms. The trypanosome's morphology can provide a presumptive species identification, but some species cannot be distinguished by this method, and unusual morphologies have occasionally been reported. Animals can be infected concurrently with more than one species of trypanosome (OIE, 2018). These techniques have variable detection limits and the BCT is considered more sensitive than HCT. Desquesnes, (2017) reports that using BCT, the lowest numbers of trypanosomes that could be detected were 2.5×10^2 *T. congolense*, 5×10^2 *T. vivax* and 5×10^3 *T. brucei.*

Polymerase Chain Reaction (PCR)

PCR assays on blood or tissue samples can identify trypanosomes at the genus, species or subspecies level. The World Organization for Animal Health (OIE) currently recommends the use of species-specific PCR tests; however, these tests may miss organisms other than *T. vivax, T. congolense* and *T. b. brucei.* The internal transcribed spacer (ITS-1) PCR technique can amplify multiple species of trypanosomes, followed by gel electrophoresis to distinguish these organisms by the size of the amplified fragment (Njiru et al., 2005). It can recognize infections with less commonly identified species, as well as the 3 major organisms. However, the OIE notes that sequencing is often necessary to confirm the organism's identity and ITS-1 PCR is not used routinely for diagnosis (Njiru et al., 2005; OIE, 1997). With the field samples, the diagnostic sensitivity varies between groups: 95% for Trypanozoon, 91% for T. congolense savannah and 77.4% for *T. vivax* (Njiru et al., 2005). Gaithuma et al. (2019) concluded that it is possible to detect and distinguish between different Trypanosome species and subspecies accurately (with the exception of *Trypanozoon* subgenus) and therefore infer prevalence of infection more precisely using a single test by comprehensive analysis of internal transcribed spacer ITS-1 amplicons.

Loop-Mediated Isothermal Amplification (LAMP) Tests

The current diagnostic methods, based on the visualization of the parasite in the blood, or on the detection of its DNA or the antibodies it triggers in the host, are not suitable for direct use in the field as they require specialized equipment and personnel (Boulange et al., 2017). For this reason, tests like the Loop-mediated isothermal amplification [LAMP] tests are in development for livestock and have been used to detect clinical cases in Africa. Findings by Gummery et al. (2020) support that LAMP is comparable to PCR when used in the field, an important tool for clinical decision making. However, results suggest repeatability is low in animals with low parasitaemia.

Antigen detection techniques have been described but were found to be unreliable and some newer antigen-detection methods in development may be promising (Davita et al., 2013).

In *vitro* cultivation can be used to isolate some species of trypanosomes, though it is not employed for routine diagnosis. Not all pathogenic species will grow in *vitro*, and nonpathogenic trypanosomes can readily overgrow the culture (Spickler, 2018). Animal inoculation of rats or mice is rarely employed for diagnosis.

Indirect Tests

Serology is mainly used in surveillance and for identifying infected animals before transport to trypanosome-free areas. The indirect fluorescent antibody test and ELISAs are generally used in cattle (Sivajothi et al., 2016). Cross-reactions can occur between the various trypanosomes causing AAT, as well as with nonpathogenic trypanosomes (e.g. *T. theileri*).

During the acute phase, there is loss of condition and anaemia, but not as severe as in the chronic phase of the disease. Miscroscopic examination of the blood will show that the haemopoietic system is actively trying to compensate for the loss of red blood cells (regenerative changes such as anisocytosis, normoblasts, Howel-Jolly bodies). The spleen and lymph nodes are enlarged while the latter is edematous too. The liver is enlarged and congested while the heart may be somewhat enlarged and may show a few hemorrhages on the muscle surface. There is also likely to be more fluid than normal in the chest, lungs, abdomen and heart sac. The kidneys are pale and swollen.

During the chronic phase, the pathological changes seen are more striking. The carcass is emaciated and often dehydrated. The skin may show pressure sores and ulcers, as a result of the animal having been recumbent for quite some time. The skin is adherent to the underlying muscles and bone and the subcutaneous fat reserves are all used up. The underlying bones are prominent and the muscles have wasted appreciably. The muscles are pale because of the anaemia and the blood is watery and pale, with an increased clotting time. The heart is often enlarged and flabby due to muscle deterioration, and its weaker pumping action may have contributed to circulatory disturbances and oedema. Unlike in acute phase of the disease, the lymph nodes are mostly normal or even hard, dry and reduced in size. The spleen is also normal in size or contracted with a drier pulp than normally seen (Galiza et al., 2011).

TREATMENT OF AAT

Chemotherapy

Chemotherapy is the treatment of disease by use of chemical drugs which are mainly curative. These drugs disrupt or block one or more of the vital processes which are essential to the invading microorganism. Certain compounds have specific effects on some enzyme system or block essential metabolic pathways. By stopping the multiplication of the trypanosome parasites, chemotherapy helps the immune system to overcome the infection. Treatment will be more effective in well-fed animals, in which the immune system is not adversely affected by stress and lack of food. Curative drugs (see table 2) are mainly used where disease incidence is low and only a limited number of animals in a herd contract the disease during the course of a year (FAO, 1992). This is often the case when the threat mainly occurs seasonally e.g. the higher risk situations that may occur during the rains, as in the case of *T. evansi* in camels, of mechanically transmitted *T. vivax*, or in marginal tsetse areas. The trypanosomosis risk may also be higher during the dry season, for instance when herds enter tsetse belts during transhumance, either accidentally or deliberately. Curative drugs are meant to cure individual infected animals, not to protect the whole herd or group for a longer period. Curative treatment is most effective in herds that are inspected and treated at regular intervals. The more frequently the visits are carried out, the more effective the treatment regimen. For instance, if loss of condition and some mortality occur when cattle are visited and treated on a three-monthly basis, the situation may be corrected if the visits are carried out on a monthly or two-monthly basis.

Block treatment of all animals in a herd should not be carried out with curative drugs, at least not with one that possesses a residual effect, because of the high risk of drug resistance developing during the period that the drug reaches a subcurative level. All curative drugs also possess some residual effect, but in the case of diminazene aceturate and Mel cy this is practically negligible.

Chemotherapy in cattle relies on two compounds; Diminazene aceturate (See Table 2) which is effective against all the three African animal trypanosomes, and Homidium bromide or chloride (Radostits et al., 2000). The isometamidium drugs are also excellent chemotherapeutic agents as are the quaternary ammonium trypanocides i.e antrycide, ethidium and prothidium (Munsterman et al., 1992). It is common practise for many owners to treat their animals with trypanocides on a presumptive diagnosis of trypanosomosis based solely on recognition of clinical signs (Leak, 1998). In such cases sub-curative treatment is often given, the effect of the trypanocide being merely sufficient to restore the host-parasite equilibrium and effective premunity. Clinical recovery often follows such treatment but the exposure of trypanosomes to low concentrations of trypanocide increases the risk of drug resistance developing (Peregrine, 1994). Nevertheless, this enables livestock to be maintained in trypanosomosis-enzootic areas at minimal cost. Combination therapy is not common but VERYL® (CEVA-Sante Animale), containing diminazene diaceturate and levamisole chloride for the control of helminths in cattle kept in trypanosomosis endemic areas is a recent introduction (Maichomo et al., 2019). This study revealed that treatment of cattle with VERYL significantly ($p < 0.001$) reduced FECs by day 7 - day 21 post-treatment, and subsequent larval recoveries compared to the control group.

Chemoprophylaxis or Chemo-Prevention

This refers to the prevention of disease by the use of chemical drugs, implying that there is a residual effect, since prevention depends on the persistence of the drug in the system of the animal. A chemo prophylactic drug persists long and therefore preventive and is also curative, but a curative drug is not necessarily prophylactic. Preventive drugs are administered at defined intervals, and care must be taken not to administer them more frequently than prescribed due to their toxicity for example, Isometamidium, will cause weight loss in poorly nourished cattle if administered at short (monthly) intervals. Equally, combination with other drugs is dangerous for example cattle treated with diminazene aceturate after several isometamidium treatments suffer from hepatic damage and may even die (Uilenberg, 1998).

Often, a depot of the drug is formed at the site of the injection where it is retained and slowly released into the circulation to maintain a concentration in the blood at a level at which no trypanosomes can exist. Other drugs are not maintained in the form of such a local depot, but are found loosely attached to blood proteins and become slowly available to act on the parasite

Prophylactic drugs are used where the risk is so high that the health of the herds cannot be maintained by individual application of curative compounds. In such a situation too many animals contract the disease too frequently. This is particularly the case in tsetse-infested areas. Although curative treatments may prevent a high death rate, the overall health of the affected herds is not satisfactory, with loss of

Table 2. Common commercial trypanocides

Generic name	Trade names	Solution for use	Dosage rate[1]	Route	Remarks
Suramin	Naganol	10%	10 mg/kg (1 ml/10 kg)	IV	Mainly used against *T. evansi* in camels
Diminazene aceturate	Berenil, Ganaseg, Trypazen, Veriben	7%	3.5–7 mg/kg (1–2 ml/20 kg)	IM	Mainly used in cattle and small ruminants
Homidium bromide	Ethidium bromide	2.5%	1 mg/kg (1 ml/25 kg)	IM	Mainly used in cattle and small ruminants. Should be dissolved in hot water. Potentially carcinogenic
Homidium chloride	Ethidium C, Novidium	2.5%	1 mg/kg (1 ml/25 kg)	IM	See above, but soluble in cold water
Quinapyramine methyl sulphate	Antrycide, Trypacide, Noroquin, Quintrycide	10%	5 mg/kg (1 ml/20 kg)	SC	Now mainly used against *T. evansi* and *T. brucei* in camels and horses
Mel cy	Cymelarsan	0.5%	0.25–0.5 mg/kg (1–2 ml/20 kg)	IM or SC	Registered only for use against *T. evansi* in camels
Isometamidium chloride	Samorin, Trypamidium	1% 2%	0.25–0.5 mg/kg (1.25–2.5 ml/50 kg) 1 .0 mg/kg (2.5 ml/50 kg)	IM	Used mainly in cattle, as a curative at lower rates, as a prophylactic at higher rates. Also contains homidium, and is therefore to be considered as potentially carcinogenic as well

[1]Dosage rate of solutions for use are given in brackets. IV-Intravascular, IM-Intramuscular, SC-Subcutaneous

condition, poor (re)productive performance, many clinical cases and some deaths, despite a relatively large number of treatments at each regular visit. Other circumstances that might warrant the use of prophylaxis may occur when the infected animals cannot be reached (for instance, inaccessibility during the rainy season, transhumant herds, trade cattle moving to distant markets and passing through tsetse belts) (Uilenberg, 1998).

General Control Methods Used for AAT in Livestock

Veterinarians and other animal health providers who encounter or suspect AAT follow their national and/or local guidelines for disease control. In Africa, trypanosomosis can be controlled by reducing or eliminating tsetse flies with traps, insecticides and other means, and by treating infected animals. The specific control methods may differ depending on the epidemiological importance of various livestock species and wildlife in a region, and the predominant species of tsetse flies.

An area-wide integrated vector management strategy is the preferred mode of controlling trypanosomosis in large tsetse infested areas (Vreysen et al., 2007). Use of deltamethrin impregnated target technology has improved the suppression of tsetse populations, with a significant impact on animal Trypanosomosis (Percoma et al., 2018). The effectiveness of traps and targets has been greatly enhanced by the addition of appropriate odour bait. Such odour baits are usually short-chain aromatic compounds such as acetone or octenol (Kuzoe & Schofield, 2004). These substances enhance trap efficiency and are mostly components of host breath, urine, and skin secretions for baiting traps. The development of novel control tools, as well as standardization of existing models, has led to the identification of cost-effective devices for tsetse management (Wamwiri & Changasi, 2016). There have been considerable advances made in the search for efficient and cost-effective control tools against riverine tsetse involved in the transmission of HAT (Simarro et al., 2008).

The integration of the insecticide-treated cattle for the control of trypanosomosis is based on the use of insecticide formulations (mainly based on pyrethroids) on a wide range of techniques (pour-on, spraying, whole body dips/ baths and sprays/ showers) (Gimonneau *et al.*, 2018). Sequential aerosol techniques have been used in Botswana, Ghana, and Ethiopia with various levels of success in eradication of T&T (Adam et al., 2013). Current methods of trypanosomosis control target the vector (the tsetse fly) and the parasite (the trypanosome). Vector control methods include the use of traps and insecticides applied on targets, or on livestock as pour-ons, while parasite control methods rely mainly on the treatment of sick animals and prophylaxis. All these initiatives are in line with the Pan African Tsetse and Trypanosomosis Eradication Campaign (PATTEC), an initiative by the African heads of States in Lome Togo in 2000 which aims to eradicate tsetse flies from Africa through use of appropriate technologies, sustainable land management and capacity building of the livestock keepers. PATTEC's eventual goal is to eliminate tsetse flies from the continent, and with them, to eliminate most animal trypanosomes. Whether this is realistic is currently uncertain; however, trypanosomes were cleared or mostly cleared from limited areas in the past.

The selection of trypanotolerant breeds (see Morbidity and Mortality) can lessen the impact of African animal trypanosomosis. The use of trypanotolerant breeds of domestic animals that require little or no chemotherapy has been viewed as a potentially sustainable approach to livestock development in the tsetse-infested parts of Africa (Trail et al., 1993). Trypanotolerance has for long been attributed to the taurine breeds of cattle in West and Central Africa, including the N'dama and the West African shorthorn (Jones & Akinbamijo, 1999). In Eastern Africa, trypanotolerance has been identified in some zebu cattle, especially the Orma Boran (OB), which is big in size and originally kept by the Orma tribe of Tana River County, Kenya. When compared with the improved Kenyan Boran or Galana Boran, this breed was found to have lower trypanosome infection rates and when untreated, showed better control of anaemia and decreased mortality (Mwangi et al., 1998).

Good nutrition and the management of concurrent illnesses are also helpful. Chemoprophylaxis is sometimes used to protect susceptible livestock. Chemoprophylaxis with isometamidium also reduced the number of clinical cases in trypanosome-naive dogs introduced to Africa, together with the use of deltamethrin impregnated collars.

Tsetse control is expensive and some of the methods are environmentally undesirable while chemotherapy has resulted in drug resistance (Kulohoma et al., 2020). Insecticide-impregnated visual targets are not effective on all species of tsetse (Green, 1994). The major constraint to the development of a viable vaccine against animal trypanosomosis is the antigenic variation exhibited by the parasite (Black & Mansfield, 2016).

FUTURE RESEARCH DIRECTIONS

Further work is required to better understand the transmission dynamics of trypanosomes between host and vectors, as well as inter-host transmission.

Technology for the tsetse vector control is available. However robust community adoption is still lacking and better methodologies are therefore required to ensure sustainability in tsetse vector control.

Trypanocidal drug resistance challenge requires that better disease and vector control methods are exploited to reduce over reliance on available drugs. Indeed, trypanotolerance is a more sustainable trypanosomosis control option and available trypanotolerant indigenous livestock breeds need better utilization through structured breeding programs.

CONCLUSION

There is good understanding of AAT among the disease control professionals and effective control methodologies are available. If well utilized, livestock productivity can be enhanced to improve livelihood, food and nutrition security of the animal keepers.

REFERENCES

Abebe, G., & Jobre, Y. (1996). Trypanosomosis: A threat to cattle production in Ethiopia. *Rev. Méd. Vét.*, *147*, 897–902.

Adam, Y., Cecchi, G., Kgori, P., Marcotty, T., Mahama, C., Abavana, M., Anderson, B., Paone, M., Mattioli, R., & Bouyer, J. (2013). The Sequential Aerosol Technique: A Major Component in an Integrated Strategy of Intervention against Riverine Tsetse in Ghana. *PLoS Neglected Tropical Diseases*, *7*(3), e2135. doi:10.1371/journal.pntd.0002135 PMID:23516662

Alingu, R., Muhanguzi, D., MacLeod, E., Waiswa, C., & Fyfe, J. (2014). Bovine trypanosome species prevalence and farmers' trypanosomiasis control methods in south-western Uganda. *Journal of the South African Veterinary Association*, *28*(85(1)). Advance online publication. doi:10.4102/jsava.v85i1.1094 PMID:25686358

Anderson, H., Brodsky, I.E., & Mangalmurti, N.S. (2018). The Evolving Erythrocyte: Red Blood Cells as Modulators of Innate Immunity. *Journal of Immunology, 201*(5), 1343–1351. doi:10.4049/jimmunol.1800565

Batista, J., Riet-Correa, F., Teixeira, M. M. G., Madruga, C. R., Simoes, S. D. V., & Maia, T. F. (2006). Trypanosomiasis by Trypanosoma vivax in cattle in the Brazilian semiarid: Description of an outbreak and lesions in the nervous system. *Veterinary Parasitology*, *143*(2), 174–181. doi:10.1016/j.vetpar.2006.08.017 PMID:16965857

Black, S., & Mansfield, J. (2016). Prospects for vaccination against pathogenic African trypanosomes. Parasite Immunology. *Parasite Immunology*, *38*(12), 735–743. Advance online publication. doi:10.1111/pim.12387 PMID:27636100

Boulange, A., Pillay, D., Chevtzoff, C., Biteau, N., Come de Graca, V., Rempeters, L., Theodoridis, D., & Baltz, T. (2017). Development of a rapid antibody test for point-of-care diagnosis of animal African trypanosomosis. *Veterinary Parasitology*, *233*, 32–38. doi:10.1016/j.vetpar.2016.11.017 PMID:28043385

Chartier, C., Itard, J., Morel, P., & Troncy, P. (2000). *Précis de parasitologie vétérinaire tropicale. Universités francophones*. Aupelf Uref.

Claxton, J., Faye, J., & Rawlings, P. (1991). Trypanosome infections in warthogs (Phacochoerus aethiopicus) in the Gambia. *Veterinary Parasitology*, *41*(3–4), 179–187. PMID:1502780

Davilla, A., & Silva, R. (2000). *Animal trypanosomiasis in South America: Current status, partnership, and information technology*. Academic Press.

Davita, P., Julien, I., Regassa, F., Philipe, B., Hermogenes, M., Luis, N., Alain, B., Momar, T. S., Jeremy, B., Grant, B. N., Cyrille, C., Virginie, C., & Theo, B. (2013). Trypanosoma vivax GM6 Antigen: A Candidate Antigen for Diagnosis of African Animal Trypanosomosis in Cattle. *PLoS One*, *8*(10), e78565. Advance online publication. doi:10.1371/journal.pone.0078565 PMID:24205263

Desquesnes, M. (2017). *Compendium of Diagnostic Protocols of the OIE Reference Laboratory for Animal Trypanosomoses of African Origin*. OIE. https://www.oie.int/nttat/Attached/A16-REC-OMPENDIUM_PROTOCOLES_TRYPANO-En.pdf

Egeru, A., Opio, J., Siya, A., Barasa, B., Magaya, J., & Namaalwa, J. J. (2020). Tsetse Invasion as an Emerging Threat to Socioecological Resilience of Pastoral Communities in Karamoja, Uganda. *Sustainability*, *12*(4), 1599. doi:10.3390u12041599

FAO. (1992). Training Manual for Tsetse control personnel (vol. 4). FAO.

Filip, C., Philipps, B., Louis, T., & Bruno, M. (2005). Trypanosoma equiperdum: Master of disguise or historical mistake? *Trends in Parasitology*, *21*(7), 316–321. doi:10.1016/j.pt.2005.05.010 PMID:15923142

Foil, L. (1989). Tabanids as vectors of disease agents. Parasitology Today, 5:88-96. *Parasitology Today (Personal Ed.)*, *5*, 88–96. doi:10.1016/0169-4758(89)90009-4 PMID:15463186

Gaithuma, A., Yamagishi, J., Martinelli, A., Hayashida, K., Kawai, N., Marsela, M., & Sugimoto, C. (2019). A single test approach for accurate and sensitive detection and taxonomic characterization of Trypanosomes by comprehensive analysis of internal transcribed spacer 1 amplicons. *PLoS Neglected Tropical Diseases*, *13*(2), e0006842. Advance online publication. doi:10.1371/journal.pntd.0006842 PMID:30802245

Galiza, G., Garcia, H., Assis, A., Oliveira, D., Pimentel, L., Dantas, A. F., Simoes, S. V., Teixeira, M. M., & Riet-Correa, F. (2011). High mortality and lesions of the central nervous system in trypanosomosis by Trypanosoma vivax in Brazilian hair sheep. *Veterinary Parasitology*, *182*(2–4), 359–363. doi:10.1016/j.vetpar.2011.05.016 PMID:21664764

Green, C. (1994). Bait methods for tsetse fly control. *Advances in Parasitology*, *34*, 229–291. doi:10.1016/S0065-308X(08)60140-2 PMID:7976751

Griffith, F. (1898). *The Petrie Papyri: Hieratic Papyri from Kahun and Gurob (Principally of the Middle Kingdom)*. Bernard Quaritch.

Gummery, L., Jallow, S., Raftery, A., Bennet, E., Rodgers, J., & Sutton, D. (2020). Comparison of loop-mediated isothermal amplification (LAMP) and PCR for the diagnosis of infection with Trypanosoma brucei ssp. In equids in The Gambia. *PLoS One*, *15*(8), e0237187. Advance online publication. doi:10.1371/journal.pone.0237187 PMID:32833981

Hendrickx, G., & Napala, A. (1997). *Le controle de la trypanosome a la carte: Une approche integree basee sur un Systeme d'Information Geographique*. Acadamie Royale des Sciences d'Oture Mer. Memoires Classe Sciences Naturelles and Medicales.

Isaac, C., Ohiolei, J., Ebhodaghe, F., Igbinosa, I., & Eze, A. (2017). Animal African trypanosomiasis in Nigeria: A long way from elimination/eradication. *Acta Tropica*, *176*, 323–331. doi:10.1016/j.actatropica.2017.08.032 PMID:28870536

Jones, T., & Akinbamijo, O. (1999). Further investigations on the role of a 33 KD antigen in the trypanotolerance of working N'dama bulls. 25th International Scientific Council for Trypanosomosis Research and Control, Mombasa, Kenya.

Kennedy, P. (2009). Cytokines in central nervous system trypanosomiasis: Cause, effect or both? *Transactions of the Royal Society of Tropical Medicine and Hygiene*, *103*(3), 213–214. doi:10.1016/j.trstmh.2008.08.013 PMID:18814893

KETRI. (1997). *Tsetse distribution map of Kenya* [Research Report]. Kenya Trypanosomiasis Research Institute.

Kristjanson, P. B. M., Rowlands, G., Kruska, R., & deLeeuw, P. (1999). Measuring the costs of African animal trypanosomosis: The potential benefits of control and returns to research. *Agricultural Systems*, *59*(1), 79–98. doi:10.1016/S0308-521X(98)00086-9

Kulohoma, B., Wamwenje, S., Wangwe, I., Masila, N., Mirieri, C., & Wambua, L. (2020). Drug resistance in Shimba Hills, Kwale County, Kenya. *BMC Research Notes*, *13*(1), 234. doi:10.118613104-020-05077-3 PMID:32349785

Kuzoe, F., & Schofield, C. (2004). *Strategic review of traps and targets for tsetse and African trypanosomiasis control* [Research report]. WHO.

Leak, S. (1998). *Tsetse Biology and Ecology. Their Role in the Epidemiology and Control of Trypanosomosis.* CABI Publishing in association with ILRI.

Leta, S., Habtamu, Y., Alemayehu, G., Ayele, B., Chanie, M., Tesfaye, S., & Masele, F. (2015). *Spatial analysis of the distribution of tsetse flies in Ethiopia using high resolution environmental datasets and Maxent modeling technique. National Symposium on Trends and Challenges in Adoption of Science, Technology and Innovation in Local Development Endeavours*, Ethiopia.

Maichomo, M., Ndung'u, J. M., Ngare, P., & Ole-Mapenay, I. M. (2005). The performance of Orma Boran and Maasai Zebu crossbreeds in a trypanosomosis endemic area of Nguruman, South-western Kenya. *The Onderstepoort Journal of Veterinary Research*, *72*(1), 87–93. doi:10.4102/ojvr.v72i1.225 PMID:15991707

Maichomo, M., Nginyi, J. M., Caruso-Varès, A., Mungube, E. O., Pagny, G., Chevtzoff, C., Njanja, J. C., Lumumba, P. A., Collin, J. F., & Isaka, N. (2019). Efficacy of VERYL® in the treatment of cattle naturally infected with gastro-intestinal nematodes in Kenya. *Tropical Animal Health and Production*, *52*(4), 1699–1705. doi:10.100711250-019-02179-4 PMID:31865537

Majekodunmi, A., Fajinmi, A., Dongkum, C., Picozzi, K., Thrusfield, M., & Welburn, S. (2013). A longitudinal survey of African animal trypanosomiasis in domestic cattle on the Jos Plateau, Nigeria: Prevalence, distribution and risk factors. *Parasites & Vectors*, *6*(1), 239. doi:10.1186/1756-3305-6-239 PMID:23958205

Malele, I. H., & Msangi, A. (2011). Factors defining the distribution limit of tsetse infestation and the implication for livestock sector in Tanzania. *African Journal of Agricultural Research*, *6*(10), 2341–2347.

Meyer, A., Holt, H., Selby, R., & Guitian, J. (2016). Past and Ongoing Tsetse and Animal Trypanosomiasis Control Operations in Five African Countries: A Systematic Review. *PLoS Neglected Tropical Diseases*, *10*(12), e0005247. Advance online publication. doi:10.1371/journal.pntd.0005247 PMID:28027299

Msangi, A., Mramba, F., Kiwia, N., Kitwika, W., Malele, I., Kasilagila G, Byamungu M, Chalo O, & Athuman J. (1999). Eradication of tsetse fly from Mafia Island after the Zanzibar success. *25th International Scientific Council for Trypanosomosis Research and Control.*

Mugunieri, L., Matete, G., Otieno, K., & Olubai, W. (2003). *Assessment of the impact of animal trypanosomosis on dairy cattle productivity in western Kenya* [Research Report].

Muhanguzi, D., Mugenyi, A., Bigirwa, G., Kamusiime, M., Kitibwa, A., Akurut, G. G., Ochwo, S., Amanyire, W., Okech, S. G., Hattendorf, J., & Tweyongyere, R. (2017). African animal trypanosomiasis as a constraint to livestock health and production in Karamoja region: A detailed qualitative and quantitative assessment. *BMC Veterinary Research*, *13*(1), 355. doi:10.118612917-017-1285-z PMID:29178951

Munsterman, S., Mbura, R., Maloo, S., & Lohr, C. (1992). Trypanosomiasis control in Boran cattle in Kenya: A comparison between chemoprophylaxis and a parasite detection and treatment. *Tropical Animal Health and Production*, *24*, 17–27. doi:10.1007/BF02357230 PMID:1306914

Murray, M., Trail, J., & D'Ieteren, G. (1990). Trypanotolerance in cattle and prospects for the control of trypanosomiasis by selective breeding. *Scientific and Technical Review Office International Des Epizooties*, *9*(2), 369–386. doi:10.20506/rst.9.2.506 PMID:2132686

Mwangi, E., Stevenson, P., Gettinby, G., Reid, S., & Murray, M. (1998). Susceptibility to trypanosomosis of three Bos indicus cattle breeds in areas of differing tsetse fly challenge. *Veterinary Parasitology*, *79*(1), 1–17. doi:10.1016/S0304-4017(98)00147-2 PMID:9777722

Mwongela, G., Kovatch, R., & Fazil, M. (2005). Acute trypanosoma vivax infection in dairy cattle in coast province, Kenya. *Tropical Animal Health and Production*, *13*(1), 63–69. doi:10.1007/BF02237893 PMID:7233561

Naessens, J. (2006). ovine trypanotolerance: A natural ability to prevent severe anaemia and haemophagocytic syndrome? *International Journal for Parasitology*, *36*(5), 521–528. doi:10.1016/j.ijpara.2006.02.012 PMID:16678182

Ngari, N., Gamba, D., Olet, P., Zhao, W., Paone, M., & Cecchi, G. (2020). Developing a national atlas to support the progressive control of tsetse-transmitted animal trypanosomosis in Kenya. *Vectors*, *13*(1), 286. doi:10.118613071-020-04156-5 PMID:32503681

Ngongolo, K., Estes, A. B., Hudson, P. J., & Gwakisa, P. S. (2019). Assessing Risk Factors for Trypanosome Infections in Cattle and Wildlife Interface Areas in Northern Tanzania. *J Infect Dis Epidemiol*, *5*(3). Advance online publication. doi:10.23937/2474-3658/1510078

Njagu, Z., Mihok, S., Kokwaro, E., & Verloo, D. (1999). Isolation of Trypanosoma brucei from the monitor lizard (Varanus niloticus) in an endemic focus of Rhodesian sleeping sickness in Kenya. *Acta Tropica*, *72*(2), 137–148. doi:10.1016/S0001-706X(98)00092-8 PMID:10206114

Njiru, Z., Constance, C., Guya, S., Crowther, K., Kiragu, J., Thompson, R., & Davila, A. (2005). The use of ITS1 rDNA PCR in detecting pathogenic African trypanosomes. *Parasitology Research*, *95*, 186–192. doi:10.100700436-004-1267-5 PMID:15619129

Odeniran, P., & Ademola, I. (2018). A meta-analysis of the prevalence of African animal trypanosomiasis in Nigeria from 1960 to 2017. *Parasites & Vectors*, *11*(1), 280. doi:10.118613071-018-2801-0 PMID:29720251

OIE. (1997). *Manual of standards for diagnostic tests and vaccines* (3rd ed.). World Organization for Animal Health.

OIE. (2012). *Manual of Diagnostic Tests and Vaccines for Terrestrial Animals*. OIE.

OIE. (2018). *OIE Terrestrial Manual. Animal trypanosomosis*. https://www.oie.int/fileadmin/Home/eng/Health_standards/tahm/3.04.16_TRYPANOSOMOSIS.pdf

Okech, G., Watson, E., Luckins, A., & Makawiti, D. (1996). The effect of experimental infection of Boran cattle in early and mid-pregnancy with Trypanosoma vivax. *The British Veterinary Journal*, *152*(4), 441–451. doi:10.1016/S0007-1935(96)80038-8 PMID:8791852

Okoth, S. (2007). *Transmission dynamics and epidemiology of Rhodesian sleeping sickness in allopatric populations of Glossina pallidipes of Kenya* [PhD Thesis]. Kenyatta University.

Onyilagha, C., & Uzonna, J. (2019). Host Immune Responses and Immune Evasion Strategies in African Trypanosomiasis. *Frontiers in Immunology*, *10*, 2738. doi:10.3389/fimmu.2019.02738 PMID:31824512

Paling, R., Leak, S., Katende, J., Kamunya, G., & Moloo, S. (1987). Epidemiology of animal trypanosomiasis on a cattle ranch in Kilifi, Kenya. *Acta Tropica*, *44*, 67–82. PMID:2884841

Paris, J., Murray, M., & McOdimba, A. (1982). A comparative evaluation of the parasitological techniques currently available for the diagnosis of African trypanosomiasis in cattle. *Acta Tropica*, *39*, 307–316. PMID:6131590

Percoma, L. A., & Pagabeleguem, S. (2018). Impact of an integrated control campaign on tsetse populations in Burkina Faso. *Parasites & Vectors*, *11*(1), 270. doi:10.118613071-017-2609-3 PMID:29703229

Peregrine, A. (1994). Chemotherapy and delivery systems: Haemoparasites. *Veterinary Parasitology*, *54*(1 3), 223 248. doi:10.1016/0304 4017(94)90092 2 PMID:7846852

Radostits, O., Blood, D., Gay, C., & Hinchcliff, K. (2000). *Veterinary Medicine: A textbook of the diseases of cattle, sheep, pigs, goats and horses*. Baillière Tindall.

Ramanatha, K., Keshavamurthy, B., & Gopal, T. (1990). Acute trypanosomiasis in zebu cattle (Bos Indicus) in epidemic form-a report. *Indian Journal of Comparative Microbiology. Immunology & Infectious Diseases*, *11*, 158–159.

Roberts, L., Wellde, B., Reardon, M., & Onyango, F. (1989). Mechanical transmission of Trypanosoma brucei rhodesiense by Glossina morsitans morsitans (Diptera: Glossinidae). *Annals of Tropical Medicine and Parasitology*, *83*(sup1), 127–131. doi:10.1080/00034983.1989.11812417 PMID:2619386

Rottcher, D., Schillinger, D., & Sweygarth, E. (1987). Trypanosomosis in the camel. *Rev. Sci. Tech. Off. Int. Epiz.*, *6*, 463–470.

Rowlands, G.J., Kyule, M.N., & Perry, B.D. (1994). *Assessing impacts of animal health interventions on productivity*. Academic Press.

Seyoum, Z., Terefe, G., & Ashenafi, H. (2013). Farmers' perception of impacts of bovine trypanosomosis and tsetse fly in selected districts in Baro-Akobo and Gojeb river basins, Southwestern Ethiopia. *BMC Veterinary Research*, *9*(1), 214. doi:10.1186/1746-6148-9-214 PMID:24139090

Simarro, P., Jannin, J., & Cattand, P. (2008). Eliminating human African trypanosomiasis: Where do we stand and what comes next? *PLoS Medicine*, *5*(2), 174–180. doi:10.1371/journal.pmed.0050055 PMID:18303943

Sivajothi, S., Rayulu, V., Malakondaiah, P., & Sreenivasulu, D. (2016). Diagnosis of Trypanosoma evansi in bovines by indirect ELISA. *Journal of Parasitic Diseases: Official Organ of the Indian Society for Parasitology*, *40*(1), 141–144. doi:10.100712639-014-0465-z PMID:27065614

Spickler, A. (2018). *African Animal Trypanosomosis*. https://www.cfsph.iastate.edu/DiseaseInfo/factsheets.php

Swallow, B. (2000). *Impacts of trypanosomiasis on African agriculture*. Academic Press.

Trail, J., D'Ieteren, G., Murray, M., Ordner, G., Yangari, G., Collardelle, C., Sauveroche, B., Maille, J., & Viviani, P. (1993). Measurement of trypanotolerance criteria and their effect on reproductive performance of N'dama cattle. *Veterinary Parasitology*, *45*(3-4), 241–255. doi:10.1016/0304-4017(93)90079-3 PMID:8447067

Uilenberg, G. (1998). *A field guide for the diagnosis treatment and prevention of African animal trypanosomosis*. FAO.

Vreysen, M. J., Seck, M., Sall, B., & Bouyer, J. (2013). Tsetse flies: Their biology and control using area-wide integrated pest management approaches. *Journal of Invertebrate Pathology*, *112*, 15–25. doi:10.1016/j.jip.2012.07.026 PMID:22878217

Vreysen, M. J. B., Robinson, A., Hendrichs, J., & Kenmore, P. (2007). Area-wide integrated pest management (AW-IPM): Principles, practice and prospects. In Area-wide control of insect pests (pp. 3–33). Springer.

Wamwiri, F., & Changasi, R. (2016). Tsetse Flies (Glossina) as Vectors of Human African Trypanosomiasis: A Review. *BioMed Research International*, *2016*, 1–8. Advance online publication. doi:10.1155/2016/6201350 PMID:27034944

WHO. (2020). *World Health Organization (WHO) fact sheet*. https://www.who.int/news-room/fact-sheets/detail/trypanosomiasis-human-african-(sleeping-sickness)

Woo, T. (1970). The haematocrit centrifuge technique for the diagnosis of African trypanosomiasis. *Acta Tropica*, *27*, 384–386. PMID:4396363

Yaro, M., Munyard, K., Stear, M. J., & Groth, D. M. (2016). Combatting African Animal Trypanosomiasis (AAT) in livestock: The potential role of trypanotolerance. *Veterinary Parasitology*, *225*, 43–52. doi:10.1016/j.vetpar.2016.05.003 PMID:27369574

Chapter 2
Epidemiology and Economic Importance of African Animal Trypanosomiasis

Esther Gwae Kimaro
Tropical Pesticides Research Institute, Tanzania

Popoola Moshood Abiola
Federal College of Animal Health and Production Technology, Ibadan, Nigeria

ABSTRACT

African animal trypanosomiasis (AAT), also called Nagana, is a vector-borne parasitic disease caused by an extracellular protozoan belonging to the genus Trypanosoma. It has serious effects on the health status and welfare of domestic mammals which considerably results in a reduction in their productivity. In this review, a comprehensive overview of the epidemiology of AAT was provided with a special focus on its general clinical aspects (the clinical signs and pathogenesis as well as its transmission cycle), the parasite (Trypanosoma spp.), the parasite life cycle and transmission, its vector (Glossina spp.), tsetse fly lifecycle and reproduction, risk factors of AAT, and economic importance of AAT in the affected countries. The present work gave a detailed account of epidemiology in the context of infestation patterns, the parasite causing it, its vector, and the economic impacts of the disease on different livestock species.

INTRODUCTION

African animal trypanosomiasis (AAT), also known as Nagana, is a vector-borne parasitic disease caused by an extracellular protozoan belonging to the genus, *Trypanosoma* (Cox, 1979). According to the Food and Agriculture Organization (FAO), it is probably the only disease that has affected the settlement and economic development of a major part of African continent. Nagana remains endemic in 37 of the 54 countries in Africa, affecting an area of approximately 10 million square kilometers of arable land, and reduces the efficiency of productivity of over 150 million cattle and 260 million sheep and goats (Leigh et al., 2015; Nyimba et al., 2015). AAT is a very significant economic and animal

DOI: 10.4018/978-1-7998-6433-2.ch002

health issue for this sub-Saharan region (Habila et al., 2012). Furthermore, the disease has an extended impact on the crop, human settlement, and welfare, because seven million square kilometers of the six region's land is rendered unsuitable for mixed crop-livestock ecosystems (Nigatu et al., 2015; Peregrine, 1994). AAT is estimated to cause annual losses of more than US$ 4.5 8 billion dollars through direct and indirect agricultural production costs (Bezie, 2015). It is not surprising, that the 21 countries where Trypanosomiasis is endemic are included in the world's 25 poorest countries (Shaw, 2009), and 32 were considered highly indebted according to IAEA, (2002).

The wide distribution of AAT is attributed to the abundance of its biological and mechanical transmitting vectors known as tsetse flies and other biting flies respectively (Diall, et al., 2017). All warm-blooded animals including wildlife species have been implicated in the transmission cycle for Nagana(Reichard, 2002). Trypanosomiasis is a parasitic disorder that multiplies in the host bloodstream, lymphatic vessels and tissues including the cardiac muscles and the central nervous system. The parasite trypanosomes invade the immune system of the host because it possesses a variable surface antigen (VSG) which prevents lying by complement alternative pathways, (Vincendeau & Bouteille, 2006).

Tsetse flies, the main vectors for trypanosomes occur exclusively in Africa over an area of approximately 10 million km ^{2}, extending on both sides of the equator from 15° N to 30° S(Cattand, 2001). Tsetse flies are of primary importance in the spread and epidemiology of Nagana (Houre, 1976; ILARD, 1990) and thus AAT continues to thrive, and losses incurred are still existing in the affected countries (Meyer, Holt, Selby, & Guitian, 2016; Muhanguzi, et al., 2017). Extensive trypanosomiasis and tsetse (T&T) control operations have been running since the colonial era, however, tsetse infestation in sub-Saharan Africa has scarcely disappeared completely. Furthermore, there is no new veterinary drugs for the treatment of AAT that have been released since 1985(Anene, Chime, Jibike, & Anika, 1991) and at the same time, there is increasing resistance to the existing trypanocides. Following this, there has been a renewed interest in funding committed to AAT control and research into drug discovery and novel control methods since 2000. In addition, the Pan-African Tsetse and Trypanosomiasis Eradication Campaign (PATTEC), whose coordinating office is supported by the African Union, has set a goal for tsetse and AAT elimination. In this case elimination of AAT has been defined as "the reduction to zero in the incidence of AAT in a defined geographical area" (Molyneux, Hopkins, & Zagaria, 2004). This chapter will address the epidemiology of AAT and its possible impacts on livestock productivity. The scientific information given in this chapter will help to inform animal health practitioners, veterinary students, researchers and all interested parties on epidemiology of African animal trypanosomiasis and its economic impacts.

GENERAL CLINICAL ASPECTS:

Clinical Signs and Pathogenesis

Trypanosomiasis in cattle is characterised by two clinical forms: acute seen particularly when animals have just been introduced to a tsetse-infested area and subacute or chronic, the form found mainly in endemic areas (Namangala et al., 2012). Acute cases may be fatal within a week. The primary clinical signs are intermittent fever, anaemia, lymphadenopathy, lacrymation and progressive emaciation. Milk yield may be decreased in dairy animals. Affected animals may show neurological signs, oedema, cardiac lesions, diarrhoea, keratitis and appetite loss (Taylor, 1998). Usually, infected animals are weak and lethargic, and hence lag behind the herd. Effects on reproduction include spontaneous abortion in pregnant animals, premature birth and perinatal losses, as well as testicular damage in males (Naylor, 1971). Deaths are common among chronically infected animals, and animals that recover clinically may relapse when stressed. Trypanosomes can also cause immunosuppression, and concurrent infections such as with *T. parva* and *Ehrlichia ruminantium* infections (Simwango et al., 2017). Coinfections may thus also complicate management of AAT (Boniface Namangala & Odongo, 2014).

The pathogenicity of trypanosome infection varies considerably depending on several factors, including parasite-related aspects (species and virulence), host (species, breed, age, immunological status, nutritional status, presence of co-infection and physical condition), vector (species, density, infection rate and host preference), epidemiological situation (endemic or epidemic) and the environment (e.g. the availability of food and water, and the season) (Cox, 1979). When an infected tsetse fly bites the host it injects the parasites into the bloodstream, and the parasites multiply and invade the body fluids and tissues. This process stimulates an immune response, causing swelling at the bite site, and multiplication and persistence of trypanosomes before dissemination to the blood. Overall, tsetse-transmitted trypanosomiasis is categorised into groups according to the sites of host-parasite interaction. This can occur as chancre, lymphadenopathy, anemia and tissue damage (Uilenberg, 1998). Trypanosomes can be transmitted through the placenta and infect the foetus in pregnant animals, causing abortion or stillbirth in cattle (Gray & Luckins, 1978).

Transmission Cycle

Transmission: Sylvatic transmission cycle is where the disease is maintained in an ecological system which includes the presence of tsetse flies, woody vegetation and wildlife. When livestock is introduced into this ecological system, tsetse fly will use the livestock as their main food source and infect them with trypanosome, (Van den Bossche & Delespaux, 2011). Transmission is through a bite of flies in the animal. The trypanosome undergoes cyclical development in flies lasting 12-35 days before they become infective.Trypanosome is complex in both the tsetse fly vector and mammalian host; trypanosomes undergo a series of transformations into different forms. Though all species of tsetse fly are capable of cyclically transmitting trypanosomes, their importance as vectors depends on feeding habit, relative infectivity and distribution in relation to domestic stock. As flies feed on animals infected with the parasite, they take up blood containing trypanosomes which then completes the life cycle (Houre, 1976; ILARD, 1990).However, *T. equiperdium*, and *T. evansi* are not transmitted sylvatic instead are transmitted cyclically by tsetse fly.

Notably, *T. vivax* infection have been found in areas free from tsetse flies Africa and South American countries like Brazil, Colombia, and Guyana, it is believed and reported that other biting flies such as Tabanidae and Stomoxys are implicated in transmitting the parasite mechanically, (Maxie, Losos, & Tabel, 1979).Whereas tsetse fly might have been introduced by*T. vivax*,the failure to trap tsetse and the presence of a large number of biting flies (*Stomoxys* and *Tabanus taenicola*) strongly suggested that in the outbreak other etiological agents were transmitted mechanically (Conner & Mulkangi, 1986).

Trypanosomes are transmitted to some wild species which serve as a reservoir of infection that may in turn infect domestic animals and people. Previous studies on the analysis of blood meals of tsetse flies collected from wildlife show that tsetse flies feed on a wide variety of wildlife host including primates (baboons, monkeys), suids (warthog, bush pig, red river hog, and giant bush pig) and various Bovidae especially antelopes. In East Africa, the most detected pathogenic trypanosomes in wildlife include *T.*

vivax, T. congolense and *T. brucei* and have been isolated from wildlife in East Africa (Anosa, 1980; Ashcroft, 1959).

THE PARASITE (*Trypanosoma* spp.)

Trypanosoma spp. is a unicellular, microscopic, elongated protozoan parasite that moves with the help of a single flagellum located at the kinetoplast. The genus comprises two main divisions, Salivaria and Stercoraria(Levine, et al., 1980). In the Salivaria group transmission occurs via saliva. The trypanosomes in this group are considered of economic importance, causing diseases in animals and humans(Vickerman, 1985). By contrast, Stercoraria transmission is via fecal contamination, and most trypanosomes in this division are non-pathogenic. The salivarian division consists of three subgenera: *Dutonella, Nannomonas* and *Trypanozoon*. The *Dutonella* consist of *Trypanosoma vivax*, while *Nannomonas* includes *T. congolense* and *T. simiae.Trypanozoon* includes *T. bruceibrucei*, also *T. b. rhodesiense and T. b. gambiense,* which are responsible for human African trypanosomiasis (sleeping sickness) (Levine, et al., 1980). Other species are *T. evansi, T. equiperdum* and *T. suis*, which are found in a diversity of animal species

Table 1. Major characteristic features of animal trypanosomiais syndromes

Parameter	Nagana	Surra	Dourine	Chaga Disease
Aetiology	*T. congolense, T. vivax, T. brucei* subspecies, *T. simiae*	*T. evansi*	*T. equiperdum*	*T. cruzi*
Transmission	Mainly by tsetse flies; other biting flies may play a role	Biting flies	By coitus, non-vector-borne,	Blood-sucking triatomine bugs
Main susceptible domestic animals	Cattle, small ruminants, pigs, horses, dogs	Camels, horses, mules, donkeys, domestic ruminants, Indian elephants, dogs, cats	Horses, donkeys	Dogs, cats, pigs, rodents
Main clinical signs	Intermittent fever, emaciation, anaemia, abortion, lymphadenopathy, oedema of the throat and ventrum, ocular discharge	As for nagana	Mucopurulent discharge from genitalia, oedema of the genitalia, surrounding tissues and ventrum, anaemia, hyperaemia, skin plaques, nervous symptoms	Sudden death, lethargy, tachycardia, exercise intolerance, fever, anorexia, lymphadenopathy, diarrhoea
Geographical distribution	Sub-Saharan Africa	North and North-East Africa, Middle and Far East, Mexico, Central and South America	In all the continents wherever large numbers of horses are reared	Mexico, Central and South America

such as horses, camels, mules and donkeys (Stephen, 1986). Nagana, mainly caused by *Trypanosoma congolense*, followed by *Trypanosoma vivax* and to a lesser extent T*rypanosoma brucei* subspecies. Table 1.1 shows AAT syndrome affecting domestic animals in sub-Saharan Africa.

Parasite Life Cycle/Transmission

Pathogenic trypanosomes are transmitted biologically through the bite of tsetse flies (*Glossina*spp.) except for *T. evansi*and *T. equiperdum,* which are transmitted through coitus and mechanical transmission respectively, (Stephen, 1986). In addition, *T. vivax* is the only species of trypanosome that possesses double transmission ability: biologically and mechanically through blood-sucking insects reported in Mauritius, and Central and South America (Desquesnes & Dia, 2003; Gonzatti, González-Baradat, Aso, & Reyna-Bello, 2014).

The trypanosome undergoes a developmental cycle within the tsetse fly involving substantial morphological, biochemical and physiological transformations to the final metacyclic stage that is infective for a new mammalian host (Fig 1.1). Once inoculated in the mammalian host through blood-feeding, these metacyclic stages undergo development and multiplication at the site of infection, causing a swelling (chancre). Eventually, trypomastigotes are released into the bloodstream through the lymphatic system (Caljon, De Vooght, & Van Den Abbeele, 2014). During their development in the host, the parasites evade the host immune system by changing their surface antigenic coat, the variable surface glycoprotein (VSG) (Hall and Plenderleith, 2014). For T*rypanosoma congolense,* the pathogen localise in capillaries and small blood vessels, where they do attach to vascular endothelial cells because of their body small size. *T.brucei* and *T. vivax* invade tissues of various organs, where they cause damage. The immune response causes the clinical signs exhibited by the vertebrate hosts. This occurs within two to three weeks of infection (Radostitiset al., 2006; Bezie et al., 2014).

Trypanosoma congolense is further grouped into three different classes: Savannah, Forest and Kilifi(Knowles, Betschart, Kukla, Scott, & Majiwa, 1988). The savanna strains of *T. congolense* strains are regarded to be the most pathogenic/virulent and widespread throughout the savannah ecosystem of sub-Saharan Africa (Bengaly Z, et al., 2002). While, *T. congolense* Kilifi reported to be more restricted to East Africa and to a lesser extent in Southern Africa and the Forest type occurs in humid forest ecosystems of West, Central and East Africa(Mamabolo MV, Ntantiso L, A, & Majiwa, 2009). In endemic areas where livestock constitutes the main reservoir of infection, low pathogenic *T. congolense* parasites circulate, resulting in generally low impact of the disease (Chitanga, Namangala, De Deken, & Marcotty, 2013; P. Van den Bossche & Rowlands, 2001). Though, epidemics of *T. congolense* AAT may occur especially following recent introduction of naïve livestock to tsetse-infested areas occurring at the livestock/wildlife interface areas that may lead to increased animal mortalities and reduced calving rates.

Trypanosoma vivax is the second most important cause of nagana to ruminants, pigs and horses in sub-Saharan Africa although is generally regarded as milder than the one caused by *T. congolense*. In West Africa, *T. Vivax* accounts for the majority of AAT cases (Adam, et al., 2012).

Trypanosoma brucei brucei affecting almost every domestic mammal including ruminants, horses and dogs and is cyclically transmitted by tsetse flies. Horses and dogs are highly susceptible to *T. brucei* infection, as well as the human-infective *T. b. rhodesiense*, resulting in acute fulminant disease (Matete, 2003; Stephen, 1986).However, the pathogenicity of *T. b. brucei* and *T. b. rhodesiense* in ruminants is considered to be relatively low. Relatively, ruminants are considered to be domestic reservoirs of these trypanosome subspecies (Fevre, et al., 2001).For *T. b. gambiense*, humans are concerned to be the main reservoirs of the parasite, although studies indicate that domestic animals such as pigs and to a lesser extent, sheep and goats in West and Central Africa are implicated to reservoirs (Cordon-Obras C, et al., 2009; Simo, Asonganyi, Nkinin, Njiokou, & Herder, 2006).

Trypanosoma simiae causes an acute fatal disease mainly in pigs and to a lesser extent in camels and sheep in the tsetse-infested sub-Saharan Africa. Although the other animal trypanosome species and subspecies may cause monolytic infections in specific domestic animals, however co-infections of a combination of two or more different trypanosome species and/or subspecies are frequently reported in the field(B. Namangala, et al., 2012).

Trypanosome evansi (subgenus Trypanozoon) causes a disease called Sura and is transmitted by Surra is the trypanosome syndrome caused by haematophagous Diptera, particularly those belonging to the Tabanidae family. Firstly this protozoa was fisrt discovered as the aetilogical agent of sura for horse in India by Griffith Evans in 1980, the local people determined the disease as the infrequent syndrome which termed as "surra" denoting anything rotten (Stephen, 1986) however surra has similar manifestation as Nagana and it affects mainly camels, horses and water buffalos. Other livestock such as donkeys, mules, goats, sheep, cattle, Indian elephants, dogs and cats are also susceptible to surra (Stephen 1986; Delafosse and Doutoum 2004; Dobson et al. 2009) (Table 1.1). The disease is confined to the northern and north-eastern part of the continent also surra is endemic in Middle and Far East, Asia, Mexico, Central and South America (Stephen 1986; Delafosse and Doutoum 2004; Dobson et al. 2009).

Figure 1. Illustration of the life cycle of trypanosome species in tsetse flies and mammalian hosts.

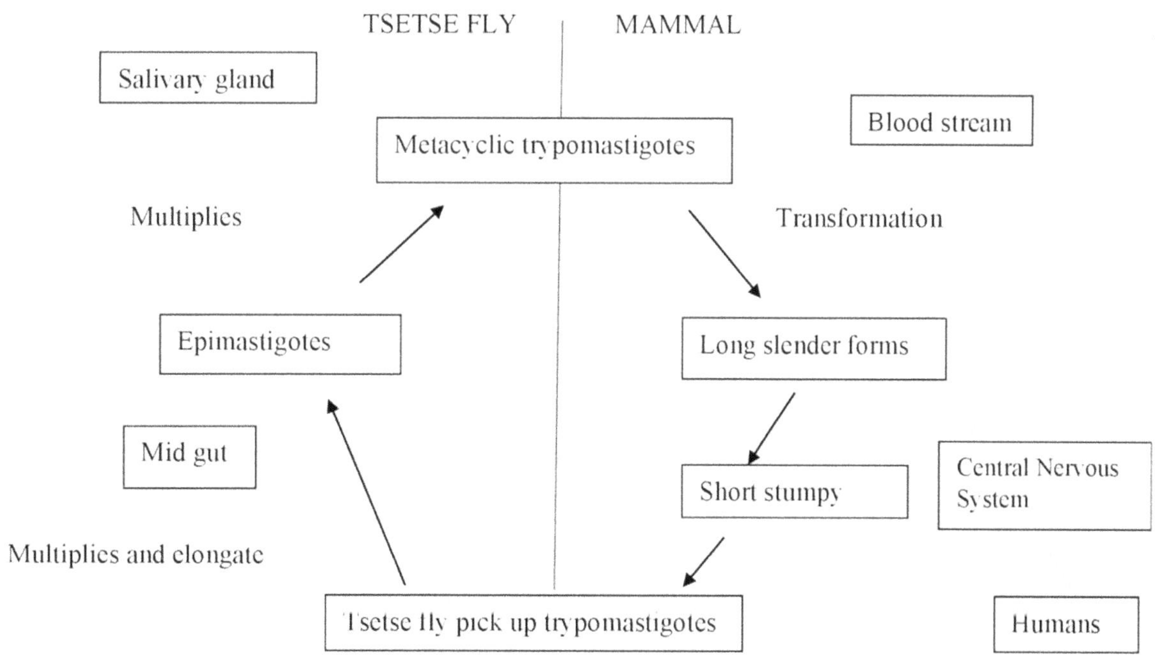

THE VECTOR (*Glossina* spp.)

Animal trypanosomiasis is transmitted through the bite of an infected tsetse fly (genus *Glossina*).Thirty-three species of tsetse fly have been described that belong to the genus *Glossina*(Elsen et al. 1990; Leak 1999).The flies are all restricted to the African continent and are limited by several natural borders, such as the Sahara and Somali deserts in the North and Kalahari and Namibian deserts in the South. Tsetse flies belong to the order of Diptera and based on their reproduction through adenotropic viviparity (Leak 1999; Gooding and Krafsur 2005), tsetse flies are categorized as members of the superfamily Hippoboscidea, the family of Glossinidae and genus of Glossina. The tsetse fly is about the size of a housefly, 6-14 mm in length, although two anatomical characteristics make them distinguishable while resting. Specifically, the tsetse fly possesses biting mouth parts: a long proboscis, which extends forward and is attached by a distinct bulb to the ventral surface of its head. Further, when at rest, one wing rests directly on top of the other over the abdomen and that has a characteristic "hatchet" cell(Pollock, 1992).

Tsetse species form three main groups that are related to different habitats (Table 1.1). The vector is present in 38 countries of sub-Saharan Africa, which covers an area of approximately 10 million km^2.

Table 2. Species of Glossina and their hosts

	Group	Species	Habitat	Host
1	Glossina Morsitans	*G. morsitans*	Mainly found in savannah areas, including open areas and thickets	Warthog, ox, buffalo, kudu and man
		G. pallidipes		Bushbuck, warthog, bushpig, buffalo
		G. swynnertoni		Warthog, buffalo, giraffe, rhinoceros
		G. austen		Bushpig, Ox, duiker
		G. morsitans morsitans		
2	Glossina Palpalis	*G. fuscipes*	Mainly riverine and lakeshore habitats	Reptiles, bushbuck, Man
		G. palpalis		Man, reptiles, Bushbuck, Ox
		G. tachnoides		Man, Ox, Porcupine
		G. palpalis gambiensis		
		G. palpalispalpalis		
3	Glossina Fusca	*G. fusca*	Mainly forest in West and Central Africa	Bushbuck, Red river hog, Ox, hippopotamus
		G. longipennis	Primary forest in lakes banks/sides and riverine	Rhinoceros, Buffalo, elephant
		G. brevipalpis		Bush pig, hippopotamus, bushbuck, buffalo
		G.fuscacongolense		
		G. schwetzi		
		G. fuscipleuris		Bush pig, Giant forest hog aardvark

The upper limit of the geographical distribution is latitude 14° North, while the lower limit is 29° South of Sahel desert, and north of the Namibian and Kalahari Desert.

Group one: Subgenus *Nemorhina* also known aspalpalis group. This group is found in western and East Africa, found in vegetation close to water source, such as forest, gallery forest, riverbanks and lakes, swamps and mangrove, coffee and coco plantations. Importantly, this group of testes fly can even adapt to environmental changes serving on peri-urban areas and in intensive agriculture. This subgenus describes the main vector of sleeping sickness (Robays J, et al., 2004).

Figure 2. Distribution and numbers of tsetse fly species belonging to the Fusca, Morsitans and Palpalis groups in Africa.
Source of figures: https://ergodd.zoo.ox.ac.uk/livatl2/tsetse.htm

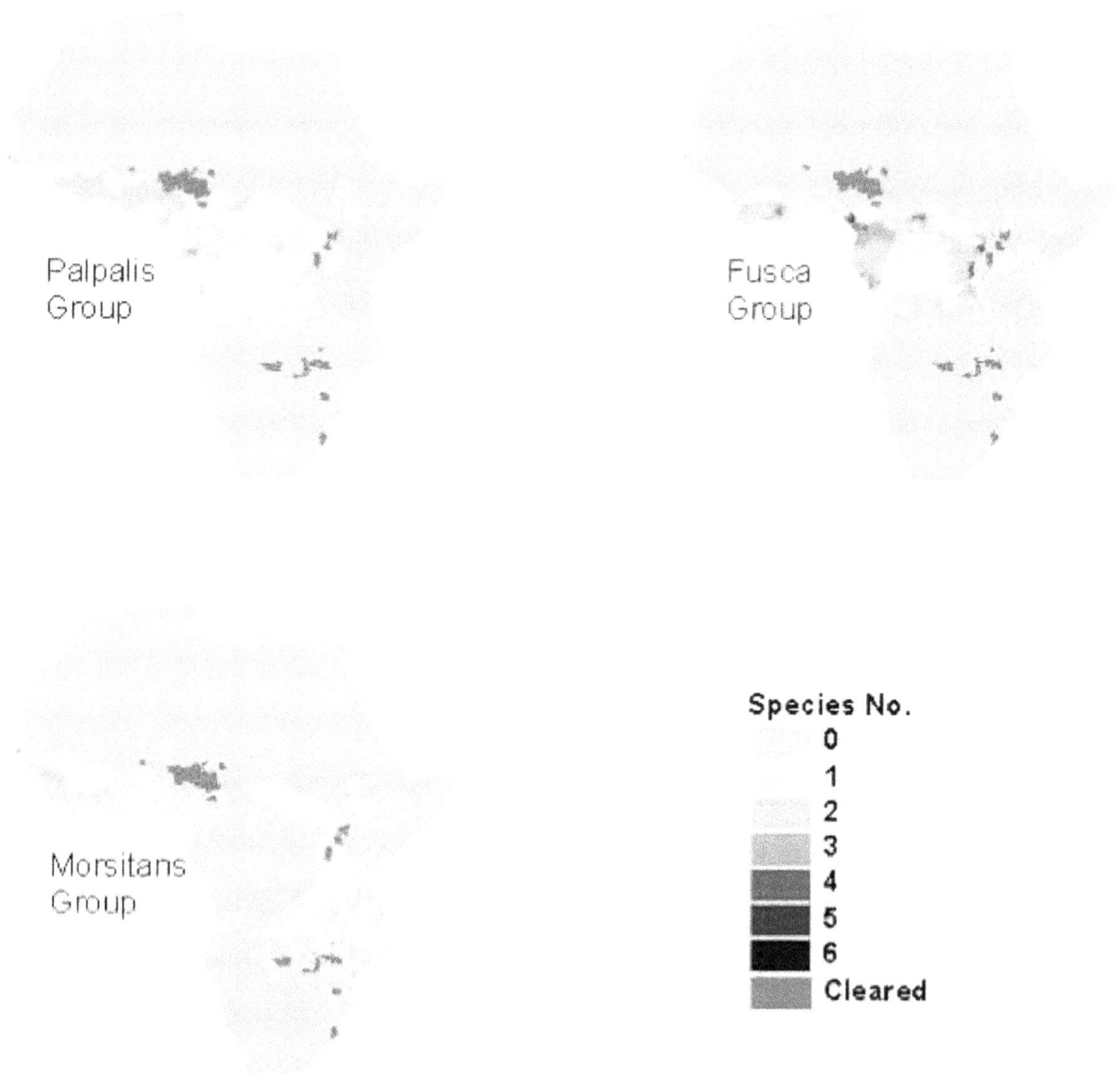

Group two: Sub genus *Glossina:* This one occurs in woodland savannah and linked to the presence of wild fauna and cattle (Table 1.2) Most of these subspecies are located in East Africa and might be involved in the transmission of *T. b rhodesiense*.

Group three: Sub genus *Austenina*: lives in primary forest belts (rainforest, savannah, or coastal forests).Growing human activities in these foreststend to make this subgenusdisappear.

Tsetse Fly Life-Cycle and Reproduction

Both male and female adult tsetse flies feed on blood from a variety of vertebrate hosts every few days and in so doing may transmit pathogenic trypanosomes (Leak, et al., 1993). After mating, the female tsetse fly develops eggs and larvae within her uterus until maturity (Fig 1.3). A full-grown larva is produced every ten days and deposited in a shady area. Once the larva is expelled from the uterus it enters the soil to pupate (Buxton, 1955). The larva burrows into the humid soil in a shady place to a depth of 2−8 cm and pupates almost immediately. The entire process of metamorphosis occurs within the puparium, leading to the adult form of the insect. The period of intrapupal development is variable according to species, sex and climatic conditions. Timing for the development of the pupa varies according to temperature but rapid pupation takes about 15 min with sufficient ground moisture of over 60% (Caljon, et al., 2014; Solano, Ravel, & de Meeûs, 2010). Depending on temperature and humidity, adult flies emerge after 30–40 days(Pollock, 1992).

Tsetse flies are already sexually mature when they emerge from the puparium and mating can occur in the following hours. In natural conditions, almost all females are fertilised as soon as they exit from the puparium. Each female can produce only one offspring at time, producing up to 12 offspring during her typical adult lifespan of 2−3 months (David Rogers, 1979). Males can mate approximately 10 times if mating is adequately spaced and mating time ranges from 30 minutes to three hours (Jordan, 1972). One mating event is generally sufficient for the female fly to produce larvae for several months; spermatozoids can survive for nearly 200 days in the spermatheca. Occasionally, a female can accept several matings to fill her spermatheca(Solano, et al., 2010).

Figure 3. The Glossina life cycle illustrating the development from an intrauterine larva into an adult fly. Adapted from Caljon et al. (2014)

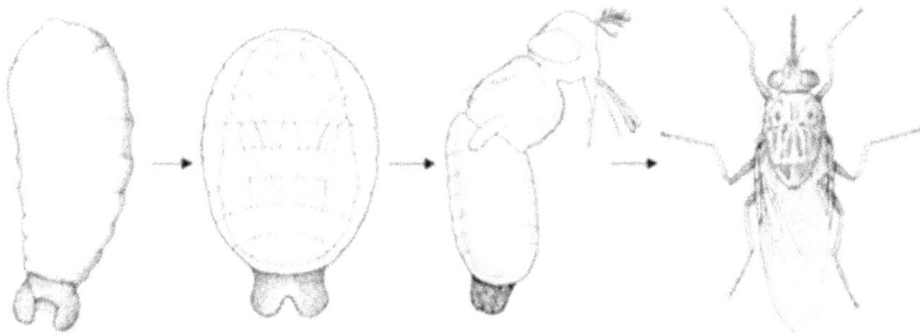

The life span of tsetse flies is highly variable depending on the season, in rainy season the fly can survive longer for 3-5 months while in the dry season it survives shorter 1-2 months and rarely exceeding 7 months.Although females tsetse fly live longer than males, being numerous in natural populations. Both males and females tsetse flies are hemotaphagus thus both are capable of transmitting infection. Normally the newly emerged flies need to immediately feed on a host and are not very discriminatory about the host and thus the chance for the tsetse flies to become infected is the highest on this first blood meal(Welburn & Maudlin, 1992). Meals are taken from 2-4 days (up to 10 days especially when the climatic conditions or the availability of hosts is diversed). Tsetse fly saliva is injected during the blood meal to avoid coagulation and to produce vasodilation if tsetse saliva contains infective trypanosomes, they are transferred to the host. If the host is infected by trypanosomes these parasites are ingested by the fly and it can develop an infection. The feeding preferences of tsetse flies to hosts is triggered bya sense of smell (odour stimuli), from up to 100 m away. Tsetse smells the host, the fly moves up-wind, which brings it closer to the host animal. Then the fly is able to see the host (at 50 m or more). Flies may land on a greater variety of host species and some hosts may prevent flies from feeding by flicking their tails and twitching the skin. Likewise non-hungry males, are attracted to hosts even though they are not going to feed. They make up the 'following swarm', in which a gathering of males tsetse settles on the ground or low vegetation close to a moving host.

Ecology of Tsetse Fly

Dispersal/Spreading

In general, tsetse flies are unable to fly for long periods and instead fly in short bursts, fly only about 15–30 minutes of each day with a relatively low capacity for active dispersal in which the speed of flight may be 3–6 m/sec (11–24 km/h = 7–15 miles/h), but it is much slower immediately after a meal. The average total distance flown per day is between 4.5 and 9 km. However, they can also be passively dispersed in vehicles, on floating vegetation and on animals, and this ability to be carried rapidly over long distances causes a risk of transmission of trypanosomosis to new areas or reintroduction to areas previously under control(Kuzoe & Schofield, 2005). After taking in a large blood meal, tsetse a short, slow flight is made away from the host animal to a resting place such as a tree trunk. In the resting, the weight of the fly is quickly reduced by primary excretion (some drops of water are passed out through the anus). More short flights may then follow, taking the fly to a safer place(FAO, 1992a).

Population Dynamics and Seasonality

The tsetse population is mainly influenced by density-independent (abiotic) factors such as temperature and humidity, which in turn depend on vegetation cover (Table 1.1). For survival, tsetse flies need a temperature between the limits of 17 °C and 36 °C. In general, a temperature above 38°C is damaging to the adults; and a temperature below about 17°C will not allow the adult to live a normal active life. For pupae, 32°C is the upper limit for normal development, and 16°C is the lower limit. The limits of tsetse distribution in the continent is due to high temperatures and dryness (in the hot dry season). But in some areas, seasonal low temperatures are more important. Cold winter months may delay the development of tsetse pupae for so long, that they cannot produce adults successfully. Temperature also affects the rate at which the fly lives (FAO, 1992a).

Increases in temperature also shorten the inter-larval development period, shorten the period spent as a pupa (pupal period), shorten the life of the adult tsetse,shorten the period for adult can last without a blood meal (Torr & Hargrove, 1999). Tsetse fly density is determined by suitable habitat and availability of hosts, while the latter is influenced by human activities such as the expansion of human settlement and agriculture, deforestation, livestock movements and habitat fragmentation (David Rogers, 1979; D. Rogers, 1985).

THE RESERVOIR AND THE HOST: THE ANIMALS IN THE CYCLE OF ANIMAL TRYPANOSOMIASIS

Tsetse are capable to find their hosts by their sense of smell, and by sight.Specific host species are preferred by tsetse flies possibly for the following reasons:

a. The hosts occupy the same habitat as the tsetse.
b. The smell and sight of the host is attractive to the tsetse.
c. The hosts remain fairly quiet and still when fed upon by tsetse.

Albeit, tsetse flies do not necessarily feed on any hosts that happen to be in the same habitat. For instance, the common wild animals in the game areas such as, zebra and wildebeest, are not fed upon by tsetse. The reasons for this are not fully understood.The dependence of tsetse species on various hosts has been reported from averaging out blood meal analysis data from various studies made in many parts of Africa. Domestic animals (cattle, pigs, goats, sheep, horses, and dogs) get infected following encroachment of humans and their animals near or into tsetse-infested areas. Tsetse flies may pick bloodstream trypomastigotes from wildlife when feeding and transmit them as metacyclics to such domestic animal. Larger animals such as cattle (mainly adult cows and oxen), possibly due to the larger odour plumes they produce, and indeed their greater availability as compared to others livestock, are more attractive to tsetse flies than smaller animals that are usually kept in fewer numbers (Simukoko, Marcotty, Phiri, Vercruysse, & Van den Bossche, 2007).

Table 1.2 indicate tsetse species and their hosts it serves as a rough guide for tsetse hosts, this information was extracted from various studies for each of thirteen tsetse species. Some species of wild animals are very important as hosts for tsetse, and other are less important. The type of host that is important may depend on the availability of the host animal, the species of tsetse, and on the season. For example, the pig family (Suidae), members of the pig family are very important as hosts oftsetse.The pig family includes the warthog, bushpig, red river hog and the giant forest hog, as well as the domestic pig. The other most important family is, the cattle (ox) Bovidae, antelope and buffalo. The bovids include a large number of important species, especially antelopes such as bushbuck, kudu and eland, as well as buffalo and cattle. In general, the larger members may be important hosts, but the smaller ones are much less important. Other important hosts for tsetse flies are primates, predisposing man. The primates are a group of mammals into that also include man, monkey and baboon. Other mammals sometimes used as hosts are rhinoceros, elephant, hippopotamus, and giraffe. Other manuals seldom used as hosts are Zebra, wildebeest, Many small antelopes. The following common wild game animals are fed upon only quite rarely: Waterbuck and Impala (FAO, 1992b).

For cattle breeds there are some evidence that significant differences in resistance to animal trypanosomiasis occur among various Zebu-Bos indicus types, these breeds are genetically resistant to the development of clinical signs, a phenomenon known as trypanotolerance. The breeds include West African Shorthorn zebu (which may also be known as Muturu, Baoule, Laguna/ Lagune, Namchi, Samba/ Somba or Dahomey), also N'Dama cattle, Djallonke . For small ruminants such as Red Maasai sheep, West African Dwarf and Small East African goats are also regarded as trypanotolerant brreds (Griffin and Allonby, 1979). Some of these trypanotolerant breeds tend to self-cure and eliminate the pathogen (Trypanosomes); while others may remain persistently infected but maintain productivity and show few or no clinical signs. Nonetheless, trypanotolerance is not absolute, and resistant animals may require some treatment in areas where the challenge intensity is very high. Trypanotolerance varies with the pathogen species, and some breeds are susceptible to the strains of *T. vivax* that cause hemorrhagic disease (Spickler 2018). In addition, there is also some evidence that significant differences in resistance to trypanosomiasis occur among various Zebu-Bos indicus types across the continent such as East African Short Horn Zebu, where there was no introduction of tolerant breeds. Most Bos indicus cattle in tsetse-infested areas require regular treatment (Njogu et al., 1985).

RISK FACTORS OF ANIMAL TRYPANOSOMIASIS

African Animal trypanosomosis is one of the major constraints to improved livestock and agriculture production in the affected countries and remains one of the major causes of livestock production loss in the livestock industry (Swallow, 1999). In cattle, trypanosomosis directly affects meat and dairy productivity and increases livestock keepers' vulnerability to economic and food insecurity (Holt et al., 2016). About 33% of the suitable grazing land of Tanzania is infested by tsetse flies, causing a considerable impact on livestock development (Daffa et al., 2013). Cattle are the main species affected by trypanosomes, due to their status as tsetse flies' feeding preference compared to other domestic animals.

There are several factors reported to influence the prevalence and incidence of AAT (Biyazen et al., 2014; Alemu and Alemneh, 2017). Adult animals are reported to have a higher incidence of the disease than young animals (Swai and Kaaya, 2012; Alemu and Alemneh, 2017; Dagnachew et al., 2017). This association relates to seasonal movements of animals, in which adult animals travel long distances in search of adequate food and water while young animals usually graze around homesteads. This means adult animals can have greater exposure than the young ones.The correlation of AAT prevalence and age of the animal is well observed in extensive grazing/endemic areas (Simukoko et al., 2007; Karimuribo et al., 2011). A link between the sex of the animal and the disease incidence is also reported (Teka et al., 2012; Alemu and Alemneh, 2017). Male animals are reported to have higher rates of infection than female animals (Swai and Kaaya, 2012; Duguma et al., 2015; Dagnachew et al., 2017). However, this association was not found to be consistent (Biyazen et al., 2014). A higher proportion of male animals are involved in draught activities and this increases their exposure rates to infected tsetse flies (Alemu and Alemneh, 2017).

Cattle breed also influences the prevalence of AAT due to the presence of trypanotolerance traits (Dumesa and Demessie, 2015).This trait describes the innate ability of an animal to control parasitaemia levels and relates to the regulation of parasite growth as well as how rapidly the immune response is triggered. This trait is generally recognised in the West African taurine breeds and the Shorthorn breeds of Central Africa, namely, the N'Dama and Muturu cattle (Murray et al 1990; ILARD, 1989). These breeds can be maintained in endemic areas where it is impossible to keep zebu cattle. In addition, some east African Zebu-*Bos indicus* types such as East African Short Horn Zebu are also trypanotolerant, albeit to a lesser degree (Dolan, 1998; Magona and Mayende, 2002). Most *Bos indicus* cattle in tsetse-infested areas require regular treatment, however (Njogu et al., 1985).

Anthropogenic Changes and the Epidemiology of AAT

Tsetse flies, like many other arthropods, rely on habitats for their survival and development with appropriate biotic and abiotic conditions (P Van Den Bossche, Rocque, Hendrickx, & Bouyer, 2010). Each subgenus of tsetse fly differs in its habitat (Table 1.1) and this diversity influences the interactions between hosts (human, livestock and/or wildlife), pathogens and vectors. Environmental changes can considerably alter tsetse fly dynamics and thus influence disease transmission, pathogenicity and impact (Reid, Kruska, Deichmann, Thornton, & Leak, 2000; Van Den Bossche, et al., 2010). The potential drivers for environmental change include climate change, agriculture development, urbanisation and changes in land use and cover, livestock-production systems and habitat fragmentation (deforestation) (Reid, et al., 2000; David Rogers, 1979). Drastic changes in the habitat of tsetse flies have in some instances resulted in a reduced population of tsetse flies and density in various tsetse infested countries including some regions of West Africa, South Africa and Tanzania (Daffa, Byamungu, Nsengwa, Mwambembe, & Mleche, 2013; Van den Bossche & Delespaux, 2011; Van Den Bossche, et al., 2010).

African Animal Trypanosomosis and the Changing Environment

The effects of anthropogenic changes have influenced the epidemiology of AAT, as illustrated by Van Den Bossche et al. (2010). Three distinct epidemiological scenarios of AAT have been identified; for example, in South Africa, where a range of long-term studies has demonstrated the effects of human pressure on the epidemiology of animal trypanosomosis. The first scenario occurs when livestock are introduced into game areas where wildlife are abundant and constitute the main source of food for tsetse flies. In these areas wildlife acts as an important reservoir of tsetse flies and trypanosomes, so the introduction of people and their livestock into such areas due to human encroachments lead to epidemics of animal trypanosomosis (Van den Bossche et al., 2000).

The second scenario involves livestock living at the edge of tsetse-infested wildlife zones. These areas create a wildlife-livestock interface between tsetse infested and non-infested areas and become a threat to livestock entering this zone. For instance, in Tanzania tsetse flies are more confined to the protected National Wildlife Parks, Game Reserves and Forest Reserves (Auty, et al., 2012; Malele, et al., 2011). During animal seasonal movements pastoralists move their cattle into proximity to the wildlife national parks in search of adequate pasture, increasing cattle-wildlife contact and exposure to tsetse flies (Gustafson, et al., 2015). A similar situation occurs in West Africa (Reid, et al., 2000).

The third scenario (also termed 'domestic form' of AAT) occurs because of human interference and habitat change, in areas where the population density of wildlife is low and livestock constitute the main source of blood meal for tsetse flies (Reid et al., 2000). The progressive clearing of natural vegetation, expansion of human settlement and agricultural development, and disappearance of large wildlife species due to natural or unnatural causes have had repercussions in the distribution, density and dispersal, and the life span of tsetse flies (Van den Bossche et al., 2000).

ECONOMIC IMPORTANCE OF AAT IN LIVESTOCK

The AAT is particularly important in cattle and it is mainly caused by *Trypanosoma congolense*, especially Savannah type, the most pathogenic form (Bengaly et al., 2002), *T. vivax*, the most prevalent, and, to a lesser extent, *T. brucei*. However, other hosts such as horses, donkeys, camels, goats, sheep, pigs, dogs and even humans may be affected.

AAT is a disease complex caused by one or several of Trypanosoma spp., transmitted mainly cyclically by the genus Glossina (tsetse flies), but also mechanically by biting flies such as tabanids and Stomoxys

AAT is one of the major constraints affecting the economy and productivity of livestock. Despite the control measures implemented for several years of livestock development projects, bovine trypanosomiasis persists and continues to be a major obstacle to improving livestock productivity (Sas et al., 2019). Trypanosome parasites usually invade the animal's lymphatic vessels, the blood circulation, and eventually the brain causing a wide range of pathologies, most commonly severe anaemia, weight loss, foetal abortion, and cachexia. It can result in the death of the host if left untreated (Stijlemans et al., 2015). Other symptoms reported for African Trypanosomiasis include infertility, sleeping disorders, psychiatric disorders, paralysis, neuroendocrine dysfunctions and coma (Steverding, 2008).

Direct and Indirect Economic Importance of AAT in Livestock

Direct Economic Importance of AAT in Livestock

A report by Swallow (2000) and Samdi et al (2010) indicated that trypanosomiasis has a direct impact on livestock productivity, reducing meat and milk off-take by 20%, calving rate by 20% increase, calf mortality by 20%, decreases both lambing and kidding rates in sheep and goat and livestock management especially the number of livestock kept by farmers, the breed and species composition of the livestock herd, the way the livestock are grazed, cost of trypanocidal drugs and cost of insecticides. Trypanosomiasis also causes livestock farmers who are majorly pastoralists to move from one place to another which ultimately leads to shift in land use and land cover. In an outbreak of trypanosomiasis that causes large losses in livestock numbers, people leave tsetse affected area to controlled area thereby changes settlement patterns and increase the population density of the area (Swallow, 2000). Abenga et al (2004) reported an outbreak of bovine trypanosomiasis in North Central Nigeria which led to livestock deaths and seasonal migration of semi-nomadic Fulanis of the area. Trypanosomiasis can also be an important constraint on migration and human settlement patterns. Reid et al. (1999) studied the impacts of an outbreak of trypanosomiasis in the Ghibe Valley of Ethiopia on land use and land cover. The main factor driving changes in land use and land cover was migration – rapid movement of large numbers of people out of the area – and changes in settlement patterns – with many households moving away from the low-laying areas near rivers. A similar study conducted in the Satiri and Bekuy areas of southern Burkina Faso did not find any evidence of an association between tsetse control and migrants' decisions of when and where to settle (Kamuanga et al., 1999). Another study conducted in Gokwe District in the Zambezi Valley of Zimbabwe found evidence of a positive impact of trypanosomiasis control on migration (Govereh, 1999). Overall it appears that trypanosomiasis can have major impacts on migration and settlement. The most rapid impacts may occur when an outbreak of trypanosomiasis occurs that causes large losses in livestock numbers. Reductions in trypanosomiasis risk have less rapid impacts on in-migration. The case study evidence suggests that the magnitude of the inmigration will depend upon the origin of the control itself. Areas cleared of tsetse are much more attractive to potential immigrants. This trend is particularly worrying since the riverine forests are a main concentration of biological diversity within the landscape (Reid et al., 1997b). In Western Senegal, Seck et al., (2010) reported that AAT remains a major problem in the area. In Southern Senegal, AAT reduces significantly animal draught power and hematocrite in trypanotolerant breeds such as Ndama (Seck et al., 2002).

Livestock have the potential to be transformative by enhancing food and nutrition security, and providing income to pay for education and other needs, livestock can enable poor children to develop into healthy, well-educated, productive adults (Smith et al., 2015). Thus, the economic value of livestock to development in food security is indispensable. The occurrence of AAT has directly affected the productivity of livestock populations especially cattle, sheep, goats and pigs. As a result of the devastating effect of the disease, animal protein output had not been able to match national demand which results in a shortage in animal protein intake (Samdi et al., 2010), hence its effect on food security.

The disease has devastating effects on the livelihoods of local farmers, for whom ruminant animals most especially cattle, sheep and goat represent not only a source of food (meat and milk), manure, and draught power but have also fundamental social roles as ‘living banks’ and are used for social obligations such as dowry and ritual use (Grace et al. 2009; Mungube et al. 2012) as well as a form of social status as found among some tribes in Sub-Saharan Africa. Tsetse flies mainly affect the rural poor and are considered a root cause of poverty in Africa because they are solely responsible for the cyclical transmission of trypanosomes for humans, it causes a disease called ‘sleeping sickness’ or Human African Trypanosomosis (HAT) is an acute and rapidly progressive disease in this way, the disease contributes to maintaining the poverty cycle in these rural communities. AAT is the greatest health constraint to increased livestock production in sub-Saharan Africa with direct annual production losses in cattle estimated at US$ 600–1200 million (Hursey and Slingenbergh, 1995; Kame-Ngasse et al., 2018).

The cost incurred is another direct effect of AAT, in terms of loss of animals which worth a huge amount of money as well as the cost of treating the disease. Annual losses in meat production are estimated at US$ billion. If losses in milk, manure production and traction could be prevented, the benefit from livestock and mixed agricultural developments in tsetse infected Africa could amount to US$50 billion annually (Agyemang, 2005; Shaw et al., 2013; Kame-Ngasse et al., 2018). African animal trypanosomiasis costs Africa US$5 billion a year and Africa spends every year at least $30 million to control cattle trypanosomiasis in terms of curative and prophylactic treatments. In Burkina Faso, livestock owners´ estimates in the Southern part of the country revealed that livestock losses ranged from 75% to 85% in individual herds between 1989 and 1992 (Kamuanga et al., 1995). In Uganda, livestock keepers meet the costs of most veterinary services (Okello and Welburn, 2014), which implies that AAT control programs will need to tailor tsetse control to livestock keepers’ budgets to be sustainable (Torr et al., 2007). According to Muhanguzi et al (2015) in a study conducted in south-eastern Uganda, the authors reported cost analyses of low-cost tsetse control options should include full delivery costs as a result the relatively low cost of a restricted insecticide application protocol for AAT control and its collateral impact on tick control make it an attractive option for livestock management by smallholder livestock keepers.

Indirect Impact of Trypanosomiasis

The AAT has indirect economic effects on livestock, crops and humans. One of the indirect losses of AAT is due to severe immunosuppression shaped by the disease leading to failure of vaccination against bacterial and viral infections (Singla et al., 2010), or by increasing susceptibility of the host to other infections. The indirect impact of trypanosomiasis mostly lies in crop production; through the availability and cost of animals that provide traction power (Swallow, 2000). Evaluation of the effect of trypanosomiasis incidence on the productivity of animals used for traction showed that relative inefficiency in the high-risk area was 38% less efficient than the animals in the low-risk area (Leak et al., 1995).

The overall impact extends to the restricted access to fertile and cultivable areas, imbalances in land use and exploitation of natural resources, and compromised growth and diversification of crop-livestock production systems (Mattioli et al., 2004). The presence of tsetse flies and animal trypanosomiasis in much of Africa south of the Sahara also had a major influence on the agricultural systems. In some Central African countries, there are still extensive areas of relatively undeveloped land such that only trypanotolerant breeds of domestic livestock can be kept in such areas without chemoprophylaxis.

Economic Importance of AAT in Livestock Species

Economic Importance of AAT in Cattle

In 2000, the Pan African Tsetse and trypanosomiasis Global Veterinaria, Eradication campaign (PAT-TEC) asserted that Tsetse (*Glossina spp.*) and AAT is an important constraint to livestock production and a threat to food security in sub-Saharan Africa. The production losses in cattle due to trypanosome infections have been estimated to be up to 20% which cuts across mortality, calving rate, draft power, meat, and milk production. According to Swallow (2000), a high tsetse-trypanosome burden constrains the use of land for livestock production, with most farmers often being more reliant on crop farming. However, trypanosomiasis also compromises crop production by reducing the availability of draft animals to plough fields and provide manure for fertilizer. Despite the yearly administration of 35 million doses of trypanocidal drugs (at US$ 1 per dose), African farmers lose 3 million cattle every year to the disease and annual direct economic losses are estimated at US$ 600 to 1200 million.

Attempts to quantify the impact of the disease have taken into account the direct (which includes mortality, fertility, milk production, animal traction and weight) and indirect effects on key productivity measures of the animal. Livestock kept under trypanosomiasis challenge has 6-20% point higher annual calf mortality, a 6-19% point lower calving rate and a 20% decrease in milk yield (Shaw, 2004). Untreated, the disease often becomes deadly. A series of indirect negative effects can exacerbate the situation, meaning that AAT impedes the use of productive exotic or cross-breeds of cattle; depresses the population growth of cattle; affects the distribution of livestock, herd size and structure; makes mixed farming impossible; affects human settlements and results in less draught power and less manure (Feldmann et al., 2005). Mariam (2006) has reported an economic loss of 70 million US dollars annually due to trypanosome infection of cattle in six northern states of Nigeria.

Economic Importance of AAT in Camel

Trypanosoma evansi is a species belonging to the subgenus *Trypanozoon* and is the causative agent of camel trypanosomiasis also known as surra. Camels get into contact with tsetse flies acquired infections, and when such camels moved to non-tsetse areas, the transmission is likely to spread the disease via haematophagous flies like those belonging to the genus *Stomoxys* and the family *Tabanidae*. Luckins (1998) asserted that *Trypanosoma evansi* originated from *Trypanosoma brucei* by adaptation to a non-cyclical mode of transmission and loss of ability to undergo growth and differentiation in the fly vector

(Luckins, 1998). Other species of trypanosomes, e.g. *Trypanosoma congolense, Trypanosoma brucei and Trypanosoma vivax* have also been isolated from camels in Sudan. According to Njiru et al (2002), it is the most important single cause of economic losses in camel rearing areas, as it causes morbidity of up to 30.0% and mortality of around 3.0%. Since its discovery in camel, studies have shown that the parasite can infect all species of domesticated livestock (Al-Rawashdeh et al., 2000). Economically, camel trypanosomiasis, pose a major constraint to camel productivity given their importance as a source of meat, milk production, transportation, and draught power, for wool, hair, skin and hides production.

In camels, the disease is manifested by elevation of body temperature which is directly associated with parasitaemia. Infected animals show progressive anaemia, marked depression, dullness, loss of condition, and often rapid death. Some camels develop oedema in their dependent parts of the body, urticaria plaques and petechial haemorrhages in serous membranes and if these animals are not treated, it may finally lead to their death. Studies have also reported various constraints this epidemic can pose to productivity - death (Tuntsuvan et al., 1997); abortion (Lohr et al., 1986); weight loss, weakness and eventually reduced draught power (Luckins, 1998) and nervous signs like circling movement and trembling, unusual aggressiveness, running aimlessly and sudden collapse in severely stressed and overworked animals (Manuel, 1998). A post mortem, necrotic foci in the liver and spleen, as well as generalised lymphoid tissue hyperplaisia, are common in camels suffering from surra (Rottcher et al., 1987).

Economic Importance of AAT in Sheep and Goats

Trypanasome congolense is one of the trypanosome species causing AAT in sheep and goats. Samdi et al., (2010) reported that trypanosomiasis decreases both lambing and kidding rates in sheep and goats. Griffin and Allonby (1979) studied the economic effects of naturally acquired trypanosomiasis in sheep and goats over 35 weeks under range conditions in Kenya, these authors reported there was a financial loss from reduced weight gain and death of sheep and goats.

IMPACT SUCCESS CASES OF AAT INTERVENTION IN SUB-SAHARAN AFRICA

Animal African Trypanosomiasis (AAT) and its primary vector, the tsetse, are among the biggest

Constraints to sustainable livestock production in Africa (Swallow, 1999). Despite the fact that extensive AAT control operations have been running since the beginning of the 20th century, tsetse infestation in sub-Saharan Africa has hardly receded. According to (Swallow, 1999). AAT incidence was very high in the 1980s-1990s, where cattle herds almost disappeared in some areas, such as the Ghibe Valley of Ethiopia and the Yale Province of Burkina Faso. From 1986 to 1995, four Southern African countries received assistance and funding from the Regional Tsetse and Trypanosomiasis Control Programme (RTTCP) to eliminate tsetse in the region. This program, mostly funded by the European Commission, targeted the common tsetse belt in Malawi, Mozambique, Zambia and Zimbabwe (Bourn, 2001). In the late 1990's in sub-Saharan Africa several programmes were established to address the issue of AAT and control the disease. In 1997, several programmes were established which was led by the FAO includes the development of an Atlas of tsetse and AAT which aims to build and regularly update a geospatial database of tsetse species occurrence and AAT at the continental level (Cecchi et al., 2014). The Farming In Tsetse Controlled Areas (FITCA) programme was implemented from 1997 to 2004 in East Africa.

Funded by the European Development Fund, it operated in Kenya, Uganda, Ethiopia and Tanzania. In Africa, there are lots of impact success cases of AAT which are discussed as follows:

Impact Success Cases of AAT Interventions in Burkina Faso

Three tsetse species are the primary vectors of bovine trypanosomiasis in Burkina Faso: *Glossina palpalis gambiensis* and *G. tachinoides* infest the riverine forests, whereas *G. morsitans submorsitans* subsists in some savannah areas (Van den Bossche et al., 2010). The Pan African Tsetse and Trypanosomiasis Eradication Campaign (PATTEC) established a campaign is to repel the northern limit of tsetse distribution in Burkina Faso gradually southwards, the strategy is called 'rolling carpet strategy' (Sow, 2013). The longstanding involvement of the Centre de Recherche sur les Trypanosomoses Animales and later the Centre International de Recherche-DeÂveloppement sur l'EÂlevage en zone Subhumide (CIRDES) in tsetse and trypanosomiasis (T&T) control in Burkina Faso has shaped the operations carried out in the last 35 years. The PATTEC and the Food and Agriculture Organisation (FAO) are two of the other organisations heavily involved in T&T control in the country. Early research-oriented field experiments such as those in SideÂradougou and YaleÂ, designed as proofs of concept, were later followed by development-oriented projects (Meyer et al., 2016).

Between 1993-1995, CIRDES-CIRAD established a programme to control AAT pandemic in Dafinso (Burkina Faso), with 100% reduction in tsetse level, the AAT epidemic was controlled, although, there was return of AAT after end of campaign but was perceived as less problematic (Grace, 2003). Also between 1994-1997, another intervention programme was established to control AAT pandemic by CIRDES-ILRI in Yalê (pastoral development zone of Burkina Faso) which resulted in 80% reduction of AAT level in the area. However, tsetse reinvasion and return of AAT cases after end of campaign (Bauer et al., 1999). A tsetse factory near Bobo-Dioulasso was initiated under the PATTEC project and is continuing with financial support from the Ministry of Animal Resources.

Impact Success Cases of AAT Interventions in Cameroun

In Cameroon two thirds of the territory and 90% of the 6 million cattle are at risk of AAT. In the country, Plateau, was the first place invaded by tsetse in 1950, and was host two species responsible for the transmission of AAT, these are: *G. morsitans submorsitans*, and *G. tachinoides* (Mamoudou et al., 2008). The Mission SpeÂciale d'Eradication des Glossines (MSEG) was established in 1979 within the Ministry for Animal Production as a specialised tsetse unit incorporating existing vector control units. It was responsible for a long-term tsetse control program in the Adamawa region which ran from 1976 to 1994. Preventive and curative trypanocidal drugs were administered to cattle by government services from 1960 to 1975 in the regions of Adamawa, East and North. In the 1970s, control operations were reoriented towards vector control. They mainly targeted the Adamawa plateau, but DDT ground spraying operations were also conducted in the Far North region of Cameroon and adjacent regions of Nigeria (Mamoudou et al., 2006). MSEG-PASSEC campaigns held in Adamawa, North and Far North to eliminate AAT was faced with delays in obtaining funding as a result the tentative regional programmes were aborted (Banipe, 2012).

Impact Success Cases of AAT Interventions in Ethiopia

Four species, *G. morsitans submorsitans, G. pallidipes, G. fuscipes fuscipes* and *G. tachinoides* are responsible for most AAT transmission in Ethiopia (Dinka, 2005). Countrywide chemotherapy operations started in cattle during the 1980s, but failed to control the disease which almost eliminated livestock in some areas, such as the Upper Didessa Valley (Jemal and Hugh-Jones 1995). Failure of chemotherapy control was partly attributed to widespread drug resistance in this area with some strains becoming resistant to all three drugs used for cattle treatment (Rowlands et al., 1999). Therefore the government, assisted by international organisations, started to implement vector control campaigns in some areas. The government branch responsible for T&T control is the National Tsetse and Trypanosomiasis Investigation and Control Centre, which comes under the Ministry of Agriculture. Besides government-lead operations, some non-governmental organisations, such as FARM-Africa, and research institutes, such as the International Livestock Research Institute (ILRI), also conducted T&T control campaigns in Ethiopia. In Konso, the FARM-Africa established a campaign programme to control AAT between 1995-2000, although there was 100% reduction of AAT prevalence, the programme was faced with low level of coverage which could be explained by treatment fee and lack of information as a result, the programme was not sustainable with possible tsetse reinvasion (Pound Jonfa 2006).

Impact Success Cases of AAT Interventions in Uganda

In Uganda, four species of flies are epidemiologically important in the country: G. *fuscipes fuscipes, G. morsitans submorsitans, G. pallipides* and *G. brevipalpis*. The two latter species were eliminated in some areas in the 1980s, following drastic reduction in their wildlife hosts. *G. fuscipes fuscipes* is an opportunistic feeder and more resilient to the anthropization of its habitat (Okoth, 1999). However, reinvasion by *G. pallipides* into several districts has been reported since the end of the 1990s, increasing the risk of AAT in livestock (Magona et al., 2005). Several control campaign operations were conducted and reported. Between 1991-1993 in Busia, Livestock Health Research Institute (LHRI) established a campaign programme to control AAT in which AAT prevalence was reduced to 92%, although the programme was faced with diminution of coverage which led to a disease upsurge (Magona et al., 2004). In Tororo between 1991-1995, similar campaign programme was established which reduced AAT to 94% but the programme was faced with persistence of mechanical transmission and there was lack of funding but AAT prevalence still lower in control area in 2000 (Magona et al., 2000). Farming In Tsetse Controlled Areas project (FITCA) in 1999-2004 at Busoga established campaign programme to control AAT and HAT which was challenged by low level of coverage and farmers' groups did not persist in time. Thus the programme was not sustainable and there was re-invasion of AAT in 2009 (Selby, 2010).

Impact Success Cases of AAT Interventions in Zambia

Four tsetse species are of epidemiological importance in Zambia: *G. morsitans morsitans, G. morsitans centralis, G. pallidipes* and *G. brevipalpis* (Symeonakis et al., 2007). During the past 50 years, the Tsetse Control Services conducted numerous SAS campaigns, covering large areas mainly in the Western Province (Allsopp and Hursey, 2004). The control activities in the Kwando-Zambezi area, which aimed at the elimination of this entire tsetse belt, are conducted under the guidance of the PATTEC. The Zambian case provides another example of control operations with impacts that have been sustainable in time. Several factors may explain the success of control operations in Zambia in terms of sustainability, when compared to other countries, including ecological, political and economic factors. First the tsetse ecological characteristics are more favourable for elimination as populations are isolated within three main tsetse belts: Kwando-Zambezi, Lower Zambezi and Eastern (also referred to as `Common') belts. Second, effective and sustained regional cooperation enabled the four countries (Angola, Botswana, Namibia and Zambia) involved in the Kwando-Zambezi Control Project to make enormous progress in the elimination of the eponym belt. Third, the Zambian government is entirely supporting the control efforts, which makes the availability of external funding less critical for the long-term viability of the control (Meyer et al., 2016).

Impact Success Cases of AAT Interventions in South Africa

The tsetse belt in South Africa is limited to north-eastern KwaZulu-Natal occupying 12,000 km^2 and is conserved mainly in nature reserves, national and private game parks and resorts (De Beer et al. 2016; Kappmeier et al., 2007). As reported by Latif et al. (2019) the first integrated tsetse and tick control was introduced in 1990, when a severe nagana outbreak occurred in the tsetse infested areas of north-eastern KwaZulu-Natal, about 10 000 cattle died of AAT and 116, 000 were treated using ethidium bromide during this outbreak (Kappmeier et al., 1998). The control measures included the use of pyrethroid dip, a chemical which has an insecticidal and acaricidal activity, on a 4-year interval to control the outbreak and challenge by the vector tsetse flies. Ethidium and novidium strategic treatment produced attractive results whereby cattle were protected for an extended period of up to 6 months. Therefore, two to three treatments per year may be sufficient to keep cattle productivity on the farm under all year tsetse and trypanosomosis challenge (Latif et al., 2019).

Impact Success Cases of AAT Interventions in Nigeria

About 70% of Nigeria landmass is infested with eleven *Glossina species* (Jordan, 1961) with four (*G. palpalis palpalis, G. tachinoides, G. morsitans submorsitans* and *G. longipalpis*) commonly seen (Federal Ministry of Agriculture, 1981; Onyiah, 1995). In Nigeria, the prevalence and occurrence of tsetse *reported G. m. submorsitans* in a series of discontinuous belt across the northern region, while *G. longipalpis* are over a wide area in central and south-western Nigeria. *Glossina palpalis* and *G. tachinoides* occur across all geopolitical zones but absent from the extreme Northeast areas with only *G. tachinoides* seen in limited areas on the southern boundary (Onyiah et al., 1983). In the bid to eradicate AAT from Nigeria, tsetse flies have to be eliminated. One method in achieving this is the use of Sterile Insect Technique (SIT) which was done by PATTEC-Nigeria to strategically approach elimination/eradication. In 1987, SIT in combination with other control methods were used to clear *G. palpalis* from an area of about 1500 km2 in North- Central Nigeria (Olandunmade et al., 1990). Ever since, SIT has been halted on account of poor funding which has led to a reinvasion of once cleared areas (Oluwafemi et al., 2008).

As soon as the disease started to cause the death of livestock and consequent economic losses, the defunct West African Institute for Trypanosomiasis Research (WAITR) in 1947 now known as Nigerian Institute for Trypanosomiasis Research (NITR) in Kaduna (North-West) was established to conduct research and development for the control and eradication of trypanosomiasis in all geo-ecological zones (Isaac et al., 2017).

Impact Success Cases of AAT Interventions in Benin

In 2002, Beninese cattle herd was estimated at 1,300,000 of which 10.50 to 45.00% of each herd may have been infected with trypanosomosis (Codjia, 1989). Beninese government has set up several Livestock Development Programs, the most recent being the Milk and Meat Sector Support Program (PAFILAV). This program aims to control various animal pathologies, including African animal trypanosomosis, which is a real constraint to the development of cattle breeding in Benin. In 1993, the Agricultural Services Restructuring Project (PRSA) was set up, led to the creation of the Ministry of Agriculture, Livestock and Fisheries (MAEP), and contributed to the development of animal production. As a result, two projects to improve cattle breeding and control of bovine trypanosomosis were set up. This is the Livestock Development Project (LDP) and the Milk and Meat Sector Support Project (PAFILAV). The implementation of numerous breeding projects, Pan African Program for the Control of Epizootics (PACE), Livestock Development Project, stage III (LDP III), Livestock Development Project in East Borgou (PDEBE), and Livestock Development Support Project in Borgou (PADEB) over the last ten years has reached cattle and small ruminants. The main activities focused on animal health, zootechnics and research and development (PDE, 2007).

DISCUSSION

While progress has been made on fighting AAT in Africa, for several years through Pan African Tsetse and Trypanosomiasis Eradication Campaign (PATTEC) for over the last 15 years, that was established by the African Union. The union has supported a series of multinational interventions to control trypanosomiasis and in partnership with national governments (Schofield & Kabayo, 2008). Despite this revitalised atmosphere of international cooperation and opportunity established through PATTEC, until today AAT persists and continues to be a major obstacle for the improved livestock productivity in Africa. In order to continue efforts to control the AAT and ensure good livestock performance, effective control measures against tsetse vectors and extension of control policy to all endemic areas should be well reinforced. There are obvious challenges that continue to hinder complete eradication of the AAT in the affected countries. For instance, when compared to diseases of livestock such as Foot-and-Mouth Disease, Contagious Bovine Pleuropneumonia and other notifiable diseases, AAT is taken casually by the authorized institutions dealing with livestock diseases control in many affected countries, and all efforts against tsetse and AAT are left within livestock keepers (Machila, Emongor, & Shaw, 2007). Lack of close monitoring of AAT by the livestock authorities, lack of funding and establishment of core capacities, also less understanding of AAT risk and impact together with less intergrated AAT management, these are obstacle undermining AAT control programmes in many affected countries in Africa. For effective control of AAT control, the strategic measures through developing a Progressive Control Pathway (PCPs) for AAT, these are stepwise approaches for the reduction, elimination, and eradication of diseases. which have been addressed(Diall, et al., 2017). These are must be well observed for effective implementation of AAT control strategies that will enhance reduction and elimination of AAT.

Another important challenge is the increase in human population which has led to encroachment of tsetse-infested areas by livestock keepers in search for adequate pastures. Likewise, climate change caused by global warming has also contributed to the raid of tsetse-infested areas especially in national parks and the surrounding communities where there is still a high density of tsetse flies and hence there is more exposure to livestock when livestock search for adequate pasture and water. This has lead to increasing migration of livestock keepers in search of pastures and water, resulting in the spread of AAT. This predicament can be prevented by promoting rangeland management to ensure the availability of pastures in livestock keeping society. Another crucial factor that has contributed to the spread of AAT beyond its known boundaries is climate change caused by global warming.

Economic analysis for AAT is crucial although it has been patchily carried out in the affected countries. It is suggested that there are should be a framework that integrates an established decision support tool, aimed at optimizing resource allocation regarding tsetse population management, with a detailed analysis of livestock productivity under different AAT control scenarios. The integration of cost-benefit studies should be supported by recent estimates of key parameters such as frequency of trypanosome infection and impact (e.g. duration of morbidity, mortality), livestock, and tsetse demographics. In addition comprehensive economic benefits to livestock keepers from intervening against livestock trypanosomiasis especially bovine trypanosomiasis is important, such kind of analysis has been carried out in some areas such as in Eastern Africa and other countries.

CONTRIBUTION TO THE BODY OF KNOWLEDGE

African animal trypanosomosis is one of the major constraints affecting the economy and productivity of livestock. This review took inventory of fixtures of the epidemiology of AAT, its measures and economic importance of AAT and possible recommendations. However, this review identifies that AAT is still a problem and enforced measures and determined strategies should be well observed if affected countries clearly want to stamp out AAT, thus this article reveals and opens perspective for AAT occurrence and its risk factors, a better understanding of these essential parameters will be benchmarks for the management of the disease. In addition, this review indicates that AAT still exists and continue to impair livestock production and thus threatens food security in sub-Saharan Africa. There is a need to revive strategies for the control of AAT s for an effective and sustainable management of AAT in the affected countries. This includes the promptness of affected countries to carry out continuous surveillance of the disease. In the end successful control of the disease will ultimately improve the livelihood of the communities where AAT is endemic.

FUTURE RESEARCH DIRECTIONS ON AAT

After many years of AAT research, the prognosis of AAT still remains ambiguous. Given that the future research towards the disease control should focus on cost-effective disease surveillance and control, improvement of vector control methods and early case detection and treatment. The followings are future researches needed on AAT:

Socio-Economic Consequences of AAT

Socio-economic consequence of AAT is difficult to assess, as the data available are frequently very approximate. Thus, the need for future and further studies on socio-economic consequences of AAT.

Molecular Epidemiological Researches on AAT

Most studies on trypanosomiasis have employed mainly parasitological and serological tools which are less accurate. Inherent shortcomings of serologic and parasitologic diagnostic methods can be overcome by molecular techniques. More studies on molecular epidemiology of AAT will give more accurate and efficient identification of trypanosome species found in the fly vectors and vertebrate hosts.

Nanotechnological Interventions for Treatment of AAT

The complete life cycle stages of trypanosomes span between insect vector and mammalian host, as a result only few drugs have been approved for treatment of trypanosomiasis. It has been discovered that trypanocidal therapy has major limitations of poor efficacy, serious side effects and drug resistance. Hence, there is need for more effective innovative strategies like nanotechnological approaches for delivery of drugs to reduce deaths associated with AAT in animals.

Compilation of Comprehensive Maps of AAT Distribution

There is need for future research on taking advantage of focal nature of the disease to design AAT atlas which will provide a valuable contribution towards informed decision-making for planning and monitoring of control activities and assessment of epidemiological trends and research activities on AAT.

CONCLUSION

The potential for increasing livestock production can only be fully realized if the animals are adequately protected against the adverse effects of diseases. Profitability of animal product demand efficient husbandry of animals, as disease remains a profit limitation factor in many tropical countries. Africa is a continent with high agropastoral activities faced by many constraints to livestock production, economic development, human and animal welfare. African Animal Trypanosomosis is one of these major constraints. AAT serves as a constraint to improved livestock and agriculture production in Africa and remains one of the major causes of livestock production loss in the livestock industry. It is probably the only disease that has affected the settlement and economic development of a major part of a continent. it has an extended impact on the crop, human settlement, and welfare. In the livestock industry, AAT is a very important and significant constraint, causing significant losses in the affected countries ranging from a decrease in milk production to death. AAT is transmitted biologically and mechanically through vectors known as tsetse flies and biting flies. Despite the control measures for tsetse and trypanosomiasis implemented for several years in livestock development projects, AAT remains to be a major constraint to improving livestock productivity in the affected countries. AAT is still a problem and will need to re-invest in tsetse and trypanosomiasis control programs to particularly targeting either sustainable reduction or elimination of AAT. For sufficient livestock production to be fully realized to meet the growing population of Africa, integrated approach to trypanosomiasis control is required to increase the present level of livestock production.

As explained by Diall et al., 2017, the adoption of the initial stages of PCPs focus on the establishment of national coordination structures, engagement of stakeholders, development of technical capacities, data collection and management, and pilot field interventions. The intermediate stage aims at a sustainable and economically profitable reduction of disease burden, while higher stages target elimination. The mixed-reports of favorable results and failures in the past efforts against AAT make the development of this PCP a high priority in this era. In line with the insights given in this review, the following are proposed:

- Sophisticated diagnostic techniques should be used in prevalence estimation and strain **Identification of AAT.**
- Biology of the parasite and interaction with host-pathogen and the vector should be studied for each specific geographical area for better interventions.
- The dynamics of mechanical transmission of AAT in endemic areas have to be thoroughly studied by including those factors contributing to occasional outbreaks.
- The PCP stages especially the initial one must-be enforced to help lifting the burden of AAT in the affected countries. In particular, to develop technical capacities and to gain a sufficient understanding of AAT distribution, risk, and impact for evidence-based planning of field activities. That will consequently lead to a sustainable and economically-profitable reduction in AAT risk and burden by creating areas of low AAT incidence and impact.

REFERENCES

Abenga, J. N., Enwezor, F. N. C., Lawani, F. A. G., Osue, H. O., & Ikemereh, E. C. D. (2004). Trypanosome prevalence in cattle in Lere Area in Kaduna State, North central Nigeria. *Revue d'Élevage et de Médecine Vétérinaire des Pays Tropicaux*, *57*(1-2), 45–48. doi:10.19182/remvt.9904

Adam, Y., Marcotty, T., Cecchi, G., Mahama, C. I., Solano, P., Bengaly, Z., & Van den Bossche, P. (2012). Bovine trypanosomosis in the Upper West Region of Ghana: Entomological, parasitological and serological cross-sectional surveys. *Research in Veterinary Science*, *92*(3), 462–468. doi:10.1016/j.rvsc.2011.04.004 PMID:21550616

Anene, B. M., Chime, A. B., Jibike, G. I., & Anika, S. M. (1991). Prevalence of trypanosomiasis in Zebu cattle at Obudu ranch— A tsetse-free zone in Nigeria. *Preventive Veterinary Medicine*, *10*(4), 257–260. doi:10.1016/0167-5877(91)90011-P

Anosa, V. O. a. O., T. U. (1980). Haematological studies on domestic animals in Nigeria. The effect of age, breed and haemoglobin type on bovine haematology and anaemia. *Bulletin of Veterinary Medicine, 27,* 773 – 788.

Ashcroft, M. T. (1959). The importance of African wild animals as a reservoir of trypanosomiasis. *East African Medical Journal*, *36*, 289–297. PMID:13672101

Auty, H. K., Picozzi, K., Malele, I., Torr, S. J., Cleaveland, S., & Welburn, S. (2012). Using molecular data for epidemiological inference: Assessing the prevalence of Trypanosoma brucei rhodesiense in tsetse in Serengeti, Tanzania. *PLoS Neglected Tropical Diseases*, *6*(1), e1501. doi:10.1371/journal.pntd.0001501 PMID:22303496

Bengaly, Z., Sidibe, I. R. G., Desquesnes, M., Boly, H., & Sawadogo, L. (2002). Comparative pathogenicity of three genetically distinct types of Trypanosoma congolense in cattle: Clinical observations and haematological changes. *Veterinary Parasitology*, *108*(1), 1–19. doi:10.1016/S0304-4017(02)00164-4 PMID:12191895

Buxton, P. A. (1955). *The natural history of tsetse flies: an account of the biology of the genus Glossina (Diptera)* (Vol. 10). London: H.K. Lewis.

Caljon, G., De Vooght, L., & Van Den Abbeele, J. (2014). The biology of tsetse-trypanosome interactions. In The Biology of Tsetse–Trypanosome Interactions. Springer.

Cattand, P. (2001). African Trypanosomiasis, human. In Encyclopedia of arthropod–transmitted infections of man and domestic animals (pp. 13–23). CABI.

Chitanga, S., Namangala, B., De Deken, R., & Marcotty, T. (2013). Shifting from wild to domestic hosts: The effect on the transmission of Trypanosoma congolense to tsetse flies. *Acta Tropica*, *125*(1), 32–36. doi:10.1016/j.actatropica.2012.08.019 PMID:23000543

Conner, R. J., & Mulkangi, D. J. A. (1986). . Concurrent outbreak of pseudo-lumpy skin disease and acute *T. vivax* infection in cattle. *Tropical Animal Health Production, 18*, 127 – 132.

Cordon-Obras, C., Berzosa, P., Ndong-Mabale, N., Bobuakasi, L., Buatiche, J. N., Ndongo-Asumu, P., Benito, A., & Cano, J. (2009). Trypanosoma brucei gambiense in domestic livestock of Kogo and Mbini foci (Equatorial Guinea). *Tropical Medicine & International Health*, *14*(5), 535–541. doi:10.1111/j.1365-3156.2009.02271.x PMID:19320872

Cox, F. E. G. (1979). Pathogenesis of animal trypanosomiasis. *Nature*, *277*(5698), 603–604. doi:10.1038/277603a0 PMID:423960

Daffa, J., Byamungu, M., Nsengwa, G., Mwambembe, E., & Mleche, W. (2013). Tsetse distribution in Tanzania: 2012 status. *Tanzania Veterinary Journal, 28*.

Desquesnes, M., & Dia, M. L. (2003). Mechanical transmission of Trypanosoma congolense in cattle by the African tabanid Atylotus agrestis. *Experimental Parasitology*, *105*(3), 226–231. doi:10.1016/j.exppara.2003.12.014 PMID:14990316

Diall, O., Cecchi, G., Wanda, G., Argiles-Herrero, R., Vreysen, M. J. B., Cattoli, G., Viljoen, G. J., Mattioli, R., & Bouyer, J. (2017). Developing a Progressive Control Pathway for African Animal Trypanosomosis. *Trends in Parasitology*, *33*(7), 499–509. doi:10.1016/j.pt.2017.02.005 PMID:28456474

FAO. (1992a). *Ecology and behaviour of tsetse*. FAO.

FAO. (1992b). *Hosts of glossina and transmission of disease*. FAO.

Fevre, E., Coleman, P., Odiit, M., Magona, J. W., Welburn, S. C., & Woolhouse, M. E. (2001). Occurrence, Clinical Manifestation and the Epidemiological Implications of Naturally Occurring Canine Trypanosomosis in Western Kenya. *Lancet*, *358*(9282), 625–628. PMID:11530149

Gonzatti, M. I., González-Baradat, B., Aso, P. M., & Reyna-Bello, A. (2014). *Trypanosoma (duttonella) vivax and typanosomosis in latin America: Secadera/Huequera/Cacho Hueco* (Vol. 9783709115565). Academic Press.

Gray, A. R., & Luckins, A. G. (1978). An extravascular site of development of Trypanosoma congolense. *Nature*, *272*(5654), 613–614. doi:10.1038/272613a0 PMID:643055

Gustafson, C. R., VanWormer, E., Kazwala, R., Makweta, A., Paul, G., Smith, W., & Mazet, J. A. K. (2015). Educating pastoralists and extension officers on diverse livestock diseases in a changing environment in Tanzania. *Pastoralism*, *5*(1), 1–12. doi:10.118613570-014-0022-5

Houre, C. A. (1976). Trypanosomes. *Veterinary Parasitology*, *42*, 9–29.

ILARD. (1990). *Problems of current control method.* Academic Press.

Jordan, A. M. (1972). The insemination potential of male Glossina austeni Newstead and Glossina morsitans morsitans Westw. (Dipt.: Glossinidae). *Bulletin of Entomological Research*, *62*(2), 319–325. doi:10.1017/S000748530004774X

Knowles, G., Betschart, B., Kukla, B. A., Scott, J. R., & Majiwa, P. A. O. (1988). Genetically discrete populations of Trypanosoma congolense from livestock on the Kenyan coast. *Parasitology*, *96*(3), 461–474. doi:10.1017/S0031182000080100 PMID:3405633

Kuzoe, F. A. S., & Schofield, C. J. (2005). *Strategic review of traps and targets for tsetse and African trypanosomiasis control.* Academic Press.

Leak, S. G. A., Mulatu, W., Authié, E., d'Ieteren, G. D. M., Peregrine, A. S., Rowlands, G. J., & Trail, J. C. M. (1993). Epidemiology of bovine trypanosomiasis in the Ghibe valley, southwest Ethiopia 1. Tsetse challenge and its relationship to trypanosome prevalence in cattle. *Acta Tropica*, *53*(2), 121–134. doi:10.1016/0001-706X(93)90024-6 PMID:8098898

Levine, N. D., Corliss, J. O., Cox, F. E. G., Deroux, G., Grain, J., Honigberg, B. M., Leedale, G. F., Loeblich, A. R., Lom, I. J., Lynn, D., Merinfeld, E. G., Page, F. C., Poljansky, G., Sprague, V., Vavra, J., & Wallace, F. G. (1980). A Newly Revised Classification of the Protozoa. *The Journal of Protozoology*, *27*(1), 37–58. doi:10.1111/j.1550-7408.1980.tb04228.x PMID:6989987

Machila, N., Emongor, R., Shaw, A., Welburn, S. C., McDermott, J., Maudlin, I., & Eisler, M. C. (2007). A community education intervention to improve bovine trypanosomiasis knowledge and appropriate use of trypanocidal drugs on smallholder farms in Kenya. *Agricultural Systems*, *94*(2), 261–272. doi:10.1016/j.agsy.2006.09.004

Malele, I. I., Magwisha, H. B., Nyingilili, H. S., Mamiro, K. A., Rukambile, E. J., Daffa, J. W., Lyaruu, E. A., Kapange, L. A., Kasilagila, G. K., Lwitiko, N. K., Msami, H. M., & Kimbita, E. N. (2011). Multiple Trypanosoma infections are common amongst Glossina species in the new farming areas of Rufiji district, Tanzania. *Parasites & Vectors*, *4*(1), 217–217. doi:10.1186/1756-3305-4-217 PMID:22093363

Mamabolo, M. V., Ntantiso, L. A. L., & Majiwa, P. A. O. (2009). Natural infection of cattle and tsetse flies in South Africa with two genotypic groups of Trypanosoma congolense. *Parasitology*, *136*(4), 425–431. doi:10.1017/S0031182009005587 PMID:19250578

Matete, G. O. (2003). Occurrence, clinical manifestation and the epidemiological implications of naturally occurring canine trypanosomosis in western Kenya. *The Onderstepoort Journal of Veterinary Research, 70*(4), 317–323. doi:10.4102/ojvr.v70i4.296 PMID:14971734

Maxie, M. G., Losos, G. J., & Tabel, H. (1979). Experimental bovine trypanosomiasis (*T. vivax and T. congolense*) 1. Symptomatology and clinical pathology. *Tropical Medicine and Parasitology, 30*, 275–282.

Meyer, A., Holt, H. R., Selby, R., & Guitian, J. (2016). Past and Ongoing Tsetse and Animal Trypanosomiasis Control Operations in Five African Countries: A Systematic Review. *PLoS Neglected Tropical Diseases, 10*(12), e0005247. doi:10.1371/journal.pntd.0005247 PMID:28027299

Molyneux, D. H., Hopkins, D. R., & Zagaria, N. (2004). Disease eradication, elimination and control: The need for accurate and consistent usage. *Trends in Parasitology, 20*(8), 347–351. doi:10.1016/j.pt.2004.06.004 PMID:15246314

Muhanguzi, D., Mugenyi, A., Bigirwa, G., Kamusiime, M., Kitibwa, A., Akurut, G. G., Ochwo, S., Amanyire, W., Okech, S. G., Hattendorf, J., & Tweyongyere, R. (2017). African animal trypanosomiasis as a constraint to livestock health and production in Karamoja region: A detailed qualitative and quantitative assessment. *BMC Veterinary Research, 13*(355), 355. doi:10.118612917-017-1285-z PMID:29178951

Namangala, B., Hachaambwa, L., Kajino, K., Mweene, A. S., Hayashida, K., Simuunza, M., Simukoko, H., Choongo, K., Chansa, P., Lakhi, S., Moonga, L., Chota, A., Ndebe, J., Nsakashalo-Senkwe, M., Chizema, E., Kasonka, L., & Sugimoto, C. (2012). The use of Loop-mediated Isothermal Amplification (LAMP) to detect the re-emerging Human African Trypanosomiasis (HAT) in the Luangwa and Zambezi valleys. *Parasites & Vectors, 5*(1), 282. doi:10.1186/1756-3305-5-282 PMID:23211002

Namangala, B., & Odongo, S. (2014). *Animal African trypanosomosis in sub-saharan Africa and beyond African borders*. Academic Press.

Naylor, D. C. (1971). The Haematology and Histopathology of Trypanosoma Congolense Infection in Cattle. Ii. Haematology (Including Symptoms). *Tropical Animal Health and Production, 3*(3), 159–168. doi:10.1007/BF02360746 PMID:4945970

Pollock, J. N. (1992). *Tsetse biology, systematics and distribution techniques*. FAO.

Reichard, R. E. (2002). Area-wide biological control of disease vectors and agents affecting wildlife. *Rev. sci. tech. Off. int. Epiz., 21*(1), 179–185. PMID:11974628

Reid, R. S., Kruska, R. L., Deichmann, U., Thornton, P. K., & Leak, S. G. A. (2000). Human population growth and the extinction of the tsetse fly. *Agriculture, Ecosystems & Environment, 77*(3), 227–236. doi:10.1016/S0167-8809(99)00103-6

Robays, J., Ebeja Kadima, A., Lutumba, P., Miaka mia Bilenge, C., Kande Betu Ku Mesu, V., De Deken, R., Makabuza, J., Deguerry, M., Van der Stuyft, P., & Boelaert, M. (2004). Human African trypanosomiasis amongst urban residents in Kinshasa: A case-control study. T. *Tropical Medicine & International Health, 9*(8), 869–875. doi:10.1111/j.1365-3156.2004.01266.x PMID:15303991

Rogers, D. (1979). Tsetse Population Dynamics and Distribution: A New Analytical Approach. *Journal of Animal Ecology, 48*(3), 825–849. doi:10.2307/4197

Rogers, D., & Randolph, S. E. (1985). Population Ecology of Tsetse. *Annual Review of Entomology*, *30*(1), 197–216. doi:10.1146/annurev.en.30.010185.001213 PMID:3882049

Schofield, C. J., & Kabayo, J. P. (2008). Trypanosomiasis vector control in Africa and Latin America. *Parasites & Vectors*, *1*(24), 24. doi:10.1186/1756-3305-1-24 PMID:18673535

Simo, G., Asonganyi, T., Nkinin, S. W., Njiokou, F., & Herder, S. (2006). High prevalence of Trypanosoma brucei gambiense group 1 in pigs from the Fontem sleeping sickness focus in Cameroon. *Veterinary Parasitology*, *139*(1), 57–66. doi:10.1016/j.vetpar.2006.02.026 PMID:16567049

Simukoko, H., Marcotty, T., Phiri, I., Vercruysse, J., & Van den Bossche, P. (2007). Heterogeneity in the trypanosomosis incidence in Zebu cattle of different ages and sex on the plateau of eastern Zambia. *Acta Tropica*, *103*(2), 98–101. doi:10.1016/j.actatropica.2007.05.011 PMID:17618594

Solano, P., Ravel, S., & de Meeûs, T. (2010). How can tsetse population genetics contribute to African trypanosomiasis control? *Trends in Parasitology*, *26*(5), 255–263. doi:10.1016/j.pt.2010.02.006 PMID:20202905

Stephen, L. E. (1986). *Trypanosomiasis: a veterinary perspective.* Pergamon/A. Wheaton & Co.

Swallow, B. M. (1999). *Impacts of trypanosomiasis on African agriculture. Report from the Programme Against African Trypanosomiasis*. Food and Agriculture Organization of the United Nations.

Torr, S. J., & Hargrove, J. W. (1999). Behaviour of tsetse (Diptera: Glossinidae) during the hot season in Zimbabwe: the interaction of micro-climate and reproductive status. *Bulletin of Entomological Research*, *89*(4), 365–379. doi:10.1017/S0007485399000504

Uilenberg, G. (1998). A field guide for the diagnosis, treatment and prevention of African animal. FAO.

Van den Bossche, P., & Delespaux, V. (2011). Options for the control of tsetse-transmitted livestock trypanosomosis. An epidemiological perspective. *Veterinary Parasitology*, *181*(1), 37–42. doi:10.1016/j.vetpar.2011.04.021 PMID:21561717

Van Den Bossche, P., Rocque, S., Hendrickx, G., & Bouyer, J. (2010). A changing environment and the epidemiology of tsetse-transmitted livestock trypanosomiasis. *Trends in Parasitology*, *26*(5), 236–243. doi:10.1016/j.pt.2010.02.010 PMID:20304707

Van den Bossche, P., & Rowlands, G. J. (2001). The relationship between the parasitological prevalence of trypanosomal infections in cattle and herd average packed cell volume. *Acta Tropica*, *78*(2), 163–170. doi:10.1016/S0001-706X(00)00182-0 PMID:11230826

Vickerman, K. (1985). Developmental cycles and biology of pathogenic trypanosomes. *British Medical Bulletin*, *41*(2), 105–114. doi:10.1093/oxfordjournals.bmb.a072036 PMID:3928017

Vincendeau, P., & Bouteille, B. (2006). Immunology and immunopathology of African trypanosomiasis. *Anais da Academia Brasileira de Ciências*, *78*(6), 645–665. doi:10.1590/S0001 37652006000400004 PMID:17143404

Welburn, S. C., & Maudlin, I. (1992). The nature of the teneral state in Glossina and its role in the acquisition of trypanosome infection in tsetse. . .. *Annals of Tropical Medicine and Parasitology*, *86*(5), 529–536. doi:10.1080/00034983.1992.11812703

Chapter 3
Overview of the Vectors and Their Role in Transmission of African Animal Trypanosomiasis

Florence Njeri Wamwiri
https://orcid.org/0000-0002-3264-1940
Biotechnology Research Institute, Kenya Agricultural and Livestock Research Organisation (KALRO), Kenya

Joanna Eseri Auma
Biotechnology Research Institute, Kenya Agricultural and Livestock Research Organisation (KALRO), Kenya

ABSTRACT

African animal trypanosomiasis (AAT) is a major constraint to livestock productivity, particularly in cattle and in camels. This chapter covers some general aspects of the arthropod vectors of animal trypanosomiasis, the tsetse flies Glossina spp., and to a lesser extent the biting flies. This chapter covers the classification, morphology, basic biology, and the eco-distribution of tsetse flies. The role of tsetse flies in disease epidemiology has also been reviewed. The elementary biology of these vectors is quite well known and elucidated. However, with advances in molecular and other biological techniques, new insights related to tsetse biology have been obtained. This chapter will revisit these basics and include some updated information emanating from research done in the recent past. The final part of the chapter is devoted to a brief discussion on biting flies, the vectors of T. evansi, which causes camel trypanosomiasis.

INTRODUCTION

African Animal Trypanosomiasis (AAT) is caused by infection due to a variety of trypanosome species. The main vector of these trypanosomes is the tsetse fly, which transmits the parasite through cyclical transmission. Outside of the tsetse-belt, trypanosomiasis is transmitted by biting flies (mainly tabanids and *Stomoxys* species) via mechanical transmission. An updated knowledge of the vector biology and

DOI: 10.4018/978-1-7998-6433-2.ch003

ecology is required by in order to devise appropriate control interventions. This chapter aims to provide a detailed review of the biology of tsetse flies and their role in the epidemiology of AAT.

TSETSE FLIES, *GLOSSINA* SPP

Classification

The tsetse flies, *Glossina* species are of great veterinary and medical importance because they transmit the trypanosome parasites that cause both the human and the animal forms of trypanosomiasis. The taxonomy of *Glossina* species is well known and any prior controversies about the taxonomic classification are considered resolved (Lane and Crosskey 1993). Tsetse flies belong to the phylum Arthropoda. The major characteristic of Arthropoda is the division of the body into three segments viz. the head, thorax and abdomen. Tsetse flies are further classified into the order Diptera (or "true flies"). Thus, tsetse flies, like other Dipterans, possess a thorax bearing a single pair of functional wings. The second pair of wings has been adapted to form halters, which help the insect to maintain in-flight balance. Under the Order Diptera, tsetse flies are placed within the superfamily Hippoboscoidea due to the morphological and reproductive similarities of tsetse flies to keds and other hippoboscid flies (McAlpine, 1989). The superfamily Hippoboscoidea contains four families including the Hippoboscidae (ked flies) and the Glossinidae, mono-genetic family with only one genus *Glossina*, wherein the tsetse flies are classified. The genus *Glossina* is divided into three distinct taxonomic groups (subgenera) based on anatomical characteristics (e.g. external genitalia of the males). The differentiation of the subgenera is also roughly co-related to the environmental territory of the flies. These three groups are the *morsitans* (savannah-dwelling), *palpalis* (riverine or lacustrine-dwelling) and the *fusca* (forest-dwelling group). One of the earliest species-specific descriptions of tsetse flies was given by Westwood in the 1800's. Since then, more than 30 different species and subspecies have been described. The most recent species to be identified

Table 1. Currently described tsetse species by subgeneric division

subgenus Morsitans - 'savannah' flies	subgenus Fusca - 'forest' flies	subgenus Palpalis - 'riverine' & 'lacustrine' flies
G. austeni Newstead, 1912 *G. morsitans* Westwood, 1850 *G. pallidipes* Austen, 1903 *G. swynnertoni* Austen, 1923 *G. morsitans submorsitans* Newstead, 1910 *G. morsitans centralis* Machado 1970 *G. morsitans morsitans* Westwood 1859	*G. fusca fusca* Walker, 1849 *G. fuscipleuris* Austen, 1911 *G. frezili* Gouteux, 1987 *G. haningtoni* Newstead and Evans, 1922 *G. longipennis* Corti, 1895 *G. medicorum* Austen, 1911 *G. nashi* Potts,1955 *G. nigrofusca nigrofusca* Newstead, 1910 *G. severini* Newstead, 1913 G. schwetzi Newstead and Evans, 1921 *G. tabaniformis* Westwood, 1850 *G. vanhoofi* Henrard, 1952 *G. nigrofusca hopkinsi* van Emden 1944 *G. brevipalpis* Newstead 1910	*G. caliginea* Austen, 1911 *G. fuscipes fuscipes* Newstead, 1911 *G. fuscipes martinii* Zumpt, 1935 *G. fuscipes quanzensis* Pires, 1948 *G. pallicera pallicera* Bigot, 1891 *G. pallicera newsteadi* Austen, 1929 *G. palpalis palpalis* Robineau-Desvoidy, 1830 *G. palpalis gambiensis* Vanderplank, 1911 *G. tachinoides* Westwood, 1850

and described is *Glossina frezili*, which infests the mangrove swamps in Congo (Gouteux, 1987). The currently known tsetse fly species and subspecies are detailed in Table 1.

Distinguishing Morphological Characteristics of Glossina

Tsetse flies, similar to other Dipterans are two-winged flies. Different species have different sizes and external appearance but generally the size ranges from about 0.5-1.5 cm and their colour is brown to greyish. Tsetse flies have a large head with distinctly separated eyes. The thorax is quite large, while the abdomen is wide rather than elongated and shorter than the wings. There are four major characteristics that may be used to distinguish adult tsetse from other flies;

a) Folded wings – when the fly is at rest, the wings fold completely one on top of the other as shown in Figure 1. In other dipterans, the wings are usually splayed at rest.

Figure 1. Male G. m. morsitans showing wings folded one on top of the other during rest

b) The second feature that distinguishes *Glossina* from other dipterans is the venation of the discal medial ("middle") cell of the wing. The wing has a characteristic hatchet shape resembling a meat cleaver or a hatchet, as shown in Figure 2

Figure 2. Wing of G. m. morsitans showing the hatchet-shaped cell

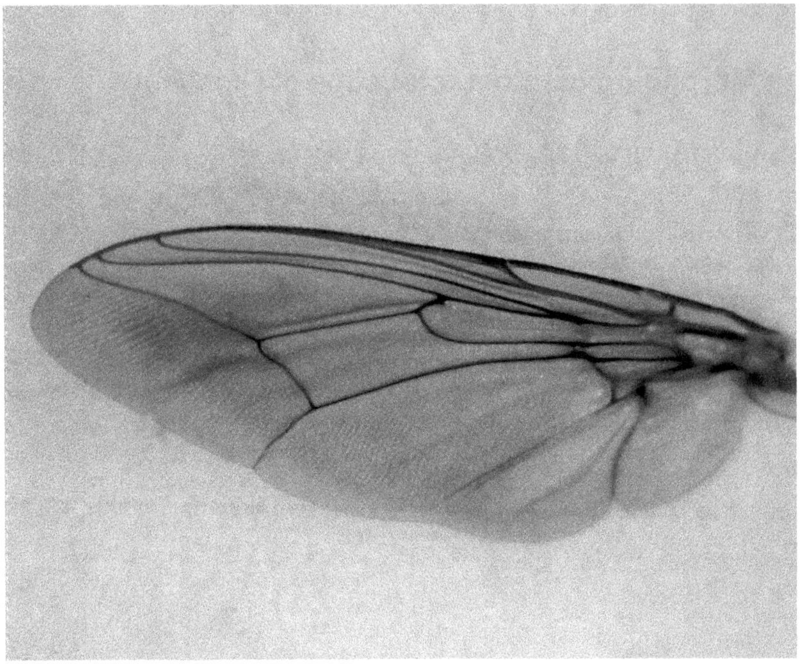

c) proboscis - tsetse have a distinct proboscis, which is the long thin structure attached to the bottom of the head. In tsetse flies, the proboscis is found characteristically pointing forward as shown in Figure 3.

Figure 3. Head of G. m. morsitans showing antennae with branched arista hairs. Note also the characteristically forward pointing proboscis

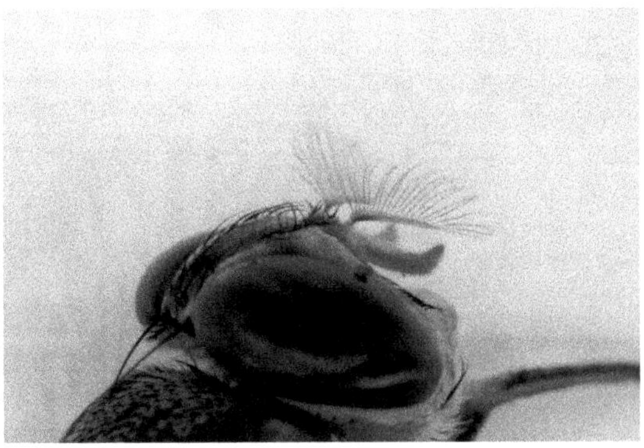

d) Branched arista hairs - the antennae have arista with hairs which are themselves branched (Figure 3).

Life Cycle of Glossina

Tsetse flies are thought to mate near to or on host animals. Females are sexually-receptive while they are relatively young which is before or at about the time of taking the first blood meal. Older males have been found to be significantly more successful in securing a female for mating than very young ones (Abila, Kiendrebeogo, Mutika, Parker, & Robinson, 2003; C. De Beer, Venter, & Vreysen, 2015). In nature, the females usually mate only once in their lives, thereafter keeping the sperm within the body for future insemination events. However, evidence of multiple mating (polyandry) has been demonstrated in some species (Bonomi et al., 2011).

Glossina have a unique method of reproduction called "adenotrophic viviparity" whereby the females retain the fertilized egg *in utero*. The progeny is nourished by maternal secretions while in the uterus and is then expelled in a live birth (Benoit, Attardo, Baumann, Michalkova, & Aksoy, 2015). In contrast, most dipterans and indeed most insects reproduce by laying eggs, termed oviparity. Female tsetse produce a single egg per gonotrophic cycle(Tobe & Langley, 1978). The egg produced from the ovary is maintained and nourished in the uterus via the milk gland tubules until it reaches the 3rd instar larval stage. It is then deposited and quickly burrows into the surrounding substrate (e.g sand or soft soil). In the following 1-2 hours, the larva becomes barrel-shaped and darkens as it changes into a pupa. The pupa incubates for 4-5 weeks after which the adult fly emerges to continue the life cycle (Tobe & Langley, 1978). The generalized life cycle of *Glossina* is detailed in Figure 4.

Figure 4. Life cycle of the tsetse fly Glossina species (adapted from Leak, 1999) Dark circles indicate stages developing in utero

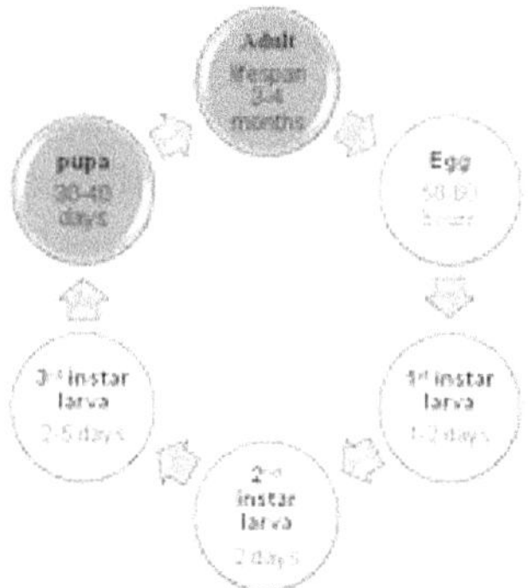

The particular reproduction pattern of *Glossina*, whereby one egg and thus potentially one offspring is produced every 10 days signals a very low rate of reproduction. Other insects have been known to produce hundreds or even thousands of eggs in a single brood. The latter are *r*-strategists and produce large numbers of eggs in order to improve the chances of survival. On the other hand, tsetse flies are *k*-strategists which produce only one offspring per cycle, but offer enhanced "maternal care" to the intrauterine larva to boost survival chances. This unique adaptation of the tsetse fly has enhanced its survival over the years.

GEOGRAPHICAL DISTRIBUTION OF TSETSE SPECIES

Glossina species were once globally distributed and there is even archeological evidence of fossil tsetse species in North America dated from the Oligocene period (Cockerell, referenced in (S. Leak, 1999). However, the present distribution is limited to a belt across tropical Africa between 15°N and 29°S, covering about 10 million km^2 of sub-Saharan Africa and extending through 37 countries(Simarro et al., 2012) . A total of 20 out of the 37 tsetse-infested countries rank among the world's poorest 25 countries (PAAT, 2008).

The first continent-wide distribution of tsetse flies was described by Ford and Katondo in the 1970s (Ford & Katondo, 1977). Since then, workers in different countries have updated this distribution map on a fine-scale basis, primarily based on presence/absence and environmental data. (Albert, Wardrop, Atkinson, Torr, & Welburn, 2015; Chikowore et al., 2017; C. J. De Beer et al., 2016; Duguma et al., 2015; Malele, Hamisi, & Atway, 2011; Shereni, Anderson, Nyakupinda, & Cecchi, 2016). Early workers used traditional methods such as man fly-rounds and stationary traps which have proved to be resource-intensive approach. Later efforts to map tsetse distribution have incorporated remote sensing, geographic information system (GIS) techniques as well as remote-sensing to generate geo-referenced datasets that help define distribution limits at different levels of resolution (Hay, Tucker, Rogers, & Packer, 1996; Kitron et al., 1996; Mahama, Koné, de la Rocque, De Deken, & Geert, 2005; Rogers, Hay, & Packer, 1996; Stanton, Esterhuizen, Tirados, Betts, & Torr, 2018).

A composite spatial distribution map of the three subgenera of tsetse combined from data sources such predicted areas of suitability (PAS) and environmental predictor variables (e.g. normalized difference vegetation index NDVI and land surface temperature) has been developed (Cecchi, Mattioli, J, & De la Rocque, 2008). Additionally, a continental Atlas of tsetse distribution, which condenses the spatially explicit data that is available in peer-reviewed scientific publications from January 1990 has been drawn up by FAO (Cecchi, Paone, Argiles, Vreysen, & Mattioli, 2015). For each input in the Atlas, details about survey period, occurrence, apparent density, trap type and density, and duration of trapping are included where available. Under the umbrella of the continental Atlas project, some countries including Sudan and Mali have updated their respective national atlases (Ahmed et al., 2016; Boucader, Modibo, Oumar, & Boubacar, 2019). The national atlases include all epidemiological data, both published and unpublished, that has been collected in the respective country. It is expected that distribution updates from the individual countries will ultimately provide a composite updated Atlas of tsetse distribution in Africa.

Changes in tsetse distribution have been driven principally by disruption of traditional tsetse habitats. This has been occasioned by a rapid increase in human population which in turn has resulted in increased human activities (settlement, farming) in areas that were formerly infested by tsetse. Indeed, simulation models have suggested that human population growth will cause a decline of savannah and forest tsetse, with possible extinction in eastern and southern Africa (De La Rocque, Augusseau, & Guillobez, 2001). This encroachment has led to fragmentation of suitable tsetse habitat and a resultant decrease in fly abundance (Ducheyne et al., 2009; Scoone et al., 2017). In these regions, tsetse populations have been confined to discrete habitats, with a high abundance in and around wildlife conservation areas such as game parks and reserves. Such conservation areas provide suitable conditions for tsetse survival and function as breeding site. Therefore, today most of the tsetse-infested areas lie in protected areas such as game parks and game reserves.

Identification of *Glossina* Species

One of the main outcomes of the long history of tsetse research in Africa is that the identities of any tsetse species occurring in any African country are generally known, as is their approximate distribution. Distribution maps, which workers can refer to, are widely available. Where this is not the case, information on the location of collection, the size and the morphology, including the genitalia of the flies, can help to distinguish the available species. This is of particular importance where two or more species occur together (sympatric populations). For example, if one is provided with unidentified specimens of *G. morsitans*, *G. palpalis*, *G. tabaniformis* and *G. longipennis,* a comparison of the ventral view of the thecal bulb can help to distinguish between them, due to their different colouration (FAO, 1982). As a subsequent step, supporting information about, for example, collection location and relative size will further help to narrow down the options. A useful tool in the morphological identification of different species is a dichotomous key. One such comprehensive key for distinguishing *Glossina* adults has been detailed in a training manual (FAO, 1982) and an updated version is also available (S. G. A. Leak, Ejigu, & Vreysen, 2008). When using a dichotomous key, it is however advisable to have two or more species for comparison because some features are highly subjective and therefore, a comparative approach is recommended.

While morphological methods form the basis of species identification, it is not always possible to use this pathway for identification. Specimens may lose important external features of their anatomy such as legs and heads, either in the sampling trap or due to degradation after long periods of preservation (e.g. museum specimens).At other times, the level of knowledge and skill required to perform the highly technical morphological identification may be lacking (Erlank, Koekemoer, & Coetzee, 2018). Alternative systems of species identification have been developed in the last few decades. These methods require specialized resources and are therefore usually more expensive to use. Wing morphometrics can be used to distinguish between species and subpopulations of tsetse flies, with the advantage of simplicity of data acquisition and low cost (Getahun, Cecchic, & Seyouma, 2014; Rohlf & Marcus, 1993; Solano et al., 1999). Mitochondrial markers, specifically cytochrome oxidase subunit II (COII) and cytochrome b (CytB).have been used in species identification (Orji, Onyeocha, Shaida, & Ella, 2015). A schematic representation of a step-by-step application of molecular tools for tsetse species identification has been described by various authors (Abd-Alla, Meki, Bourtzis, & Parker, 2018; Augustinos et al., 2018). Here, the authors describe the steps necessary to identify live specimens as well as preserved specimens. A non-destructive method is applied for the live specimens by using a single leg for the assay and therefore maintaining the live specimen. This method uses a combination of nuclear markers, microsatellites and *Wolbachia*-infection status to distinguish DNA from different tsetse species.

Host-Location and Feeding

Tsetse flies are obligate blood-feeders, feeding on a variety of vertebrate hosts. Host choice is dependent upon various factors such as temperature, humidity, light intensity and ultimately on the availability of hosts. Early workers concluded that the observed species-specific patterns feeding they observed maybe due to, at least in part, genetics (Weitz, 1963). Host location in tsetse is driven by visual and olfactory cues. It is thought that the fly first "smells" a potential host via an odour plume. The fly then follows that odour plume until it visualizes the potential host. Upon landing on the host, the fly is stimulated by the host body heat to probe and commence feeding. It has been shown that different species exhibit different olfactory characteristics, which has ultimately impacts host choice (Obiero et al., 2014) .

The three subgenera have varying affinities for different hosts, which is roughly related to the ecological niche they inhabit. For example, the *palpalis* group tsetse which inhabit riverine or lacustrine habitats, live close to reptiles such as crocodiles and monitor lizards which are their preferred hosts (Omolo et al., 2009). In West Africa, *palpalis-* group tsetse feed predominantly on humans and suidae, a key pathway in the epidemiology of human trypanosomiasis. The *morsitans* group tsetse which infest mainly the savannah grasslands, preferentially feed on wild bovids such as buffalo, eland and bushbuck. Where cattle are available they form a much-favoured source of bloodmeals for *morsitans*-group tsetse(S. Leak, 1999). The *fusca*-group tsetse mostly feed on forest-dwelling animals such as wild pigs and rhinoceros (Sasaki, Kang'ethe, & Kaburia, 1995).

It should however be noted that the most abundant host is not always the most preferred host. In the Serengeti, *G. swynnertoni* feeds preferentially on warthog, buffalo and giraffe despite the proportionally low population of these species (Auty et al., 2016). This begs the question whether tsetse flies are opportunistic feeders which feed on any vertebrate available or if they are selective when presented with options. It is suggested that host size is an important deciding factor, presumably because visual cues are a critical complement to olfaction, as does host availability (Muturi et al., 2011). It has been shown that different species exhibit different olfactory characteristics, which has ultimately impacts host choice.

Artificial bait technologies, such as traps and insecticide-impregnated targets, have been designed based on the characteristics that influence tsetse host location. Natural odours such as urine and ox breath are known to increase fly catches. This has led to the isolation, identification and synthesis of artificial odours which have a similar olfactory profile as natural odours. These so-called attractants which enhance fly catches have been used successfully in tsetse control interventions (Opiyo, Njogu, & Omuse, 1990; Rayaisse et al., 2015; Vale, Lovemore, & Flint, 1988). Similarly, some natural odour cues such as those from the waterbuck, have been shown to have a repellent effect on flies. Synthetic analogues of such repellents have been incorporated into control technologies using this repellence effect (Saini et al., 2017).

Tsetse Gut Microbiota

Many species of arthropods are host to various bacterial microorganisms. These microorganisms have traditionally been classified as mutualists when beneficial, parasitic when harmful, or as commensal when neutral. Insects that feed on a single restricted diet throughout their lives such as plant sap, wood or animal blood often harbor mutualists which provide additional nutrients that may be unavailable through their normal diet, and which they are unable to synthesize(S. Aksoy, Maudlin, Dale, Robinson, & O´Neill, 2001; Moran & Baumann, 2000)

Tsetse flies harbor at least three bacterial endosymbionts, some of which contribute to their nutrition by synthesis of vitamins such as folic acid and vitamin B complex metabolites (Nogge, 1981). The most important is *Wigglesworthia glossinidia,* which is the primary (obligate) endosymbiont. *Wigglesworthia* is transmitted through the maternal milk gland secretions and has obligate functions in host biology (S Aksoy, 1995). *Wigglesworthia* is responsible for maintaining fecundity in the adult female (Hill & Campbell, 1973). Thus, it is found in all populations of tsetse flies where it forms the most abundant bacterial species in the tsetse gut (E. Aksoy et al., 2014; Griffith et al., 2018 ; Ngoune et al., 2019).

The tsetse gut also contains secondary endosymbionts which are not obligatory to host biology but may provide host benefits depending on environmental context, such as during periods of heat stress, parasitoid attack or towards the utilization of particular host plant substrates (Snyder, Adkins, & Rio, 2011). The secondary endosymbiont *Sodalis glossinidius* is thought to have an effect on vector competence (Dennis et al., 2014) . However, the association between trypanosome prevalence and *S. glossinidius* presence in tsetse flies may be confounded by the specifics of the population tested at different locations (Channumsin et al., 2018). This bacteria has been detected in several fly species albeit at low prevalence levels in West Africa(Kame-Ngasse et al., 2018 ; Kanté, Farikou, Njiokou, & Simo, 2018) East Africa (E. Aksoy, et al., 2014; Griffith, et al., 2018 ; Wamwiri et al., 2013) and in southern Africa (Dennis, et al., 2014). It should be noted that most of the fore-mentioned studies used conventional PCR for the diagnosis. The density of *S. glossinidius* in tsetse in nature is known to be very low, a fact which may contribute to the low prevalence rate observed. In contrast, a very high prevalence rate (upto 100%) was detected when using highly sensitive methods such as 16S rRNA deep sequencing(E. Aksoy, et al., 2014).

The third endosymbiont present in some tsetse species is the parasitic *Wolbachia pipientis* (Cheng et al., 2000; Rio, Wu, Filardo, & Aksoy, 2006) which is a common non-obligatory endosymbiont of arthropods. *Wolbachia* is usually maternally-transmitted through the mother's milk gland secretions into the developing larvae(Cheng & Aksoy, 1999) but horizontal transmission is thought to occur in rare instances (Rousset, Bouchon, Pintureau, Juchault, & Solignac, 1992). The functions of *Wolbachia* are not well known but its presence has been shown to cause cytoplasmic incompatibility, resulting to lack of offspring, in tsetse flies (Alam et al., 2011). The prevalence rates of *Wolbachia* infections are heterogeneous between species and even populations. The infections in populations reared under colony conditions may reach levels of 100%, (Cheng, et al., 2000). This is also the case in field populations of species such *as G. austeni* (Wamwiri, et al., 2013). The large variations obtained in *Wolbachia* prevalence levels may be attributed to low-titre infections and/or the sensitivity fo the diagnostic test used (Ouedraogo et al., 2018). For example, in the case of *G. p. palpalis* different workers report either an absence or a relatively low rate of *Wolbachia* infection, depending on the method use of infection (Cheng et al., 2000 ; Doudoumis et al., 2012; Kanté, et al., 2018).

Several low-density organisms have been isolated from the guts of tsetse flies, using culture-dependent and culture-independent methods. These organisms include species from the genera *Serratia, Staphylococcus, Acinetobacter, Spiroplasma, Bacillus, Pseudomonas and Corynebacterium* (Jacob et al., 2017; Lindh & Lehane, 2011; Malele et al., 2018). In general, these microbes account for less than 1% of tsetse's entire enteric microbiota (Rio, et al., 2006). Current knowledge on the potential effect of these low-density organisms on tsetse fly vector competence is still scanty. There is little experimental evidence that suggests that environmentally acquired bacteria mediate trypanosome infection outcomes in tsetse (Griffith, et al., 2018). However, similar studies conducted using other insects, such as mosquitoes and sandflies, have shown that microbial infections significantly altered the rates of infection with *Plasmodium* and *Leishmania* parasites respectively. This field therefore requires more investigation in order to be fully understood.

Role of Tsetse Flies in Disease Transmission

Trypanosomes

Trypanosomes are unicellular of hemoflagellate protozoan organisms. They are kinetoplastids classified under family Trypanosomatidae and genus *Trypanosoma*. They have a flat body surrounded by a strong but thin skin membrane. The kinetoplast, located near the posterior end of the trypanosome contains the mitochondrial DNA of the parasite. There is a flagellum that originates from the kinetoplast and runs along the body length connected with an undulating membrane. Part of this flagellum remains free at the anterior end. The flagellum has both locomotive and sensory functions. Figure 5 shows bloodstream *T. b. rhodesiense* parasites in a blood smear from mice.

There are two distinct genera of trypanosomes divided according to the method by which the parasite leaves the insect vector to be transmitted to the vertebrate host. The Stercoraria includes species whose development is terminated in the rear part of the digestive tract of the vector i.e. the posterior station. The infective forms for these species (e.g. *Trypanosoma cruzi*, *T. rangeli* and *T. theleri*) are found in faecal material. The second *Trypanosoma* genus is the Salivaria where development in the vector is terminated in the anterior station i.e. the salivary glands. The salivarian trypanosomes are further divided into four subgenera *Trypanozoon, Duttonella, Nannomonas* and *Pycnomonas*. This division is based loosely on the developmental cycle of the parasite in the vector.

The subgenus *Trypanozoon* species develop successively in the tsetse gut, proboscis, and salivary glands is divided into 3 species T. *brucei, T. evansi and T. equiperdum,* based on pathogenicity, distribution and host range. *T. brucei* is subdivided further into three morphologically indistinguishable subspecies viz. *T. b. gambiense*, and *T. b. brucei*. The first two are the aetiological agents of human African trypanosomiasis whereas *T. b. brucei* causes infections in ungulates. *T. evansi* is the causative agent of surra in camels, while. *T. equiperdum* causes dourine in horses and donkeys.

The subgenus *Dutonella* is composed of two main species *T. vivax* and *T. uniforme*. Development of parasites in this subgenus (also called *T.vivax*-group) is restricted to the mouthparts. This characteristic enables both cyclically and mechanically transmission of *T. vivax,* which has led to its spread beyond the tsetse belt in Africa, and into South America and Asia.

The *Nannomonas* species, which develop successively in the gut and the mouthparts include the animal-infective *T. simiae, T. godfreyi* and *T. congolense*. The two first species are mainly infective to mammals belonging to the Suidae family (domestic pigs, warthogs) while *T. congolense* has a broader range of hosts including livestock and game animals. The fourth subgenus *Pycnomonas* includes *T. suis*, a parasite which is closely related to *T. brucei* and which affects mainly Suidae.

Figure 5. Bloodstream T. b. rhodesiense parasites in a blood smear from mice (Courtesy Kariuki Nd'ungu).

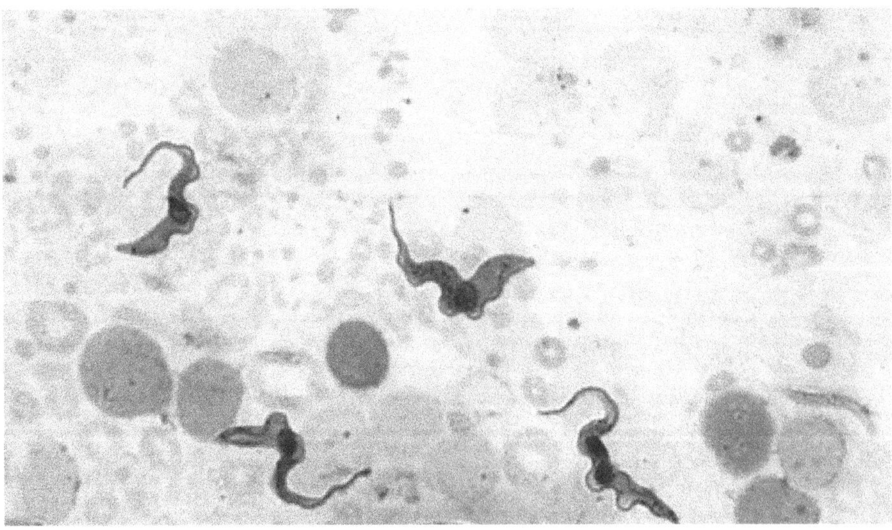

Transmission Cycle of Trypanosomes

The transmission cycle begins when a tsetse fly picks up infective trypomastigotes in a bloodmeal from an infected host animal. In the midgut, the parasites multiply by binary fission over a period of 2-3 weeks and transform into procyclic trypomastigotes. The procyclics may either die in the midgut of refractory individuals or survive to yield persistent procyclic infections in susceptible insects. Some procyclics may leave the midgut, transform into "stumpy form" epimastigotes and migrate into the salivary glands, where they transform further and mature into metacyclic trypomastigotes (Roberts and Janovy, 2000). The metacyclics are the only forms that are infective to mammalian hosts. When the infected fly feeds upon an uninfected host, metacyclic trypomastigotes are injected into the skin. The parasites enter the lymphatic system and pass into the bloodstream, multiply and finally transform into bloodstream trypomastigotes (O'Neill et al., 1992). These can then be picked up by another uninfected fly and the cycle continues.

The completion of this developmental cycle includes several steps of proliferation, migration and differentiation. It varies with trypanosome and fly species, and may take from a few days for *T. vivax* up to 3 weeks for *T. brucei*(Rotureau & Van Den Abbeele, 2013). The transmission of trypanosomes after they undergo a cycle of development culminating in the production of infective metacyclic trypanosomes is referred to as cyclical transmission, and is more common than mechanical transmission, which requires that blood containing infectious trypanosomes be transferred from one animal to another.

Complex interactions of factors contribute to the successful establishment, survival and maturation of the parasite in the fly. These include tsetse-dependent factors (age of fly at infective feed, source of bloodmeal, fly species), ecological factors (e.g. climate) as well as parasite/host factors. Certain species are better vectors of trypanosomes than others. All these inter-dependent factors contribute to varying levels of vector competence among different fly species.

Figure 6. Typical transmission cycle of trypanosomes: (H) - stages taking place in the host, (V) - stages taking place in the vector

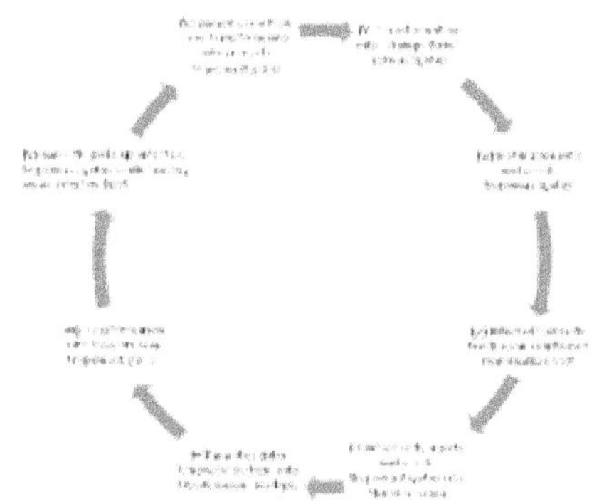

THE ROLE OF BITING FLIES IN AAT EPIDEMIOLOGY

Biting flies are the vectors of *Trypanosoma evansi* and to a lesser extent, *T. vivax*, outside the tsetse belts. *T. evansi* is often found infecting camels, whereas *T. vivax* may infect livestock in general. The tabanids are dipteran flies belonging to the suborder Brachycera and the family Tabanidae. Stomoxes are dipterans belonging to the family Muscidae and the sub-family Stomoxyinae (Dia & Desquesnes, 2004). The biting flies implicated in trypanosome transmission primarily belong to the families Tabanidae (also known as "horse flies"), Hippoboscidae and Muscidae.Tabanids have a wide distribution across the world and more than four thousand species have been described (Baldacchino et al., 2014). The tabanid genera often encountered include the *Ancala, Tabanus, Atylotus, Chrysops*(deer flies), *Philoliche* and *Haematopota* (clegs).Among the hippoboscids,*Hippobosca camelina* (camel flies)are the most important vector, especially for *surra* (camel trypanosomiasis). From the Muscidae, the most important species are the cosmopolitan *S. calcitrans* (stable fly) as well as *Haematobia* (horn flies). The latter usually congregates in very high densities on camels. Several species of biting flies have been confirmed, either in nature or by experimental methods, to transmit trypanosomes. These include camel flies (or camel keds) *Hippobosca camelina* (Kidambasi, Masiga, Villinger, Carrington, & Bargul, 2019), *Haematobia minuta* (Oyieke, 1989), *Philoliche* spp (Dirie, Wallbanks, Aden, Bornstein, & Ibrahim, 1989), *Stomoxys* spp (Sumba, Mihok, & FA, 1998) and *Tabanus* spp (Odeniran, Macleod, Ademola, & Welburn, 2019). The density and distribution of tabanid flies is highly seasonal (Baldacchino et al., 2013). Thus, although flies are caught throughout the year, there are often seasonal variations in abundance and diversity which corresponds to the end of the rains and the beginning of the dry season.

Mechanical transmission of parasites occurs when the feeding process is interrupted on one infected host and thereafter re-started on a new uninfected host. The actual transmission occurs through either contamination of the mouthparts or by regurgitation of digestive tract contents. The successful transmission of trypanosomes depends on their survival in the mouthparts or digestive tract of the vector. Therefore, the shorter the interval of feeding between an infected and an uninfected animal, the greater the success of parasite transmission(Aregawi, Agga, Abdi, & Büscher, 2019). This mode of transmission is more efficient inside a group of animals (intra-herd). However, transmission between spatially distinct groups of animals (inter-herd) can also occur. Tabanids and *Stomoxys* have different feeding frequencies related perhaps to their relative sizes and feeding habits. It has been shown that *Stomoxys* feed more frequently at intervals of less than 72 hours, whereas tabanids may feed about every 5 - 7 days For this reason, it is thought that *Stomoxys* are involved in both intra- and inter-herd transmission of *T. evansi*, whereas the more persistent tabanids are usually implicated in intra-herd transmission (Desquesnes, Dargantes, & Lai, 2013). Successful parasite transmission is influenced by a variety of host-related and vector-related factors. Host-related factors include the level of parasitaemia and herd dynamics, for example, the relative proportion of infected and non-infected hosts in the herd (Baldacchino, et al., 2013).On the other hand vector-related aspects, for instance, the survival period of the pathogen in the vector and the time the insect takes between subsequent feedings has also been shown to influence mechanical transmission of parasites(Schofield & Torr, 2002).

FUTURE RESEARCH DIRECTIONS

More than one hundred years have been applied to research into tsetse and trypanosomiasis. Despite this, the disease continues to have a negative impact on human and animal health in many parts of Africa. The basic biology and ecology of *Glossina* has been studied very thoroughly and has formed a strong background for research that is increasingly more applied than basic. Generally, the research emphasis has shifted along the line from basic - strategic - applied - adaptive. Thus current research themes are focused on the tsetse-trypanosome interface in the quest for transmission blocking applications and/or the development of refractory populations. In addition, there is an increased need for a multi-disciplinary approach to vector and disease control, and in particular the application of disease modeling to translate GIS data into practical and useable vector and disease control information.

CONCLUSION

The fact that trypanosomiasis continues to be a constraint to livestock productivity even after more than a century of research, is an indicator that there are gaps in scientific knowledge that must be resolved in order to fully control or eradicate this disease. This chapter has covered various aspects related to the insect vectors of trypanosomes, both the tsetse flies and, to a lesser extent, the biting flies. A summary of the current state of knowledge on tsetse biology and ecology has been provided. Research on some aspects covered is continually evolving as more intricate molecular tools are developed. However, the information here provides a strong baseline for researchers interested in this field and contributes to the ongoing efforts to control trypanosomiasis.

REFERENCES

Abd-Alla, A., Meki, I., Bourtzis, K., & Parker, A. (2018). *Standard Operating Procedures for Identification of Tsetse Species from Wild Populations and Laboratory Colonies*. FAO/IAEA.

Abila, P., Kiendrebeogo, M., Mutika, G., Parker, A., & Robinson, A. (2003). The effect of age on the mating competitiveness of male Glossina fuscipes fuscipes and G. palpalis palpalis. *Journal of Insect Science*, *3*(13), 1–8. doi:10.1673/031.003.1301 PMID:15841229

Ahmed, S., Rahman, A., Hassan, M., Salih, S., Paone, M., & Cecchi, G. (2016). An atlas of tsetse and bovine trypanosomosis in Sudan. *Parasites & Vectors*, *9*(1), 194. doi:10.118613071-016-1485-6 PMID:27056678

Aksoy, E., Telleria, E., Echodu, R., Wu, Y., Okedi, L., Weiss, B., Aksoy, S., & Caccone, A. (2014). Analysis of multiple tsetse fly populations in Uganda reveals limited diversity and species-specific gut microbiota. *Applied and Environmental Microbiology*, *80*(14), 4301–4312. doi:10.1128/AEM.00079-14 PMID:24814785

Aksoy, S. (1995). *Wigglesworthia* gen. nov. and *Wigglesworthia glossinidae* sp. nov., taxa consisting of the mycetocyte-associated primary endosymbionts of tsetse flies. *International Journal of Systematic Bacteriology*, *45*(4), 848–851. doi:10.1099/00207713-45-4-848 PMID:7547309

Aksoy, S., Maudlin, I., Dale, C., Robinson, A. S., & O'Neill, S. L. (2001). Prospects for control of African trypanosomiasis by tsetse vector manipulation. *Trends in Parasitology*, *17*(1), 29–35. doi:10.1016/S1471-4922(00)01850-X PMID:11137738

Alam, U., Medlock, J., Brelsfoard, C., Pais, R., Lohs, C., Balmand, S., Carnogursky, J., Heddi, A., Takac, P., Galvani, A., & Aksoy, S. (2011). Wolbachia symbiont infections induce strong cytoplasmic incompatibility in the tsetse ñy Glossina morsitans. *PLoS Pathogens*, *7*(12), e1002415. Advance online publication. doi:10.1371/journal.ppat.1002415 PMID:22174680

Albert, M., Wardrop, N., Atkinson, P., Torr, S., & Welburn, S. (2015). Tsetse Fly (G.f. fuscipes) Distribution in the Lake Victoria Basin of Uganda. *PLoS Neglected Tropical Diseases*, *9*(4), e0003705. doi:10.1371/journal.pntd.0003705 PMID:25875201

Aregawi, W. G., Agga, G. E., Abdi, R. D., & Büscher, P. (2019). Systematic review and meta-analysis on the global distribution, host range, and prevalence of Trypanosoma evansi. *Parasites & Vectors*, *12*(1), 67. Advance online publication. doi:10.118613071-019-3311-4 PMID:30704516

Augustinos, A. A., Meki, I. K., Demirbas-Uzel, G., Ouédraogo, G. M. S., Saridaki, A., Tsiamis, G., Parker, A. G., Abd-Alla, A. M. M., & Bourtzis, K. (2018). Nuclear and Wolbachia-based multimarker approach for the rapid and accurate identification of tsetse species. *BMC Microbiology*, *18*(1), 147. doi:10.118612866-018-1295-4 PMID:30470190

Auty, H., Cleaveland, S., Malele, I., Masoy, J., Lembo, T., & Bessell, P. (2016). Quantifying Heterogeneity in Host-Vector Contact: Tsetse (Glossina swynnertoni and G. pallidipes) Host Choice in Serengeti National Park, Tanzania. *PLoS ONE, 11*, e0161291. doi: .pone.0161291 doi:10.1371/journal

Baldacchino, F., Desquesnes, M., Mihok, S., Foil, L., Duvallet, G., & Jittapalapong, S. (2014). Tabanids: Neglected subjects of research, but important vectors of disease agents. *Infection, Genetics and Evolution*, *28*, 596–615. doi:10.1016/j.meegid.2014.03.029 PMID:24727644

Baldacchino, F., Muenworn, V., Desquesnes, M., Desoli, F., Charoenviriyaphap, T., & Duvallet, G. (2013). Transmission of pathogens by Stomoxys flies (Diptera, Muscidae): A review. *Parasite (Paris, France)*, *20*, 20–26. doi:10.1051/parasite/2013026 PMID:23985165

Benoit, J. B., Attardo, G. M., Baumann, A. A., Michalkova, V., & Aksoy, S. (2015). Adenotrophic viviparity in tsetse flies: Potential for population control and as an insect model for lactation. *Annual Review of Entomology*, *60*(1), 351–371. doi:10.1146/annurev-ento-010814-020834 PMID:25341093

Bonomi, A., Bassetti, F., Gabrieli, P., Beadell, J., Falchetto, M., Scolari, F., Gomulski, L. M., Regazzini, E., Ouma, J. O., Caccone, A., Okedi, L. M., Attardo, G. M., Guglielmino, C. R., Aksoy, S., & Malacrida, A. R. (2011). Polyandry is a common event in wild populations of the Tsetse fly Glossina fuscipes fuscipes and may impact population reduction measures. *PLoS Neglected Tropical Diseases*, *5*(6), e1190–e1190. doi:10.1371/journal.pntd.0001190 PMID:21666797

Boucader, D., Modibo, D., Oumar, D., & Boubacar, B. (2019). A national atlas of tsetse and African animal trypanosomosis in Mali. *Parasites & Vectors*, *12*(1), 466. doi:10.118613071-019-3721-3 PMID:31597558

Cecchi, G., Mattioli, R. J. S., & De la Rocque, S. (2008). Land cover and tsetse fly distributions in sub-Saharan Africa. *Medical and Veterinary Entomology*, *22*(4), 364–373. Advance online publication. doi:10.1111/j.1365-2915.2008.00747.x PMID:18785934

Cecchi, G., Paone, M., Argiles, H. R., Vreysen, M., & Mattioli, R. (2015). Developing a continental atlas of the distribution and trypanosomal infection of tsetse flies (Glossina species). *Parasites & Vectors*, *8*(1), 284. doi:10.118613071-015-0898-y PMID:25994757

Channumsin, M., Ciosi, M., Masiga, D., Mable, B., Michael, C., & Turner, R. (2018). Sodalis glossinidius presence in wild tsetse is only associated with presence of trypanosomes in complex interactions with other tsetse-specific factors. *BMC Microbiology*, *18*(S1), 163. doi:10.118612866-018-1285-6 PMID:30470184

Cheng, Q., & Aksoy, S. (1999). Tissue tropism, transmission and expression of foreign genes *in vivo* in midgut symbionts of tsetse flies. *Insect Molecular Biology*, *8*(1), 125–132. doi:10.1046/j.1365-2583.1999.810125.x PMID:9927181

Cheng, Q., Ruel, T., Zhou, W., Moloo, S., Majiwa, P., O'Neill, S., & Aksoy, S. (2000). Tissue distribution and prevalence of Wolbachia infections in tsetse flies, Glossina spp. *Medical and Veterinary Entomology*, *14*(1), 44–50. doi:10.1046/j.1365-2915.2000.00202.x PMID:10759311

Cheng, Q., Ruel, T. D., Zhou, W., Moloo, S. K., Majiwa, P., O'Neill, S. L., & Aksoy, S. (2000). Tissue distribution and prevalence of *Wolbachia* infections in tsetse flies, *Glossina* spp. *Medical and Veterinary Entomology*, *14*(1), 44–50. doi:10.1046/j.1365-2915.2000.00202.x PMID:10759311

Chikowore, G., Dicko, A., Chinwada, P., Zimba, M., Shereni, W., Roger, F., Bouyer, J., & Guerrini, L. (2017). A pilot study to delimit tsetse target populations in Zimbabwe. *PLoS Neglected Tropical Diseases*, *11*(5), e0005566. doi:10.1371/journal.pntd.0005566 PMID:28467409

De Beer, C., Venter, G., & Vreysen, M. (2015). Determination of the optimal mating age of colonised Glossina brevipalpis and Glossina austeni using walk-in field cages in South Africa. *Parasites & Vectors*, *8*(467), 467. Advance online publication. doi:10.118613071-015-1073-1 PMID:26380975

De Beer, C. J., Venter, G. J., Kappmeier Green, K., Esterhuizen, J., De Klerk, D. G., Ntshangase, J., Vreysen, M. J. B., Pienaar, R., Motloang, M., Ntantiso, L., & Latif, A. A. (2016). An update of the tsetse fly (Diptera: Glossinidae) distribution and African animal trypanosomosis prevalence in north-eastern KwaZulu- Natal, South Africa. *The Onderstepoort Journal of Veterinary Research*, *1172*(1). Advance online publication. doi:10.4102/ojvr.v83i1.1172 PMID:27380653

De La Rocque, S., Augusseau, X., & Guillobez, S. (2001). The changing distribution of two riverine tsetse flies over 15 years in an increasingly cultivated area of Burkina Faso. *Bulletin of Entomological Research*, *91*, 157–166. PMID:11415469

Dennis, J., Durkin, S., Horsley, D., Hamill, L., Anderson, N., & MacLeod, E. (2014). Sodalis glossinidius prevalence and trypanosome presence in tsetse from Luambe National Park, Zambia. *Parasites & Vectors*, *7*(1), 378. doi:10.1186/1756-3305-7-378 PMID:25138709

Desquesnes, M., Dargantes, A., Lai, D.-H., Lun, Z.-R., Holzmuller, P., & Jittapalapong, S. (2013). *Trypanosoma evansi* and Surra: A Review and Perspectives on Transmission, Epidemiology and Control, Impact, and Zoonotic Aspects. *BioMed Research International*, *2013*, 1–20. Advance online publication. doi:10.1155/2013/321237 PMID:24151595

Dia, M., & Desquesnes, M. (2004). Evaluation of a New Trap for Tabanids and Stomoxyines. *Bulletin Société royale belge d'Entomologie Koninklijke Belgische Vereniging voor Entomologie, 140*, 72-81.

Dirie, M., Wallbanks, K., Aden, A., Bornstein, S., & Ibrahim, M. (1989). Camel trypanosomiasis and its vectors in Somalia. *Veterinary Parasitology*, *32*(4), 285–291. doi:10.1016/0304-4017(89)90039-3 PMID:2781715

Doudoumis, V., Tsiamis, G., Wamwiri, F., Brelsfoard, C., Alam, U., Aksoy, E., Dalaperas, S., Abd-Alla, A., Ouma, J., Takac, P., Aksoy, S., & Bourtzis, K. (2012). Detection and characterization of Wolbachia infections in laboratory and natural populations of different species of tsetse flies (genus Glossina). *BMC Microbiology*, *12*(Suppl 1), S3. doi:10.1186/1471-2180-12-S1-S3 PMID:22376025

Ducheyne, H., Mweempwa, C., De Pus, C., Vernieuwe, H., De Deken, R., Hendrickx, G., & Van den Bossche, P. (2009). The impact of habitat fragmentation on tsetse abundance on the plateau of eastern Zambia. *Preventive Veterinary Medicine*, *91*(1), 11–18. doi:10.1016/j.prevetmed.2009.05.009 PMID:19523702

Duguma, R., Tasew, S., Olani, A., Damena, D., Alemu, D., Mulatu, T., Alemayehu, Y., Yohannes, M., Bekana, M., Hoppenheit, A., Abatih, E., Habtewold, T., Delespaux, V., & Duchateau, L. (2015). Spatial distribution of Glossina sp. and Trypanosoma sp. in south-western Ethiopia. *Parasites & Vectors*, *8*(1), 430. doi:10.118613071-015-1041-9 PMID:26286484

Erlank, E., Koekemoer, L. L., & Coetzee, M. (2018). The importance of morphological identification of African anopheline mosquitoes (Diptera: Culicidae) for malaria control programmes. *Malaria Journal*, *17*(1), 43. Advance online publication. doi:10.118612936-018-2189-5 PMID:29357873

FAO. (1982). *Tsetse Biology, Systematics and Distribution: Techniques* (Vol. 1). Food and Agriculture Organisation of the United Nations.

Ford, J., & Katondo, K. (1977). Maps of tsetse flies (Glossina) distribution in Africa, 1973 according to sub-generic groups on a scale of 1: 5 000 000. *Bulletin of Animal Health and Production in Africa*, *25*, 187–193.

Getahun, M., Cecchic, G., & Seyouma, E. (2014). Population studies of Glossina pallidipes in Ethiopia: Emphasis on cuticular hydrocarbons and wing morphometric analysis. *Acta Tropica*, *138*, S12–S21. doi:10.1016/j.actatropica.2014.04.015 PMID:24751419

Gouteux, J. (1987). Une nouvelle glossine du Congo: Glossina (Austenina) frezili sp. nov. (Diptera; Glossinidae). *Tropical Medicine and Parasitology*, *38*, 97–100. PMID:3629143

Griffith, B., Weiss, B., Aksoy, E., Mireji, P., Auma, J., Wamwiri, F., Echodu, R., Murilla, G., & Aksoy, S. (2018). Analysis of the gut-specific microbiome from field-captured tsetse flies, and its potential relevance to host trypanosome vector competence. *BMC Microbiology*, *18*(S1), 146. doi:10.118612866-018-1284-7 PMID:30470178

Hay, S., Tucker, C., Rogers, D., & Packer, M. (1996). Remotely sensed surrogates of meteorological data for the study of the distribution and abundance of arthropod vectors of disease. *Annals of Tropical Medicine and Parasitology*, *90*(1), 1–19. doi:10.1080/00034983.1996.11813021 PMID:8729623

Hill, P. D. S., & Campbell, J. A. (1973). The production of symbiont-free *Glossina morsitans* and an associated loss of female fertility. *Transactions of the Royal Society of Tropical Medicine and Hygiene*, *67*(5), 727–728. doi:10.1016/0035-9203(73)90051-5 PMID:4798390

Jacob, F., Melachio, T., Njitchouang, G., Gimonneau, G., Njiokou, F., Abate, L., Christen, R., Reveillaud, J., & Geiger, A. (2017). Intestinal bacterial communities of trypanosome-infected and uninfected Glossina palpalis palpalis from three human African trypanosomiasis foci in Cameroon. *Frontiers in Microbiology*, *8*, 1464. doi:10.3389/fmicb.2017.01464 PMID:28824591

Kame-Ngasse, G., Njiokou, F., Melachio-Tanekou, T., Farikou, O., Simo, G., & Geiger, A. (2018). Prevalence of symbionts and trypanosome infections in tsetse flies of two villages of the "Faro and Déo" division of the Adamawa region of Cameroon. *BMC Microbiology*, *18*(S1), 159. doi:10.118612866-018-1286-5 PMID:30470177

Kanté, T. S., Farikou, O., Njiokou, F., & Simo, G. (2018). Prevalence of Sodalis glossinidius and different trypanosome species in Glossina palpalis palpalis caught in the Fontem sleeping sickness focus of the southern Cameroon. *Parasite (Paris, France)*, *25*, 44. doi:10.1051/parasite/2018044 PMID:30117802

KidambasiK.MasigaD.VillingerJ.CarringtonM.BargulJ. (2019). Detection of blood pathogens in camels and their associated ectoparasitic camel biting keds: the potential application of keds in xenodiagnosis of camel haemopathogens *Hippobosca camelina* AAS Open Research *164*. doi:10.12688/aasopenres.13021.1

Kitron, U., Otieno, L., Hungerford, L., Odulaja, A., Brigham, W. U., Okello, O. O., Joselyn, M., Mohamed-Ahmed, M. M., & Cook, E. (1996). Spatial analysis of the distribution of tsetse flies in the Lambwe valley, Kenya, using Landsat TM satellite imagery and GIS. *Journal of Animal Ecology*, *65*(3), 371–380. doi:10.2307/5883

Leak, S. (1999). *Tsetse biology and ecology: their role in the epidemiology and control of trypanosomosis*. ILRI.

Leak, S. G. A., Ejigu, D., & Vreysen, M. J. B. (2008). *Collection of entomological baseline data for tsetse area-wide integrated pest management programmes*. Joint FAO/IAEA Programme of Nuclear Techniques in Food and Agriculture.

Lindh, J., & Lehane, M. (2011). The tsetse fly Glossina fuscipes fuscipes (Diptera: Glossina) harbours a surprising diversity of bacteria other than symbionts. *Antonie van Leeuwenhoek*, *711*(3), 20–29. doi:10.100710482-010-9546-x PMID:21203841

Mahama, C., Koné, A., de la Rocque, S., De Deken, R., & Geert, S. (2005). Prediction of the distribution of Glossina tachinoides (Diptera: Glossinidae) in the Volta basin of northern Ghana. *Bulletin of Entomological Research*, *95*(1), 63–67. doi:10.1079/BER2004340 PMID:15705216

Malele, I., Hamisi, N., & Atway, M. (2011). Factors defining the distribution limit of tsetse infestation and the implication for livestock sector in Tanzania. *African Journal of Agricultural Research*, *6*, 2341–2347.

Malele, I., Nyingilili, H., Lyaruu, E., Tauzin, M., Ollivier, B., Cayol, J., & al., e. (2018). Bacterial diversity obtained by culturable approaches in the gut of Glossina pallidipes population from a non-sleeping sickness focus in Tanzania: preliminary results. *BMC Microbiology, 18*, 164.

McAlpine, J. (1989). Phylogeny and classification of the Muscomorpha. In J. McAlpine & D. Wood (Eds.), *Manual of Nearctic Diptera* (Vol. 3, pp. 1397–1518). Agriculture Canada.

Moran, N. A., & Baumann, P. (2000). Endosymbionts in animals. *Current Opinions in Microbiology*, 270–275.

Muturi, C., Ouma, J., Malele, I., Ngure, R., Rutto, J., Mithöfer, K., Enyaru, J., & Masiga, D. K. (2011). Tracking the Feeding Patterns of Tsetse Flies (Glossina Genus) by Analysis of Bloodmeals Using Mitochondrial Cytochromes Genes. *PLoS One*, *6*(2), e17284. doi:10.1371/journal.pone.0017284 PMID:21386971

Ngoune, J. M. T., Reveillaud, J., Sempere, G., Njiokou, F., Melachio, T. T., Abate, L., ... Geiger, A. (2019). The composition and abundance of bacterial communities residing in the gut of *Glossina palpalis palpalis* captured in two sites of southern Cameroon. *Parasites & Vectors*, *12*(1), 151. Advance online publication. doi:10.118613071-019-3402-2 PMID:30940213

Nogge, G. (1981). Significance of symbionts for the maintenance of an optimal nutritional state for successful reproduction in haematophagous arthropods. *Parasitology Today (Personal Ed.)*, *82*, 101–104.

Obiero, G. F., Mireji, P. O., Nyanjom, S. R. G., Christoffels, A., Robertson, H. M., & Masiga, D. K. (2014). Odorant and Gustatory Receptors in the Tsetse Fly Glossina morsitans morsitans. *PLoS Neglected Tropical Diseases*, *8*(4), e2663. doi:10.1371/journal.pntd.0002663 PMID:24763191

Odeniran, P., Macleod, E., Ademola, I., & Welburn, S. (2019). Molecular identifcation of bloodmeal sources and trypanosomes in *Glossina* spp., *Tabanus* spp. and *Stomoxys* spp. trapped on cattle farm settlements in southwest Nigeria. *Medical and Veterinary Entomology*, *33*(2), 269–281. doi:10.1111/mve.12358 PMID:30730048

Omolo, M., Hassanali, A., Mpiana, S., Esterhuizen, J., Lindh, J., Lehane, M., Solano, P., Rayaisse, J. B., Vale, G. A., Torr, S. J., & Tirados, I. (2009). Prospects for developing odour baits to control Glossina fuscipes spp., the major vector of human African trypanosomiasis. *PLoS Neglected Tropical Diseases*, *3*(5), e435. doi:10.1371/journal.pntd.0000435 PMID:19434232

Opiyo, E. A., Njogu, A. R., & Omuse, J. K. (1990). Use of impregnated targets for control of Glossina pallidipes in Kenya. *International Journal of Tropical Insect Science*, *11*(03), 417–425. doi:10.1017/S1742758400012844

Orji, C. U., Onyeocha, I. O., Shaida, S., & Ella, E. (2015). Molecular identification of tsetse fly (Diptera: Glossinidae) species based on mitochondrial DNA (COII and CytB) sequences. *African Journal of Biotechnology*, *14*, 1605–1613. doi:10.5897/AJB2015.14411

Ouedraogo, G. M. S., Demirbas-Uzel, G., Rayaisse, J.-B., Gimonneau, G., Traore, A. C., Avgoustinos, A., Parker, A. G., Sidibe, I., Ouedraogo, A. G., Traore, A., Bayala, B., Vreysen, M. J. B., Bourtzis, K., & Abd-Alla, A. M. (2018). Prevalence of trypanosomes, salivary gland hypertrophy virus and Wolbachia in wild populations of tsetse flies from West Africa. *BMC Microbiology*, *18*(1), 153. doi:10.118612866-018-1287-4 PMID:30470187

Oyieke, F. (1989). Studies on Occurrence and Transmission of Camel Trypanosomiasis in Northern Kenya. PhD, University of Nairobi, Kenya.

PAAT. (2008). *On Target Against Poverty - The Programme Against African Trypanosomiasis 1997-2007*. PAAT Information Service.

Rayaisse, J., Salou, E., Courtin, F., Yoni, W., Barry, I., Dofini, F., Kagbadouno, M., Camara, M., Torr, S. J., & Solano, P. (2015). Baited-boats: An innovative way to control riverine tsetse, vectors of sleeping sickness in West Africa. *Parasites & Vectors*, *8*(1), 236. doi:10.118613071-015-0851-0 PMID:25928366

Rio, R., Wu, Y., Filardo, G., & Aksoy, S. (2006). Dynamics of multiple symbiont density regulation during host development: Tsetse fly and its microbial flora. *Proceedings. Biological Sciences*, *273*(1588), 805–814. doi:10.1098/rspb.2005.3399 PMID:16618673

Rogers, D. J., Hay, S. I., & Packer, M. J. (1996). Predicting the distribution of tsetse flies in West Africa using temporal Fourier processed meteorological satellite data. *Annals of Tropical Medicine and Parasitology*, *90*(3), 225–241. doi:10.1080/00034983.1996.11813049 PMID:8758138

Rohlf, F., & Marcus, L. (1993). A revolution in morphometrics. *Trends in Ecology & Evolution*, *8*(4), 129–132. doi:10.1016/0169-5347(93)90024-J PMID:21236128

Rotureau, B., & Van Den Abbeele, J. (2013). Through the dark continent: African trypanosome devel- opment in the tsetse ñy. *Frontiers in Cellular and Infection Microbiology*, *3*, 53. doi:10.3389/fcimb.2013.00053 PMID:24066283

Rousset, F., Bouchon, D., Pintureau, B., Juchault, P., & Solignac, M. (1992). *Wolbachia* endosymbionts responsible for various alterations of sexuality in arthropods. *Proceedings. Biological Sciences*, *250*(1328), 91–98. doi:10.1098/rspb.1992.0135 PMID:1361987

Saini, R. K., Orindi, B. O., Mbahin, N., Andoke, J. A., Muasa, P. N., Mbuvi, D. M., Muya, C. M., Pickett, J. A., & Borgemeister, C. W. (2017). Protecting cows in small holder farms in East Africa from tsetse flies by mimicking the odor profile of a non-host bovid. *PLoS Neglected Tropical Diseases*, *11*(10), e0005977–e0005977. doi:10.1371/journal.pntd.0005977 PMID:29040267

Sasaki, H., Kang'ethe, E., & Kaburia, H. (1995). Blood meal sources of Glossina pallidipes and G. longipennis (Diptera: Glossinidae) in Nguruman, southwest Kenya. *Journal of Medical Entomology*, *32*(3), 390–393. doi:10.1093/jmedent/32.3.390 PMID:7616534

Schofield, S., & Torr, S. (2002). A comparison of the feeding behaviour of tsetse and stable flies. *Medical and Veterinary Entomology*, *16*(2), 177–185. doi:10.1046/j.1365-2915.2002.00361.x PMID:12109712

Scoone, I., Dzingira, V., Anderson, N., MacLeod, E., Mangwanya, L., Matawa, F., ... Welburn, S. C. (2017). People, Patches, and Parasites: The Case of Trypanosomiasis in Zimbabwe. *Human Ecology*, *45*(5), 643–654. doi:10.100710745-017-9929-y PMID:29170590

Shereni, W., Anderson, N. E., Nyakupinda, L., & Cecchi, G. (2016). Spatial distribution and trypanosome infection of tsetse flies in the sleeping sickness focus of Zimbabwe in Hurungwe District. *Parasites & Vectors*, *9*(1), 605. doi:10.118613071-016-1879-5 PMID:27884172

Simarro, P., Cecchi, G., Franco, J., Paone, M., Diarra, A., Ruiz-Postigo, J. A., & Jannin, J. G. (Eds.). (2012). *Estimating and Mapping the Population at Risk of Sleeping Sickness*. doi:10.1371/journal.pntd.0001859

Snyder, A. K., Adkins, K. Z., & Rio, R. V. M. (2011). Use of the Internal Transcribed Spacer (ITS) Regions to Examine Symbiont Divergence and as a Diagnostic Tool for *Sodalis*-Related Bacteria. *Insects*, *2*(4), 515–531. doi:10.3390/insects2040515 PMID:26467831

Solano, P., De La Rocque, S., Cuisance, D., Geoffroy, B., De Meeus, T., Cuny, G., & Du Vallet, G. (1999). Intraspecific variability in natural populations of Glossina palpalis gambiensis from West Africa, revealed by genetic morphometric analysis. *Medical and Veterinary Entomology*, *13*(4), 401–407. doi:10.1046/j.1365-2915.1999.00189.x PMID:10608229

Stanton, M. C., Esterhuizen, J., Tirados, I., Betts, H., & Torr, S. J. (2018). The development of high resolution maps of tsetse abundance to guide interventions against human African trypanosomiasis in northern Uganda. *Parasites & Vectors*, *11*(1), 340. doi:10.118613071-018-2922-5 PMID:29884213

Sumba, A., Mihok, S., & Oyieke, F. A. (1998). Mechanical transmission of Trypanosoma evansi and T. congolense by Stomoxys niger and S. taeniatus in a laboratory mouse model. *Medical and Veterinary Entomology*, *12*(4), 417–422. doi:10.1046/j.1365-2915.1998.00131.x PMID:9824826

Tobe, S., & Langley, P. (1978). Reproductive physiology of *Glossina*. *Annual Review of Entomology*, *23*(1), 283–307. doi:10.1146/annurev.en.23.010178.001435 PMID:343707

Vale, G., Lovemore, D., Flint, S., & Cockbill, G. F. (1988). Odour-baited targets to control tsetse flies, Glossina spp. (Diptera: Glossinidae) in Zimbabwe. *Bulletin of Entomological Research*, *78*(1), 31–49. doi:10.1017/S0007485300016059

Wamwiri, F., Alam, U., Thande, P., Aksoy, E., Ngure, R. M., Aksoy, S., Ouma, J. O., & Murilla, G. A. (2013). *Wolbachia, Sodalis* and trypanosome co-infections in natural populations of *Glossina austeni* and *Glossina pallidipes*. *Parasites & Vectors*, *6*(1), 232. doi:10.1186/1756-3305-6-232 PMID:23924682

Weitz, B. (1963). The feeding habits of *Glossina*. *Bulletin of the World Health Organization*, *28*, 711–729. PMID:13999790

ADDITIONAL READING

Aksoy, S., Weiss, B., & Attardo, G. (2008). Paratransgenesis Applied for Control of Tsetse Transmitted Sleeping Sickness. In S. Aksoy (Ed.), *Transgenesis and the Management of Vector-Borne Disease* (pp. 35–48). Springer. doi:10.1007/978-0-387-78225-6_3

Chapter 4
Role of Kenya Tsetse and Trypanosomiasis Eradication Council (KENTTEC) in Control of African Animal Trypanosomiasis (AAT)/Nagana

Daniel O. Gamba
Kenya Tsetse and Trypanosomiasis Eradication Council, Kenya

Pamela Akinyi Olet
Kenya Tsetse and Trypanosomiasis Eradication Council, Kenya

Monicah W. Maichomo
Kenya Agriculturla and Livestock Research Organizaion, Kenya

Sylvia Muthama Korir
Kenya Tsetse and Trypanosomiasis Eradication Council, Kenya

Isaiah Ndaburu Kiteto
https://orcid.org/0000-0003-1868-0801
Kenya Tsetse and Trypanosomiasis Eradication Council, Kenya

ABSTRACT

The role of the Kenya Tsetse and Trypanosomiasis Eradication Council (KENTTEC) in the control of animal trypanosomiasis is premised on the fact that a large proportion of animal trypanosomiasis in Kenya is tsetse transmitted. Tsetse distribution in Kenya is characterized by eight discontinuous belts defined by topographical, environmental and land-use. KENTTEC's strategy for control of African animal trypanosomiasis is based on use of community based organizations for spraying of livestock, control of the vector using various devices such as targets and traps, and development of strategies and policies for use of land after the intervention. The council has developed linkage with research institutions for

DOI: 10.4018/978-1-7998-6433-2.ch004

adaptive and operational research. The council has initiated the development of national atlas by mapping tsetse and animal trypanosomiasis distribution in collaboration with stakeholders at the national, regional, and international levels.

INTRODUCTION

The role Kenya Tsetse and Trypanosomiasis Eradication Council (KENTTEC) in the control of animal trypanosomiasis is premised on the fact that a large proportion of the animal disease in Kenya is tsetse transmitted. In Kenya, tsetse is distributed in eight discontinuous belts defined by topographical, environmental, and land-use systems. Tsetse and trypanosomiasis (T&T) control programs in Kenya began during colonial era, was characterized by epidemics in humans and their livestock.

KENTTEC is a state agency mandated to control tsetse and trypanosomiasis in Kenya, is a successor to Pan African Tsetse and Trypanosomiasis Eradication Campaign (PATTEC) program initiated by African Heads of States in Lomé Togo in 2000 to eliminate T&T in Africa. The PATTEC initiative generated a continental plan of action and funding of six African countries by African Development Bank (ADB). Kenya implemented the PATTEC campaign (2005-2011) in four of the tsetse belts (22,000Km^2), and subsequently expanded this to five belts (24,000km^2) (North rift tsetse belt, central Kenya belt, Lake Victoria, Coastal and Western tsetse belts) under the KENTTEC program. The PATTEC Kenya chapter was autonomous from the ministry and had a Project coordination management unit and five regional coordinators with staff deployed specifically for the tsetse control activities. The model of devolving the project from the main ministry enabled the project in Kenya to achieve major milestones in implementation of the project. Leveraging on achievements of the PATTEC initiative, KENTTEC based its approach on:Partnering with communities, through inclusive community-based organizations for spraying of livestock; control of the vector using various devices such insecticide treated targets and traps; Development of strategies and policies of land use after tsetse intervention; linkage with county governments for capacity development; linkage with research institutions for adaptive and operational research such as push-pull technology; development of national atlas by mapping tsetse and animal trypanosomiasis distribution; advocacy and sensitization of communities on African animal Trypanosomiasis control, linkage with the Directorate of Veterinary Services (DVS) on determination of usage of trypanocide to minimize drugs resistance; and policy intervention such as the introduction of Progressive control pathway for planning control programs and promotion of standards and guidelines for tsetse control.

The achievements of KENTTEC in control of African Animal Trypanosomiasis (AAT) (Nagana) are hinged on application of strategies specific to predominant tsetse species, economic benefits, and return on investments, environmental and public good. Integrated vector management is the guiding principle in the control of trypanosomiasis in pastoral and agro-pastoral areas. Current control methods are geared towards use of communities in spraying livestock as mobile targets, and insecticide-impregnated targets. Use of novel attractants and repellants have greatly improved the performance of the traps and targets in the control of tsetse flies. The deployment of Novel target screens, and rearing of trypanotolerant cattle have also reduced incidence of AAT in endemic areas and permitted economic utilization of land for agriculture. KENTTEC in collaboration with Kenya Agricultural and Livestock Research Organization (KALRO) and the Directorate of Veterinary Services (DVS) has instituted animal and human disease surveillance in the intervention areas. The data obtained through active and passive surveillance is used to inform the relevant control intervention. The information is directed into appropriate database for preparation of distribution maps and Progressive control pathway for planning and provide policy direction.

BACKGROUND ON TSETSE AND ANIMAL DISEASE CONTROL IN KENYA

The T&T control in Kenya began in early 1900 spearheaded by the defunct Tsetse Control Section in the Directorate of Veterinary Services (DVS). The DVS role focused on AAT while the Division of Vector-Borne Diseases within the Ministry of Health focused on Human African Trypanosomiasis (HAT) (KENTTEC, 2011).

It is reported that in the early century that G. *longipennis, G. pallidipes,* and *G. brevipalpis* were dispersed to livestock -farming areas beyond the natural limit of the fly belt by trains passing through tsetse-infested areas between Mombasa and Nairobi in Kenya. This necessitated the de-flying of passenger and goods trains in Sultan Hamud and Emali (Lewis *et al,* 1950). In early 1950s East African Trypanosomiasis Research Organization (EATRO) was established to carry out research and develop technologies for effective control of trypanosomiasis, a key institutional capacity development of the country (Murilla, *et al.,* 2014). Subsequent control efforts included bush clearing and settlement of farmers in Shimba Hills in1959-1962

The Construction of tsetse fly barrier between Trans-Mara and chepalungu by bush clearing in 1950 was one of the reported tsetse control efforts (Otieno, 2010). The Alego tsetse control campaign in Siaya to contain the HAT epidemic in 1964 was followed by aerial spraying in Lambwe Valley under World Health Organization/Food and Agriculture Organization/United Nations Development Program (WHO/FAO/UNDP) operation in 1968-1972,(Wellde *et al* 1989). In 1960 tsetse control camps were establishment in tsetse 'Hot Zones' of Kiboko, Kilifi, Ukunda, Aitong, Kiburine, Sisenye, Nambale, Luhano, Nyamonye, Manyuanda, Kaloka, Homa Bay, Sindo, Karungu, and Oyugis (Wellde *et al,* 1989). The collapse of the East African Community (EAC) in 1977 led to the dissolution of EATRO and establishment of Kenya Trypanosomiasis Research Institute (KETRI), which was located in Muguga, Kenya. KETRI piloted the Use of insecticide-impregnated targets in Lambwe Valley from 1988 to 1997 for T&T control. In 1999 Farming in Tsetse Controlled Areas (FITCA) project was launched in Busia, Western Kenya and ended in 2004 with varying levels of success in the control of AAT and HAT (FITCA, 2004)

Despite these noble efforts, the problem of tsetse and trypanosomiasis remained unabated not only in Kenya but also in the rest of the continent. In the year 2000, the African Heads of State and Government, vide Decision AHG/156 (XXXVI) of the 36th Assembly of the African Union held in Lome, Togo, unanimously agreed to eradicate tsetse and trypanosomiasis in Africa. This resulted in the creation of the Pan-African Tsetse and Trypanosomiasis Eradication Campaign (PATTEC) project. Kenya was one of the six African countries that implemented PATTEC under a local chapter called PATTEC Kenya. Following the promulgation of the new constitution in 2010 and the reorganization of state corporations, PATTEC-Kenya was transformed into Kenya Tsetse and Trypanosomiasis Eradication Council (KENTTEC) in 2012, as a State Corporation mandated to coordinate tsetse and trypanosomiasis eradication in Kenya.

ROLE OF KENTTEC IN CONTROL OF AFRICAN ANIMAL TRYPANOSOMIASIS

Tsetse control is an effective approach and constitutes the cornerstone in trypanosomiasis suppression (Kettle, 1990; Elliot *et al*, 1978). KENTTEC has promoted and up scaled the use of a variety of techniques to control tsetse populations, including insecticide spraying, trap and target technology, livestock protective fencing (LPF) and sustainable land use management in tsetse-freed areas. The effective application of these methods is contingent on the emergence of reported cases of resistance of trypanocides and environmental concerns. The development of bait technology and repellants offer opportunities for environmentally friendly techniques and improved effectiveness of the current methods. These methods use a combination of visual and odor cues on traps or targets (Leak, *et al* 2008). Use of sterile insect techniques in Zanzibar and subequent trials in Lambwe valley through IAEA project had introduced a biological control option. KENTTEC has employed an integrated vector management principle in the control of animal and Human African trypanosomiasis. The mandate of KENTTEC is as given in Legal Notice no 77 of July 2012

Integration of Vector Control Strategies in tsetse control

a) Targets and Traps

Traps use blue and black cloth in a shape that attracts the flies and then funnels them upwards into a netting trap – usually in the form of a monoconical (pyramidal) or biconical shape. KENTTEC has been using biconical, NGU and H traps in the Bogoria- Mara, Meru Mwea, Nyanza, Western and Coast regions to conduct baseline and monitoring entomological surveys over the years. Traps are usually deployed following this method as described by (Stephen *et al* 2008). A Global Positioning System (GPS) receiver is used to record the exact location of the traps. The area for trap deployment is based on ease of access, vegetation type and proximity to rivers or streams and vegetation. The height of the traps above ground level and the light direction is such that the trap is shaded from direct sunlight and the base of the blue cloth is not more than 20 cm above the ground. The wind direction and trap entrance orientation ensures that the odor plumes go to the right direction. The trapped flies are protected from predators by applying grease to the pole at 10 cm above the ground. The flies are trapped over a period of 48 to 96hours to enable the calculation of fly apparent density; flies per trap per day (FTD). Details on the data recording sheets include target or trap code, date of deployment, detailed GPS coordinates of each target/ trap deployed, altitude, acetone, phenols, target condition (cloth and metal frames) and remarks. As for the servicing of targets, the data recording sheets include target code, detailed GPS coordinates of the targets serviced, cloth replacement, acetone refilling, phenol replacement, metal frame replacement, missing target information, clearing of encroaching canopy, clearing of undercover and comments.

Traps give information on fly populations and indicate the presence or absence of tsetse and other hematophagous species. Targets consist of suspended screen of blue and black cloth, impregnated with a biodegradable pyrethroid insecticide such as deltamethrin. The effectiveness of traps and targets are enhanced by the addition of appropriate odour bait. Such odour baits are usually short-chain aromatic compounds such as acetone or octenol (Kuzoe *et al*, 2004). These substances enhance trap efficiency and are mostly components of host breath, urine and skin secretions for baiting traps.

The development of novel control tools, as well as standardization of existing models, has led to the identification of cost-effective devices for tsetse management (Wamwiri & Changasi, 2016). There have been considerable advances made in the search for efficient and cost-effective control tools against riverine tsetse involved in the transmission of HAT (Simarro *et al,* 2008). This search culminated in the development of "tiny targets" (Lindh *et al,* 2009). These insecticide-treated targets are much smaller (25cm×50cm) than the traditional standard 1×1 m targets. Despite their size, tiny targets have proved to be quite effective for the control of riverine tsetse, specifically *G. fuscipes spp.* and *G. palpalis spp.* (Esterhuizen *et al,* 2011; Rayaisse *et al,* 2011). They cost much less due to their smaller size, hence reduced costs for impregnation and materials. Besides, their lightweight, they can be easily deployed by foot, or using bicycles and motorcycles (Shaw *et al,* 2015). Further, tiny targets when mounted on pirogues moving along a river are effective in reducing tsetse density (Rayaisse *et al,* 2015). An example of a trap NGU trap figure 1 below .

Figure 1. Picture of NGU trap

b) Push-Pull Technology in Tsetse Control

Bait technology based on attractive olfactory and visual cues identified from tsetse hosts has shown promise in tsetse control in eastern Africa. Relative low cost, community acceptability (Mangwiro et al, 1999; Green, 1994; Vale et al, 1988; Allsopp, 1984), as well as high specificity and minimal environmental contamination (Allsopp, 1984) have shown potential for this technology in tsetse control (Nagel, 1995). Recent characterization of repellent constituents in the odour of tsetse refractory waterbuck has opened up novel tools for protecting livestock (Wachira *et al*, 2016; Bett *et al*, 2015; Gikonyo *et al*, 2003; 2002) and minimizing competition with baited traps/targets. The two possible technologies can be deployed in behavioral control of tsetse: one that is based on improved bait attractant ('pull') tactic for reducing tsetse population in a targeted area; and the other based on the repellence ('push') tactic which can be used to protect individual livestock animals or herds. The 'push' and 'pull' olfaction together with visual attraction tactics can be spatially integrated into 'push-pull' modes to facilitate a rapid and efficient reduction of target tsetse populations and suppression of disease transmission. The relative efficiency of control initiative can significantly influence the overall cost of control (duration) operations and vector abundance.

c) Insecticide-Treated Cattle (ITC)

This method is based on the treatment of cattle with insecticide formulations (mainly based on pyrethroids) using a wide range of techniques (pour-on, spraying, whole body dips/ baths and sprays/ showers), (Gimonneau, *et al* 2018). The development and introduction of pyrethroids, insecticides of low mammalian toxicity, opened the way to their use in cattle dips. Since pyrethroids have acaricidal properties, protection of cattle from ticks, tsetse, and other biting flies is achieved (Grant, 2001). Cattle treated with the formulation act as attractive and lethal baits for tsetse and ticks due to their odour, movement, and size (Gimonneau, *et al,* 2018). Use of baited animals as Insecticide Treated Cattle (ITC) is a common practice adopted by farmers to control tsetse (Malele, 2011). ITC also has important spillover benefits on other livestock and human health problems (Shaw, *et al* 2015).

Trypanosomiasis control involves financial input from livestock owners and their active participation (Hargrove *et al*, 2000). In the areas where KENTTEC is implementing its activities, there are three main types of cattle rearing systems. These are trans-human rearing systems, sedentary and zero-grazing units where animals are fed with harvested forage or agro-industrial sub-products (Bouyer *et al,* 2013).

Previous studies has shown that insecticide treatment of a high proportion livestock of prevents reinvasion of tsetse flies if they are evenly distributed in a whole area (Warnes *et al*, 1999). In the KENTTEC intervention areas bordering wildlife conservation areas (high tsetse challenge), ITC is complemented with tsetse barriers, as ITC alone cannot prevent re-invasion.

In their review on cheap and safe tsetse control for livestock production, (Bourn *et al* 2005) concluded that wider use of insecticide-treated cattle has the potential for not only increasing overall crop and livestock production and enhancing rural livelihoods but also for improving food security through mixed farming. Where individual villages, or even more so, only individuals within villages treat their cattle, higher numbers will need to be treated to achieve the same level of tsetse control (Shaw et al, 2015). To cushion farmers from the costs of using ITC as a method in the tsetse infested area, KENTTEC in its areas of intervention supports this method through the issuance of insecticides to livestock spraying groups and dipping groups. Spraying records are maintained and groups are trained on the safe use of insecticides to mitigate against the development of resistance and environmental degradation. Figure 2 and 3 illustrates.

Figure 2. spraying of livestock in crushpens

Figure 3. Dipping of livestock

d). Livestock Protective Fences

Livestock protective fences consist of a one-meter high black netting that is treated with insecticide, attached to the perimeter wall of a livestock pen. KENTTEC supports dairy farmers in the installation of livestock protective fence.

KENTTEC promotes the use of the Zero fly screen, which is manufactured by Vestergaard Group, South Africa is made up of 100% polyethylene and incorporated with deltamethrin at the rate of 4.0mg-deltamethrin/kg of textiles. The product has a UV protection incorporated to extend durability during outdoor usage. Zero fly nets, when used alone, are effective in controlling biting and nuisance flies and remain effective in hot and rainy weather (Nagagi *et al,* 2017). KENTTEC has installed 2000 LPF across the regions in the last five years in collaboration under the FAO funded project. The zero grazing unit is shown in figure 4 below.

Figure 4. Zero grazing unit with Livestock protective fence

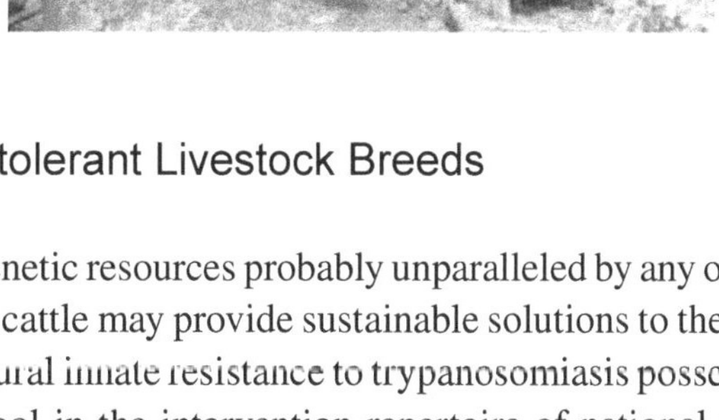

e). Adoption of Trypanotolerant Livestock Breeds

Africa possesses livestock genetic resources probably unparalleled by any other continent. The resources in N'Dama and Orma Boran cattle may provide sustainable solutions to the problem of trypanosomiasis (Abenga *et al,* 2005) The natural innate resistance to trypanosomiasis possessed by these breeds of cattle is an important additional tool in the intervention repertoire of national and regional disease control programs.

The improvement of the more productive susceptible breeds' trypanotolerance will inherently be a solution to the problem of trypanosomiasis in Africa. It has been observed that the better performance of the Orma Zebu is an attribute that could be exploited in the adoption of the trypanotolerance genotype, as a sustainable trypanosomiasis control program (Maichomo *et al,* 2005).

KENTTEC in collaboration with KALRO-BioRI is collaborating on a project for developing strategies to enhance the utilization of the improved trypanotolerant Orma Boran cattle breed in KenyaThe project funded by National research fund is being implemented in collaboration with the Kwale county government and the local farmers. The project has mapped out the preference of the improved trypanotolerant breed by farmers in Kwale.The project is ongoing in the county.

Collaborative Frameworks

a). Linkage With Stakeholders

Among the functions of KENTTEC as stipulated in the legal notice are; co-ordinate activities of Government departments, agencies, and other stakeholders at the National and County levels in matters related to tsetse and trypanosomiasis eradication, and promote collaboration among stakeholders on tsetse and trypanosomiasis eradication (Kenya Law, 2012). The key stakeholders involved in the tsetse and trypanosomiasis eradication efforts include; Ministry of Agriculture, Livestock and Fisheries, County Governments, Community/ farmers, Research Institutions (KALRO, ILRI, ICIPE, and Universities), International and Regional Organizations, Ministry of Health, Kenya Wildlife Services, Kenya Forestry Services, Media, and Ministry of Interior and Coordination of National Government (KENTTEC, 2019). These have different roles and all need to play an active role in tsetse control. Their roles could be through policy formulations, funding, harmonization and regulation and enforcement of policy and laws related to the implementation of vector control plans; coordination of tsetse control at ecological interface areas, promotion and involvement of all gender groups because tsetse effects cut across gender groups (Malele, 2015). During the 2018-2023 Strategic planning period, this is expected is to be undertaken through consultative forums with stakeholders, development of memoranda of understanding (MoU), public-private-partnerships (PPPs), joint planning with stakeholders, and lobbying for the establishment of sub-regional committees for cross-border tsetse and trypanosomiaisis control. The arduous journey from research results to feasibility studies, to pilot testing, to eventual successful field operations requires close collaboration between researchers, extension specialists, community leaders, and the agriculture, natural resources, and public health communities (Vreysen *et al* 2007).

The need for stakeholder's involvement and collaboration in the fight against Human African Trypanosomiasis (HAT) and Animal trypanosomiaisis (AAT) under One health approach is key in avoiding the resurgence of the disease in Western Kenya. The mechanisms of collaboration and coordination among stakeholders were established within the WHO network for HAT elimination, and the meeting held in 2014 looked at procedures needed for monitoring and evaluation of the elimination process as well as confirmation of outcomes (Holmes, 2015). In creating linkages with stakeholders in the tsetse and trypanosomiasis eradication works, participatory approaches (PAs) in all the field activities is important as it combines strengths of the various Institutions to achieve the same goals.

Income-generating activities for rural populations are invariably linked to an improvement of animal production, which is threatened by the presence of tsetse flies and other arthropod vectors (Bauer *et al* 2011). Harnessing the interests of community/ farmers is important, as their support is beneficial in ensuring sustainability of the projects. KENTTEC is working closely with the farmers in the tsetse-infested areas in traps making, deployment/ servicing of insecticide-impregnated targets, and use of Insecticide Treated Cattle (ITC).

KENTTEC is currently working on the development of an Atlas of AAT in collaboration with Food and Agriculture Organization (FAO) and AU-PATTEC. Developing contemporary global maps of AAT and tsetse requires substantial efforts, which can only be justified and sustained in the light of the broader objective of the Atlas initiative, which means full engagement of, and collaboration with the national animal health authorities and research institutes are crucial (Cecchi *et al* 2014). Lastly, effective engagement of the mass media is important in creating awareness on neglected diseases for the general lay public and policymakers. Good understanding of how the mass media assigns news value to an issue and timing are the main success factors in getting mass media publicity (Olet &Othieno, 2015).

b). Farmers Sensitization, Training, and Capacity Building

The pivotal role played by farmers in control of AAT, disease surveillance, and sustainability of the control interventions requires their active involvement The Council prioritized farmer sensitization and training to promote partnerships with the communities on T and T management and adoption of techniques and technologies (Geofrey *et al,* 2018).

The farmers are trained and engaged in maintenance of traps and targets, trap making, construction of crush pens, and protecting them from vandalism and theft. For community ownership, the farmers identify target attendants who are trained on deployment, replacement of traps and targets, tsetse identification and AAT incidence reporting (KENTTEC 2019). The farmers are organized into community livestock spraying groups for ease of coordination and management. The groups are trained on group dynamics, animal disease control and production techniques, sustainable land management and alternative income generation approaches (KENTTEC 2019)

c). Active and Passive Monitoring of Animal Trypanosomiasis

The KENTTEC conducts periodic baseline surveys and monitoring missions known as active surveillance to establish disease prevalence and incidence in known tsetse hotspots and suspected areas. Active surveillance is both field and laboratory-based. The Field-based approach begins by mobilizing farmers through their local area administration and community barazas to present their livestock for screening and in some instances taken to the laboratory (KENTTEC-parasitological survey report 2016, Pate Island)

The KENTTEC conducts continuous passive surveillance through its field staff, local veterinary officers, private animal health practitioners, and community leaders who form a reporting network for collecting AAT data and generating reports for action by the technical division of the council.

ACHIEVEMENTS

Mapping and Distribution of African Animal Trypanosomiasis (AAT)

Geographic Information System (GIS) has been used by KENTTEC to map the status of AAT in the Country. During monitoring of tsetse population and the disease, the site of the monitoring is geo-referenced to obtain their respective coordinates, which is then combined with a data of the area and livestock herd for mapping.

Mapping also helps in predicting the trends of the disease by geospatial modeling and use of remote sensing data while interfacing Habitat suitability (Dicko *et al,* 2015).Mapping of the control intervention helps in visualizing the tsetse control activities being undertaken. When over-laid with the tsetse population or disease prevalence layers, the impact of the intervention activities on tsetse and trypanosomiasis in a given area is demonstrated.

The map of tsetse and trypanosomiasis distribution Kenya is over 20years old, KENTTE C in an effort to bridge this gap launched the development of a national database and atlas for tsetse and AAT supported by FAO. The map/atlas is linked to the development of the Progressive Control Pathway (PCP) for AAT (Diall et al., 2017). PCPs are staged approaches to plan and evaluate progress in the reduction, elimination and eradication of a range of diseases (Fao/Oie, 2015; Sumption et al., 2012) which requires a national level information system to advance in it. PCP helps to inform in decision-making as well as for undertaking self-assessment.

The national databases has helped to display and compare pre-intervention and post invention status. The database has been used to assess the impact of interventions, by focusing on the two areas where tsetse control was particularly successful [i.e. Mwea National Reserve (central Kenya) and Pate island (coastal Kenya – Lamu Archipelago)]. In these two selected areas, tsetse flies were reduced below detectable levels, while AAT was reduced but not eliminated (Ngari *et al.,* 2020).

Central Management of Data

Kenya has been using reference materials on tsetse distribution in the country that data as far back as the 1970s and 1990s (Ford &Katondo, 1977) though a lot of work that has been undertaken in the country on tsetse and trypanosomiasis control by KENTTEC and various stakeholders

KENTTEC in its mandate to manage tsetse and trypanosomiasis data; is in the process of developing a national database of tsetse and trypanosomiasis. This involves collating data from partnering institutions and creating a centralized database. The institutions have been trained on how to organize and clean their respective data, therefore, helping the institutions to manage their data too. Once the data from the other institutions have been collated, it will be incorporated into the KENTTEC database to create an integrated database for both AAT and tsetse in the country. The integrated data will show the impacts that different interventions have on tsetse and trypanosomiasis as well as temporal and spatial gaps over the years. This will enhance the future planning of tsetse and trypanosomiasis control.

POLICY LEVEL ADVICE ON AFRICAN ANIMAL TRYPANOSOMIASIS

Strategy for the Control of Tsetse and Trypanosomiasis in the Country

The council has developed the strategy for eradication of tsetse and trypanosomiasis in Kenya. The 10-year strategy has been implemented since 2011 and the council has taken lead in reviewing the strategy to be in line with the constitution of Kenya 2010 to embrace the devolved governance system.

Progressive Control Pathway (PCP) for the Planning of Eradication in the Country

The PCP tool provides a systematic framework for planning and evaluating field tsetse and trypanosomiasis eradication activities. The KENTTEC in partnership with FAO/WHO has used the PCP tool for the control of AAT in Kenya in collaboration with key stakeholders.

Contribution to Policy Development in Related Sectors

The council participates in the development of legislation that facilitates the development of livestock resources in the country. To enable KENTTEC to play its role in tsetse and trypanosomiasis eradication and that a legislative framework is in place for that purpose, tsetse and trypanosomiasis eradication matters have been anchored in other policies and legislation including the National Veterinary Policy and the reviewed National Livestock Policy. Both policy documents provide that KENTTEC will be strengthened for tsetse and trypanosomiasis eradication.

The Synergy of HAT and AAT Control in HAT Endemic Area in Kenya (One Health Approach)

A one health approach in the control of trypanosomiasis in the country has been advocated by KENTTEC in Kenya. The approach is being implemented in the HAT endemic areas of Western, Lake Victoria, and transmara area. The teams from the Ministry of Health, KENTTEC, KALRO, and County government of the respective areas are involved in the active and passive surveillance of the animal and human disease.

Advocacy in Control of African Animal Trypanosomiaisis

The Council has employed several advocacy strategies in creating awareness among the policymakers and the general public at large. The goal is to achieve a favorable policy framework that supports Tsetse and trypanosomiaisis and to gain the goodwill of citizens and subsequently citizen advocacy. This is due to the neglected nature of the T& T problem.

The Council has use three strategies namely; program advocacy, mass media advocacy, and citizen advocacy. The Council utilized available opportunities to sensitize policymakers on the problem of tsetse and trypanosomiasis, including senior government officials in the executive and members of parliament. The Council engaged especially members of parliament from tsetse-infested areas and those in the agriculture subcommittees.

The main challenge of this approach has been the time limitation as dictated by the electoral system that leads to changes every five years, forcing the council to begin all over again. This is not only time consuming but also requires financial resources. The introduction of devolved government after the passage of a new constitution further made this approach more complex as every county is a unit of advocacy. The Council has engaged tsetse infested counties to prioritize T& T problems in their respective counties.

Mass media advocacy is another strategy that the Council has employed. The mass media plays a watchdog role over the government and is always a trusted source of news. The Council is probably the only one in Africa with a Communication Division. Through this division, the council has undertaken mass media engagement and sensitization on Tsetse and Trypanosomiasis (Olet & Othieno, 2015). The

Council sensitized journalists on the problem of Tsetse and Trypanosomiaisis to enable them to write news and feature articles in the papers and to make TV news and features. The Council developed and aired four documentaries on T and T on national TVs (Olet & Othieno, 2015).

On citizen advocacy, the Council has worked with local communities. The Council has sensitized these communities to push their leaders to prioritize T and T. This method has especially worked well with the devolved systems where budget making is participatory in nature. The Council has also prepared Information Education and Communication (IECs) materials targeting various audience groups. These materials have been prepared, pre-tested, and disseminated to audiences for education purposes.

THE ROLE OF KENTTEC IN T&T CONTROL

Case Study of Lake Bogoria Area

The Lake Bogoria ecosystem comprises of Nakuru and Baringo counties where tsetse flies species (mainly *G. pallidipes*) are concentrated in and around the Lake Bogoria Game Reserve . The area is predominantly rocky and hilly, the rocky hills are dominated by the Acacia spp, shrubs and bushes of different plant species suitable for browsers. The low-lying areas are covered by *Acacia mellioflora*, and savannah grassland. The presence of wild life species preferred by tsetse for blood meals like antelopes and greater gazelles, coupled with thick vegetation makes this reserve a breeding ground for tsetse. Information from key informant interviews conducted before the baseline survey also indicated that the tsetse menace intensified with the gazettment of the park in 1977. Livestock owners are allowed to graze their cattle in the reserve during dry seasons, which increases exposure to animal trypanosomiasis.

Baseline data on tsetse population density and animal trypanosomiasis prevalence was collected in October 2008. Two hundred trapping sites were identified and geo referenced along 10 transects. Biconical traps (Challier & Laveissiere, 1973) baited with acetone and 8:4:1 phenol (Torr, 1997.) were placed in the identified sites checked and harvested after 48hrs. In subsequent monitoring exercises, traps were placed in the same sites and the catches counted and recorded as follows;

Two control methods, namely insecticide treated cattle and deployment of stationary targets impregnated with a formulation of deltamethrin (Glossinex, Roussel Uclaf Ltd, France) produced commercially for impregnation of targets for tsetse control were employed. A concentration of 0.6% w/v of active ingredient (a i) was used for the initial impregnation of standard targets. Re-impregnation was carried out at quarterly intervals during the rainy season and at monthly intervals during the dry season at half the original concentration (0.05% w/v a.i.). They were deployed at approximately 100- metre intervals in suitable tsetse habitat, mainly along the fringing Forest. 3,227 Standard Tsetse targets had been deployed in Lake Bogoria Game Reserve and its environs since 2008 till 2018. In December, 2019, 1,020 Standard targets were deployed in the Game Reserve. In February, 2020, 38 targets were deployed along the Hot springs and 1,004 additional Standard targets were deployed in April, 2020, a total of 5,316. Target deployment is captured in figure 5 below.

Cattle were sprayed or dipped in Alphacypermethrin which KENTTEC distributed to community dips and crush pens. These were organized into groups, which were trained on various aspects of group management and T& T control. KENTTEC has supported 52 communal dips and 25 crush pens in the area so far.

Figure 5. Target deployment exercise, Feb, 2020

Figure 6. trends in tsetse and trypanosomiaisis in Baringo county

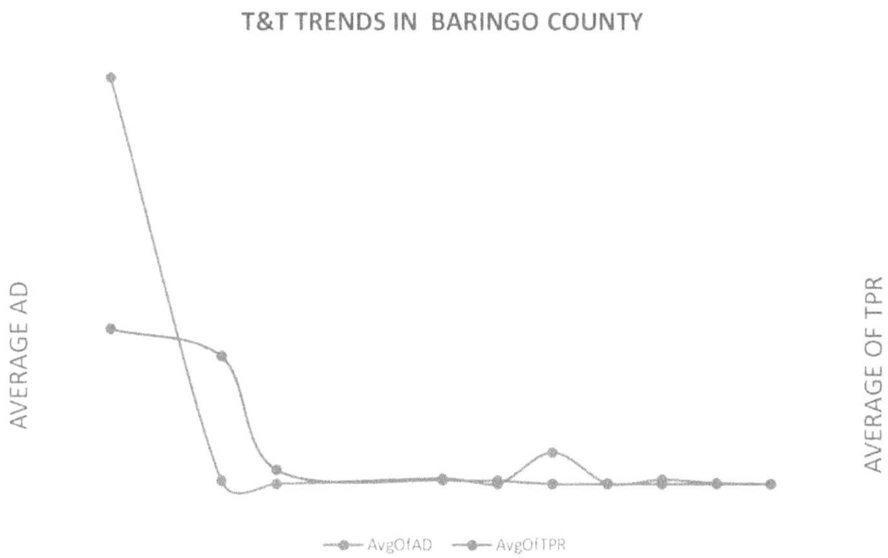

The trends indicate a steady reduction of the tsetse population from February 2008 to August 2010, where the population reached zero is shown in Figure 6 above. The level of suppression was maintained for the subsequent years and only a rise was evident in the monitoring in October 2014 and 2016 when there was insecurity in the areas leading to destruction of the targets by vandals. The increased movement of wild life in the area from Laikipia to the Lake Bogoria reserve also contributed to the increase of the tsetse population.

One hundred cattle were selected for sampling from each of the ten predetermined sites that surround the Lake Bogoria game reserve. The sites were selected based on the number of cattle and the geographical distribution of the crush pens used for the livestock spraying.

The result showed that the mean trypanosome prevalence in cattle during the baseline in 2008 was 5.44%, reduced to 0.12 in 2014 and leveled out to zero in 2018. The import of the intervention in the Lake Bogoria reserve has been the increase in the livestock population in the areas adjoining the park and the improved status of the wildlife.

Case Study of Tsetse and Trypanosomiasis Control in Mwea National Reserve and its Environs

Mwea National Reserve is one of the protected areas located in Meru/ Mwea tsetse flies belt together with Meru National Park, Ol Donyo Sabuk National Park and Mwingi North National Reserve. Mwea National Reserve in Mbeere South sub-county Figure 7 below, Kenya is located between latitude 00 45¢N and 00 52¢N, and longitude 370 35¢E and 370 40¢E. The reserve is 48 Km2 and is situated between the confluence of the Thiba River and the Tana Rivers, which are characterized by a beautiful scenery and long river frontage (Kaitho, *et al* 2013). The Mwea National Reserve which is found in Mbeere South area in Embu County (then Embu District) and its environs is an area covering approximately 200 Km2 and 48 Km2 being the National Reserve. Apart from the tsetse flies problem, the area has had a problem of human-elephant conflict that resulted to an average of 3deaths yearly but this was solved by use of physical barrier in form of electric fence completed in 1996 (Omondi, *et al* 2003).

Figure 7. Map of Mwea National Reserve and its environs
(Source:- Kenya Wildlife Services)

Due to the tsetse flies infestation, farmers abandoned their farm lands and there was shortage of food which resulted to food aid. The farmers lost their animals due to African Animal Trypanosomiasis (AAT)/ Nagana and those that survived were wasted and couldn't be used for draught power in farming. The number of visitors visiting the Reserve diminished due to tsetse flies infestation, which had an effect in revenue collection.

In January 2007, a team comprising of the Veterinary Service Department, Pan African Tsetse and Trypanosomiasis Eradication Campaign (PATTEC), ICIPE and local community was formed to continue with tsetse and trypanosomiasis eradication that included tsetse monitoring, traps servicing, crush pens establishments for use in spraying of livestock, monitoring of emergency activities and outreach conservation programme. A total of 330 NGU traps were deployed inside the Reserve and 216 deployed in the environs of the Reserve. The community was trained on servicing of the traps. Even though the traps were being deployed in the periphery and on the roads including Thiba circuit, Thiba fence line road, Tana Circuit and Central road in the Reserve that left most of the areas uncontrolled, there were was reduction of tsetse flies from Flies trapped per Trap per Day (FTD) of 70.1 to 33.5 by December 2007. The Animal African Trypanosomiasis (AAT) remained high with prevalence of 18.7% in 2004 and 12.9% in 2008.

The establishment of crush pens involved registered groups as entry point. The farmers were sensitized about importance and effectiveness of tsetse flies and tick control using spraying of livestock in the crush pens. The sprayed livestock with synthetic pyrethroids became mobile targets as whenever tsetse flies bit them they were knocked down.

The PATTEC project then changed the control technology in January 2008 from use of NGU traps in the Reserve and its environs, to use of static insecticide impregnated targets in and outside the Reserve. In total, 334 targets were deployed in all the roads inside the Reserve, dry riverbeds, along the fence and river circuit. The targets were geo-referenced for ease in locating them during servicing of the targets. This resulted to reduction of FTD to 0.2 by August 2008. By May 2009, the FTD dropped to 0 which was recorded to the end of the PATTEC project in June, 2012. The trypanosomiasis prevalence during implementation of PATTEC dropped to 0%.

When PATTEC project ended, KENTTEC continued with the PATTEC activities from 2012. In 2013 and 2014, Grid based Entomological monitoring carried out in wet sand dry seasons where Odour baited Biconical traps, NGU traps and sticky panels deployed in each square kilometer. The results showed absence of tsetse flies, only Tabanus, Chrysops, Haematopota and Stomoxys species trapped. In 2014, there was a rise of trypanosomiasis prevalence to 0.25%, which was attributed to introduction of migrant cattle in the area. Trends of tsetse and trypanosomiasis trends in Embu County is shown in figure 8 below.

Figure 8. Graph showing the trends of tsetse and trypanosomiasis trends in Embu County

The tsetse and trypanosomiasis interventions in Mwea National Reserve and its environs is one of the successful PATTEC/ KENTTEC interventions where tsetse flies were reduced below detectable levels, while AAT appears to have been reduced but not eliminated (Ngari, *et al* 2020). As a result of tsetse and trypanosomiasis interventions that started in 2003, the tsetse flies population has been 0 since May, 2009 with a few cases recorded of trypanosomiasis. The effect of that is farmers adjacent to Mwea National Reserve, who were used to Government food aid are currently exporting their farm produce e.g. cereals and horticulture to other areas within and outside Embu County. Acreage under crop agriculture in Mbeere South has increased due to increased use of animal draught power to open up more agricultural land where land area under cereals alone increased from 7,020 Ha in 2009 to over 21,000 Ha in 2019. This has directly contributed to improved food security and incomes for families in these areas. Over 1,000 exotic dairy animals have been introduced into tsetse freed areas by the communities. Commercial farms have come up in areas previously infested; for example, Thiba Farm in Mwea has over 100 dairy cattle and goats employing over 200 youth and women from the area (KENTTEC, 2020).

DISCUSSION

Kenya tsetse control efforts dates back to the 1900s through the tsetse control division in the Directorate of veterinary services. The Kenya government has its shown commitment to eradicate tsetse and trypanosomiasis through various control approaches spearheaded by the government, partners and multilateral agencies.(Adungo *et al* 2020)

The Kenya Tsetse and Trypanosomiasis Eradication Council (KENTTEC) is a successor to the Pan African Tsetse and Trypanosomiaisis Eradication campaign (PATTEC) initiative formed following the African Union (AU) Heads of State and Government Decision AHG/156 (XXXVI) of the 36th Assembly, which urged member states to re-double their efforts to eradicate tsetse and trypanosomiasis from Africa by 2025. Kenya was among the six African countries that implemented the first phase of PATTEC. The continental coordination of the tsetse eradication was aimed at implementing Area wide integrated pest management system that would commence on one country and rolled to other countries. The success of the continental initiative was premised on the decision by all countries to mobilize resources, joint planning, coordination and implementation of the projects. The linkages at the borders of the countries with shared tsetse belts would allow for the development of joint control projects for intervention at the borders. This would inform the development of common strategies and methodologies for the eradication of the species of tsetse flies along the borders. Kenya and Tanzania have developed the Mara Serengeti ecosystem project that tackles the *G.swynertoni* population along the common border, whereas the Kenya –Uganda border there is a project to secure the common border.

Integrated tsetse control strategy in Kenya incorporates key methods of that comprise of insecticide treated cattle, Insecticide treated targets, Traps, insecticide application by either ground or aerial spraying. Sterile insect technique in Kenya has not being tried given the low number in the *G. pallidipes* in the colony at KALRO. The methods that have been outlined have drawbacks, as the residual populations are difficult to mop-up.

A one-health approach in the control of trypanosomiasis in the country has been advocated by KENTTEC in Kenya. The approach has been implemented in the HAT endemic areas of Western, Lake Victoria, and transmara area. The teams from the Ministry of Health, KENTTEC, KALRO, and County government of the respective areas are involved in the Surveillance of the animal and human disease. These approaches significantly influence the resource sharing and mobilization at the intervention stage.

The aggressive control programmes together with climatic, land-use, and socio-economic changes immensely contribute to the reduction of African trypanosomiasis in Kenya

FUTURE RESEARCH DIRECTION

Research in tsetse and animal trypanosomiasis will focus on two critical areas of vector control and development of better diagnostics and treatment of livestock. The vector management component will focus on the exploiting of the knowledge in tsetse genetics, improvement of odors, repellants, insecticide application, understanding behavior in the field and well as in the colony.

On animal trypanosomiasis, the focus would be for the development of better molecues for treatment of the disease and improvement of animal genetics in terms of trypanotolerance,

CONCLUSION

The lessons' learnt in the control tsetse and trypanosomiaisis from the various programs and projects inform policy and implementation of future programs.

REFERENCES

Abenga, J. N., & Vuza, D. (2005). About factors that determine trypanotolerance and prospects for increasing resistance against trypanosomosis. *African Journal of Biotechnology*, *4*(13), 1563–1567.

Adungo, F., Mokaya, T., Makwaga, O., & Mwau, M. (2020). Tsetse distribution, trypanosome infection rates, and small-holder livestock producers' capacity enhancement for sustainable tsetse and trypanosomiasis control in Busia, Kenya. *Tropical Medicine and Health*, *48*(1), 62. doi:10.118641182-020-00249-0 PMID:32760194

Allsop, R. (1984). Control of tsetse flies (Diptera: Glossinidae) using insecticides: a review and future prospects. *Bulletin of Entomological Research*, *74*(1), 1–23. doi:10.1017/S0007485300009895

Bauer, B., Holzgrefe, B., Mahama, C. I., Baumann, M. P., Mehlitz, D., & Clausen, P. H. (2011). Managing tsetse transmitted trypanosomosis by insecticide-treated nets-an an affordable and sustainable method for resource-poor pig farmers in Ghana. *PLoS Neglected Tropical Diseases*, *5*(10), e1343. doi:10.1371/journal.pntd.0001343 PMID:22022625

Bett, M. K., Saini, R. K., & Hassanali, A. (2015). Repellency of tsetse-refractory waterbuck (Kobus defassa) body odour to Glossinapallidipes (Diptera: Glossinidae): assessment of the relative contribution of different classes and individual constituents. *Acta Tropica*, *146*, 17–24. doi:10.1016/j.actatropica.2015.02.017 PMID:25746973

Bourn, D., Reid, R., Rogers, D., Snow, B., & Wint, W. (2001). Environmental change and the autonomous 476 control of tsetse and trypanosomiasis in sub-Saharan Africa: case histories from Ethiopia, The 477 Gambia, Kenya, Nigeria, and Zimbabwe. Oxford, UK: Environmental Research Group.

Bourn, D. A. V. I. D., Grant, I., Shaw, A., Torr, S., & Protection, C. E. (2005). Cheap and safe tsetse control for livestock production and mixed farming in Africa. *Aspects of Applied Biology*, *75*, 81.

Bouyer, J., Bouyer, F., Donadeu, M., Rowan, T., & Napier, G. (2013). Community-and farmer-based management of animal African trypanosomosis in cattle. *Trends in Parasitology*, *29*(11), 519–522. doi:10.1016/j.pt.2013.08.003 PMID:24012356

Cecchi, G., Paone, M., Argiles Herrero, R., Vreysen, M. J., & Mattioli, R. C. (2015). Developing a continental atlas of the distribution and trypanosomal infection of tsetse flies (Glossina species). *Parasites & Vectors*, *8*(1), 284. doi:10.118613071-015-0898-y PMID:25994757

Cecchi, G., Paone, M., Feldmann, U., Vreysen, M. J., Diall, O., & Mattioli, R. C. (2014). Assembling a geospatial database of tsetse-transmitted animal trypanosomosis for Africa. *Parasites & Vectors*, *7*(1), 39. doi:10.1186/1756-3305-7-39 PMID:24447638

Chira, R. M. (2003). *Changes in wildlife habitat and numbers in Embu and Mbeere districts*. Eastern Province.

Diall, O., Cecchi, G., Wanda, G., Argiles-Herrero, R., Vreysen, M. J. B., Cattoli, G., Viljoen, G. J., Mattioli, R., & Bouyer, J. (2017). Developing a Progressive Control Pathway for African Animal Trypanosomosis. *Trends in Parasitology*, *33*(7), 499–509. doi:10.1016/j.pt.2017.02.005 PMID:28456474

Dicko, A. H., Percoma, L., Sow, A., Adam, Y., Mahama, C., Sidibe, I., Dayo, G.-K., Thévenon, S., Fonta, W., Sanfo, S., Djiteye, A., Salou, E., Djohan, V., Cecchi, G., & Bouyer, J. (2015). A Spatio-temporal Model of African Animal Trypanosomosis Risk. *PLoS Neglected Tropical Diseases*, *9*(7), e0003921. doi:10.1371/journal.pntd.0003921 PMID:26154506

Elliot, M., Janes, N.F., & Potter, C. (1978). The future of pyrethroids in insect control. *Annal rev Entomology, 23*, 443-469.

Esterhuizen, J., Rayaisse, J. B., Tirados, I., Mpiana, S., Solano, P., Vale, G. A., Lehane, M. J., & Torr, S. J. (2011). Improving the cost-effectiveness of visual devices for the control of riverine tsetse flies, the major vectors of human African trypanosomiasis. *PLoS Neglected Tropical Diseases*, *5*(8), e1257. doi:10.1371/journal.pntd.0001257 PMID:21829743

FAO. (1992). *Use of attractive devices for tsetse survey and control* (Vol. 4). FAO.

FAO. (1993). *Insecticides for tsetse and trypanosomiasis control using attractive bait techniques* (Vol. 5). FAO.

FAO/OIE. (2015). *Global Strategy for the Control and Eradication of PPR*. FAO/OIE.

FITCA. (2004). *Farming in Tsetse Controlled Controlled Areas Kenya Project 1999-2004 Lessons Learned*. FAITCA.

Ford, J. (1971). *The role of the trypanosomiases in African ecology: A study of the tsetse fly problem*. Clarendon Press.

Ford, J., & Katondo, K. M. (1977). Maps of tsetse flies (Glossina) distribution in Africa, 1973 according to sub-generic groups on scale of 1:5 000 000. *Bulletin of Animal Health and Production in Africa*, 187–193.

Ford, J., & Katondo, K. M. (1977b). Maps of tsetse flies (Glossina) distribution in Africa, 1973 according to sub-generic groups on scale of 1:5 000 000. *Bulletin of Animal Health and Production in Africa*, 187–193.

Franco, J. R., Cecchi, G., Priotto, G., Paone, M., Diarra, A., Grout, L., Simarro, P. P., Zhao, W., & Argaw, D. (2018). Monitoring the elimination of human African trypanosomiasis: Update to 2016. *PLoS Neglected Tropical Diseases*, *12*(12), e0006890. doi:10.1371/journal.pntd.0006890 PMID:30521525

Franco, J. R., Simarro, P. P., Diarra, A., & Jannin, J. G. (2014). Epidemiology of human African trypanosomiasis. *Clinical Epidemiology*, *6*, 257–275. doi:10.2147/CLEP.S39728 PMID:25125985

Gikonyo, N. K., Hassanali, A., Njagi, P. G., Gitu, P. M., & Midiwo, J. O. (2002). Odor composition of preferred (buffalo and ox) and nonpreferred (waterbuck) hosts of some Savanna tsetse flies. *Journal of Chemical Ecology*, *28*(5), 969–981. doi:10.1023/A:1015205716921 PMID:12049234

Gikonyo, N. K., Hassanali, A., Njagi, P. G., & Saini, R. K. (2003). Responses of Glossina morsitans morsitans to blends of electroantennographically active compounds in the odors of its preferred (Buffalo and ox) and nonpreferred (waterbuck) hosts. *Journal of Chemical Ecology*, *29*(10), 2331–2345. doi:10.1023/A:1026230615877 PMID:14682515

Gimonneau, Rayaisse, & Bouyer. (2018). Integrated control of trypanosomosis. *Pests and vector-borne diseases in the livestock industry*, 147 - 174. doi:10.3920/978-90-8686-863-6_6

Gimonneau, G., Rayaisse, J. B., & Bouyer, J. (2018). Integrated control of trypanosomosis. In *Pests and vector-borne diseases in the livestock industry* (pp. 351–357). Wageningen Academic Publishers. doi:10.3920/978-90-8686-863-6_6

Grant, C., Anderson, N., & Machila, N. (2015). Stakeholder narratives on trypanosomiasis, their effect on policy, and the scope for One Health. *PLoS Neglected Tropical Diseases*, *9*(12), e0004241. doi:10.1371/journal.pntd.0004241 PMID:26658646

Grant, I. F. (2001). Insecticides for tsetse and trypanosomiasis control: Is the environmental risk acceptable? *Trends in Parasitology*, *17*(1), 10–14. doi:10.1016/S1471-4922(00)01848-1 PMID:11137734

Green, C. H. (1994). Bait methods for tsetse fly control. *Advances in Parasitology*, *34*, 229–291. doi:10.1016/S0065-308X(08)60140-2 PMID:7976751

Hargrove, J. W., Omolo, S., Msalilwa, J. S. I., & Fox, B. (2000). Insecticide-treated cattle for tsetse control: The power and the problems. *Medical and Veterinary Entomology*, *14*(2), 123–130. doi:10.1046/j.1365-2915.2000.00226.x PMID:10872856

Holmes, P. (2015). On the road to elimination of rhodesiense human African trypanosomiasis: First WHO meeting of stakeholders. *PLoS Neglected Tropical Diseases*, *9*(4), e0003571. doi:10.1371/journal.pntd.0003571 PMID:25836650

Kaitho, T., Ndeereh, D., & Ngoru, B. (2013). An outbreak of anthrax in endangered Rothschild's giraffes in Mwea National Reserve, Kenya. *Veterinary Medicine: Research and Reports*, *4*, 45–48.

KENTTEC. (2019). https://www.kenttec.go.ke/wp-content/uploads/2019/08/Parasitological-and-Entomologuical-monitoring-report_Pate_Nov_Dec-2016.pdf

KENTTEC. (2019). *Draft KENTTEC Strategic Plan*. Retrieved April 27th, 2020 from https://www.kenttec.go.ke/wp-content/uploads/2019/08/Draft-KENTTEC-Strategic-Plan.pdf

KENTTEC. (2011). *Strategy for tsetse and trypanosomiasis eradication in Kenya 2011-2021*. Ministry of Livestock Development Dovs2011. 449 8.

Kenya Law. (2012). *State Corporations Act (Kenya Tsetse and Trypanosomiasis Eradication Council) Order, 2012*. Retrieved April 27th, 2020 from, www. http://kenyalaw.org/kl/index.php?id=784

Kenya Tsetse and Trypanosomiasis Eradication Council (KENTTEC). (2020). *Achievements*. https://www.kenttec.go.ke/achievements/

Kettle, D. S. (1990). *Medical and Veterinary Entomology*. C.A.B. International.

Kuzoe, F. A. S., & Schofield, C. J. (2004). Strategic review of traps and targets for tsetse and African trypanosomiasis control. WHO/ TDR.

Leak, S. G. E., Ejigu, D., & Vreysen, M. J. B. (2008). *Collection of entomological baseline data for tsetse area-wide integrated pest management programmes*. Food and Agriculture Organization of the United Nations.

Lewis, E. (1950). Tsetse Flies carried by Railway Trains in Kenya Colony. *Bulletin of Entomological Research*, *40*(4), 511–531. doi:10.1017/S0007485300024421

Lindh, J. M., Torr, S. J., Vale, G. A., & Lehane, M. J. (2009). Improving the cost-effectiveness of artificial visual baits for controlling the tsetse fly Glossina fuscipes fuscipes. *PLoS Neglected Tropical Diseases*, *3*(7), e474.

Maichomo, M.W., Ndung'u, J.M., Ngare, P.M., & Ole-Mapenay, I.M. (2005). The performance of Orm-aBoran and Maasai Zebu crossbreeds in a trypanosomosis endemic area of Nguruman, southwestern Kenya. *Onderstepoort Journal of Veterinary Research, 72*, 87–93.

Malele, I. I. (2011). Fifty years of tsetse control in Tanzania: Challenges and prospects for the future. *Tanzania Journal of Health Research, 13*(5).

Mangwiro, T. N. C., Torr, S. J., Cox, J. R., & Holloway, M. T. P. (1999). The efficacy of various pyrethroid insecticides for use on odour-baited targets to control tsetse. *Medical and Veterinary Entomology*, *13*, 315–323.

Moloo, S.K. (1993). The distribution of Glossina species in Africa and their natural hosts. *Insect Science 480 and its Application, 14*, 511-28.

Murilla, G. A., Ndung'u, K., Thuita, J. K., Gitonga, P. K., Kahiga, D. T., Auma, J. E., Ouma, J. O., Rutto, J. J., & Ndung'u, J. M. (2014). Kenya Trypanosomiasis Research Institute cryobank for human and animal trypanosome isolates to support research: Opportunities and challenges. *PLoS Neglected Tropical Diseases*, *8*(5), e2747. https://doi.org/10.1371/journal.pntd.0002747

Nagel, P. (1995). *Ecological side effect of tsetse control. Abstracts of VIII International Conference of Institutions of Tropical Veterinary Medicine*, Berlin, Germany.

Neave, S. (1912). Notes on the blood-sucking insects of Eastern Tropical Africa. *Bulletin of Entomological Research*, *3*(3), 275–323. doi:10.1017/S0007485300001942

Ngari, N. N., Gamba, D. O., Olet, P. A., Zhao, W., Paone, M., & Cecchi, G. (2020). Developing a national atlas to support the progressive control of tsetse-transmitted animal trypanosomosis in Kenya. *Parasites & Vectors*, *13*(1), 1–12.

Njiru, Z. K., Makumi, J. N., Okoth, S., Ndungu, J. M., & Gibson, W. C. (2004). Identification of trypanosomes in Glossina pallidipes and G. longipennis in Kenya. *Infection, Genetics and Evolution*, *4*(1), 29–35.

Olet, P., & Othieno, J. (2015). How to do mass media publicity for a neglected disease. Lessons from Tsetse and Trypanosomiasis in Kenya. *Journal of Science Communication, 14*(3), Y03.

Omondi, P., Bitok, E., & Kagiri, J. (2004). Managing human–elephant conflicts: The Kenyan experience. *Pachyderm, 36*, 80–86.

Otieno, A. (2010). 'Reordering and Counterordering': Forestry Preservation, Bush Clearing and the Sociophysical Mapping of Chepalungu, Kericho District, Kenya, 1930–1963. *Environment and History, 16*(4), 381-408. Retrieved April 22, 2020, from www.jstor.org/stable/25799095

Pollock, J. N. (1982). Training manual for tsetse control personnel – volume 1: tsetse biology, systematics and distribution; techniques. FAO.

Rayaisse, J., Salou, E., & Courtin, F. (2015). Baited-boats: An innovative way to control riverine tsetse, vectors of sleeping sickness in West Africa. *Parasites & Vectors, 8*(1), 236.

Rayaisse, J., Salou, E., & Courtin, F. (2015). Baited-boats: An innovative way to control riverine tsetse, vectors of sleeping sickness in West Africa. *Parasites & Vectors, 8*(1), 236.

Rayaisse, J. B., Tirados, I., Kaba, D., Dewhirst, S. Y., Logan, J. G., Diarrassouba, A., & Esterhuizen, J. (2010). Article. *PLoS Neglected Tropical Diseases, 4*, 632–638.

Rebeski, D. E. (1999). Improved Methods for the Diagnosis of African Trypanosomosis. *Mem. Inst. Oswaldo Cruz, 94*(2), 249-253.

Rutto, J., Osano, O., Thuranira, E. G., Kurgat, R. K., & Odenyo, V. A. O. (2013). Socio-economic and cultural determinants of human African trypanosomiasis at the Kenya—Uganda transboundary. PLoS Neglected Tropical Diseases, 7(4).

Shaw, A. P. M., Cecchi, G., Wint, G. R. W., Mattioli, R. C., & Robinson, T. P. (2014). Mapping the economic benefits to livestock keepers from intervening against bovine trypanosomosis in Eastern Africa. *Preventive Veterinary Medicine, 113*(2), 197–210.

Shaw, A. P. M., Wint, G. R. W., Cecchi, G., Torr, S. J., Mattioli, R. C., & Robinson, T. P. (2015). Mapping the benefit-cost ratios of interventions against bovine trypanosomosis in Eastern Africa. *Preventive Veterinary Medicine, 122*(4), 406–416.

Simarro, P. P., Jannin, J., & Cattand, P. (2008). Eliminating human African trypanosomiasis: Where do we stand and what comes next? *PLoS Medicine, 5*(2), 174–180.

Simarro, P. P., Jannin, J., & Cattand, P. (2008). Eliminating human African trypanosomiasis: Where do we stand and what comes next? *PLoS Medicine, 5*(2), 174–180.

Sumption, K., Domenech, J., & Ferrari, G. (2012). Progressive control of FMD on a global scale. *The Veterinary Record, 170*(25), 637.

Takken, W., & Weiss, M. (1990). The Sterile Insect Technique for Tsetse Flies in Africa. *International Atomic Energy Agency Bulletin, 20*(3).

Torr, S. J., Holloway, M. T. P., & Vale, G. A. (1992). Improved persistence of insecticide deposits on targets for controlling Glossinapallidipes (Diptera: Glossinidae). *Bulletin of Entomological Research*, *82*, 525–533.

Vale, G. A. (1977). Feeding responses of tsetse flies (Diptera: Glossinidae) to stationary hosts. *Bulletin of Entomological Research*, *67*, 635–649.

Vale, G. A., Hall, D. R., & Gough, A. J. E. (1988). The olfactory responses of tsetse flies Glossina spp. (Diptera:Glossinidae) to phenols and urine in the field. *Bulletin of Entomological Research*, *78*, 2.

Vreysen, M.J. (2001). *Principles of Area Wide Integrated Tsetse Fly Control Using the Sterile Insect Technique*. Academic Press.

Vreysen, M. J. B., Robinson, A. S., Hendrichs, J., & Kenmore, P. (2007). Area-wide integrated pest management (AW-IPM): principles, practice and prospects. In *Area-wide control of insect pests* (pp. 3–33). Springer.

Vreysen, M. J. B., Robinson, A. S., Hendrichs, J., & Kenmore, P. (2007). Area-wide integrated pest management (AW-IPM): principles, practice and prospects. In *Area-wide control of insect pests* (pp. 3–33). Springer.

Wachira, B. M., Mireji, P. O., Okoth, S., Ng'ang'a, M. M., William, J. M., Murilla, G. A., & Hassanali, A. (2016). Responses of Glossinapallidipes and Glossinamorsitansmorsitans tsetse flies to analogues of δ-octalactone and selected blends. *Acta Tropica*, *160*, 53–57.

Wamwiri, F. N., & Changasi, R. E. (2016). Tsetse Flies (Glossina) as Vectors of Human African Trypanosomiasis: A Review. *BioMed research international, 2016*, 6201350. doi:10.1155/2016/6201350

Warnes, M. L., Van den Bossche, P., Chihiya, J., Mudenge, D., Robinson, T. P., Shereni, W., & Chadenga, V. (1999). Evaluation of insecticide-treated cattle as a barrier to re-invasion of tsetse to cleared areas in northeastern Zimbabwe. *Medical and Veterinary Entomology*, *13*(2), 177–184.

Wellde, Chumo, Waema, Reardon, & Smith. (1989). A history of sleeping sickness in Kenya. *Annals of Tropical Medicine & Parasitology, 83*(S1), 1-11. DOI: doi:10.1080/00034983.1989.11812405

WHO. (2012). *Research Priorities for Chagas Disease, Human African Trypanosomiasis and Leishmaniasis, WHO Technical Report Series*. WHO.

WHO. (2012). *Research Priorities for Chagas Disease, Human African Trypanosomiasis and Leishmaniasis, WHO Technical Report Series*. WHO.

Wilson, S. G. (1953). The control of Glossina palpalis fuscipes Newstead in Kenya colony. *Bulletin of Entomological Research*, *44*, 711–728.

Chapter 5
Use of Trypanotolerant Breeds:
The Case of the Orma Boran

Monicah Maichomo
Kenya Agricultural and Livestock Research Organization, Kenya

Caleb Oburu Orenge
Department of Veterinary Anatomy and Physiology, Faculty of Veterinary Medicine and Surgery, Egerton University, Kenya

Samuel Mwanzia Mbuku
Kenya Agriculture and Livestock Research Organization, Kenya

ABSTRACT

This chapter describes the meaning of trypanotolerance and its implication as used with trypanotolerant livestock that are known and studied for many years in West Africa. This trait is observed in many West African breeds of cattle that survive in areas of tsetse fly challenge where the humped zebu cannot. Small ruminants and wildlife also exhibit increased resistance to infection with trypanosomes. Trypanotolerance is viewed as a more sustainable and environmentally friendly approach to control of trypanosomiasis given the limitations of the present methods of control and lack of vaccine. Little work has been conducted with East African cattle breeds. The Kenyan Orma Boran (OB) has been studied since 1983 to enhance its conservation and utilization in East African region. Nucleus breeding herd of superior genotypes was multiplied in Galana Agricultural Development Corporation (ADC) and disseminated to trypanosomosis endemic areas of Kenya. On evaluation, performance of the crosses was comparable to that of other Borans. Challenges of adoption in pastoral area are presented.

INTRODUCTION

The African continent has diverse livestock species and breeds with unique characteristics and more importantly, adapted to the environment through natural selection over time. They are useful assets supporting human livelihoods either directly or indirectly. This chapter focuses on cattle and trypanotolerance

DOI: 10.4018/978-1-7998-6433-2.ch005

as a sustainable control option for trypanosomosis which is a major constraint in livestock production. Trypanotolerant livestock breeds are more utilized in West Africa where tsetse vector and trypanosomosis infection are more prevalent than the rest of sub-saharan Africa. Thus a lot of knowledge available on trypanotolerance is derived from breeds like N'Dama, Baoule and West African short horn among others. These have made cattle rearing a possibility in the forested West Africa. The Orma Boran has potential for exploitation in the tsetse fly belts of East Africa but more work is required, from structured multiplication and within breed selection, to community based multiplication schemes. The reader is introduced to the African cattle types and trypanotolerant breeds with a focus on the East African Orma Boran breed.

OVERVIEW OF AFRICAN INDIGENOUS CATTLE BREEDS

The contribution of indigenous cattle genetic resources to food production in Africa cannot be overemphasized. With the demand for animal foods projected to triple in Africa by the year 2050 (Food and Agriculture Organization of the United Nations, 2011), and the predicted negative effects of climate change impacts in the tropics, cattle breed like Ankole, Fulani, Orma Boran, Nguni, Mashona, Tswana and Tuli that are less demanding in terms of feeds, while able to cope with higher temperatures and humidity are vital for ensuring future food and nutrition security for African people. These unique African indigenous cattle breeds provide milk, meat, insurance against economic bottlenecks, and social capital products. Their origin, characteristics and distribution are well documented in literature (DAD-IS, 2020; DAGRIS, 2020; Mwai et al., 2015; Rege et al., 2001; Rege & Tawah, 1999).

Africa is uniquely rich in cattle diversity with approximately 152 African cattle breeds or populations recognized. The two main morphological classes of African cattle genotypes are humped *Bos indicus*, commonly referred to as Zebu, and humpless *Bos taurus* cattle. Additionally, there are two other defined breed combinations; (i) Sanga (stable crosses of *B. indicus* × *B. taurus*) and Zenga (stable crosses of Sanga × Zebu) (Hanotte, 2002). The Zebu cattle include the N'Dama from West Africa, Orma Boran (eastern Africa), Kuri (western/central Africa), Borana (eastern/central Africa), Fulani (west Africa), Ankole (Great Lakes Region), and Maure (Mali). Others include the Nguni (southern Africa), Tonga (Zambia), Tuli (Zimbabwe) and Afrikaner (South Africa) (Rege & Tawah, 1999). Rege et al. (2001) has described the East African Zebu (EAZ) as either "Small" or "Large" type based on their body sizes. The taurine breeds are geographically distributed in the sub-humid and humid regions of West Africa, which are largely infested with the tsetse fly (Kim et al., 2017). Zenga are found in eastern Africa while Sanga cattle are predominantly found in central Africa around the Great Lakes region and on the southern Africa region.

Today, Zebu cattle have wide geographical spread in Africa and adapted to dry climatic conditions such as high temperature and humidity. They are more tolerant to tropical cattle diseases like East Coast Fever compared to their exotic counterparts. However, majority of Zebu cattle are usually more susceptible to trypanosomosis except N'Dama, several West African taurine breeds, Orma Boran which have comparative considerable levels of trypanotolerance traits (Kim et al., 2017). Although the productivity of these indigenous genotypes is lower compared to the exotic cattle, they offer critical livelihood options to millions of farmers and livestock keepers domiciled in low input production systems, where exotic breed struggle to perform under the prevailing traditional management circumstances. Therefore, a wide portfolio of indigenous cattle is crucial to adapting and developing our livestock production systems.

Climate change underlines the need to retain this adaptive capacity. Studies to unravel their admixtures may provide critical information into the genetic mechanisms underlying their adaptation to various agro-ecosystems and tolerance to various tropical diseases in Africa.

Although African cattle genetic diversity remains large, several breeds continue to face extinction. It is even more worrying that unrecorded indigenous genotypes are being lost before their characteristics can be studied and their potential evaluated. Some of these endangered breeds include Kavirondo and Kikuyu Zebu (Kenya), Ghana Dwarf Muturu, Pare (Small East African Zebu), Chagga (Small East African Zebu), Baria (Small East African Zebu), Nkone (South African sanga), Pedi (South African sanga) and Shangan (South African sanga) (Mwai et al., 2015). The influx of exotic genotypes due to market demand for increased production has been a threat to these African indigenous cattle genotypes. Indiscriminate and generally uncontrolled crossbreeding has continuously prevailed in Africa, with minimal within breed selection program thus eroding the integrity of some of these cattle populations including for adaptive and tolerance traits. For example, loss of trypanotolerance traits in Orma Boran would be tragic for Kenya and other tsetse fly belts in sub-Saharan Africa where trypanosomosis is one of the major constraints on livestock production. Tsetse-transmitted trypanosomiasis affects productivity in cattle, with direct and indirect losses estimated in billions of dollars annually in sub-Saharan Africa. There is little hope that a conventional, anti-infection vaccine will be produced in the near future. While eradication of the disease from East Africa region is an unrealistic goal, at least for now, considerable effort and investment is required towards exploitation of the genetic resistance exhibited by the Orma Boran. The fact that Orma Boran also possess considerable production potential, offers at a stroke, an unparalleled opportunity to improve other indigenous cattle genotypes production in the vast areas of East Africa infested by tsetse flies.

Definition of Trypanotolerance

Trypanotolerance is defined as the relative capacity of an animal to control the development of the trypanosome parasite once infected and to limit its pathological effects, the most prominent of which is anaemia the major cause of death associated with the disease (Murray et al., 1990; J Naessens, 2006). Such animals are capable of maintaining packed cell volume and limiting peak level parasitaemia and eventually control, reduce or even eliminate the parasite, hence remaining productive with or without minimal aid of trypanocidal drugs (d'Ieteren et al., 1998; J Naessens, 2006). These livestock breeds demonstrate innate ability to survive and remain productive under trypanosome challenge (Yaro et al., 2016). This trait is observed in many West African breeds of cattle such as N'Dama, Muturu and West African Shorthorn that survive in areas of tsetse fly challenge where the East African zebu cannot. Even though the zebu is viewed as the most susceptible of the African breeds, some zebu have developed a degree of tolerance such as those around the shores of Lake Victoria in East Africa that have survived continuous exposure to tsetse (Cunningham, 1966). Small ruminants and wildlife also exhibit increased resistance to infection with trypanosomes (Mortelmans & Kageruka, 1976; Radcliffe & Khairani, 2019; Tilki et al., 2016). Trypanotolerance in small ruminants is less pronounced than in cattle and should be considered as resilience rather than resistance (Chiejina et al., 2009). The West African Dwarf (WAD) goat is known for its trypanotolerance as well as resistance to *Haemonchus contortus* but seems to be less trypanotolerant than Djallonke sheep (Geerts et al., 2009). Considering their potential in trypanosomosis endemic areas of Africa, there is need to upgrade their genetic purity for trypanotolerance.

A trypanotorelent animal can either resist infection mediated primarily by the immune system to reduce pathogen load following infection. Essentially, this distinct humoral immune response to trypanosome infection is the major feature of bovine trypanotolerance. Alternatively, the animal can tolerate infection by reducing the deleterious effects of pathogen in the body (Medzhitov et al., 2012). Superiority of the N'Dama cattle is associated with its ability to mount a more effective secondary immune response compared to the susceptible zebu (Desowitz, 1959).

The tolerance is not absolute and some infected cattle can die while most survive in a poor state of production with wasting, stunting, abortion, high calf mortality and low grade anaemia (Berthier et al., 2015). Thus trypanotolerance is a relative trait that is affected by stressors such as nutritional deficiencies, intense work and concurrent infections. Continous exposure to trypanosomes is therefore required in order to maintain this trait in the host. This has implications in crossbreeding programs where selection is done for maximum trypanotolerance while retaining required adaptive characteristics of the non-trypanotolerant animal.

Use of Trypanotolerant Livestock in Control of African Animal Trypanosomosis (AAT)

Trypanotolerance has a complex genetic basis and is thought to have evolved in tsetsefly infested areas of Africa by natural selection of the more resistant livestock breeds such as Djallonke sheep, west African dwarf goat and Taurine cattle of West Africa (Abenga & Vuza, 2005; Mwai et al., 2015; Yaro et al., 2016). The trait was described in indigenous East African breeds of cattle as early as 1913, for example the Zebu in Sudan, the Lugware of Zaire, Uganda and Sudan and the Zebu around the shores of Lake Victoria (Cunningham, 1966; Yaro et al., 2016).

Use of trypanotolerant animals in trypanosomosis endemic areas of Africa optimizes production, a practice widely associated with certain *Bos taurus* breeds of West Africa (Geerts et al., 2009; Murray et al., 1990; Naessens et al., 2002). These taurine cattle include the Muturu, Baoule, Laguna, Samba, Dahomey and the N'dama (Agyemang et al., 1997). Susceptibility studies have shown the N'dama to be the most resistant to trypanosome infection in The Gambia (Leperre & Claxton, 1994). Further work with the N'dama confirmed the positive correlation between key indicators of trypanotolerance (control of anaemia, parasitaemia and body weight) and growth. A review by Mortelmans & Kageruka (1976) details excellent performance of N'Dama cattle in the tsetse fly infested region of southwestern Zaire in view of their tolerance to trypanosomes, at a time when the government and livestock enterprises struggled to establish cattle farms in the forest zone where no other bovines could survive. N'Dama cattle breed has been used to improve hardiness and trypanotolerance of the larger, heavier and less resistant breeds of western and central Africa (Hanotte et al., 2003; Namangala, 2012). The tolerant Dahomey cattle breed although small was found very hardy, well adapted to poor regions and useful as slaughter stock with a relatively high dressing percentage.

These and other studies have recommended use of trypanotolerance as the most economically sustainable and environmentally friendly approach to control of anaimal trypanosomosis given the limitations of the present methods of control such as chemotherapy and vector control, and lack of vaccine (Jan Naessens, 2006; Orenge et al., 2011; Yaro et al., 2016). Their small size notwithstanding, the productivity of trypanotolerant livestock compared to the more susceptible breeds may be much higher than previously assumed and hence hold promise for efficient land utilization in tsetse infested areas in Africa (Maichomo et al., 2005; Mwai et al., 2015; Orenge et al., 2012). In addition, trypanotolerant cattle have other genetic advantages that contribute to their potential for use in livestock development programs in the tropics (d'Ieteren et al., 1998). (Mwangi et al., 1998) documented resistance of the Orma Boran to ticks,Tick-borne diseases (TBD's) and helminths. N'dama cattle are less susceptible to *Amblyoma* and *Hyallomma* tick species than Gobra zebu cattle, with significantly lower prevalence of *Anaplasma marginale* infection, but are however susceptible to East Coast fever (ECF). An economic study by Itty et al (1995) on the production of N'dama cattle in the Democratic Republic of Congo demonstrated their profitability and successful integration into the farming system. Wosu & Nwanta (1989) found the Fulani zebu of Southern Nigeria to have better reproductive performance and healthier than their Northern counterparts despite having high incidence of trypanosomosis.

Despite these recommendations, trypanotolerance has been less exploited as a control measure for AAT (Dolan, 1987; Kosgey & Okeyo, 2007) Leak (1998) estimated that trypanotolerant cattle accounted for about 5% of Africa's cattle population, indicating under utilization of this useful AAT control strategy. Indeed, in spite of these early reports, limited work has been carried out in East Africa on trypanotolerance unlike the case of West African breeds. However, from the 1980s, some efforts have been made to investigate the nature and extent of trypanotolerance amongst cattle breeds of East Africa (Dolan, 1998; Maichomo, 2008; Njogu et al., 1985). It is worth noting that wider application of trypanotolerance as a control strategy for AAT can positively impact on food and nutrition security in Sub-saharan Africa.

Use of Trypanotolerant Orma Boran breed for Control of African Animal Trypanosomosis

Over 60% of Kenya's total landmass, which supports over half of the country's national cattle herd, is tsetse infested, and could benefit from introduction of trypanotolerant breeds. The Orma Boran (OB) was the focus of studies conducted by the defunct Kenya Trypanosomiasis Research Institute (KETRI), since 1983 to enhance its conservation and utilization (Dolan, 1998; Mwangi et al., 1998; Njogu et al., 1985). This work is currently continuing with the Kenya Agricultural and Livestock Research Organization (KALRO) – Beef Research Institute. Orma Boran is categorized as the large type East African Zebu (EAZ) (Rege et al., 2001). The breed was selected from source in Tana River County and the nucleus breeding herd multiplied under ranching conditions in Galana Agricultural Development Coorporation (ADC) in order to avail breeding stock for use in Tsetse and trypanosomosis infested areas of Kenya. Effects of Climate Change has caused variations in the tsetse vector, with shrinking of some belts and extension into new areas. According to Kenya Tsetse and Trypanosomosis Eradication Council (KENTTEC), the country currently has 6 defined tsetse fly belts whose livestock production can benefit immensely by using the OB breed (KENTTEC, 2020). The belts include (i) Lake Victoria basin fly belt (ii) Narok – Kajiado fly belt (iii) Lake Bogoria – Baringo – Kerio Valley – Koibatek – Turkwell fly belt (iv) Central Kenya fly belt (v) Isiolo- Samburu fly belt, and, (vi) Coastal fly belt. The Orma Boran breed will also be very useful outside the tsetse fly belt where mechanical transmission occurs.

HISTORY OF ORMA BORAN BREEDING IN KENYA

A review of the Orma boran (OB) breeding herd in Galana ranch is given by Dolan (1998). This herd was established in 1983 using a foundation stock of best genotypes obtained from the Orma people of Tana River County (Dolan, 1998; Njogu et al., 1985). Through to 1997, consequent generations formed the

Figure 1. Breeding structure for the Orma Boran cattle at the Galana Agricultural Development Coorporation ranch and satellite centres.

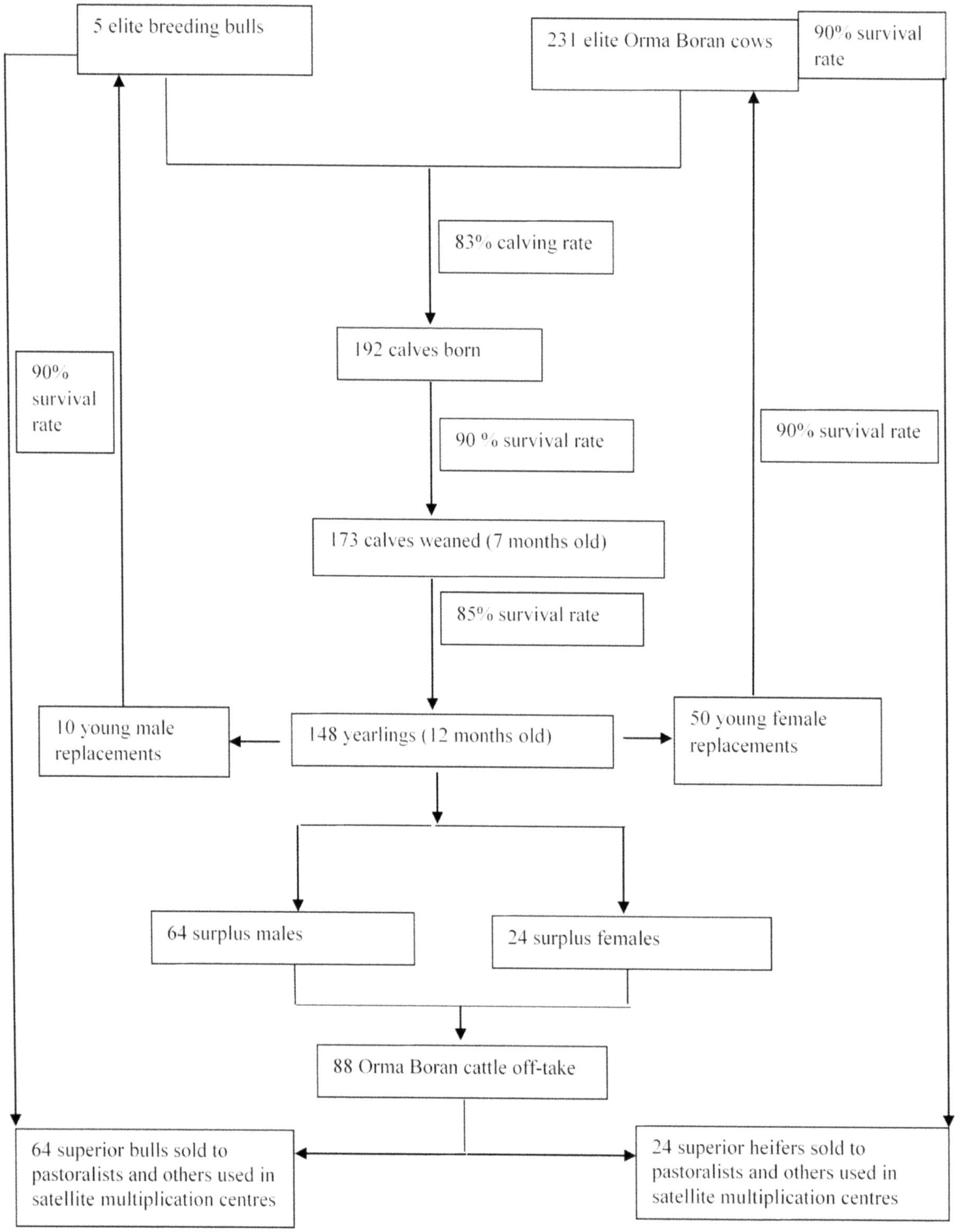

basis for the OB breeding herd and selection program aimed at improving growth rate characteristics (Fig 1). Males with above average growth rate were selected for breeding, the the trait of choice being post-weaning growth rate under high tsetse challenge, a trait with medium to high heritability and frequently used in beef production breeding program as it responds well to selection. Over time, improvement was observed on productivity parameters such as birth weight values, daily weight gain and age at first calving (Dolan, 1998). Repeatability values for cow PCV, trypanosome infection rates and body weight were estimated at 26%, 40% and 50% respectively. Heritability estimates of calves born at Galana ranch for birth weight, weaning weight and PCV were medium to high, which is a suitable selection criterion

The Orma Boran was registered with the Kenya stud book but limited numbers of improved animals hindered widespread adoption by pastoralists (Dolan, 1998). Besides, there was need to evaluate its performance in other tsetse infested areas of Kenya. The breeding program therefore selected superior bulls for multiplication and dissemination purposes to tsetse infested areas of Kenya. Semen was also collected from superior bulls and stored with the Kenya Animal Genetic Resources Centre (KAGRC) for further multiplication using Artificial Insemination in order to meet the growing demand (Maichomo et al., 2005). Between 1999 and 2001, a breeding stock selected from Galana ranch was established in Kenya Agricultural Research Institute (KARI), Kiboko range research station in Makueni County for further multiplication. Some bulls from this station were disseminated to Kajiado and Busia counties which are endemic for tsetse and trypanosomosis, for adoption and multiplication by the resident communities.

Community Conservation of Orma Boran Cattle in Tana River County

The Orma Boran (OB) cattle have lived under intense natural tsetse challenge for centuries, first in the Borana province of Ethiopia and then in Tana River delta of coastal Kenya. Although the delta provides abundant pastures for dry season grazing, the area is heavily infested with tsetse flies, which limits intensive exploitation. In Kenya, the Orma Boran are kept by the Orma community who are predominantly pastoralists keeping cattle, sheep and goats in Tana River county (Irungu, 2000). This community has excellent cattle keeping skills and selects their cattle for breeding and commercial purposes based on their milk yield, body size, resistance to trypanosomosis and white to grey coat colour (Irungu, 2000).. The selection is aimed at supporting pastoral livelihoods by supplying milk, meat and money. The breed preference criterion for large size and high milk production seems to be the same between the Orma, Maasai and perhaps other pastoral communities (Maichomo et al., 2005). Colour choice is based on the fact that white coloured animals are less frequently bitten by tsetse flies compared to black or brown colours. This indigenous knowledge by the host Orma community informed selection of foundation breeding stock used at Galana ranch which were predominantly white in colour (Dolan, 1998).

Despite lack of adequate comparative data, information from Tana River county showed that the growth rates of Orma Boran cattle under traditional management were lower than that under ranch management, and possibly higher than that of other indigenous breeds in sub-Saharan Africa (Irungu, 2000). Village level assessment of weaning weight was 78kg compared to 107kg reported at Galana ranch after 12 years of breed selection program (Dolan, 1998), and 82kg at 12 months of several traditional cattle production systems in Sub-Saharan Africa (De Leeuw, 1990). Bulls and cows of over three years weighed 269kg and 256kg respectively. Average daily milk off-take was 1.6 litres per cow compared to 1 and 1.1 litres by Bekure (1991) and Roderick (1995) respectively for the Maasai Zebu in Kajiado County. Despite the infrastructural challenges encountered in Tana River county, the Orma pastoralists were found efficient and knowledgeable in cattle keeping and with ability to select for economic traits.

Extra foundation stock for ongoing Orma Boran multiplication in KALRO-Beef Research Institute is therefore being sourced from this community.

Performance of Orma Boran Bulls Disseminated to Kajiado and Busia Counties

The dissemination of Orma Boran breed (OB) was aimed at enhancing adoption of trypanotolerant cattle thereby unlocking the huge potential for income generation and food security in tsetse infested areas. Its inclusion in structured community breeding schemes to introgress the trypanotolerant alleles into local breeds can improve productivity in arid and semi-arid lands (ASALs) that are home to pastoral communities.

Kajiado County located in South West Kenya has three main agro-ecological zones (AEZ; IV-8%, V-55%, VI-37). The rainfall pattern is bimodal (mean 440mm, range 400-600), temperatures range between 18.4^{0}C and 34^{0}C and relative humidity vary from 35-56% (Agnew, 1996). Thus, livestock keeping is the only sustainable agriculture in over 90% of this region whose vegetation provides suitable habitat for tsetse vector. Communal grazing that allows free movement of animals is common and wildlife is prevalent. This coupled with poor infrastructure constrains effective delivery of animal health services (Warinda, 2001). Pastoralists in Kajiado County have a long history of upgrading their Maasai zebu with the Sahiwal and Boran breeds from 1985 through Maasai Rural Training Centre (MRTC) to breed for large body size and higher milk yiel (Mosi & Mbuuza, 1996). The resulting Zebu admixture is large sized with high market value, but with poor adaptation evidenced by high disease susceptibility and cost of treatment. Twenty-six (26) OB bulls disseminated to Kajiado County between 1996 and 2003 were subjected to performance testing to compare its adaptive and productive attributes in view of the communities breeding targets (Fig 2). The adoption was demand driven and delivery was based on a cost-sharing and partial ownership agreement between pastoralists and the research institution (KETRI) to facilitate research data collection.

Figure 2. Mature Orma Boran bull in Kajiado County

Figure 3. Orma Boran females in Busia County

Busia County in Western Kenya borders the Republic of Uganda and the common practice is mixed crop-livestock agriculture. Altitude ranges from 1,300m- 1500m above sea level, annual rainfall ranges from 1,270 - 2000mm and is bimodal, temperature ranges between 26°C-30°C. Nine (9) Orma Boran bulls were disseminated to mixed crop-livestock farmers in November 2001 and subjected to same management as the local Teso zebu cattle (Gachohi et al., 2009). The local community was sensitized on the presence and availability of the OB bulls for crossbreeding with the local Teso zebu cattle by the KETRI research team. The offspring were reared in Chakol, Amagoro and Amukura villages in Teso sub-county, Kenya.

i) Orma Boran Bulls Disseminated to Kajiado and Busia Counties and Duration of Follow-Up

Table 1 shows the number of OB bulls that were monitored for health, adaptation and production parameters in their new environment. Duration of follow-up differed between farmers and Counties. Consistent data was thus only available for fourteen bulls in Kajiado county that were followed up for 21 months

Table 1. Number of OB bulls disseminated between 1996 and 2003 and their offspring's

County	Bulls bought	Year	Mode of disposal for those absent	Approximate O/Z offspring born	Approximate O/Z offspring monitored
Kajiado	12	1996	Death (7), slaughter (3), sale (2)	300	100
Kajiado	5	2002	Death (ECF, cloth ingestion), sale	150	100
Kajiado	9	2003	Death (ECF and starvation, unknown), sale	200	190
Busia	9	2001	Death (ECF-4, bloody diarrhea-1), sale (3)	70	62

Source: (Gachohi et al., 2009; Maichomo, 2008; Maichomo et al., 2001, 2005)

(range 14-30) from November 2002 to December 2004, which was 3 months less the intended observation period (Maichomo, 2008). This difference in retention period was significant (p=0.023) between bulls observed in low (27 months) and high (18.5 months) tsetse density areas. Only 4/14 (29%) bulls completed the anticipated 2-year retention period. Nevertheless, the OB bulls gave sufficient progeny for follow-up in both Counties where sixty-three percent (63%; 452/720) of offsprings were monitored to assess their productivity attributes.

Livestock commercialization in Kajiado County is vibrant with pastoralists improving their livestock for more meat and milk production, thus contributing to better livelihoods. This trend is responsible for the continous breed improvement with the sahiwal, improved Kenyan boran and Orma boran breeds, and is in line with the economic pillar enshrined in the Kenya Government vision 2030 and the draft Livestock bill 2020. Adaptation challenges observed during adoption of the OB bulls were within the norm in ASALs. Results showed that OB withstood drought and disease challenges better than the Sahiwal breed (Maichomo, 2008). The OB bulls were easily manageable than the Sahiwal which was majorly affected by recurrent severe droughts. The advantage attached to the Sahiwal is its superior body weight and higher milk production achieved with good plane of nutrition, as well as its "attractive brown colour" which is preferred by the Maasai. However, considering the recurrent dry spells experienced in ASALs, this spirited opinion may not solve the immediate needs of the pastoralists and the Orma Boran is recommended to bridge this gap. Indeed, one of the farmers who retained his bull to the end affirmed its hardiness as he had larger herd size than his counterparts during extremely dry spell.

(ii) Challenges Observed in Adopting the Orma Boran Breed

It was a challenge loosing many bulls to follow-up for various reasons as shown in Table 1. Some were lost to death, salvage sale from drought and income generation, culling due to disease (lameness, blindness) while two had given sufficient progeny (Table 2). Two bulls sold for breeding in Tanzania earned income comparable to that of Sahiwal and improved Boran. The high culling rate was mainly attributed to drought related starvation that nearly decimated population of immunocompromised livestock that succumbed to infectious diseases in the ASALs (Maichomo et al., 2001). Challenges of technology adoption were observed as some farmers withdrew after losing interest in the project. Given a choice, the pastoralists preferred to cull the OB instead of Sahiwal or improved Kenyan boran bull, an indicator of the value attached to the latter breeds. The value of healthy breeding bulls was demonstrated by the culling of those with deformities (overgrown hooves and blindness), that limited their reproduction despite possessing desirable traits.

Responses by Maasai pastoralists on performance and adaptation of the OB bulls were generally positive regarding trypanotolerance but were dissatisfied with their size and colour (Maichomo et al., 2001). Sixty-seven percent (67%) of the pastoralists interviewed were willing to purchase the bulls for breeding but 33% were non-committal, citing their inability to withstand harsh environmental conditions or preserve endemic stability. Issues of economic costs, apathy and lack of education partly contributed to the ambivalence felt by the Maasai pastoralists regarding the trypanotolerant OB bulls (Maichomo et al., 2001).

Table 2. Factors associated with PCV in the Orma Boran crosses in Olkiramatian and Shompole group ranches, Kajiado District, January 2004 to July 2005 (Maichomo, 2008)

Variable	Level	Estimate (b)	SE(b)	t value	P-value
Animal level factors					
Sex (1=M;0=F)		1.1289	.04923	2.29	.0220
Trypanosomosis (Yes=1;no=0)		4.2622	.7174	5.94	<.0001
Age (absolute values)		-.5047	.07219	-6.99	<.0001
Body weight (absolute values)		.03377	.01012	3.34	.0009
Climatic factors					
Season (short rains= 1;others=0)					0.0006
	Long rains	-1.1796	0.5019	-2.35	0.0189
	Dry 1	-0.2568	0.5281	-0.49	0.6269
	Dry 2	1.0143	0.0807	1.26	0.2091
Geographic factors					
Location (shompole= 1;others=0)					<.0001
	Daraja	5.4641	2.2398	2.44	0.0149
	Ledama	4.1214	1.8337	2.25	0.0248
	Lengobey	3.0266	1.1914	2.54	0.0112
	Mugunani	3.1506	1.0398	3.03	0.0025
	Ngejomony	3.7980	1.3248	2.87	0.0042
	Ntigiya	3.0920	6.6421	1.17	0.2421
	Olkiramatian	1.5269	2.1349	0.72	0.4746
	Oloibortoto	4.1216	2.0825	1.98	0.048
	Oloisinyai	5.7143	2.0876	2.74	0.0063
	Pakase	2.8982	3.0842	0.94	0.3476

iii) Performance Parameters for the Orma Boran Bulls in Kajiado County

(a): Packed Cell Volume (PCV) and Body Weight

Amongst the OB bulls introduced to Magadi region of Kajiado County, the mean PCV was 27% (19-31%). The highest PCV of 31% was achieved during and after the short rains when pastures were abundant and plane of nutrition high. Trypanosomosis infection significantly lowered PCV to an average of 25.2% compared to negative ones at27.8% (p-0.0016) (Fig 4). However, infection with tick-borne diseases (TBDs) had no significant effect on PCV (p=0.58). Bulls whose PCV dropped below 18% were treated with trypanocides even when trypanosome parasites were not detected, considering the low sensitivity of buffy coat technique used under field conditions (Maichomo, 2008; Maichomo et al., 1999). The bulls maintained consistent body weights after introduction to Kajiado County and even gained weight when pastures were good. However, significant weight loss was also observed during extreme dry spells.

(b). Prevalence of Trypanosomosis and Tick-Borne Diseases (TBDs)

Parasitic diseases limit livestock production in most of tropical Africa and contribute to considerable economic loss (Muraguri et al., 2005). Tick-borne diseases (TBD), trypanosomosis and helminthosis are endemic and often occur concurrently in the low-input livestock production system (Rubaire-Akiiki et al., 2004). They are economically most important in the ASAL's due to nutritional stress which compromises immune status of the animals (Norval et al., 1992). It was therefore important to know how the OB bulls performed in their new environment where all these infections are common.

Prevalence of trypanosomosis infection was 20% with a parasitaemia score of 3.6 (Fig. 4). Trypanosomes were never detected in 2 bulls during the entire follow-up period but 1 bull was treated 3 times with diminazene aceturate at 7mg/kg bwt when its' PCV fell below 18% which was considered life threatening (Maichomo, 2008). There was a significant difference in trypanosomosis prevalence between group ranches (χ^2=3.9, p=0.0482) with 20.6% and 9.7% of bulls found positive in Olkiramatian (wetter) and Shompole (drier) group ranches respectively. *T. congolense, T. vivax and mixed T.c/T.v* parasites were responsible for 63%, 37% and 6% of infections in the OB bulls. Tick-borne infections including Anaplasmosis, ECF and babesiosis were also prevalent but their effect on PCV was insignificant. The mean prevalence of ECF, anaplasmosis and babesiosis infections were 9%, 47% and 7%, respectively (Maichomo, 2008). High prevalence of TBDs particularly as mixed infections corresponded with low PCV (Fig 4).

Figure 4. Mean PCV, prevalence of trypanosomosis, anaplasmosis, ECF, and babesiosis in OB bulls at Olkiramatian and Shompole group ranches of Kajiado County (November 2002 - December 2004) (Maichomo, 2008).

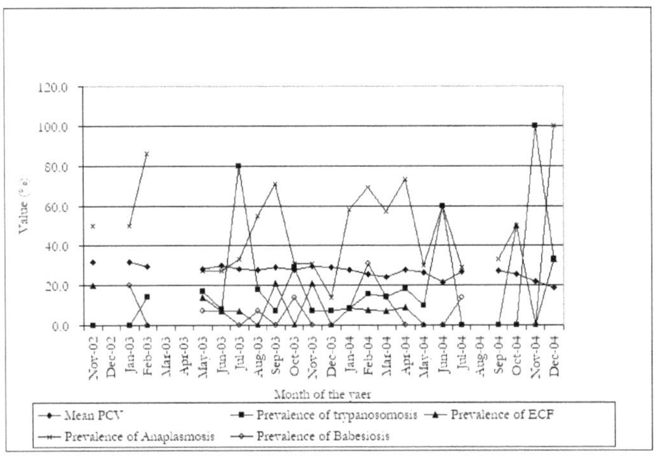

As observed earlier, trypanosomosis infection was the most important parasitaemic infection observed in the OB bulls due to its negative effect on PCV. Results indicated that trypanosomosis positive bulls lost weight especially when plane of nutrition was poor necessitating treatment but self-cure was noted when pasture was plenty (Maichomo et al., 1999). However, the two bulls that were never detected with trypanosomes for 30 months may be a true case of trypanotolerance. This compares well with findings by Njogu et al. (1985) in Galana Ranch who reported that the OB was superior in its response to tsetse and trypanosomosis when compared to other Boran cattle, and when left untreated, the OB showed better control of anemia and decreased mortality (Mwangi et al., 1998; Njogu et al., 1985). This superiority of the OB was more apparent under *T. vivax* challenge, a phenomenon also observed in N'dama cattle (Murray et al., 1981). However, like the N'Dama, the Orma Boran in Kajiado county was also found to be overwhelmed by trypanosomosis under intense tsetse challenge necessitating treatment to keep it productive.

Anaplasmosis infection though highly prevalent had no effect on PCV, a clear demonstration of endemic stability. Concurrent infection with TBDs necessitates proper use of acaricides with the trypanotolerant Orma boran cattle and their crosses. The economic benefit of Orma boran to the pastoralists is reduced cost of trypanosomosis control when compared to the predominant Sahiwal/zebu crosses (Maichomo et al., 2009).

Propagation and Performance of Orma Boran Crosses in Kajiado County

Crosses between the Orma Boran and Maasai Zebu shown in Table 1 and Fig 5 were monitored to address the following issues: (i) comparison of disease resistance and production performances of the Orma cattle with the small East African zebu under the same environment, (ii) resistance to other disease parasites and, (iii) adaptability to different environments. Data was collected on the incidence of trypanosomosis, TBD's, helminthosis, and their effect on growth rate in 400 Orma Boran crosses. Thirteen herds purposively selected were followed up for 1½ years between January 2004 and July 2005 in a cohort study design. The Maasai Zebu (Fig 6) are basically admixtures with imprints of Sahiwal and Improved Kenyan Boran in an attempt to improve milk production for household consumption (Mosi & Mbuuza, 1996). However, no breeding program for the same is in place and the pastoralists use indiscriminate crossbreeding (Maichomo et al., 2008).

Figure 5. A proud pastoralist with OB cross calve

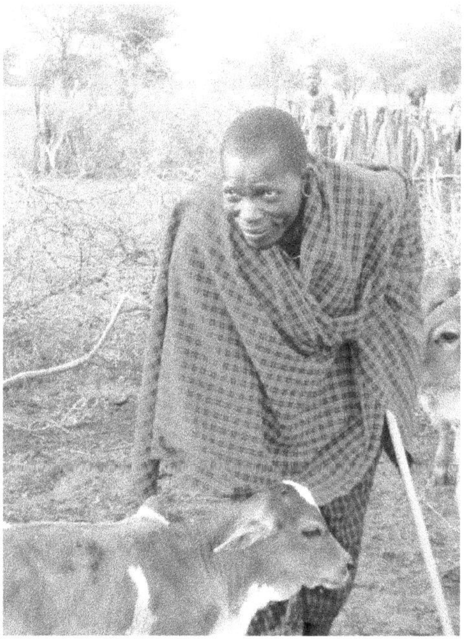

i). PCV Trends in the Orma Boran Crosses

PCV trend was relatively constant as described by Maichomo (2008). Variation was however observed with age where calves had significantly higher PCV (31.7%; P=0.000) compared to weaners (29%) and adults (29.6%). PCV was also significantly high post the rainyseason when pastures were plenty (32.9%; P=0.000) compared to other seasons. While there were no obvious differences in PCV between animals infected with tick-borne diseases (TBDs (p>0.05), the differences were significant for those infected with trypanosomosis and helminthosis. Sex, trypanosomosis infection, body weight and presence of tsetse vector were significantly associated with PCV (Table 2). Trypanosomosis infected calves and with PCV <18% were treated with Diminazene aceturate (Veriben®, CEVA at recommended dose rate, while those with PCV > 20% were left untreated.

ii). Incidence and Risk Factors of Trypanosomosis, Tick-Borne Diseases and Helminthosis in the Orma Boran Crosses

Trypanosomosis incidence was higher in adults (9.9%, p<0.001) compared to calves (2.3%) and weaners (5.2%), and an average parasitaemia score of 3.9. Incidence of anaplasmosis, East Coast Fever (ECF), babesiosis and helminthosis was 57.9%, 22.1%, 10.9% and 27.9%, respectively (Figs 5 & 6). Anaplasmosis infection was observed to increase with age which may be attributed to husbandry practices and development of endemic stability. Given the endemicity of trypanosomosis, TBDs and the importance of concurrent infections on health and growth of Orma Boran crosses, their effective control is recommended as they can significantly constraint livestock productivity in Kajiado County.

Figure 6. Maasai cattle herd in Kajiado County

Significant associations were observed between trypanosomosis infection, group ranch and age where it was higher in older calves (Table 3). Trypanosomosis was also significantly associated with PCV (inversely proportional) and concurrent infections with babesiosis and anaplasmosis (Maichomo,

Figure 7. PCV and incidence of trypanosomosis and helminthosis in Orma Boran crosses in Olkiramatian and Shompole group ranches of Kajiado County, January 2004 to July 2005 (Maichomo, 2008)

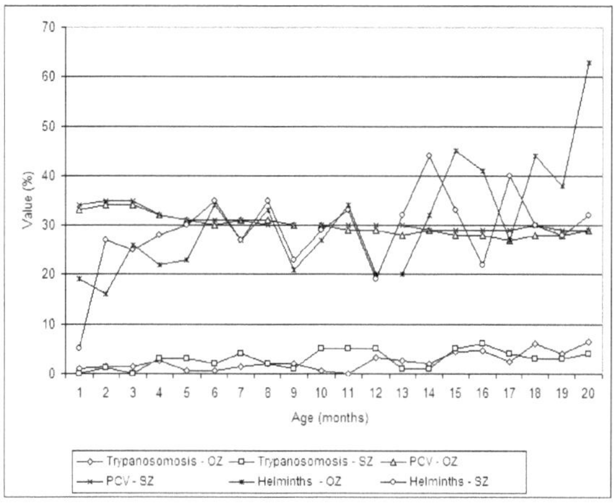

Figure 8. Incidence of anaplasmosis, ECF, babesiosis and trypanosomosis infections in Orma Boran crosses in Olkiramatian and Shompole group ranches of Kajiado County, January 2004 to July 2005 (Maichomo, 2008)

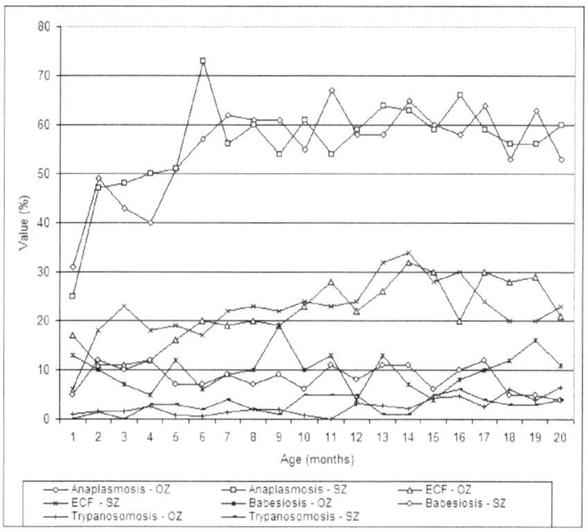

2008). Variations observed in trypanosomosis incidence between seasons and tsetse vector habitats were insignificant (p>0.05). Varying risk factors significantly associated with TBDs and helminthosis were age, sex, concurrent infections and season (Table 3). Interaction between cattle and wildlife in the dry season grazing reserves used for wildlife conservation directly impacts on the epidemiology of TBDs and trypanosomosis to which wildlife are reservoirs. Apparently, livestock wildlife interaction is excellent in Olkiramatian and Shompole group ranches where this work was done. Presence of high helminthosis incidence in high tsetse density areas is in agreement with observation by Dwinger et al (1994).

Table 3. Factors associated with trypanosomosis, TBDs and helminthosis from the GLMM in the Orma Boran crosses in Kajiado County (January 2004 – July 2005) (Maichomo, 2008)

Disease	Variable	Parameter estimate (b)	SE (b)	Wald statistic (χ^2)	P-Value
Anaplasmosis	Animal level factors				
	Age (absolute values)	0.0604	0.011	30.28	<.001
	Babesiosis (Yes=1;no=0)	1.4545	0.2190	44.12	<.001
	Sex (male=1;female=0)	0.3114	0.1198	6.76	0.010
	Season (Dry 1=0;others=1)			54.67	<.001
	§Dry 2	-1.4349			
	§Long rains	-0.7622			
	§Short rains	-1.1728			
Trypanosomosis	PCV (absolute values)	-0.1619	0.0174	86.81	<.001
	Babesiosis (Yes=1;no=0)	1.2006	0.2289	27.51	<.001
	Anaplasmosis (Yes=1;no=0)	-0.6969	0.2025	11.84	<.001
	Age (absolute values)	0.0604	0.0181	11.17	<.001
	Geographic factors				
	Location (Daraja=0;others=1)			22.64	<.001
	§Ledama	-0.148			
	§Lengobey	0.000			
	§Mugunani	-0.466			
	§Ngejomony	9.755			
	§Ntigiya	-6.079			
	§Olk iramatian	-3.099			
	§Oloibortoto	0.516			
	§Oloisinyai	0.251			
	§Pakase	-9.165			
	§Shompole	-3.237			
ECF	Animal level factors				
	Helminthosis (Yes=1;no=0)	-0.3121	0.1369	5.19	0.023
	Babesiosis (Yes=1;no=0)	0.3427	0.1680	4.16	0.041
	Climatic factors				
	Season (dry 1=1;others=0)			9.96	0.019
	§dry 2	0.6244			
	§Long rains	-0.0135			
	§Short rains	-0.1574			
Helminthosis	Age (absolute values)	0.0324	0.0105	15	<.001
	ECF	-0.3168	0.1367	5.85	0.016
	PCV (absolute values)	-0.0178	0.0102	5.95	0.015
	Climatic factors				
	Season (dry 1=1;others=0)			26.9	<.001
	§dry 2	-0.2938			
	§Long rains	0.5809			
	§Short rains	0.1537			
	Group ranch	-0.2757	0.9532	7.47	0.072

continued on following page

Table 3. Continued

Disease	Variable	Parameter estimate (b)	SE (b)	Wald statistic (χ^2)	P-Value
Babesiosis	Trypanosomosis (Yes=1;no=0)	1.005	0.3168	10.06	0.002
	ECF (Yes=1;no=0)	0.3235	0.1602	4.08	0.044
	Anaplasmosis (Yes=1;no=0)	1.4845	0.2083	50.8	<.001
	Season (dry 1=1;others=0)			20.02	<.001
	§dry 2	0.3492			
	§Long rains	-0.6214			
	§Short rains	-0.0978			

Figure 9. Body weight changes of Orma Boran crosses in Olkiramatian and Shompole group ranches, Kajiado County, January 2004 to July 2005 (Maichomo, 2008)

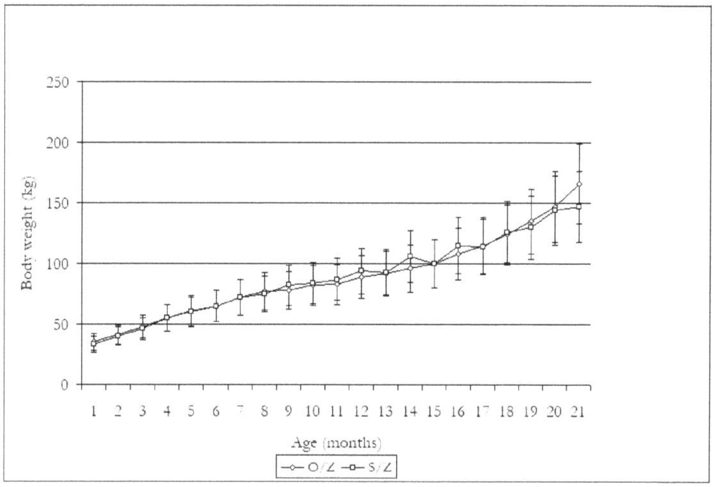

Figure 10. Normalized Difference Vegetation Index (NDVI) for Kajiado County, January 2004 to July 2005 (Maichomo, 2008)

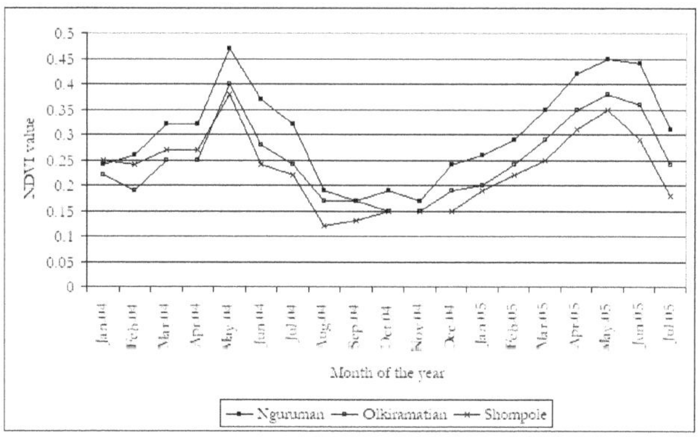

Figure 11. ADWG for 225 Orma Boran crosses in Olkiramatian and Shompole group ranches, Kajiado County, January 2004 to July 2005 (Maichomo, 2008)

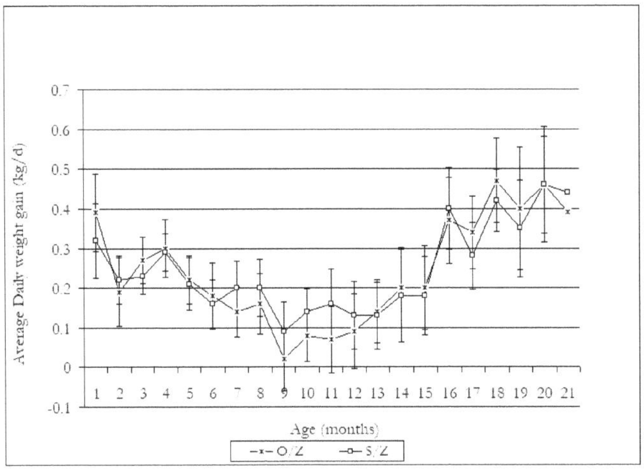

Table 4. Factors associated with ADWG of Orma Boran crosses in Kajiado County, January 2004 to July 2005 (Maichomo, 2008)

Variable	Level	Estimate (b)	SE(b)	t value	P-value
Animal level factors					
Helminthosis (Yes=1;no=0)		0.04101	0.0156	2.63	0.0087
Anaplasmosis (Yes=1;no=0)		-0.04034	0.01533	-2.63	0.0086
Body weight (absolute values)		0.00181	0.00040	4.53	<.0001
Climatic factors					
Season (short rains= 1;others=0)	short rains				<.0001.
	Long rains	0.2426	0.02489	9.75	<.0001
	Dry 1	0.1362	0.02625	5.19	<.0001
	Dry 2	0.2547	0.04040	6.3	<.0001
Geographic factors					
Location (shompole= 1;others=0)	Shompole				0.0464
	Daraja	0.00052	0.09007	-0.01	0.9954
	Ledama	0.03903	0.07957	-0.49	0.6238
	Lengobey	0.08696	0.04748	1.83	0.0672
	Mugunani	-0.0953	0.03949	-2.41	0.016
	Ngejomony	0.0326	0.04313	0.76	0.45
	Ntigiya	-0.1483	0.1159	-1.28	0.2009
	Olkiramatian	-0.039	0.0881	-0.44	0.6581
	Oloibortoto	-0.03563	0.08669	-0.41	0.6811
	Oloisinyai	0.01385	0.09043	0.15	0.8783
	Pakase	-0.03719	0.1144	-0.33	0.7452

iii). Growth Pattern of the Orma Boran (OB) Crosses in Olkiramatian and Shompole Group Ranches, Kajiado County

Fig 11 shows the growth rate of 225 OB crosses from birth to 20 months. Body weight at 2 weeks was 35kg while average daily weight gain (ADWG) was 0.1702 kg/d and comparable to that of the Maasai Zebu (Fig 9 & 11). ADWG was significantly higher during the long rain season that corresponded to observed Normalized Difference Vegetation Index (NDVI) of more than 0.25 for Kajiado County. Further, the NDVI value was higher in Olkiramatian (0.45) compared to Shompole group ranch (0.37) (Fig 10). Results further showed that the weaning weight of OB crosses at 8 months (72kg) was higher than that of the Maasai Zebu admixture (64kg), while at 18 months, the OB body weight exceeded that of the Maasai Zebu by 40kg (Maichomo et al., 2005). Better results were even achieved during high rainfall and good plane of nutrition when high tsetse density and trypanosomosis prevalence was significantly different between the OB crosses and Maasai Zebu. These crosses had higher trypanosomosis prevalence (61% Vs 39%) and PCV, a clear indication of trypanotolerance. It is worth noting that the ADWG of the OB exceeded that of the Maasai Zebu when pasture situation was good. The apparent superior performance of the OB crosses with its' perceived psychological satisfaction is an attribute that could be exploited in the adoption of trypanotolerance genotype, as a sustainable trypanosomosis control strategy (Maichomo et al., 2001, 2005). Other factors that were significantly associated with ADWG are helminthosis and anaplasmosis, body weight, season and herd grazing location (Table 4).

Performance of the Orma Boran crosses during dry season was however different where significant variation in ADWG associated to geographic, climatic, management and animal level factors was observed. Breed difference was not significant probably due to poor nutrition and low trypanosomosis burden. This notwithstanding, the study animals showed a remarkable ability to withstand adverse effects of drought, which was in agreement with observation by Rege et al (2001). Observed mortality rate of 6% compared well with that reported by ILCA (Bekure, 1991). Regarding adaptation of the breed, pastoralists alluded that the Orma Boran crosses and Maasai Zebu admixtures were comparable in terms of disease morbidity, growth rate, mortality rates and economic benefits (Maichomo et al., 2009). The OB were hardy with ability to withstand harsh conditions of pasture shortage, heat stress, and tolerated diseases better than the Maasai Zebu admixtures. As such, a higher proportion of them survived, resulting in high herd turnover (Maichomo et al., 2005). Therefore, selection for superior indigenous genotypes with high productivity may reverse the loss of genetic resource being experienced in the country. Cost of treatment during the study was lower than that normally spent by pastoralists, pointing to unnecessary drug use that lowers profit margins. Many pastoralists in Kajiado county have embraced commercial livestock keeping necessitated by changing lifestyle and high cost of living. Knowledge transfer to pastoralists should emphasize the role of animal health in risk management. Besides the monetary value, the Maasai pastoralists demonstrated that other attributes of a breed such as colour and physical appearance are important in breed adoption.

Propagation and Performance of Orma Boran Crosses in Busia County

From nine Orma Boran (OB) bulls that were disseminated to Busia County, 86 crosses were available for follow-up and were monitored monthly for health and productivity parameters. The observation period varied depending on the length of time each offspring was under surveillance for trypanosomosis and TBDs. The performance of OB crosses in Busia County is described by Gachohi et al. (2009). Ana-

Table 5. Generalized linear and latent mixed models (GLLAM) (accounting for clustering) for all response variables from Orma-Teso zebu offspring surveillance data collected in Teso district, western Kenya, January 2003-June 2005

Variable	Levels	Trypanosomosis		Anaplasmosis		All response variables combined	
		OR[a] (95%CI[b])	p	OR[a] (95%CI[b])	p	OR[a] (95%CI[b])	p
Sex	Male	1.0	0.73	1.0	0.71	1.0	0.56
	Female	1.3 (0.3, 5.1)		0.9 (0.4, 1.7)		0.8 (0.5, 1.5)	
Season	Wet	1.0	0.04	1.0	0.02	1.0	0.02
	Dry	5.4 (1.1, 25.9)		1.9 (1.1, 3.6)		1.9 (1.1, 3.2)	
Age		1.1 (1.0, 1.3)	0.04	1.0 (0.9, 1.1)	0.62	1.0 (0.9, 1.1)	0.33
BWT[c]		0.9 (0.9, 1.0)	0.15	0.9 (0.9, 1.0)	0.79	0.9 (0.9, 1.0)	0.53
Variance	Animal	0.33 (s.e[f]. 1.21)		0.44 (s.e[f] 0.4)		0.21 (s.e[f] 0.29)	
	Herd	0.00		0.00		0.00	
	Village	0.00		0.12 (s.e[f] 0.17)		0.07 (s.e[f] 0.14)	
lnL[d]		-45.99		-170.34		-188.93	
Obs.[e]		399		444		444	

Source: Gachohi et al., 2009 - *a: Adjusted odds ratio; b: Confidence interval; c: body weight; d: log likelihood value; e: number of observations; f: standard error.*

plasmosis was more prevalent (15.3%) than trypanosomosis (2.3%) and piroplasmosis (1%). Infection with *Trypanosoma vivax (T.v)* species was more common than with *Trypanosoma congolense* (T.c) in a ratio of 2.7:1 (Tv:Tc). Mean PCV of the crosses was 30.1% and was dependent on season, locality, age and infection status. Like in Kajiado County, trypanosomosis prevalence was found to be inversely proportional to PCV. Mean PCV for parasitaemic calves was 29.3% compared to 33% for those aparasitaemic. Gachohi et al. (2009) observed that the difference in PCV by sex and village were significant ($p<0.001$) but that by season was not significant ($P=0.71$). Trypanosomosis and piroplasmosis infections did not appear to significantly affect PCV ($p>0,001$) compared to anaplasmosis ($p<0.001$). Season was the only variable that was significantly associated with the occurrence of trypanosomosis ($p=0.03$) and anaplasmosis ($p=0.02$). Burden of trypanosomosis and anaplasmosis were higher in the dry than the wet season. Age was also found to be significantly associated with trypanosomosis infection, which was higher in older crosses (Table 5).

Prevalence of trypanosomosis in the Orma Boran (OB) crosses in Busia County of 2.3% was relatively lower than that reported for Teso zebu (2 - 10%) in the same area and period but is comparable to that reported in the OB crosses (3%) (Karanja, 2006; Maichomo et al., 2005) in the pastoral production system of Kajiado County. Karanja (2006) observed that the low tsetse density in the area (<1fly/trap/day), calves' confinement in the homesteads, aggressive tsetse vector control by "Farming in Tsetse Controlled Areas, FITCA" coupled with use of trypanocides may have led to the low prevalence of trypanosomosis in the Orma-Teso zebu offspring.

Trypanosomosis infection had no effect on PCV in the OB crosses in Busia County compared to Kajiado County. This observation appears to confirm the weak association between control of parasitaemia and anaemia in trypanotolerant animals or alludes to better control of anaemia by the infected OB crosses. This is further in agreement with the decision to base treatment of trypanotolerant animals on low PCV (<20%) regardless of parasitaemia. Variance in PCV association with trypanosomosis infection between Busia and Kajiado counties perhaps pointed to differences in pasture availability and production systems. Season was also found to be a significant predictor with dry season being associated with higher trypanosomosis prevalence. Busia County has a predominant mixed crop-livestock system and animals are grazed near the homesteads during the wet season but do access the riverine tsetse-infested areas common in this region during dry periods (Gachohi et al., 2009). This contrasts the observation with OB crosses in Kajiado county where trypanosomosis infection was reported to increase during the wet season and peaking into the moderately dry season in OB crosses probably because of unrestricted grazing patterns (Bett et al., 2004; Maichomo et al., 2005). This observation underpins the importance of farming systems in influencing the infection challenge to which livestock are subjected and is central to the epidemiology and planning of trypanosomosis control particularly in regions and production systems where dissemination and breeding of trypanotolerant livestock are earmarked. Increasing age with increased trypanosomosis prevalence was consistent with other studies (Dolan, 1998) and appear to be influenced by livestock management systems (Bett et al., 2004). In both counties, concurrent infections were common but anaplasmosis was the most prevalent TBD that significantly affected PCV. Trypanosomosis self-cure was observed in 72% of cases.

Pre-weaning and post-weaning ADWG of 199g/d and 194g/d respectively were recorded for the OB crosses in Busia County. Gachohi et al. (2009) also observed that crosses detected parasitaemic had on average lower growth rates (122g/d) than those aparasitaemic (208g/d). ADWG dropped during the wet season, was highest in young calves and lowest in yearlings. This could perhaps be due to post weaning stress, also observed in Kajiado County (Maichomo et al., 2005).

Long-Term Orma Boran Breeding Program

The future utilization of the genetic resource represented by trypanotolerant cattle to harvest the biomass of much of tsetse-infested Africa depends on their conservation, promotion and improvement (Bradley, 1995; Kim et al., 2017; Mwai et al., 2015; Orenge et al., 2012). Just like the many trypanotolerant breeds of West Africa face extinction due to introgression by other large breeds, pastoralists in Kenya also outcross their smaller fully adapted breeds with larger ones that are perceived to be more productive (Agyemang et al., 1997; Mwai et al., 2015; Rege et al., 2001). The danger of indiscriminate cross breeding is that unique genetic resources could be depleted with time (Mwacharo & Drucker, 2005), considering that world agriculture is currently observed to rely on few animal species and fewer breed lines (Delgado et al., 2001). Over time, there has been extensive but indiscriminate breed improvement in Kajiado County with admixtures of the Maasai zebu, Sahiwal, improved Kenyan Boran and Orma Boran (Maichomo, 2008; Maichomo et al., 2008). This observation is in agreement with Rege et al. (2001) and is aimed at boosting meat and milk production in the pastoral production system. Without proper record keeping, parent-offspring relationship identification is uncertain and more so among pastoralists practicing communal grazing system. For such, relationship identification process is only subject to memory or guess work as observed in Kajiado County where commingling of bulls is common.

To leverage on gains of the trypanotolerant trait, KALRO-Beef Research Institute is currently hosting an OB nuclear breeding program for the purpose of improving, conserving and supplying tsetse infested counties in Kenya with improved OB bulls and heifers in order to improve beef production. With remnants of the female breeding stock from KARI-range research station, artificial insemination with OB

Figure 12. Farmers workshop in Busia County

semen from KAGRC is ongoing to multiply the herd on-station. Additional stock to boost the nucleus herd has been sourced from Tana River county. The purpose is to establish a structured long term breeding program and increase population of improved OB cattle to effectively reclaim the six tsetse belts in Kenya, and hopefully in the East Africa region. The low drug requirement of OB increases resilience for the cattle keepers adopting and utilizing this technology while mitigating effects of antimicrobial resistance to deliver safe products (milk, meat) for human consumption. Breeding and conservation of OB cattle for beef production characteristics while maintaining their levels of trypanotolerance is a step towards sustainable domestic red meat self-sufficiency in Kenya. The program envisages community engagement for mass multiplication and breed conservation.

FUTURE RESEARCH DIRECTION

To enhance utilization of the Orma Boran breed, more information is required:

1. Assessment of genetic and environmental factors affecting trypanotolerance using quantitative approach
2. Understanding the nature of exhibited trypanotolerance, and particularly demonstrating the breeds ability to tolerate East African strain of haemorrhagic *T. vivax* infection.
3. More work is required to refine genetic purity of trypanotolerant breeds of cattle and small ruminants in order to harness their potential in trypanosomosis endemic areas of Africa.

CONCLUSION

Qualitative responses on perception of the OB breed by farmers in Kajiado and Busia Counties was generally positive. In the drier Kajiado county, only the OB crosses distinguished one pastoralist from others during drought due to their high survival rate. In Busia County, OB crosses were reportedly born bigger and vigorous, required less acaricide wash, produced more milk and fetched higher market prices than their Teso Zebu contemporaries.

Indiscriminate crossbreeding can cause loss of trypanotolerance traits in Orma Boran. This would be tragic for tsetse infested sub-Saharan Africa where trypanosomosis is a major constraint to livestock production. Thus the OB breeding program will promote within breed selection for preferred productivity traits by communities while maintaining integrity of the adaptive traits. This is further supported by deeming hopes of getting a vaccine and challenges facing eradication of trypanosomosis. Improved OB breed offers hope to improve indigenous cattle types in tsetse infested areas of East Africa and is also a good bet for use in tsetse belts where suppression is achieved considering the risk posed by residual vector population.

REFERENCES

Abenga, J. N., & Vuza, D. (2005). About factors that determine trypanotolerance and prospects for increasing resistance against trypanosomosis. *African Journal of Biotechnology*, *4*(13), 13.

Agnew, A. (1996). *The graze and browse resource at Olkiramatian Group Ranch. Fourth report to KETRI/NRI on landscape/vegetation monitoring at Olkiramatian group ranch. Report to the ODA on a Visit to the Kenya Trypanosomosis Research Institute (Nguruman Station).* Academic Press.

Agyemang, D. R. H., Little, D. A., & Rowlands. (1997). *Village N'Dama cattle production in West Africa. Six years of research in the Gambia*. ILRI. https://agris.fao.org/agris-search/search.do?recordID=QM1998000001

Bekure, S. (1991). *Maasai herding: An analysis of the livestock production system of Maasai pastoralists in eastern Kajiado District, Kenya* (Vol. 4). ILRI (aka ILCA and ILRAD).

Berthier, D., Peylhard, M., Dayo, G.-K., Flori, L., Sylla, S., Bolly, S., Sakande, H., Chantal, I., & Thevenon, S. (2015). A Comparison of Phenotypic Traits Related to Trypanotolerance in Five West African Cattle Breeds Highlights the Value of Shorthorn Taurine Breeds. *PLoS One*, *10*(5), e0126498. doi:10.1371/journal.pone.0126498 PMID:25954819

Bett, B., Orenge, C., Irungu, P., & Munga, L. K. (2004). Epidemiological factors that influence time-to-treatment of trypanosomosis in Orma Boran cattle raised at Galana Ranch, Kenya. *Veterinary Parasitology*, *120*(1–2), 43–53. doi:10.1016/j.vetpar.2003.12.011 PMID:15019142

Bradley, D. (1995). *Genetic characterisation of cattle in West and Central Africa*. Department of Genetics, Trinity College.

Chiejina, S. N., Behnke, J. M., Nnadi, P. A., Ngongeh, L. A., & Musongong, G. A. (2009). The responses of two ecotypes of Nigerian West African Dwarf goat to experimental infections with Trypanosoma brucei and Haemonchus contortus. *Small Ruminant Research*, *85*(2–3), 91–98. doi:10.1016/j.smallrumres.2009.07.004

Cunningham, M. P. (1966). Immunity in bovine trypanosomiasis. *East African Medical Journal*, *43*(9), 394–397. PMID:5965864

d'Ieteren, G. D., Authié, E., Wissocq, N., & Murray, M. (1998). Trypanotolerance, an option for sustainable livestock production in areas at risk from trypanosomosis. *Revue Scientifique et Technique (International Office of Epizootics)*, *17*(1), 154–175. doi:10.20506/rst.17.1.1088 PMID:9638808

DAD-IS. (2020). *Domestic Animal Diversity Information System (DAD-IS) | Food and Agriculture Organization of the United Nations*. Food and Agriculture Organization of the UN. http://www.fao.org/dad-is/en/

DAGRIS. (2020). *Domestic Animal Genetic Resources Information System (DAGRIS)*. International Livestock Research Institute. http://www.dagris.ilri.cgiar.org

De Leeuw, P. N. (1990). *Interactive effects of environment, management and mortality on cattle productivity in livestock systems in sub-Saharan Africa*. Academic Press.

Delgado, C., Rosegrant, M., Steinfeld, H., Ehui, S., & Courbois, C. (2001). Livestock to 2020: The Next Food Revolution. *Outlook on Agriculture*, *30*(1), 27–29. doi:10.5367/000000001101293427

Desowitz, R. S. (1959). Studies on Immunity and Host-Parasite Relationships: I.—The Immunological Response of Resistant and Susceptible Breeds of Cattle to Trypanosomal Challenge. *Annals of Tropical Medicine and Parasitology*, *53*(3), 293–313. doi:10.1080/00034983.1959.11685927 PMID:13849587

Dolan, R. (1998). *The Orma Boran: A trypanotolerant east African breed: Fifteen years of research on Galana Ranch in Kenya*. Kenya Trypanosomiasis Research Institute.

Dolan, R. B. (1987). Genetics and trypanotolerance. *Parasitology Today (Personal Ed.)*, *3*(5), 137–143. doi:10.1016/0169-4758(87)90197-9 PMID:15462937

Dwinger, R. H., Agyemang, K., Snow, W. F., Rawlings, P., Leperre, P., & Bah, M. L. (1994). Productivity of trypanotolerant cattle kept under traditional management conditions in the Gambia. *The Veterinary Quarterly*, *16*(2), 81–86. doi:10.1080/01652176.1994.9694424 PMID:7985361

Food and Agriculture Organization of the United Nations. (Ed.). (2011). *FAO in the 21st Century: Ensuring Food Security in a Changing World*. FAO.

Gachohi, J. M., Bett, B., & Murill, G. A. (2009). Factors influencing prevalence of trypanosomosis in orma boran (trypanotolerant) and teso zebu (trypanosusceptible) crosses in Teso district, western Kenya. *Bulletin of Animal Health and Production in Africa*, *57*(4), 4.

Geerts, S., Osaer, S., Goossens, B., & Faye, D. (2009). Trypanotolerance in small ruminants of sub-Saharan Africa. *Trends in Parasitology*, *25*(3), 132–138. doi:10.1016/j.pt.2008.12.004 PMID:19200783

Hanotte, O. (2002). African Pastoralism: Genetic Imprints of Origins and Migrations. *Science*, *296*(5566), 336–339. doi:10.1126cience.1069878 PMID:11951043

Hanotte, O., Ronin, Y., Agaba, M., Nilsson, P., Gelhaus, A., Horstmann, R., Sugimoto, Y., Kemp, S., Gibson, J., Korol, A., Soller, M., & Teale, A. (2003). Mapping of quantitative trait loci controlling trypanotolerance in a cross of tolerant West African N'Dama and susceptible East African Boran cattle. *Proceedings of the National Academy of Sciences of the United States of America*, *100*(13), 7443–7448. doi:10.1073/pnas.1232392100 PMID:12805560

Irungu, P. (2000). Cattle Keeping Practices Of The Orma People A Household Survey In Tana River District. Academic Press.

Itty, P., Rowlands, G. J., Minengu, M., Ngamuna, S., Van Winkel, F., & d'Ieteren, G. D. M. (1995). The economics of recently introduced village cattle production in a tsetse affected area (I): Trypanotolerant N'Dama cattle in Zaire. *Agricultural Systems*, *47*(3), 347–366. doi:10.1016/0308-521X(95)90747-H

Karanja, S. M. (2006). *Epidemiology and importance of trypanosomosis, helminthosis and tick-borne diseases on the performance of cattle in Busia district* [PhD Thesis].

KENTTEC. (2020). Tsetse and Trypanosomiasis Problem in Kenya. *KENTTEC*. https://www.kenttec.go.ke/tsetse-and-trypanosomiasis-problem-in-kenya/

Kim, J., Hanotte, O., Mwai, O. A., Dessie, T., Bashir, S., Diallo, B., Agaba, M., Kim, K., Kwak, W., Sung, S., Seo, M., Jeong, H., Kwon, T., Taye, M., Song, K.-D., Lim, D., Cho, S., Lee, H.-J., Yoon, D., ... Kim, H. (2017). The genome landscape of indigenous African cattle. *Genome Biology*, *18*(1), 34. doi:10.118613059-017-1153-y PMID:28219390

Kosgey, I. S., & Okeyo, A. M. (2007). Genetic improvement of small ruminants in low-input, smallholder production systems: Technical and infrastructural issues. *Small Ruminant Research*, *70*(1), 76–88. doi:10.1016/j.smallrumres.2007.01.007

Leak, S. (1998). *Tsetse Biology and Ecology. Their Role in the Epidemiology and Control of Trypanosomosis*. CABI Publishing in association with ILRI.

Leperre, P., & Claxton, J. R. (1994). Comparative study of trypanosomosis in zebu and N'Dama cattle in the Gambia. *Tropical Animal Health and Production*, *26*(3), 139–145. doi:10.1007/BF02241070 PMID:7809985

Maichomo, M. (2008). *Comparative productivity and performance of Orma/Zebu and Sahiwal/Zebu cross breed cattle under trypanosomosis, tick borne diseases, and anaplasmosis and helminthosis challange in Kajiado district, Kenya* [PhD Thesis]. University of Nairobi.

Maichomo, M., Gitau, G., Gathuma, J., Ndungú, J., Kosura, W., & Hanotte, O. (2008). Extent and implications of incorrect offspring-sire relationships in pastoral production system in Kajiado District, Kenya. *Livestock Research for Rural Development*, *20*, . http://www.lrrd.org/lrrd20/5/maic20080.htm

Maichomo, M., Kosura, W., Gathuma, J., Gitau, G., Ndungú, J., & Nyamwaro, S. (2009). Economic assessment of the performance of trypanotolerant Orma/Zebu cross cattle in a pastoral production system in Kenya. *Journal of South African Veterinary Association*, *80*(3), 157–162.

Maichomo, M., Mwendia, C. M. T., & Ole-Mapenay, I. M. (1999). Preliminary performance of the Orma bull, a trypanotolerant breed at Nguruman. *The Kenya Veterinarian*, *24*, 129–131.

Maichomo, M., Ndung'u, J. M., Ngare, P., & Ole-Mapenay, I. M. (2005). The performance of Orma Boran and Maasai Zebu crossbreeds in a trypanosomosis endemic area of Nguruman, South-western Kenya. *The Onderstepoort Journal of Veterinary Research*, *72*(1), 87–93. doi:10.4102/ojvr.v72i1.225 PMID:15991707

Maichomo, M., Olubai, W., Mwendia, C. M. T., & Mapenay, I. M. (2001). Farmer responses on acceptability of the Orma Boran, a trypanotolerant breed of cattle introduced into Nguruman, kajiado District in Kenya. *OAU/STRC Publication, 120*, 293–295.

Medzhitov, R., Schneider, D. S., & Soares, M. P. (2012). Disease Tolerance as a Defense Strategy. *Science, 335*(6071), 936–941. doi:10.1126cience.1214935 PMID:22363001

Mortelmans, J., & Kageruka, P. (1976). Trypanotolerant cattle breeds in Zaire. *World Animal Review, 19*, 24.

Mosi, R., & Mbuuza, F. (1996). Animal recording, characterization and improvement schemes in East Africa. Proceedings of Livestock Improvement Association, 75th Anniversarry, 10–16.

Muraguri, G. R., McLeod, A., McDermott, J. J., & Taylor, N. (2005). The incidence of calf morbidity and mortality due to vector-borne infections in smallholder dairy farms in Kwale District, Kenya. *Veterinary Parasitology, 130*(3–4), 305–315. doi:10.1016/j.vetpar.2004.11.026 PMID:15885914

Murray, M., Trail, J., & D'Ieteren, G. (1990). Trypanotolerance in cattle and prospects for the control of trypanosomiasis by selective breeding. *Scientific and Technical Review Office International Des Epizooties, 9*(2), 369–386. doi:10.20506/rst.9.2.506 PMID:2132686

Mwacharo, J. M., & Drucker, A. G. (2005). Production objectives and management strategies of livestock keepers in south-east Kenya: Implications for a breeding programme. *Tropical Animal Health and Production, 37*(8), 635–652. doi:10.100711250-005-4253-8 PMID:16619880

Mwai, O., Hanotte, O., Kwon, Y.-J., & Cho, S. (2015). African Indigenous Cattle: Unique Genetic Resources in a Rapidly Changing World. *Asian-Australasian Journal of Animal Sciences, 28*(7), 911–921. doi:10.5713/ajas.15.0002R PMID:26104394

Mwangi, E., Stevenson, P., Gettinby, G., Reid, S., & Murray, M. (1998). Susceptibility to trypanosomosis of three Bos indicus cattle breeds in areas of differing tsetse fly challenge. *Veterinary Parasitology, 79*(1), 1–17. doi:10.1016/S0304-4017(98)00147-2 PMID:9777722

Naessens, J. (2006). ovine trypanotolerance: A natural ability to prevent severe anaemia and haemophagocytic syndrome? *International Journal for Parasitology, 36*(5), 521–528. doi:10.1016/j.ijpara.2006.02.012 PMID:16678182

Naessens, J. (2006). Bovine trypanotolerance: A natural ability to prevent severe anaemia and haemophagocytic syndrome? *International Journal for Parasitology, 36*(5), 521–528. doi:10.1016/j.ijpara.2006.02.012 PMID:16678182

Naessens, J., Teale, A. J., & Sileghem, M. (2002). Identification of mechanisms of natural resistance to African trypanosomiasis in cattle. *Veterinary Immunology and Immunopathology, 87*(3–4), 187–194. doi:10.1016/S0165-2427(02)00070-3 PMID:12072233

Namangala, B. (2012). Contribution of Innate Immune Responses towards Resistance to African Trypanosome Infections: Innate Mechanisms of Resistance to Trypanosomiasis. *Scandinavian Journal of Immunology, 75*(1), 5–15. doi:10.1111/j.1365-3083.2011.02619.x PMID:21916913

Njogu, A. R., Dolan, R. B., Wilson, A. J., & Sayer, P. D. (1985). Trypanotolerance in East African Orma Boran cattle. *The Veterinary Record*, *117*(24), 632–636. doi:10.1136/vr.117.24.632 PMID:4090214

Norval, R. A. I., Perry, B. D., & Young, A. S. (1992). *The epidemiology of theileriosis in Africa. Academic Press.*

Orenge, C., Munga, L., Kimwele, C., Kemp, S., Korol, A., Gibson, J., Hanotte, O., & Soller, M. (2011). Expression of trypanotolerance in N'Dama x Boran crosses under field challenge in relation to N'Dama genome content. *BMC Proceedings*, *5*(Suppl 4), S23. doi:10.1186/1753-6561-5-S4-S23 PMID:21645303

Orenge, C. O., Munga, L., Kimwele, C. N., Kemp, S., Korol, A., Gibson, J. P., Hanotte, O., & Soller, M. (2012). Trypanotolerance in N'Dama x Boran crosses under natural trypanosome challenge: Effect of test-year environment, gender, and breed composition. *BMC Genetics*, *13*(1), 87. doi:10.1186/1471-2156-13-87 PMID:23075408

Radcliffe, R. W., & Khairani, K. O. (2019). Health of the Forest Rhinoceros of Southeast Asia. In *Fowler's Zoo and Wild Animal Medicine Current Therapy* (Vol. 9, pp. 707–715). Elsevier., doi:10.1016/B978-0-323-55228-8.00100-4

Rege, J. E. O., Kahi, A., Okomo-Adhiambo, M., Mwacharo, J. M., & Hanotte, O. H. (2001). *Zebu cattle of Kenya: Uses, performance, farmer preferences, measures of genetic diversity and options for improved use*. International Livestock Research Institute. https://cgspace.cgiar.org/handle/10568/4196

Rege, J. E. O., & Tawah, C. L. (1999). The state of African cattle genetic resources II. Geographical distribution, characteristics and uses of present-day breeds and strains. *Animal Genetic Resources/Resources Génétiques Animales/Recursos Genéticos Animales, 26*, 1–25.

Roderick, S. (1995). *Pastoralist cattle productivity in a tsetse infested area of south west Kenya* [PhD Thesis]. University of Reading.

Rubaire-Akiiki, C., Okello-Onen, J., Nasinyama, G. W., Vaarst, M., Kabagambe, E. K., Mwayi, W., Musunga, D., & Wandukwa, W. (2004). The prevalence of serum antibodies to tick-borne infections in Mbale District, Uganda: The effect of agro-ecological zone, grazing management and age of cattle. *Journal of Insect Science*, *4*(1), 1–8. doi:10.1673/031.004.0801 PMID:15861224

Tilki, T., Dagnachew, H. M., Yohanis, M., Biniyam, M., & Shawit, K. (2016). Epidemiology of trypanosomal infections of caprine in Nono district, south west showa zone Ethiopia. *Livestock Research for Rural Development*, *28*(10). http://www.lrrd.org/lrrd28/10/mulu28182.html

Warinda, E. (2001). *A socio-economic survey and economic analysis of land use options in Shompole and Olkiramatian group ranches–Magadi division. Report*. African Conservation Centre.

Wosu, L. O., & Nwanta, J. A. (1989). Incidence of trypanosomes in settled Fulani cattle in the Anambra State, southern Nigeria. *Archives Roumaines de Pathologie Experimentales et de Microbiologie*, *48*(4), 367–372. PMID:2520674

Yaro, M., Munyard, K., Stear, M. J., & Groth, D. M. (2016). Combatting African Animal Trypanosomiasis (AAT) in livestock: The potential role of trypanotolerance. *Veterinary Parasitology*, *225*, 43–52. doi:10.1016/j.vetpar.2016.05.003 PMID:27369574

Chapter 6
Non-Tsetse-Transmitted Animal Trypanosomosis (NTT):
Trypanosomosis Due to *Trypanosoma evansi* (Surra), *T. vivax* (Duttonella), and *T. equiperdum*

William Olaho-Mukani
Makerere University, Uganda

ABSTRACT

The non-tsetse-transmitted animal trypanosomoses are infections caused by three main mammalian trypanosome species, namely Trypanosoma evansi, T. equiprdum, and T. vivax. Their global distribution is much wider than tsetse-transmitted trypanosomoses and includes Africa and Latin America, Asia, and Euro-Asia. These trypanosomoses affect a very wide range of domestic animals and game and negatively impact on the development of the animal industry, thus directly affecting national economies and people's livelihoods.

INTRODUCTION

Animal African trypanosomoses are diseases of animals of similar aetiology and epidemiology. The causative agents of the diseases are protozoan parasites of the genus *Trypanosoma* that live and multiply extracellularly in blood and tissue fluids of their mammalian hosts and are transmitted by the bite of infected tsetse flies (*Glossina* sp.); others by biting flies or by coitus.

Animal trypanosomoses comprise: nagana (caused by *T. congolense, T. vivax and T. brucei spp* and *T. simiae*), surra caused by (*T. evansi*) and dourine caused by (*T. equiperdum*). In wild animals, these parasites may cause relatively mild infections while in domestic animals they cause a severe, often fatal disease. All domestic animals can be affected in one way or another.

DOI: 10.4018/978-1-7998-6433-2.ch006

Background

The non-tsetse transmitted animal trypanosomoses are caused by three main trypanosome species, namely: *Trypanosoma evansi, T. equiprdum* and *T. vivax.* Their global distribution is wider than tsetse-transmitted trypanosomoses and includes Africa and Latin America, Asia and Euro-Asia. *Trypanosoma evansi* and *T. vivax* are transmitted mechanically by biting flies. However, *T. vivax* is also transmitted by tsetse flies, while *T. equiprdum* is transimitted by coitus. Infections by these trypanosomoses negatively impact the development of the animal industry, directly affecting national economies and people's livelihoods (Radwanska, 2018).

TRYPANOSOMOSIS DUE TO *TRYPANOSOMA EVANSI* (SURRA)

Trypanosoma (Trypanozoon) evansi (Steel 1885) Balbiani, 1888, was the first pathogenic mammalian trypanosome to be described by Evans (1880-1881-1882) who associated it with an endemic disease known as "Surra" affecting equines and camels in Dera Ismail Khan, in the Punjab district of India (Hoare, 1972).

Epidemiology

Transmission: *Trypanosoma evansi* is transmitted mechanically by a number of haematophagous flies of the genera *Tabanus, Haematopota, Chrysops* and *Stomoxys spp,* although, *Haematobia, Atylotus* and possibly *Musca* can are also be its vector. In Latin America, the vampire bat *(Desmodus rotundus*) is a unique example of an animal that can simultaneously be a host, reservoir, and a vector because of its ability to mechanically transmit the parasite in its saliva and also act as a true reservoir, maintaining the parasite in the bat colony in the absence of other hosts species affected (Hoare, 1965). Carnivores may become infected after ingesting infected meat. There is evidence of transmission through the ingestion of milk and during coitus (Luckins, 1998; OIE, 2013).

Hosts: Most mammalian species are susceptible to *T. evansi* infection and this makes the latter to have one of the widest host ranges among the mammalian trypanosomes. Originally, its principal host was believed to be the camel, but to date, it has been reported in equids, cattle, goats, sheep, buffaloes (Luckins, 1988; Desquesnes et al., 2013) and humans (Joshi et al., 2005) as well as in a wide range of other wild animals (Aregawi et al., 2019).

Prevalence: *Trypanosoma evansi* occurs beyond the African tsetse fly belt. Along with *T. equiperdum* and *T. vivax*, it is one of the pathogenic trypanosomes with the widest geographical distribution Fig.1. Available reports on surra in selected countries show varying prevalences: Nigeria (27%), Chad (30%), Mauritania (24%), Niger (29%), Kenya (48-95%), Ethiopia (21%), Sudan (33%), Somalia (58%), Morocco (34.5% in Tafilalet and 43.3% in Ouarzazate provinces), Egypt (81%), Jordan (33%), Iraq (47%), Iran (10%), United Arab Emirates (8-14%), India (22%), Pakistan (14%), South America (27%) (Enwezor & Sackcy, 2015) and in Uganda prevalences ranging from 0%o to 100% have been reported (Olaho-Mukani et al., 1998).

Higher prevalence rates have been reported in the camel, followed by the buffalo and cattle. The prevalence values within each species depend on the diagnostic method used and the geographical region (Aregawi et al., 2019). For example in the camel, parasitological prevalence ranged from 1% in Spain and Mauretania to 50% in United Arab Emirates, while the molecular prevalence ranged from 0% in Iran to 40% in Sudan. Similarly, parasitological prevalence in the buffalo was reported to vary between 2% in Vietnam and 68% in The Philippines. The overall molecular prevalence in the buffalo was 28% with only 3% in The Philippines, but 51% in India. In cattle the prevalence of surra has mainly been studied in India and Thailand and rates of a 2% and 6% pooled parasitological prevalence, respectively have been observed. One study in Venezuela reported an exceptionally high parasitological prevalence of 18% which is characteristic for an outbreak situation (Aregawi et al., 2019).

Horses are considered to be very susceptible to surra and show an acute disease and high mortality, while donkeys and mules are less susceptible to the disease. Prevalence values for equines are estimated from studies in the horse, donkey and the mule.

Small ruminants might play a role as reservoir of *T. evansi*, e.g. in camel rearing areas of Eastern Africa where small ruminants and camels are herded together. Parasitological and molecular prevalence of surra in this region, vary from 10% to 57% and cross-reactions with other possibly non-pathogenic trypanosomes might lead to higher prevalence rates. Dogs might be carriers of *T. evansi* for a short period before they succumb to the infection and are not considered important reservoirs but rather,they are epidemiological dead-end hosts. Investigations of *T. evansi* in wildlife have been carried out mainly in South America (Brazil) and showed overall 15%, with a parasitological prevalence, 22% seroprevalence and 13% molecular prevalence.

Geographical Distribution: *Trypanosoma evansi* has the widest geographical distribution among the mammalian trypanosomes, covering all the continents of the world (Fig.1). Its geographical distribution includes the Middle East ((Desquesnes et al., 2013), Southeast Asia, including India, China, Mongolia, parts of Russia, Bhutan, Nepal, Myanmar, Laos, Vietnam, Cambodia, Thailand, Malaysia, the Philippines, and Indonesia (Luckins, 1988). In Africa, it is present in all countries where camels are found, but it is limited southwards by the presence of tsetse flies (Mahmoud and Gray, 1980). In Europe, there are reports of introductions of *T. evansi* in the Canary Islands (Spain) (Guteirrez et al., 1988), the Spanish mainland (Tamarit et al., 2010) and Bulgaria (Desquesnes et al., 2013). There are no reports of its presence in North America, northern Europe, and northern Russia (Desquesnes et al., 2013).

Pathogenesis

The clinical signs of *T. evansi* infection can vary dramatically depending on host species and disease stage. Even within a host species, there can be variability in clinical signs in different geographical areas. In general, disease signs include fever, anaemia, loss of appetite, weight loss, nervous signs, abortion, cachexia, and death. In camels, signs of illness begin on appearance of parastaemia (Fig.2) with intermittent fever (41°C) for about one week. The animals appear dull and listless and become progressively weaker with loss of appetite and weight, develop oedema (particularly of the udder or scrotum), become anaemic, and develop petechial or ecchymotic haemorrhages. Nervous signs are sometimes observed, including periodic convulsions. The disease can be fatal within a few months however, it is more often chronic and can last 2-3 years (Desquesnes et al., 2013).

In horses, the incubation period is 1-4 weeks (sometimes up to 8 weeks), after which fever develops with high peaks (up to 44°C) corresponding to parasitaemia levels. Other signs, such as weakness, lethargy, anaemia, and severe weight loss follow. Additionally, petechial haemorrhages on the eyelids, vulvar and vaginal mucosa can occur, including the anterior chamber of the eye. As in camels, abortion, nervous signs, and oedema of the abdomen, testicles or udder are common (Desquesnes et al., 2013).

Anaemia in *T. evansi* infection is predominantly due to haemolytic crisis brought about by decreased haematocrit due to the shortened life span of erythrocytes and extensive erythrophagocytosis (Habila et al., 2012). The onset of anaemia in most instances is related to parasitaemia and the severity is related to the level and initial wave of parasitaemia. Haemolytic factors such as free fatty acids, immunologic mechanisms, haemodilution, coagulation disorders, depression of erythrogenesis and the release of trypanosomal substances such sialidase, have all been implicated in the development of anaemia (Habila et al., 2012). These substances induce haemolysis due to the action of sialidase, which is believed to cleave sialic acids on the erythrocyte surface and expose galactosyl residues. Sallau, et al., (2008) observed that these residues are recognized by D-galactose-specific lectins on macrophages, leading to macrophage and complement activation, erythrophagocytosis and subsequently anaemia. Trypanosomes may liberate proteolytic ferments, which digest proteins and liberate toxic products.

The observed haemolytic anaemia is macrocytic and may be associated with leucopaenia or leucocytosis and thrombocytopaenia. The predominant lesion is found in the connective tissue of various organs, blood and lymphoid system. Trypanosomes usually localise in connective tissue and are associated with mononuclear cell aggregation and exudation in body cavities. Wide spread distribution of parasites in tissues leads to subcutaneous oedema of the head, limbs, genitalia, urticarial plaques, keratitis, opthalmitis, enlargement of lymph nodes, spleen and accumulation of fluid in the body cavities. Destruction of red blood cells and platelets may also be due to cytotoxic reaction associated with immune complexes formation ultimately causing disturbances in clotting mechanism and acute glomerulonephritis. Utilisation of tryptophan by trypanosomes in the brain has been proposed as basis for functional disturbances of the Central Nervous System (CNS) malfunction, due to the generation of phenyl pyruvate and indole–3–ethanol by the trypanosomes in the brain. Mechanism of immunological injury is by delayed hyper-sensitivity reaction which also generates substances causing direct inflammatory changes.

Immune exhaustion due to antigenic variation results into polyclonal B cell activation and IgM hypergammaglobinaemia. This results into immunosuppression at both cellular and humoral levels. Immuno-suppression due to *T. evansi* results in failure of vaccination against bacterial and viral diseases (Singla et al., 2010) and secondary/concurrent infection with other pathogens (Gupta et al., 2009). Other biochemical changes in blood chemistry and cellular component include increased serum enzyme levels of sorbitol dehydrogenase, alanine aminotransferase aspartate amino transferase and decrease in serum alkaline phosphatase which is responsible for tissue damage. Hypoglycaemia results in the breakdown of the liver function and fatal intoxication. Impairment of blood brain barrier and increased intracerebral immunoglobulin biosynthesis are other observations reported (Singla et al., 1998). The lysosomal secretory proteinases, phospholipases and other hydrolytic enzymes of trypanosomes are considered potentially important factors in the development of disease. The proteases released into the blood stream may degrade the host tissue proteins and contribute to the pathogenesis. Death may occur due to endotoxins that may be released by the lysis of the parasites and hypoglycaemia which results from the consumption of large amount of blood glucose by trypanosomes. Furthermore, asphyxia due to increased blood lactic acid may interfer with the normal absorption of oxygen by the haemoglobin and may also occur due to mechanical blocking of the lung capillaries by agglutinated trypanosomes, which may cause pulmonary oedema and result into fatal termination. Erythrocyte production is inhibited by toxins. Apart from progressive anaemia and intravascular coagulation, toxins and proteolytic enzymes liberated by the parasite are considered to be the cause of death.

Post-mortem findings: The lesions are not pathognomonic of the disease. They include pale mucous membranes with emaciated body, anaemia and petechiae on some internal organs. Oedema of lungs with accumulation of sero- sanguineous fluid in peritoneal, pericardial and thoracic cavities are common findings in chronic surra. Yellow gelatinous exudates are seen under skin. Sub endocardial haemorrhages with enlarged lymph node, splenomegaly and hepatomegaly and some times ulceration and patechiation of the gastrointestinal tract, serosa and mucosa are common post mortem findings. The parasites can be detected in histopathological specimens of the brain, lung, spleen and liver.

Clinical signs of *T. evansi* infection are not pathognomonic. Therefore, laboratory methods have been developed to improve diagnosis. The direct parasitological methods for the diagnosis include: microscopic examination of blood or lymph node material. These are of low sensitivity but useful, particularly in poor countries. The application of various blood concentration procedures has significantly improved their sensitivity and added advantage of assessing the degree of anaemia (Woo, 1970; Murray et al., 1979; OIE, 2012). Sub-inoculation of fresh blood or buffy-coat preparations from suspected animals into laboratory rodents has been used to detect subclinical infections at levels of as few as 1.25 trypanosomes per ml of blood.

Specific DNA probes have been developed and used to detect trypanosome DNA in infected blood or tissues and numerous primers for both standard and real-time PCRs have been developed and evaluated for the detection of *T. evansi* DNA (Masiga et al, 2006; OIE, 2012).

Serological tests have also been developed to detect specific antibodies to *T. evansi* antigens. These include direct and indirect agglutination tests (Olaho-Mukani, 1981), complement fixation tests, indirect fluorescence antibody tests, and the trypanolysis test (OIE, 2012). However, many of these tests are either useful only for small-scale surveys or are no longer used. In recent years these tests have been replaced by the more sensitive and more easily standardisable techniques such as: enzyme linked immunosorbent assay (ELISA) (Olaho-Mukani et al., 1993; Luckins, 1992, 1997), the card agglutination test (CATT) and Latex agglutination test (Suratex®). The latter two, have the advantage over the former in that they can be used on any mammalian species. Evaluations of ELISA, CATT and Suratex have been carried out in camels, horses, cattle, buffaloes, and pigs (Desquesnes et al., 2009; Olaho-Mukani et al., 1996; Verloo et al., 2000).

Impact

The total world camel population is approximately 35 million animals (FAO, 2019) and surra is considered the most important single cause of morbidity and mortality affecting camels (OIE, 2013). Of all the protozoal diseases of livestock, *T. evansi* infection has the highest economic impact and livelihoods of the people in Africa, Middle East, Euro-Asia S.E Asia and South America. The resources spent on treatment and managemet of the disease, the losses due to mortalities and production are quite substantial and can be a serious challenge to policy makers and communities in affected areas. However, there is paucity of information and statistics on its actual impact. It affects a number of economically important animals, including camels, horses, buffaloes, and cattle (OIE, 2012). Furthermore, *T. evansi* and other livestock trypanosomes threaten more than 48 million cattle in 37 African countries and are responsible for major losses in the production of milk, meat, and manure fertilizer (Desquesnes et al., 2013). Surra is on of the OIE list of multispecies notifiable diseases and therefore affects international trade in livestock and livestock products.

Prevention and Control

Due to its wide geographic range, its control has attracted international attention, with a focus on improving diagnosis, treatment, formulating and implementing effective strategies aimed at increasing productivity and achieving a decrease in morbidity and mortality (Enwezor and Sackey, 2005). Effective prevention and control of surra are hampered by: the wide range of vectors and hosts affected, and the lack of vector specificity (Aregawi, 2019). Chemotherapy is the main method used to control *T. evansi* infections. Prophylactic drugs include quinapyramine dimethylsulfate (Triquin), prothridium and isometamidium chloride, and in some cases suramin. Quinapyramine sulfate, melarsomine (Cymelarsan), suramin and diminazene aceturate can be used for curative treatment (OIE, 2013). Only authorized formulations of diminazene aceturate should be used, as the latter could be very toxic to camels. Many of the drugs used for cattle are either not curative or are too toxic for camels. Overdosing of quinapyramines, diminazene aceturate and isometamidium chloride, can cause toxicity in camels and lead to death (Wernery and Kaaden, 2000). Drug resistance is common and should be considered in refractory cases. Monitoring for drug resistance is important and should be employed when suspicion of resistance arises. In most endemic countries, effective drugs and diagnostics are not available to those who need them. Control measures should focus on the host rather than the vector by applying parasitological detection of the parasites, treatment of infected animals, prophylactic cover of susceptible animals and protection of animals from biting flies. In Latin America, the vampire bats (OIE, 2013) are difficult to control. Though the elimination of *T. evansi* from infected herds can be challenging, systematic detection of circulating trypanosomes or their antigens followed by immediate treatment with curative drugs may completely eliminate the infection (Olaho-Mukani et al., 1993).

TRYPANOSOMOSIS DUE TO *TRYPANASOMA (DUTONELLA) VIVAX*

Trypanosoma (Dutonella) vivax is a unique trypanosome in that it can be transmitted both cyclically and mechanically. In Africa it is endemic and is transmitted cyclically by Glossina spp. and mechanically by biting flies. *Trypanosoma vivax* is also endemic in Latin America where its transmission is exclusively mechanical through biting flies. It infects several species of domestic and wild ungulates. This trypanosome was introduced into the South America around 1830, when Zebu cattle from Senegal were exported to French Guiana and the Antilles, although it is also believed that such introduction might have occurred in 1733 directly from the African continent and Cap Green Islands, as well as via the Caribbean islands and later, the parasite spread to other parts of the region (Hoare, 1972).

Epidemiology

Transmission: In Africa mechanical and cyclical transmission co-exist in the field, but only mechanical transmission can explain the permanent presence of *T. vivax* outside the tsetse belt (Hoare,1972). Mechanical transmission is mainly by lies of the genera *Tabanus, Haematopota, Haematobia, Chrysops and Stomoxys*. Some mosquitoes have also been implicated as potential vectors for *T. vivax* in countries like Venezuela and Cuba (Cordoves et al., 1982). The degree of parasitaemia in mammalian hosts and the frequency of biting fly feeding affect the rate of mechanical transmission of *T. vivax* and a direct link between the level of parasitaemia and the success of transmission has been observed (Sinshaw et al.,

2006). There are indications that *T. vivax* may sometimes persist at a low level, because of mechanical transmission, after tsetse flies have been eradicated from an area, for example the Island of Zanzibar in East Africa (FAO, 2006). The unregulated movement of infected animals across national and international borders is probably the principal reason why the parasite spreads to new areas.

Hosts and Geographic distribution: *Trypanosoma vivax* is found in an expansive area of Africa that is habitat for tsetse flies. Outside this tsetse-infested area, the parasite is also found in most of the western, eastern, and central regions of the continent (Hoare, 1972) (Fig.3). In the Sudan for example, *T. vivax* infection is now the second most important non-tsetse transmitted tyrpanosome of camels (Mossaad et al., 2017). Several species of ungulates in Africa are hosts of *T. vivax*, including: domestic animals such as cattle, sheep goats, equids, camels and wild animals (Hoare, 1972). In South America, *T. vivax* occurs in bovines and bubalines in areas subject to flooding and also in capybaras and wild ungulates such as the white- tailed deer, or cariacou (Odocoileus gymnotis). The parasite has been reported in Mauritius and the Arabian Peninsula and although it has been said that wild ungulates may serve as reservoirs, no conclusive evidence has been drawn (Desquesnes & Gardiner, 1993). Unlike in Africa, *T. vivax* has not been observed in South American horses.

Prevalence

The prevalence of *T. vivax* shows considerable variation according to geography, abundance of tsetse or biting flies and host species. Within the tsetse belt the overall *T. vivax* prevalence is reported within the range of 5-15% and accounts for up to half of total trypanosome prevalence (Kalu et al., 2001). However, outside of the tsetse belt, *T. vivax* prevalence has been reported in the range of 2–10% and dependends on the local and seasonal variation in fly abundance. Where tsetse flies are absent *T. vivax* accounts for all trypanosome infections in African livestock (Delafosse, et al., 2006).

True prevalence and incidence rates may be much higher than those commonly reported due to the limited sensitivity of the microscopy- based techniques (Delafosse et al., 2006). Faye et al. (2001) observed a seven times higher prevalence of bovine trypanosomosis when comparing PCR-based identification with the traditional microscopy technique in regions of high and low tsetse challenge in The Gambia. Similarly, Pinchbeck et al. (2008) also reported a significantly higher detection of prevalence in clinical equine samples from The Gambia when comparing PCR detection with microscopy, reporting a *T. vivax* prevalence of 87% from PCR, compared with a total trypanosome prevalence (*T. vivax, T. congolense and T. brucei*) of only 18% with microscopy. In a study conducted in Chad the prevalence of *T. vivax* infection was 1.6% using Buffy Coat Test (BCT) and 42.3% with indirect ELISA in herds of cattle from tsetse free areas around Lake Chad (Delafoss et al., 2006).

Pathogenesis

Following infection, the prepatent period of infection by *T. vivax varies* depending on the host and the pathogenicity of the strain. In cattle, it ranges from 9 to 14 days for virulent isolates, and from 9 to 59 days for less pathogenic strains (Hoare, 1972). In sheep and goats, on the other hand, the incubation period can last from 4 to 12 days (Hoof et al., 1948). The pattern of parasitaemia (Fig. 4) exhibits irregular fluctuations and this may last for several months before trypanosomes become undetectable in the blood, followed by periods of reocurrence. In some areas *T. vivax* infection may show a mortality rate of 100% three to four months after exposure to tsetse flies (Hoare, 1972), while in others such as the Congo (Hoof et al., 1948), the disease can present a sub-acute picture in calves for a period of up to two months, after which trypanosomes may disappear from the bloodstream and the animals recover spontaneous. The typical symptoms of *T. vivax* infection, include: cachexia, oedema, anaemia and nervous symptoms. Anaemia appears with progressing parasitaemia and there is lysis of large numbers of red blood cells resulting in a drop in PCV% (Murray et al., 1979). In cattle, the anaemia is described as normochromic- normocytic, with a tendency to be normochromic- macrocytic. Macrocytosis is due to erythrogenesis that takes place two weeks after the onset of infection, at which time, immature erythrocytes are released into the bloodstream (Anosa and Isoun, 1980). The antigenic variation as reported in other trypanosomal infections leads to the exhaustion of the antibody production by the host leading to immunosuppression (Nantulya et al., 1982). In addition, there is enlargement of lymph nodes and splenomegally associated with plasma cell hyperplasia and hypergammaglobulinaemia (Urquart et al., 1995).

Four stages of *T. vivax* infection can be distinguished (Van DenIngh et al., 1976): (i) A prepatent period of about 1 week. (ii) An acute or fulminating disease characterized by an almost continuous, fluctuating parasitaemia (Fig.4), a decrease of the number of thrombocytes, blood serotonin level, Packed Cell Volume (PCV) and haemoglobin (Hb), serum albumin and -globulin fractions, a low serum total lipid content and an initial rise with subsequent decrease of the blood bradykinin level. (iii) A critical stage of about 4-5 weeks post infection during which the animal may develop a progressive parasitaemia and may die from disseminated intravascular coagulation associated with microthrombi, oedema and haemorrhages in the various tissues, pulmonary lobular infarction, patches of necrosis in liver, spleen and adrenals and nephrotic changes in the kidneys. (iv) A post-critical stage characterized by an ameliorating condition, with no or few trypanosomes in the peripheral blood alternating with relatively high peaks of parasitaemia. This may be associated with increased blood bradykinin levels, an increase of the PCV and Hb, serum albumin and γ-globulin fractions, number of thrombocytes and blood serotonin level.

In the aparasitaemic phases, trypanosomes can be found extravascularly. The parasites may be found in the lymph nodes, eyes, the choroid plexus and aqueous humour (Hoere, 1972; Whitelaw, et al., 1988). Affected animals exhibit fever, anaemia, weight loss, hypoglycaemia, increased serum levels of aspartate aminotransferase and nervous signs (Batista, et al., 2009).

Certain African isolates of *T. vivax* can cause acute disease accompanied by a haemorrhagic syndrome. This has been observed during periods of high biting fly activity and typical features of this syndrome include high, persistent parasitaemia, fever, very pronounced anaemia and generalized visceral and mucosal haemorrhages, particularly in the gastrointestinal tract and bleeding from the orifices. In the field, the disease affecting adult cattle can be severe enough to lead to death or abortion, even before diagnosis is reached and treatment can be started (Mwongela et al., 1981; Magona et al., 2008). Severe acute anaemia and thrombocytopaenia are associated with the onset of parasitaemia (ILRAD, 1989). This haemorrhagic syndrome has not been observed outside Africa (Desquesnes, 2004).

Diagnosis

In terms of identifying animals with active infections, clinical diagnosis could be too unreliable to be used for diagnosis and the control of the disease so other means of diagnosis have to be employed.

Parasitological Diagnosis: Light microscopy is a readily and widely applied method for diagnosis of trypanosomal infection in the field. The examination of fresh or stained blood smears, has been modified to improve diagnostic sensitivity by concentrating the blood through haematocrit centrifugation technique (HCT) (Woo, 1970) or the dark ground buffy coat technique (DG) (Paris et al., 1982). The HCT is one of the most widely used parasitological methods in which motile trypanosomes can be viewed between the buffy-coat and the plasma. The technique has been reported to detect trypanosomes in the blood six to ten days before they are detected in a fresh drop or from thick smears (Molyneux, 1975). However, these parasitological diagnostic techniques also have limitations in sensitivity and therefore there is a need for more reliable methods.

Immunological Techniques: Serological techniques such as Enzyme Linked Immunosorbent Assay (ELISA) and Immunofluorescence Assay (IFA) can be used for the direct and indirect diagnosis of trypanosomosis in infected livestock. Both indirect IFA (Katende et al., 1987) and indirect ELISA (Luckins, 1977; Desquesnes, 1997) are useful techniques for epidemiological investigations. These tests have been applied for antibody detection but they have inherent problems related to specificity due to cross-reaction with other parasites of the same trypanosome genus, especially in endemic areas where mixed infections occur (Luckins,1972, 1977). A modification of the ELISA, based on an antigen capture assay which enables detection of circulating trypanosomal antigens in the blood of infected animals was deveped (Nantulya & Lindqvist, 1989). The species specificity of the assay was improved following the development of monoclonal antibodies as capture antibodies that recognized antigens present in *T. brucei, T. vivax and T. congolense* (Nantulya & Lindqvist, 1989), but is adoption for field application received some practical challenges.

Molecular Techniques: Biochemical and Molecur diagnostic techniques have been applied to study *T. vivax*. These include isoenzyme analysis and molecular techniques based on nucleic acids (Desquesnes & Tresse, 1996). Compared with other diagnostic techniques, the polymerase chain reaction (PCR) has yielded more excellent results in all regions where *T. vivax* infection occurs. These tests are not only suitable for detecting parasites in the mammalian hosts, but also in the insect vectors and can be used to evaluate the success of therapeutic interventions.

Post-Mortem Findings: The post-mortem findings are not pathognomonic (FAO, 2006). The liver may be enlarged and congested. The heart may be enlarged, flaby because of muscle deterioration and may show a few haemorrhages on the muscle surface and serous membranes. There is increased fluid in the chest, lungs, abdomen and pericardium. The kidneys are pale and swollen. In acute cases the spleen and lymphnodes may be enlarged. Subcutaneous oedemas may be present particularly in horses and sheep (FAO, 2006). In chronic infections, the pathological changes are more striking, but without being typical. The carcass is emaciated and often dehydrated. The skin may show sores and ulcers. The fat reserves under the skin are depletd and the skin is closely adherent to the underlying muscles and bone. The muscles are wasted and pale and the underlying bones are prominent. The lymph nodes are mostly normal or even hard, dry and reduced in size. The spleen may be normal in size or contracted, with a drier pulp than normally seen (FAO, 2006).

Prevention and Control

Control should combine restricted movement control, treatment of the diseased animals, vector control and epidemiological monitoring of the distribution and severity of the disease. The application of trypanocidal drugs has been the most widely practiced means of controlling trypanosomosis in domestic livestock since the early 1950s, either as curative or prophylactic application. Three antitrypanosomal compounds, namely: sometamidium chloride, homidium chloride or bromide and diminazene aceturate (Whitelaw et al., 1989; Desquesnes et al. 1995; Geerts et al., 1997) are used for curative and prophylaxis and in the integrated control of animal trypanosomois the use of these drugs will continue to play an important role because they are more appealling to local animal keepers, as they provide affodable means for protecting their animals. However, this still has low impact on overall animal production especially in high-challenge areas and where farmers use trypanocides therapeutically rather than prophylactically. The development of resistance to these trypanocides by the parasite is a continuous challenge to their sustainable use. Therefore, the rational use of the existing trypanocidal drugs and good animal husbandry methods that do not expose animals to the vectors will be essential until new drugs and methods of treatment emerge. In areas where resistance to trypanocides has not yet been demonstrated, the probability of its development should influence the selection of an appropriate control strategy (Bossche et al., 2000).

Although potential immunological targets within the parasite have been identified, the antigenic complexity of trypanosomes has for many years frustrated all efforts to develop a vaccine (Nantulya, 1986) and no vaccine is anticipated to be commercially available in the near future. In this regard the option is to explore alternative approaches e.g. the use of trypanotolerant breeds of African origin (Murray et al. 1990).

Economic Impact

African trypanosomoses cause negative impact on human livelihoods, resulting in income loss estimated to the tune of US $1.3 billion per year (Shaw, 2004). In Africa, losses caused by bovine trypanosomosis alone have been estimated at roughly 5 million USD annually not mentioning losses in milk production and secondary products, such as leather (ILRAD, 1989), draught-power and manure (ILRAD, 1989). African farmers spend about 35 million USD per year on trypanocidal drugs to protect and cure their cattle. Losses in meat production, milk yield and traction power and manure are estimated to cost more than US $ 500 million annually. The total losses due to tsetse and non-tsetse transmitted trypanosomoses in interms of agricultural Gross Domestic Product, has been valued at more than USD 4.75 billion per year (FAO, 2004).

The impact of *T. vivax* infection in Latin America has not been satisfactorily estimated, although it is thought to rank third among parasitic diseases in economic importance, after tickborne diseases and fasciolosis. Further economic studies are necessary in order to obtain reliable data for the assessment of real impact. The inapparent losses due to subclinical infections and reproductive disorders (Seidl, 1999), should also be considered in assessing economic impact in the *T. vivax* affected regions.

TRYPANOSOMOSIS DUE TO *TRYPANOSOMA EQUIPERDUM* (DOURINE)

Introduction

Trypanosoma equiperdum is closely related to *Trypanosoma evansi, T. rhodesiense, T. gambiense* and *T. brucei*, which constitute the *Trypanozoon subgenus*. However, recent phylogenetic analyses show that *T. equiperdum* and *T. evansi* are not monophyletic and should therefore be considered as subspecies of *T. brucei*. *Trypanosoma equiperdum* is monomorphic, but sometimes exhibits pleomorphism during subpassage in rodents (Blacklock and Yorke, 1913; Wang, 1988). Dyskinetoplastic forms of *T. equiperdum* have also been reported in all known strains maintained in rodents or continuous in vitro culture.

Epidemiology

Dourine is a World Organisation for Animal Health (OIE) notifiable disease and is the only trypanosomosis that is not transmitted by an invertebrate vector and also differs from other trypanosomes in that it is primarily a tissue parasite that rarely invades the blood. Average mortality associated with acute disease approaches 50%.

Transmission: *Trypanosoma equiperdum* may be found in vaginal secretions of infected mares, and seminal fluid, mucous exudate of the penis, and sheath of stallions. Dourine is transmitted during breeding or infected mares may occasionally pass infection to foals. Natural transmission occurs directly from animal to animal during coitus, mainly from stallion to mare, but may also be transmitted from mare to stallion, facilitated by the presence of the parasite in the seminal fluid, mucous exudates of the penis and its sheath and vaginal mucus, respectively, (OIE, 2013). *Trypanosoma equiperdum* can pass through intact mucous membranes and it is possible for foals to acquire infection by contamination of nasal or conjunctival membranes with the vaginal discharge during parturition. Transmission to foals by ingestion of infected colostrum or milk is considered rare (William and Steven, 2007), however, these foals that ingest colostrum from infected mares become seropositive due to passive transfer of antibodies, though these are usually seronegative by 4 to 7 months of age (William and Steven, 2007). The presence of trypanosomes in the mammary secretions may support the view that the infection can occasionally pass to foals during suckling (Pascucci et al., 2013). Parasitaemia is usually low. Transmission of *T. equiperdum* by tsetse fly or other vector has not been reported.

Hosts: The main hosts are horses, mules and donkeys and there is no known natural reservoir of the parasite other than infected equids. Rats, mice, rabbits and dogs can be infected experimentally; rodents are used to maintain strains of the parasite indefinitely and to prepare antigen for diagnostic tests. Establishment of *T. equiperdum* in the blood of laboratory animals is extremely difficult, but once a strain of it is adapted to rodents, it can be maintained for a long time, by serial passages in the same manner as *T. evansi* (Haig and Lund, 1948). Domestic animals such as sheep and goats infected with the murine- adapted strains of *T. equiperdum* have been reported to produce a disease similar to dourine.

Geographical Distribution: Dourine is endemic in (Fig.5) Asia, Africa, southeastern Europe, and Central America. It has been eradicated from North America, and strict control measures have reduced the incidence to a low level in most parts of Europe. It occurred in Italy in 2011. The disease is endemic in parts of Ethiopia and Namibia and is rarely reported in other parts of sub-Saharan Africa. It has not been reported in Latin America for over 20 years. It is possible that lack of reporting in some countries may be caused by very strict international regulations that tend to discourage official notification of the disease.

Prevalence

The prevalence of *T. equiperdum* has declined generally because the horse is no longer that important militarily, economically, and agriculturally, and because of strict control measures in many countries. Therefore, in most countries, dourine now occurs only sporadically. In Ethiopia, the disease is more prevalent during the breeding season from June to September. A survey of 237 horses in an endemic area of Ethiopia (Fikru, 2010) showed that infection rates varied with the method of examination. The rates were 4.6% based on standard parasitologic methods, 27.6% on serology, and up to 47.6% on DNA detection by PCR. This was the first time in more than 30 years that a fresh strain of *T. equiperdum* was isolated from clinical cases of dourine. Case mortality varies; in Europe, it may be as high as 50% to 70%, but it is much lower elsewhere, although many animals may have to be destroyed as a means of control.

Pathogenesis

The incubation period of *T. equiperdum* infection may vary from a few weeks to several years. The manifestations of dourine include: oedematous lesions of the genitalia (Fig. 6), the involvement of the nervous system, progressive emaciation and death in most cases. Classical cutaneous lesions, from which the disease derives its name dourine, have been described as circular elevated plaques of thickened skin ranging in size from one to ten centimeters in diameter, resembling money or douros (Claes et al., 2005). The constant antigenic variations of the parasite during the course of infection, results in the release of a large amount of biochemically active products and the formation of immune complexes, which are key factors in triggering a variety of clinical and pathological changes (Zwart, 1989). Some of the clinical signs, which include genital oedema, weight loss, skin lesions known as silver dollar plaques, and neurological signs, may be absent in the early stages or during latent infections (Luckins and Dwinger, 2004; Claes et al. 2005). Postmortem examination reveals gelatinous exudates under the skin. In the stallion, the scrotum, sheath, and testicular tunica appear thickened and infiltrated and the testes are embedded in a tough mass of sclerotic tissue to a level that is unrecognizable. In the mare, the vulva, vaginal mucosa, uterus, bladder, and mammary glands may be thickened with gelatinous infiltration (Fig.6). The lymph nodes, especially in the abdominal cavity, appear hypertrophied, softened, and, in some cases, haemorrhagic. The spinal cord of animals with paraplegia is often soft, pulpy, and discolored, more so in the lumbar and sacral regions (OIE, 2013).

Histological examination of tissue samples, the disease shows haemosiderin deposition in the spleen, the iliac, supramammary, and popliteal lymph nodes with hyperplasia of the plasma cells, a sign of increased leucocytosis. Pastular dermatitis, particularly more pronounced around the oedematous plaques has been observed with evidence of severe inflammation and vacuolar degeneration extending to the deepest layers of the skin, with involvement of the cutaneous adnexa and perivascular plasma cell inflammation (Pascucci et al., 2013). In the central nervous system, especially in the facial and lingual nerves of the infected horses, there is evidence of neuro-degenerative lesions and inflammatory vasculitis with oedematous infiltration. In the affected mares, the udders show severe interstitial inflammation accompanied by strong supramammary lymph node reactivity and the presence of Russell's bodies. The liver may show multifocal areas of hepatitis while the kidneys depict plasma cell inflammation of the renal pelvis. There is periglandular inflammation in the vulva, vagina, uterus, and clitoris in the infected mares and the finding of parasites in the the lymph of iliac and supramammary lymph nodes and the positive histological examination seem to support the thinking that the parasite spreads mainly through the lymphatic system (Pascucci et al., 2013).

Depigmentation around the perineum and the vulval skin of infected mares is often described as pathognomonic of clinical cases of dourine (Claes et al. 2003). The probable cause of depigmentation could be due to severe necrostic degeneration of melanocytes (Yonas, 2015) (Fig. 3)

Diagnosis

Clinical signs of dourine can provide a strong indication of the presence of the disease, but confirmatory diagnosis may be needed (Claes et al. 2005). However, diagnosis of dourine is a challenge, due to limited knowledge about the parasite and host-parasite interaction following infection. In practice, diagnosis is based on clinical evidence supported by serology (Alemu et al. 1997; Hagos et al. 2010). Diagnosis of dourine, therefore, requires laboratory confirmation by parasitological, serological, and molecular techniques.

Parasitological diagnosis: It is extremely difficult to detect *T. equiperdum* parasite in the body fluids of infected horses (Claes et al., 2005). The HCT which is a blood concentration technique (Woo, 1970), is the most frequently applied concentration technique and has better sensitivity than direct microscopic examination (Reid et al. 2001). The mini anion-exchange centrifugation technique (mAECT), which is an improment on mHCT, consists of separating the trypanosomes from blood via anion-exchange chromatography and finally concentrating them at the bottom of a plastic collector tube, followed by microscopic examination. The mAECT is more sensitive but quite tedious and time consuming (Buscher et al., 2009). Repeated attempts to use animal inoculation method were unsuccessful (Alemu et al., 1997; Clausen et al., 1999).

Serological techniques: The complement Fixation test (CFT)is the officially recognised test for dourine by the OIE and depends on the demonstration of trypanosomal antibodies in the serum (Bishop et al., 1995). Despite its usefulness and universal acceptance, the CFT gives some discrepancies, requires careful continuous titration of numerous labile agents, is not reliable with sera having anticomplementary activity and is only specific for the subgenus *Trypanozoon* (Luckins, 1994). The CFT is also less sensitive than ELISA and IFAT (Bishop et al., 1995). *T. equiperdum*-infected laboratory animals and horses suspected of dourine also react positive on the Card Agglutination trypanosomiasis (CATT)/*T. evansi* test. The CATT/*T. evansi* test is fast, uses a standardised antigen, and can be performed in the field without the need of a highly qualified staff, a well-equipped laboratory and could be a good test for equine trypanosomosis, regardless whether the causative agent is *T. evansi* (surra) or *T. equiperdum* (dourine) (Claes et al., 2003).

Molecular techniques: *Trypanozoon* subgenus-specific primers are available and can be used in PCR for the detection of T. equiperdum DNA. A highly sensitive real-time PCR for *Trypanozoon* subgenus was applied on tissues and fluid samples from a naturally dourine-infected horse, enabling the detection of low numbers of the parasites (Pascucci et al., 2013). However, these tests are expensive and cannot be applied for mass screening and negative results do not necessarily exclude the possibility of an infection with *T. equiperdum*.

Prevention and Control:

Treatment is not recommended for dourine because animals remain refractory after treatment and remain carriers of the parasite (OIE 2013). However, suramin and neoarsphenamine or quinapyramine sulfate have been used in the past and evidence from in vitro drug sensitivity indicates that suramin, diminazene, quinapyramine, and cymelarsan are effective (Brun and Lun, 1994) effective against *T. equiperdum*. Prevention of dourine should be based on the establishment of freedom from infection based on testing blood for the precence of antibodies against *T. equiperdum*. Any introductions of horses from endemic areas or areas suspected of dourne should be isolated, quarantined and the blood tested for antibodies by complement fixation test (Sidney et al. 2013). According to OIE (OIE, 2013) the control of dourine depends on compulsory notification, slaughter of infected animals, and movement control and enforcement of the veterinary legislation in most countries. In non-endemic areas dourine should be eradicated by movement control and enforcing and testing of all in contact animals and the euthanasia of infected and seropositive horses (Sidney et al. 2013). However, this policy of testing and slaughter may not be economcally feasible in developing countries. Therefore, in dourine endemic areas, based on the result of the in vivo drug sensitivity study, a revised strategy of the sequential testing of horses for trypanosome antigens or DNA followed by treatment of positive cases using curative drugs like cymelarsan (Olaho-Mukani et al., 1995) could be a more viable option in eliminating dourine. This needs to be further explored and compared with the testing and slaughter eradication approach recommended by the OIE.

FUTURE RESEARCH DIRECTION FOR NTT TRYPANOSOMOSES

Infections due to NTT Tyrpanosomes are of major economic importance in many countries of Africa, Asia, Euro-Asia and South America. Although they have been recognised for many years as important causes of economic losses, they largely remain as neglected diseases at national, regional and international levels. Therefore, the future research trust should focu on the following:

(i) Improvement of diagnosis by developing and validating, standardized and approved tests for pen-side diagnosis and tests that differentiate between T. evansi and T. equiperdum; (ii) improvement of the understanding of the dynamics of mechanical transmission in endemic areas by dentifying the key factors and major fly species responsible for transmission in different ecological settings (Abera et al., 2015); (iii) monitoring the efficacy of different trypanocidal drugs and the extent of drug resistance constantly by using molecular tests like Antigen-ELISA and PCR as well as ascertaining the causes of relapses during chemotherapeutic interventions; (iv) determination of the factors that precipitate epidemic outbreaks and the role of wild animals in the epidemiology of T. evansi and T. vivax (Enwenzor & Sackey, 2005); (v) focus on the understating the molecular mechanisms that may lead to vaccine development for NTT trypanosomoses; (vi) understanding the factors that lead to the haemorrhagic syndrome in some T. vivax infections; (vii) study the interaction of NTT trypanaosomes with other mammalian trypanosomes during mixed infections; (viii) generation of information on the ecology and natural history of NTT trypanosomoses to provide better understanding of the interplay of the various factors that affect transmission, the maintenance of infection and the dynamics of mechanical transmission by biting flies; and finally (ix), more research is needed to elucidate the nature and magnitude of the zoonotic potential of T. evansi, its spread in space and time, and how it can be controlled.

CONCLUSION

Despite the impact caused by NTT Trypanosomes, there is paucity of quantitative data on disease distribution, prevalence of infection and economic impact on different livestock species in different production systems. There is a need for information on the ecology and natural history of non-tsetse-transmitted trypanosomiases to provide greater understanding on the interplay of the factors that affect transmission, maintenance of infection and the dynamics of mechanical transmission by biting flies. More research is needed to show evidence that wild animals play an important part in the ecology of disease transmission to domesticated livestock, by developing and applying validated and standardized diagnostic tests in order to provide more accurate prevalence rates to assess the economic impact of NTT trypanomoses at the national and regional levels.

REFERENCES

Abera, Z., Usmane, A., & Zelalem, A. Z. (2015). Review on Camel Trypanosomosis: Its Epidemiology and Economic Importance. *Acta Parasitologica Globalis*, *6*(2), 117–128.

Alemu, T., Luckins, A. G., Philips, L. P., Reid, S. W. J., & Holmes, P. H. (1997). The use of ELISA to investigate the prevalence of T. equiperdum in Ethiopian horses. *Veterinary Parasitology*, *71*, 239–250. doi:10.1016/S0304-4017(96)01139-9 PMID:9299693

Anosa, V. O., & Isoun, T. T. (1980). Haematological studies on Trypanosoma vivax infection of goats and intact and splenectomized sheep. *Journal of Comparative Pathology*, *90*(1), 155–168. doi:10.1016/0021-9975(80)90036-5 PMID:7391277

Aregawa, W. G., Agga, G. E., Abdi, R. D., & Burcher, P. (2019). Systematic review and meta-analysis on the global distribution, host range and prevalence of Trypanosoma evansi. *Parasites & Vectors*, *12*(1), 67–92. doi:10.118613071-019-3311-4 PMID:30704516

Batista, J. S., Riet-Correa, F., Teixeira, M. M., Madruga, C. R., Simões, S. D., & Maia, T. F. (2007). Trypanosomosis by Trypanosoma vivax in cattle in the Brazilian semiarid: Description of an outbreak and lesions in the nervous system. *Veterinary Parasitology*, *143*(2), 174–181. doi:10.1016/j.vetpar.2006.08.017 PMID:16965857

Bishop, P., Rae, P. F., Philips, L. P., Boid, R., & Luckins, A. G. (1995). T. equiperdum: Detection of Trypanosomal antibodies and antigen by enzyme-linked immunosorbent assay. *The British Veterinary Journal*, *151*(6), 715–720. doi:10.1016/S0007-1935(95)80152-9 PMID:8605584

Blacklock, B., & Yorke, W. (1913). The trypanosomes causing dourine (Mal de Coitor Beschalseuche). *Proceedings of the Royal Society of London. Series B, Containing Papers of a Biological Character*, *87*(593), 89–96. doi:10.1098/rspb.1913.0063

Bossche, V. D. P., Shumba, W., & Makhambera, P. (2000). The distribution and epidemiology of bovine trypasonomosis in Malawi. *Veterinary Parasitology*, *88*(3-4), 163–176. doi:10.1016/S0304-4017(99)00222-8 PMID:10714455

Brun, R., & Lun, Z. R. (1994). Drug sensitivity of Chinese T. evansi and T. equiperdum isolates. *Veterinary Parasitology*, *52*, 37–46. doi:10.1016/0304-4017(94)90033-7 PMID:8030186

Buscher, P., Ngoyi, D. M., Kabore, J., Lejon, V., Robays, J., Jamonneau, V., Bebronne, N., Van der Veken, W., & Bieler, S. (2009). Improved models of mini Anion exchange centrifugation technique (MAECT) and modified single centrifugation (MSC) for sleeping sickness diagnosis and staging. *PLoS Neglected Tropical Diseases*, *3*(11), e471. doi:10.1371/journal.pntd.0000471 PMID:19936296

Claes, F., Agbo, E. C., Radwanska, M., Te Pas, M. F. W., Baltz, T., De Waal, D. T., Goddeeris, B. M., Claassen, E., & Büscher, P. (2003). How does T. equiperdum fit into the Trypanozoon group? A cluster analysis by RAPD and multiplex-endonuclease genotyping approach. *Parasitology*, *126*(5), 425–431. doi:10.1017/S0031182003002968 PMID:12793646

Claes, F., Büscher, P., Touratier, L., & Goddeeris, B. M. (2005). Trypanosoma equiperdum: Master of disguise or historical mistake? *Trends in Parasitology*, *21*(7), 316–321. doi:10.1016/j.pt.2005.05.010 PMID:15923142

Clausen, P. H., Gebreselassie, G., Abditcho, S., Mehlitz, D., & Staak, C. (1999). Detection of Trypanosoma DNA in serological positive but aparasitemic horses suspected of dourine in Ethiopia. *The Tokai Journal of Experimental and Clinical Medicine*, *23*, 303–308. PMID:10622626

Cordoves, C. O., Fernandez, C., Garcia, A. J., & Gonzalez, B. R. (1982). Trypano- soma vivax, Ziemann, 1905. Lista de transmisores mecanicos en Cuba. *Revista Cubana de Ciencias Veterinarias*, *13*, 219–221.

Delafosse, A., Thébaud, E., Desquesnes, M., & Michaux, Y. (2006). Epidemiology of Trypanosoma vivax infection in cattle in the tse-tse free area of Lake Chad. *Preventive Veterinary Medicine*, *74*(2-3), 108–119. doi:10.1016/j.prevetmed.2005.10.006 PMID:16321453

Desquesnes, M. (1997). International and regional standardization of immunoenzyme tests: Methods, concerns and limitations. *Scientific and Technical Review*, *16*, 809–823. PMID:9567307

Desquesnes, M. (2004). *Livestock trypanosomoses and their vectors in Latin America*. OIE World Organization for Animal Health.

Desquesnes, M., De La Rocque, S., & Peregrine, A. S. (1995). French Guyana stock of Trypanosoma vivax resistant to diminazine aceturate but sensitive to isometamidium chloride. *Acta Tropica*, *60*, 133–136. doi:10.1016/0001-706X(95)00117-W PMID:8610541

Desquesnes, M., & Gardiner, P. R. (1993). Épidémiologie de la trypanosome bovine (Trypanosoma vivax) en Guyane Française. *Revue d'Élevage et de Médecine Vétérinaire des Pays Tropicaux*, *46*, 463–470. PMID:7910693

Desquesnes, M., & Holzmuller, P. (2013). Trypanosoma evansi and surra: A review and perspectives on origin, history, distribution, taxonomy, morphology, hosts, and pathogenic effects. *BioMed Research International*, 1–22.

Enwezor & Sackey. (2005). *Camel trypanosomosis, Nigeria institute for Trypanosomosis Research (NITR)*. Department of veterinary surgery medicine. Ahmadu Bello University (ABU) Zaria, Kaduna State.

Evans, G. (1880). Report on "Surra" disease in the Dera Ismail Khan District of India. *Punjab Government Miltary*, (493), 446.

FAO. (2004). The State of Food Insecurity in the World (SOFI). Food and Agriculture Organization of the United Nations.

FAO. (2006). *Corporate Document Repository. A field guide for the diagnosis, treatment and prevention of African animal trypanosomosis*. Food and Agriculture Organization of the United Nations. www.fao.org/DOCREP/2006

FAO. (2019). *R*. FAOSTAT.

Faye, D., Pereira, P. J., de Almeida, B., Osaer, S., Ndao, M., Berkvens, D., Speybroeck, N., Nieberding, F., & Geerts, S. (2001). Prevalence and incidence of trypanosomosis in horses and donkeys in The Gambia. *Veterinary Parasitology*, *101*(2), 101–114. doi:10.1016/S0304-4017(01)00503-9 PMID:11587839

Fikru, R., Gari, Hagos, A., Alemu, T., Goddeeris, B. M., & Claes, F. (2010). Comparative diagnosis of parasitological, serological, and molecular tests in dourine-suspected horses. *Tropical Animal Health and Production*, *42*, 1649–1654.

Geerts, S., Holmes, P. H., Eisler, M. C., & Diall, O. (2001). African bovine trypanosomiasis: The problem of drug resistance. *Trends in Parasitology*, *17*(1), 25–28. doi:10.1016/S1471-4922(00)01827-4 PMID:11137737

Gupta, M. P., Kumar, H., & Singla, L. D. (2009). Trypanosomosis concurrent to tuberculosis in black bucks. *The Indian Veterinary Journal*, *86*, 727–728.

Gutiérrez, C., Montoya, J. A., Padron, M., Corbera, J. A., Juste, M. C., & Molina, J. M. (1998). Description of a case of trypanosomiasis due to Trypanosoma evansi in a dromedary in the Canary Islands. *Medicina Veterinária*, *15*(6), 356–357.

Habila, N., Inuwa, M. H., Aimola, I. A., Udeh, M. U., & Haruna, E. (2012). Pathogenic mechanisms of Trypanosoma evansi infections. *Research in Veterinary Science*, *93*(1), 13–17. doi:10.1016/j.rvsc.2011.08.011 PMID:21940025

Hagos, A., Abebe, G., Buscher, P., Goddeeris, B. M., & Claes, F. (2010). Serological and parasitological survey of dourine in the Arsi–Bale highlands of Ethiopia. *Tropical Animal Health and Production*, *42*(4), 769–776. doi:10.100711250-009-9485-6 PMID:19924557

Haig, D. A., & Lund, A. S. (1948). Transmission of the South African strain of dourine to laboratory animals. *Onderstepoort Journal of Veterinary Science*, *23*, 59–61. PMID:18863432

Hoare, C. (1972). The Trypanosomes of Mammals, a Zoological Monograph. Blackwell Scientific Publications.

Hoare, C. A. (1965). Vampire Bats as Vectors and Hosts of Equine and Bovine Trypanosomes. *Acta Tropica*, *22*(3), 204–216. PMID:4379528

Hoof, L., Van Henrard, C. & Peel, E. (1948). Quelques observations sur les trypanosomes del grands mammifères au Congo Belge. *Acta Tropica, 5*, 327-344.

ILRAD. (1989). *Annual report of the International Laboratory for Research on Animal Diseases*. ILRAD.

Joshi, P. P., Shegokar, V. R., Powar, R. M., & Herder, S. (2005). Human trypanosomiasis caused by Trypanosoma evansi in India: The first case report. *The American Journal of Tropical Medicine and Hygiene*, *73*(3), 491–495. doi:10.4269/ajtmh.2005.73.491 PMID:16172469

Kalu, A. U., Oboegbulem, S., & Uzoukwu, M. (2001). Trypanosomosis in small ruminants maintained by low riverine tsetse population in central Nigeria. Small Ruminant Research. *The Journal of the International Goat Association*, *40*(10), 2015–2017.

Katende, J. M., Musoke, A. J., Nantulya, V. M., & Gooderis, B. M. (1987). A new method for fixation and preservation of trypanosomal antigens for use in the indirect immunofluorescence antibody test for diagnosis of bovine trypanosomosis. *Tropical Medicine and Parasitology*, *38*, 41–44. PMID:3299658

Luckins, A. G. (1972). Studies on bovine trypanosomiasis. Serum immunoglobulin levels iin zebu callte exposed to natural infection in East Africa. *The British Veterinary Journal*, *128*(10), 523–528. doi:10.1016/S0007-1935(17)36737-4 PMID:4630575

Luckins, A. G. (1977). Detection of antibodies in trypanosome-infected cattle by means of a microplate enzyme-linked immunosorbent assay. *Tropical Animal Health and Production*, *9*(1), 53–62. doi:10.1007/BF02297393 PMID:333677

Luckins, A. G. (1988). Trypanosoma evansi in Asia. *Parasitology Today (Personal Ed.)*, *4*(5), 137–142. doi:10.1016/0169-4758(88)90188-3 PMID:15463067

Luckins, A. G. (1992). Methods for diagnosis of trypanosomosis in livestock. *World Animal Review*, *70*(71), 15–20.

Luckins, A. G. (1994). Equine trypanosomiasis. Exotic disease series. *Equine Veterinary Education*, *6*(5), 259–262. doi:10.1111/j.2042-3292.1994.tb01149.x

Luckins, A. G. (1998). Epidemiology of Surra: Unanswered Questions. *Journal. Protozoological Research*, *8*, 106–119.

Luckins, A. G., & Dwinger, R. H. (2004). Non-tsetse-transmitted animal trypanosomiases. In The Trypanosomiases. Epidemiology and Diagnosis. Cabi Publishing.

Magona, J. W., Walubengo, J., & Odimin, J. T. (2008). Acute haemorrhagic syndrome of bovine trypanosomosis in Uganda. *Acta Tropica*, *107*(2), 186–191. doi:10.1016/j.actatropica.2008.05.019 PMID:18599006

Mahmoud, M. M., & Gray, A. R. (1980). Trypanosomosis due to Trypanosoma evansi (Steel, 1885) Balbiani, 1888. A review of recent research. *Tropical Animal Health and Production*, *12*(1), 35–47. doi:10.1007/BF02242629 PMID:7370098

Masiga, D. K., Ndung'u, K., Tweedie, A., Tait, A., & Turner, C. M. R. (2006). Trypanosoma evansi: Genetic variability detected using amplified restrictionfragment length polymorphism (AFLP) and random amplified polymorphic DNA (RAPD) analysis of Kenyan isolates. *Experimental Parasitology*, *114*(3), 147–153. doi:10.1016/j.exppara.2006.03.002 PMID:16624308

Molyneux, D. H. (1975). Diagnostic methods in animal trypanosomiases. *Veterinary Parasitology*, *1*(1), 5–17. doi:10.1016/0304-4017(75)90003-5

Mossaad, E., Salim, B., Suganuma, K., Musinguzi, P., Hassan, M. A., Elamin, E. A., Mohammed, G. E., Bakhiet, A. O., Xuan, X., Satti, R. A., & Inoue, N. (2017). Trypanosoma vivax is the second leading cause of camel trypanosomosis in Sudan after Trypanosoma evansi. *Parasites & Vectors*, *10*(1), 176–186. doi:10.118613071-017-2117-5 PMID:28403897

Murray, M., Clifford, D. J., & Mcintyre, W. I. M. (1979). Diagnosis of African trypanosomosis in the bovine. *Transactions of the Royal Society of Tropical Medicine and Hygiene*, *73*(1), 120–121. doi:10.1016/0035-9203(79)90148-2 PMID:442172

Murray, M., Trail, J. C. M., & D'Ieteren, G. D. M. (1990). Tripanotolerance in cattle and prospects for the control of trypanosomiasis by selective breeding. *Scientific and Technical Review*, *9*, 369–386. PMID:2132686

Mwongela, G. N., Kovatch, R. M., & Frazil, M. A. (1981). Acute Trypanosoma vivax infection in dairy cattle in Coast Province, Kenya. *Tropical Animal Health and Production*, *13*(1), 63–69. doi:10.1007/BF02237893 PMID:7233561

Nantulya, V. M. (1986). Immunological approaches to the control of ani- mal trypanosomiasis. *Parasitology Today (Personal Ed.)*, *2*(6), 168–173. doi:10.1016/0169-4758(86)90148-1 PMID:15462812

Nantulya, V. M., & Lindqvist, K. J. (1989). Antigen-detection enzyme immunoassays for the diagnosis of Trypanosoma vivax, T. congolense and T. brucei infections in cattle. *Tropical Medicine and Parasitology*, *40*, 267–272. PMID:2617031

OIE. (2012). Trypanosoma evansi infection (Surra). In Manual of Diagnostic Tests and Vaccines for Terrestrial Animals. Paris, France: OIE.

OIE. (2013). *Trypanosoma evansi infections (including surra)*. OIE.

Olaho-Mikani, W., Nyang'ao, J. M. N., & Ouma, J. (1995). Elimination of drug-resistant Trypansoma evansi infection in a group of camels by sequential treatment with Melarsomine. *Journal of Camel Practice and Research*, *2*(2), 83–85.

Olaho-Mukani, W. (1981). *Studies on the epidemiology of Trypanosoma evansi in selected areas of Kenya* (Thesis MSc). University of Nairobi.

Olaho-Mukani, W., Kakaire, D., Matovu, E., & Enyaru, J. (1998). Prevalence of surra in dromedary camels in Uganda. *The Journal of Protozoology Research*, *8*(3), 120–125.

Olaho-Mukani, W., Munyua, W. K., Mutugi, M. W., & Njogu, A. R. (1993). Comparison of antibodyand antigen detection enzyme immunoassays for the diagnosis of Trypanosoma evansi infection in camels. *Veterinary Parasitology*, *45*(3-4), 231–240. doi:10.1016/0304-4017(93)90078-2 PMID:8447066

Olaho-Mukani, W., Nyanga'o, J. M. N., & Ouma, J. O. (1996). Use of Suratex for field diagnosis of patent and non-patent Trypanosoma evansi infections in camels. *The British Veterinary Journal*, *152*(1), 109–111. doi:10.1016/S0007-1935(96)80091-1 PMID:8634859

Paris, J., Murray, M., & McOdimba, F. (1982). A comparative evaluation of the parasitological techniques currently available for the diagnosis of African trypanosomosis in cattle. *Acta Tropica*, *39*, 307–316. PMID:6131590

Pascucci, I., Andrea, D., Cesare, C., Gabriella, D., Paolo, C., Manuela, T., Nicola, F., Massimo, S., & Vincenzo, C. (2013). Diagnosis of dourine in outbreaks in Italy. *Veterinary Parasitology*, *193*(1-3), 30–38. doi:10.1016/j.vetpar.2012.12.006 PMID:23298562

Pinchbeck, G., Morrison, L. J., Tait, A., Langford, J., Meehan, L., Jallow, S., Jallow, J., Jallow, A., & Christley, R. M. (2008). Trypanosomosis in The Gambia: Prevalence in working horses and donkeys detected by whole genome amplification and PCR, and evidence for interactions between trypanosome species. *BMC Veterinary Research*, *4*(1), 7–13. doi:10.1186/1746-6148-4-7 PMID:18289378

Radwanska, M., Vereecke, N., Deleeuw, V., Pinto, J., & Magez, S. (2018). Salivarian Trypanosomosis: A Review of parasites involved, their global distribution and their interaction with the innate and adaptive mammalian host immune system. *Frontiers in Immunology*, *9*(2253), 1–14. doi:10.3389/fimmu.2018.02253 PMID:30333827

Sallau, A. B., Ibrahim, M. A., Salihu, A., & Yusuf, I. A. (2008). Bloodstream form of Trypanosoma evansi contains galactosidase. *Middle East Journal of Scientific Research*, *3*, 49–52.

Seidl, A., Dávila, A. M. R., & Silva, R. A. M. S. (1999). Estimated financial impact of Trypanosoma vivax on the Brazilian Pantanal and Bolivian lowlands. *Memorias do Instituto Oswaldo Cruz*, *94*(2), 269–272. doi:10.1590/S0074-02761999000200027 PMID:10224541

Shaw, A. P. (2004). Economics of African trypanosomosis. In The Trypanosomiases. CABI Publishing.

Sidney, R., Andrew, McG., James, C. & Richard, N. (2013). Dourine – an emerging venereal threat to European horses. AHT / BEVA/DEFrA. *Equine Quarterly Disease Surveillance report, 6*, 7.

Singla, L. D., Juyal, P. D., & Sharma, N. S. (2010). Immune responses to haemorrhagic septicaemia (HS) vaccination in Trypanosoma evansi infected buffalo-calves. *Tropical Animal Health and Production*, *42*(4), 589–595. doi:10.100711250-009-9461-1 PMID:19784863

Singla, L. D., Juyal, P. D., & Sharma, N. S. (2010). Immune responses to haemorrhagic septicaemia (HS) vaccination in Trypanosoma evansi infected buffalo-calves. *Tropical Animal Health and Production*, *42*(4), 589–595. doi:10.100711250-009-9461-1 PMID:19784863

Sinshaw, A., Abebe, G., Desquesnes, M., & Yon, W. (2006). Biting flies and Trypanosoma vivax infection in three highland districts bordering Lake Tana, Ethiopia. *Veterinary Parasitology*, *142*(1-2), 35–46. doi:10.1016/j.vetpar.2006.06.032 PMID:16890359

Tamarit, A., Gutierrez, C., Arroyo, R., Jimenez, V., Zagalá, G., Bosch, I., Sirvent, J., Alberola, J., Alonso, I., & Caballero, C. (2010). Trypanosoma evansi infection in mainland Spain. *Veterinary Parasitology*, *167*(1), 74–76. doi:10.1016/j.vetpar.2009.09.050 PMID:19864069

Urquart, I. G. M., Armour, J., Dunncan, J. L., Dunn, A. M., & Jennings, F. W. (1995). Veterinary Parasitology. The University of Glasgow.

Van den Ingh, T. S., & Neijs-Bakker, M. H. (1979). Pancarditis in Trypano- soma vivax infection in cattle. *Tropenmedizin und Parasitologie*, *30*, 239–243. PMID:483387

Verloo, D., Holland, W., My, L. N., Thanh, N. G., Tam, P. T., Goddeeris, B., Vercruysse, J., & Büscher, P. (2000). Comparison of serological tests for Trypanosoma evansi natural infections in water buffaloes from north Vietnam. *Veterinary Parasitology*, *92*(2), 87–96. doi:10.1016/S0304-4017(00)00284-3 PMID:10946132

Wang, Z. L. (1988). The similarities and differences of the characteristics between T. equiperdum and T. evansi. *Bulletin of Veterinary College*, *8*, 300–303.

Wernery, U., & Kaaden, O. R. (2002). *Infectious diseases in camelids* (2nd ed.). Wiley-Blackwel. doi:10.1002/9780470699058

Whitelaw, D. D., Bell, I. R., & Holmes, P. H. (1989). Immunological clearance of Se-labeled Trypanosoma brucei in goats. *Acta Tropica*, *46*(2), 101–106. doi:10.1016/0001-706X(89)90003-X PMID:2565070

William, B. L., & Steven, H. S. (2007). Infectious diseases of breeding stallion in. *Current Therapy in Large Animal Veterinary Research*, *55*, 117–119.

Woo, P. T. K. (1970). The haematocrit centrifuge technique for the diagnosis of African Trypanosomosis. *Acta Tropica*, *27*, 384–386. PMID:4396363

Yonas, G. (2015). *Clinical signs, serum biochemical profiles and repro- ductive pathological lesions in horses naturally infected with dourine in Western Arsi zone, Ethiopia* (Unpublished MSc thesis). Addis Ababa University.

Zwart, D. (1989). Aspects of comparative pathology and pathogenesis of Trypanosomal infections in Africa. *Annales de la Société Belge de Medecine Tropicale*, *69*, 105–112.

KEY TERMS AND DEFINITIONS

Anaemia: A reduction in the number of circulating red blood cells or in the quantity of haemoglobin.

Antibodies: Proteins that are found in blood or other bodily fluids of vertebrates, produced by the cells of the immune system in response to invasion by foreign substance., such as parasites and other microorganisms.

Antigens: Substances (usually proteins or polysaccharides) that can stimulate an immune response.

Endemic: A disease that is native to a particular geographic region.

Enzyme Linked Immunosorbent Assay (ELISA): Is an immunological test widely used to measure antibodies, antigens, proteins, glycoproteins and drugs in biological samples.

Epidemiology: is the study of the distribution and determinants of health-related states or events in specified populations, and its application to the control of health problems.

Polymerase Chain Reaction (PCR): Is a technique used to make numerous copies of a specific segment of DNA quickly and accurately to enable investigators obtain large quantities of DNA that are required for various experiments and procedures in molecular biology, forensic analysis, evolutionary biology and biomedical diagnosistics.

Trypanosomosis: Is the name of several diseases in animals caused by parasitic protozoan trypanosomes of the genus *Trypanosoma.*

Section 2
Tick-Borne Diseases (TBDs)

Chapter 7
Epidemiology and Economic Importance of Tick-Borne Diseases of Cattle in Africa

Donald Lubembe Mukolwe
Faculty of Veterinary Medicine and Surgery, Egerton University, Kenya

Charles Byaruhanga
National Agricultural Research Organisation, Uganda

Fredrick Ojiambo Obonyo
School of Agriculture and Food Science, Meru University of Science and Technology, Kenya

ABSTRACT

The tropical and sub-tropical climate in Africa favours multiplication and maintenance of tick vectors and transmission of various pathogens to cattle. Key challenges including acaricide resistance, policy issues, transboundary animal movements, and inadequate veterinary services compromise effective control of tick-borne diseases (TBDs). This chapter discusses important host, pathogen, climatic, and management factors that impact the control of TBDs among cattle in Africa, and which affect the productivity and overall contribution to economic development. The economic losses in cattle production attributed to tick infestation and TBDs in Africa are also reviewed. The use of a sustainable integrated control approach, including vaccination, strategic tick control, surveillance for acaricide resistance, and multi-stakeholder involvement is also evaluated.

INTRODUCTION

Ticks and tick-borne diseases (TBDs) cause significant economic and health losses to livestock, especially in the tropical and sub-tropical regions (Jongejan and Uilenberg, 2004). Three genera of ticks, *Rhipicephalus*, *Amblyomma* and *Hyalomma* are of economic importance in cattle in Africa (Walker *et al.*, 2014b; Jongejan and Uilenberg, 2004).

DOI: 10.4018/978-1-7998-6433-2.ch007

The use of acaricides is the main method of control of ticks in Africa. However, inappropriate use and control, coupled with decentralization and privatization of veterinary services have resulted in the emergence and spread of acaricide resistance, which is threatening productivity of livestock and the livelihoods of livestock keepers (Abbas *et al.*, 2014; Vudriko *et al.*, 2016). Climate change and uncontrolled transboundary animal movements are other crucial factors that have contributed to the introduction of new tick species in some regions (Nyangiwe *et al.*, 2018). A high level of tick-borne pathogen diversity has been observed (Hove *et al.*, 2018; Salih *et al.*, 2018), with possible negative effects on the efficacy of vaccines. The presence of tick-borne pathogens in wildlife further complicates control at the wildlife-livestock interface (Walker *et al.*, 2014a). In addition, climatic change and cattle production systems in Africa, as well as unlevelled socio-economic contexts, contribute to the limited success in the control of ticks and TBDs. This chapter discusses recent findings on the epidemiological dynamics and economic impact of the most important tick species and TBDs of cattle in Africa, and evaluates feasible control options considering the aforementioned challenges, in order to reduce economic losses in the livestock sector.

BACKGROUND

Important Tick-Borne Diseases of Cattle in Africa

Tick-borne diseases that mostly affect cattle in Africa are anaplasmosis, heartwater, babesiosis and theileriosis. Anaplasmosis is the most prevalent TBD of cattle, mainly caused by *Anaplasma marginale*, which is distributed in tropical and sub-tropical regions in all six continents (reviewed by Aubry and Geale, 2011). *Anaplasma centrale* causes a mild form of anaplasmosis, and is used as a live vaccine to protect against *A. marginale* infection (reviewed by Kocan *et al.*, 2010). Heartwater, caused by *E. ruminantium* affects cattle in almost all of sub-Saharan Africa as well as on the islands of Reunion, Mauritius, Madagascar, Grande Comore and São Tomé, and several islands in the Caribbean (Allsopp, 2010). In Africa, bovine babesiosis is caused mainly by *B. bigemina* and *B. bovis* (De Vos *et al.*, 2004). *Babesia bovis* causes the more virulent form of the disease, while *B. bigemina* is more prevalent on the continent. Theileriosis is commonly caused by *Theileria parva*, which affects cattle in eastern central and southern Africa (reviewed by Bishop *et al.*, 2020), and was recently found for the first time in West Africa (Cameroon) (Silatsa *et al.*, 2020). It was also reported in the Comoros Island (De Deken *et al.*, 2007). *Theileria parva* causes East Coast fever (ECF) (Lawrence *et al.*, 2004b), Corridor disease (Lawrence *et al.*, 2004a) and January disease (Lawrence *et al.*, 2004c).

Transmission and Tick Vectors

Infective stages of tick-borne parasites are released in the saliva of feeding ticks into the feeding site of the host, from which they infect bovine cells. Apart from tick vectors, *A. marginale* is also transmitted mechanically by biting flies (*Stomoxys, Tabanus* and various mosquito species) or through blood-contaminated fomites, and transplacentally from dam to foetus (reviewed by Aubry and Geale, 2011). The most important tick species of cattle in Africa, and the pathogens they transmit are described below.

Rhipicephalus Appendiculatus and *Rhipicephalus Zambeziensis*

Both *R. appendiculatus* and *R. zambeziensis* are three-host ticks that transmit *T. parva*. *Rhipicephalus appendiculatus* is the principal vector, widely distributed in the warmer, more humid areas of southern, eastern and central Africa (Walker *et al.*, 2014b). The cattle-derived *T. parva* causes ECF, while the buffalo-derived *T. parva* causes Corridor disease. *Rhipicephalus zambeziensis* is found in the savanna and steppe climatic regions of Namibia, Mozambique, Zambia, Botswana, South Africa, Zimbabwe, Tanzania and Angola. In Zimbabwe, the tick transmits *T. parva* that causes January disease, *Theileria taurotragi* that causes benign bovine theileriosis, and *Anaplasma bovis* that causes tropical bovine theileriosis (Walker *et al.*, 2014b). *Rhipicephalus duttoni*, the other vector of *T. parva*, occurs in Angola.

Hyalomma Species

The most important *Hyalomma* species that affect cattle in Africa are *Hy. anatolicum* (three-host tick) and *Hy. detritum* (two-host tick), which transmit *T. annulata*, the cause of tropical theileriosis in North Africa, as well as *Hy. rufipes* (two-host tick) and *Hy. truncatum* (two-host tick) which transmits *A. marginale* and causes sweating sickness, respectively (Walker *et al.*, 2014b; CABI, 2020). *Hyalomma anatolicum* is found in Ethiopia, Algeria, Libya and Sudan, while *Hy. detritum* is found in Morocco, Tunisia and Sudan (Walker *et al.*, 2014b). *Hyalomma rufipes* and *Hy. truncatum* are distributed in most parts of Africa (Walker *et al.*, 2014b).

Amblyomma Species

Amblyomma species transmit *E. ruminantium*, the cause of heartwater, which is fatal in cattle, sheep and goats, as well as *Theileria mutans*, which causes mild theileriosis. The most important *Amblyomma* species that affect cattle in Africa are *A. variegatum* and *A. hebraeum*. Others are *A. lepidum* and *A. pomposum*. *Amblyomma variegatum* is the most widely distributed *Amblyomma* species in Africa, found in the West, Central and East Africa, and part of southern Africa. The tick also transmits *T. velifera* and *Anaplasma bovis* (Walker *et al.*, 2014b). *Amblyomma hebraeum* has got limited distribution, occurring mainly in the northern part and east coast of South Africa, Swaziland, and the east of Botswana. It is also found in southern Mozambique and Zimbabwe where is has been found to be spreading northwards. Because of this spread, the tick has established itself on the central plateau of Highveld in South Africa (Norval, 1983). *Amblyomma hebraeum* and *A. variegatum* are reported to be scarcely present in some parts of northern Zimbabwe. This could either be because of interspecific mating resulting in sterile instars, or the climatic requirements of both ticks are not similar (Estrada-Peña *et al.*, 2008).

Rhipicephalus (*Boophilus*) Species

Rhipicephalus (*Boophilus*) species are one-host tick vectors for protozoal and rickettsial pathogens that cause bovine babesiosis and anaplasmosis, which are the most widely distributed TBDs. The two most important tick species are *R. microplus* and *R. decoloratus*. *Rhipicephalus decoloratus* transmits *B. bigemina* and *A. marginale*, and is distributed throughout Africa south of the Sahara, and rarely occurs in the drier areas of Namibia, South Africa and Botswana (Walker *et al.*, 2014b). The tick has not been reported in Madagascar (Walker *et al.*, 2014a). *Rhipicephalus microplus* transmits both *B. bigemina* and *B. bovis*, as well as *A. marginale*. The tick is distributed in South Africa, Tanzania, Kenya and Mozambique, parts of the eastern and central provinces of Zambia, throughout Malawi and to the east and north of Lake Malawi. Where favorable climatic conditions exist, the tick competes and displaces populations of *R. decoloratus*, and this is attributed to the slightly shorter life-cycle of *R. microplus*, the sterile progeny of interspecific breeding, a higher level of acaricide resistance in *R. microplus* (Tønnesen *et al.*, 2004), as well as presence of its larvae on vegetation during winter (Nyangiwe *et al.*, 2011). *Rhipicephalus microplus* has extended to West Africa, including Ivory Coast, Nigeria, Benin, Burkina Faso, Mali and Togo (Kamani *et al.*, 2017; reviewed by Nyangiwe *et al.*, 2018), where it is regarded as the most successful invasive tick species. The tick has also been reported in South Sudan (Kivaria *et al.*, 2012), Zimbabwe and Namibia (reviewed by Nyangiwe *et al.*, 2018), and more recently in Uganda (Muhanguzi *et al.*, 2020), Kenya (Kanduma *et al.*, 2020), Cameroon (Silatsa *et al.*, 2019) and Angola (Gomes and Neves, 2018). The recent introduction and spread of *R. microplus* are attributed to trans-boundary cattle movements for grazing and trade.

EPIDEMIOLOGY

Factors related to management practices, socio-economic status, climate, vertebrate host and pathogen factors influence the distribution and prevalence of tick-borne pathogens. However, the epidemiology of TBDs is getting complex due to climate change, wildlife-livestock interactions, pathogen diversity and unregulated animal movements. Understanding the epidemiology of TBDs in different production systems and geographical regions is key in determining the choice of control approaches.

Climate and Environmental Factors

The type of vegetation determines the presence, density, persistence of tick species, and the chances of successful host acquisition (Cumming, 2002). All species of ticks require an area covered by vegetation which provides suitable micro-habitats for survival and moulting or egg laying. Areas that experience high rainfall with adequate vegetation have got more ticks than those with poor vegetation and erratic rainfall. However, even within a generally suitable region, differences in the suitability of habitats have been suggested based on vector tick counts on animals. For example, in Kenya, warmer and more humid areas characterized by a mixture of grass and tree cover (Savannah woodland) are found in the Lake Victoria basin, coastal region and parts of central and eastern highlands (Gachohi *et al.*, 2012), and are suitable for vector development. Areas that are harsh, hot and dry with sparse vegetation and open grasslands have been found to be unsuitable for vector development. The prevalence of *R. appendiculatus* is low in the semi-arid areas, and in regions where overgrazing and environmental degradation has occurred (Lawrence *et al.*, 1992). Savannah grasslands that are not covered with bushes or tree cover have been found to be less favorable for the survival of *Amblyomma* ticks.

Temperature and humidity influence the reproduction and survival of ticks, especially the larvae. *Rhipicephalus appendiculatus* therefore thrives better in warmer and high precipitation areas in eastern, central and southern Africa, but not in the horn of Africa, due the warmer and more humid climate in the former (Norval *et al.*, 1992). Also, the bimodal rainfall in eastern Africa versus the unimodal rainfall in southern Africa results in a higher risk of ECF in the former, because up to three generations of all instars of *R. appendiculatus* occur each year in many areas (Norval *et al.*, 1992). *Rhipicephalus ap-*

pendiculatus occurs in areas where the annual rainfall is over 500 mm and an altitude ranging from the sea level to over 8000 feet (Norval *et al.*, 1992). This tick is rare at altitudes higher than 1200 meters above sea level (Gachohi *et al.*, 2012). However, the absence of *R. appendiculatus* in areas that appear climatically and ecologically suitable for its survival (Obara *et al.*, 2020) requires that other factors that may constrain occurrence of the tick be investigated. In the sub-tropical regions of Africa, for example in South Africa, the low temperatures in winter slow down the development and questing activity of ticks. A large number of free-living ticks then appears in spring, and this is driven by the rise in temperature which promotes larval hatching, as observed with *R. microplus* in Eastern Cape Province, South Africa (Nyangiwe *et al.*, 2011). Then, the high temperatures in summer facilitate death of questing or moulting stages. In Tanzania, the shifts in the distribution of *R. microplus* and *R. decoloratus* were attributed to different climate requirements for the two tick species, i.e., less than ~700 mm annual rainfall and colder than 20°C for *R. decoloratus*, and higher values for *R. microplus* (Lynen *et al.*, 2008). It has been found that engorged nymphs of *A. variegatum* that infest their hosts in the beginning of the dry season cannot move more than two meters to find suitable microhabitats (Stachurski, 2000). During the rainy season, engorged females usually drop off but can only move only up to a few meters to find a suitable place to lay eggs (Barré, 1989). Most *Amblyomma* tick species will survive well in areas that experience at least 500 mm of annual rainfall, with at least four months of continuous rainfall and a temperature of 15 - 35°C to enhance the development of the eggs (Barré, 1989). Higher temperatures have been found to the increase the rate of development of engorged stages (Pegram and Banda, 1990).

Climate change comes with a wide range of consequences, including change in rainfall patterns, increased intensity of droughts, loss of biodiversity, change in land use and this can modify habitats and affects the viability of ecosystems (Martin *et al.*, 2008). These changes therefore affect the dynamics of ticks and their hosts, as well as pathogen biology, and this can favour expansion of the distribution and abundance of tick species, and the transmission and pathogenicity of tick-borne pathogens.

Climate change is expected to increase transmission of infections due to extended questing season for ticks, increased number of susceptible immature stages, and the number of nymphs co-feeding on an animal (Nah *et al.*, 2020). On the other hand, increased temperatures in tropical and subtropical zones can have a deleterious effect on some plant species, thereby affecting tick habitat suitability and forcing some tick species to spread to new areas. In southern Africa, for example, it has been predicted that rising temperatures will decrease the suitability of habitats for *R. decoloratus*, *A. hebraeum*, *R. appendiculatus* and *H. truncatum* (Estrada-Peña, 2003). Elsewhere, Olwoch *et al.* (2008) predicted a reduction in suitability and range of *R. appendiculatus* due to increase in temperatures in the western arid areas of Angola, southern DRC and Namibia, but increased suitability in Zambia, some areas of Botswana, eastern DRC, Eastern and Northern Cape Provinces in South Africa, due to increased rainfall and an increase in the minimum temperatures. Increased concentrations of atmospheric carbon dioxide are likely to increase biomass of vegetation, which will provide more suitable shelter for free-living stages of ticks. Pastoralists often adopt to prolonged droughts by migrating with their stock, a phenomenon that contributes to the introduction of tick species, e.g. *R. microplus* and *R. pulchellus* in Uganda, in areas where they did not exist previously (Muhanguzi *et al.*, 2020).

Gradients of agro-ecological zones (AEZs), which are determined by differences in climatic conditions give rise to different epidemiological states of TBDs. For instance, endemic stability for ECF is expected to occur in areas that favour tick development, while endemic instability is envisaged to occur in marginal areas and regions where the tick can rarely survive (Gachohi *et al.*, 2012). In the Kenyan highlands, *T. parva* prevalence and the associated ECF morbidity, mortality and case-fatality have been demonstrated to vary greatly depending on AEZ. The prevalence of the disease has been reported to increase with lower elevation and in areas with unrestricted grazing (Gachohi *et al.*, 2012). In Uganda, Rubaire-Akiki *et al.* (2004) noted that the occurrence of antibodies to *A. marginale* and *B. bigemina* in cattle showed significant difference across AEZs.

Management Systems

The livestock production system influences the rate of exposure of cattle to different ecological conditions. In Uganda, Rubaire-Akiki *et al.* (2004) observed that the probability of cattle being infested with ticks especially *R. appendiculatus* was greater in cattle that graze on field pastures than those the under zero-grazing system. However, continuous exposure of extensively grazed indigenous cattle in endemic areas in Africa can facilitate the establishment of endemic stability, thereby decreasing incidence of TBDs (Gachohi *et al.*, 2012). On the other hand, intensive acaricide application in zero grazing herds significantly reduces exposure to tick-borne pathogens, which leaves them naïve and susceptible to infection (Gachohi *et al.*, 2012). On the other hand, transhumance in African drylands is thought to contribute to local tick diversity and is perceived to cause rapid spread of ticks and related pathogens in West Africa (Kamani *et al.*, 2017).

Cattle grazing in close proximity with wildlife affects ticks' dynamics and transmission of tick-borne pathogens amongst cattle (Oura *et al.*, 2011). Wild animals, in particular the African buffalo (*Syncerus caffer*), are persistently infected and serve as reservoirs of important tick-borne parasites of cattle, as observed for *A. marginale* and *T. parva* in the four national parks in Uganda (Oura *et al.*, 2011), *T. parva* in the four national parks in South Africa and Mozambique (Chaisi *et al.*, 2011), *T. parva* and *E. ruminantium* in Kruger National Park in South Africa (Allsopp *et al.*, 1999) and *B. bovis* in Okavango delta in Botswana (Eygelaar *et al.*, 2015). Because of this, Corridor disease caused by buffalo-derived *T. parva* causes high mortalities in cattle when they come into contact with African buffalo, in the presence of *R. appendiculatus*.

Pathogen Factors

The diversity of tick-borne pathogens affects pathogenicity, immune responses and effectiveness of vaccines. *Amblyomma variegatum* populations vary in infection rates with different *E. ruminantium* genotypes, showing higher susceptibility to a western African and a Caribbean isolate than to two isolates from southern Africa (Mahan *et al.*, 1995). However, the susceptibility of *A. hebraeum* to *E. ruminantium* was found to be similar in all isolates tested, and this may explain why heartwater is generally a more serious problem in areas where *A. hebraeum* is the main vector (Allsopp *et al.*, 2004). Also, *E. ruminantium* strains vary in infectivity, antigenicity and pathogenicity, and this has led to a lack of cross-protection by vaccines in ruminants (Allsopp *et al.*, 2004).

Various studies have demonstrated antigenic and genetic diversity of *T. parva* (Hemmink *et al.*, 2018;Sitt *et al.*, 2019; Lubembe *et al.*, 2020), and strains from eastern Africa have shown more persistent infections for tick and vertebrate hosts than those from southern Africa (reviewed by Bishop *et al.*, 2020). The antigenic diversity may have implications on effectiveness of vaccination. In South Sudan, high levels of diversity were attributed to multiple migrations of *T. parva*-infected cattle infested with *R. appendiculatus* into various regions in the country (Salih *et al.*, 2018).

Anaplasma marginale strains that differ in protein sequences, antigenicity and infectivity for ticks have been identified in various regions, including South Africa (Hove *et al.*, 2018). An important aspect in the epidemiology of babesiosis is that *Babesia* strains are polyclonal and therefore difference in virulence of subpopulations occurs. Antigenic differences also occur amongst isolates and strains of *B. bigemina* and *B. bovis*, and therefore cattle become more resistant to infection with homologous isolates than with the heterologous ones (Bock *et al.*, 2004). For the two species, the tick infection rates, and therefore the rate of transmission and likelihood of endemic stability developing are higher for *B. bigemina* than *B. bovis* infections, in regions of Africa where both species occur (Bock *et al.*, 2004).

Host Factors

Generally, younger animals have been found to suffer lower tick burdens than adult animals, due to capability to protect themselves by non-specific and cell mediated immunity (Okello-Onen *et al.*, 1999). Increasing age has been associated with increased seroprevalence of most TBDs and this is attributed to exposure time as antibodies become persistent in circulation in the host up to six months (Gachohi *et al.*, 2012).

The severity of infection with *Anaplasma marginale* has been reported to vary with the age of cattle (Aubry and Geale, 2010). Calves have been found to exhibit low susceptibility to clinical disease with the disease rarely reported in animals of up to six months of age, and a mild to acute disease occurs in older cattle (Kocan *et al.*, 2003). In Uganda, animals within the age groups of 0 - 3 and 4 - 12 months were reported to be at lesser risk of infestation with *R. appendiculatus* than animals above eighteen months.

The genetic make-up of an animal determines its ability to resist tick infestation, or the ability to develop an effective immunological response to infestation. Resistance related to the breed of the animal varies with tick species (Mattioli *et al.*, 2000). Amongst indigenous cattle breeds in South Africa, the Nguni cattle were reported to be less susceptible to tick infestation than Bonsmara cattle. This was attributed to short hair and smooth light-colour (Nyangiwe *et al.*, 2011), and delayed hypersensitivity (Marufu *et al.*, 2013) in Nguni as compared to the long hair, rough darker coat colour and immediate hypersensitivity in Bonsmara. When compared with exotic breeds, indigenous breeds of cattle have been shown to have a higher degree of genetic resistance to both ticks and TBDs (Norval *et al.* 1992).

Human movements and transportation of animals have a large effect on tick distribution and pattern of tick-borne pathogens both within and across countries (Barré and Uilenberg, 2010; Nyangiwe *et al.*, 2018). The emergence of bovine babesiosis and anaplasmosis in Nigeria was attributed to unrestricted movement of livestock from Benin which led to the introduction of *R. microplus* (Kamani *et al.*, 2017). The tick species was also introduced to Namibia from South Africa in the same way (Nyangiwe *et al.*, 2018). The spread of *A. variegatum* from mainland Africa to Madagascar, Mascarene Islands and the West Indices is also attributed to human movements and migratory populations of cattle egrets (*Bubulcus ibis*), which are hosts for the immature stages (Barré and Uilenberg, 2010). The recent detection of *T. parva* in Cameroon, where it had not been reported suggests anthropogenic movement of carrier cattle from endemic areas (Silatsa *et al.*, 2020), and therefore further research is required to assess the movement patterns responsible for the spread, and determine if the Muguga cocktail vaccine is involved.

ECONOMIC LOSSES DUE TO TICKS AND TICK-BORNE DISEASES IN AFRICA

There are few recent and comprehensive reports on the economic impact of TBDs in Africa (Mukhebi *et al.*, 1992; Norval *et al.*, 1992; Manjauw and McLeod, 2003; Kivaria, 2006). Economic losses are attributed to mortality, weight loss and reduction in milk production, and costs incurred for treatment, vaccination, and application of acaricides (Kivaria, 2006, Jonsson, 2006; Ocaido *et al.*, 2009). For instance, in Uganda, ticks and TBDs are responsible for 75.4% of losses in cattle production (Ocaido *et al.*, 2009), while in Tanzania, the total annual national loss due to TBDs is approximately USD 364 million (Kivaria, 2006). Cattle mortality due to TBDs accounted for 71.4% of losses in cattle production in Tanzania for a 13-year period (1981-1993), with theileriosis, anaplasmosis, heartwater and babesiosis accounting for 43.5%, 16.5%, 6.3% and 5.1%, respectively (Kivaria, 2006). In other studies, pastoralists in north-eastern Uganda indicted that TBDs were the biggest disease constraint to cattle production, due to mortality, treatment costs and weight loss (Byaruhanga *et al.*, 2015). In Benin, the impact of tick infestation on cattle as reported by the farmers was mainly due to weight loss, skin infections and milk production (Adehan *et al.*, 2018). Losses in milk production have been estimated to account for 23% of the total losses related to TBDs in cattle herds (Regitano and Prayaga, 2011).

Control of ticks and TBDs comprises the largest component (up to 88%) of the total economic costs (Ocaido *et al.*, 2009). For instance, in Kenya, Uganda and Tanzania, in which application of acaricide is the primary means of tick control, the cost was estimated to range from USD 6 to 36 per adult animal (Minjaw and McLeod, 2003). In another study in Uganda, the mean annual cost of ticks and TBDs control per animal in pastoral and ranch herds was estimated to be USD 4.54 (Ocaido *et al.*, 2009). There are also other financial losses attributed to ticks, for example damage to hides and skins, which depreciates their commercial value (Biswas, 2003), as well as wounds, loss of blood, injury to tissues during feeding, weakness and death of calves (Taylor *et al.*, 2016). Ticks also cause irritation, hypersensitivity and inflammation, and the lesions caused by tick species predispose to dermatitis, secondary bacterial infection and invasion by myiasis flies (Taylor *et al.*, 2016). Some tick species inject toxins into the host, causing paralysis or 'sweating sickness' (Minjauw and McLeod, 2003). Other losses are attributed to income lost due to inefficiencies in production by utilizing resistant but low producing animal breeds instead of the susceptible but highly productive breeds, and restriction of trade in animals and animal products (Hurtado and Giraldo-Ríos, 2018). Furthermore, the use of acaricides in animals leaves residues in milk and meat (Hurtado and Giraldo-Ríos, 2018).

East Coast fever is the most important TBD in central, eastern and southern Africa (Norval *et al.*, 1992), and the estimated annual losses in cattle due to the disease were USD 168 million (Mukhebi *et al.*, 1992). The cost of ECF infection-and-treatment immunization protocol was estimated to be approximately USD 7 and 40 per animal in the pastoralist and small-scale cross-bred dairy systems respectively in Tanzania, which is out of reach for the poor farmers (Lynen *et al.*, 2012). In Kenya and Tanzania, Minjauw and McLeod (2003) estimated the annual cost of ECF to be USD 54.4 and 34.1 million in the small-holder dairy and traditional systems, respectively. In Sudan, the economic losses arising from treatment cost for ECF were estimated to range between USD 20 per calf and USD 40 per cow using Butalex®, and USD 15 per calf and USD 30 per cow using Parvexon® (Marcellino, 2004). Another study in Sudan, the estimated morbidity rate for ECF in two cattle camps was 14.4% and 32%, and the case fatality rate ranged between 26.7% and 38.7%, while milk losses were at USD 4185 in one cattle camp (Salih *et al.*, 2011). *Theileria annulata* is the important species in northern Africa, and the total losses reported due to an outbreak of tropical theileriosis in a dairy farm in Khartoum State, Sudan was approximated at

USD 62,000 (Siddig *et al.*, 2003), while Latiff (1994) estimated losses of USD 4-6 million annually for *Theileria annulata* in Khartoum. Economic losses due to heartwater have also been documented; for example, in Zimbabwe, the national annual loss due to the disease in cattle was estimated to be USD 5.6 million (Mukhebi *et al.* 1999), and in another study on commercial farms in the same country, the loss was USD 6.5 million (Meltzer *et al.* 1995), mainly attributed to dipping (Jongejan and Uilenberg, 2004).

More studies are required to establish the actual impact of ticks and TBDs in different production systems in Africa, especially in countries where data is lacking. This will provide evidenced-based information for national governments to invest in control strategies.

CONTROL OF TBDs AMONG CATTLE HERDS IN AFRICA

Approaches for control of TBDs include tick management, vaccination, and the use of chemotherapeutic and chemoprophylactic drugs, or integration of these approaches.

The use of acaricides is the most common method for control of ticks in livestock in Africa. The advantage of acaricides is that they control all tick species. The main factor that drives the choice and quantity of acaricides by cattle keepers is price, and therefore due to economic reasons, cattle keepers often use understrength acaricides and less quantity (Byaruhanga et al., 2015. Dip tanks were built by European settlers in East, central and southern Africa at the beginning of the twentieth century, but except for a few countries, they are no longer functional, due to governments' policy of privatization and decentralization of veterinary services, and yet communities have less capacity or interest to maintain the cattle dips. Re-introduction or strengthening these dip services by local governments for compulsory communal dipping of cattle is required, and this will facilitate regulation of acaricide use and maintain efficacy. A strategic dipping regime was evaluated in Zambia, and demonstrated a reduction in tick challenge, cattle mortality rate and acaricide use, with no disruption to endemic stability (De Meneghi *et al.*, 2016). In Burkina Faso, acaricide footbaths were used to control ticks before permanent attachment to their predilection sites, and this reduced *A. variegatum* infestation and treatment costs (De Meneghi *et al.*, 2016). In both cases, there was also minimal environmental impact (De Meneghi *et al.*, 2016).

Some drawbacks in the use of acaricides include the requirement for large volumes of water, and at wild-livestock interfaces, multi-host tick and pathogen dynamics can negatively affect the effectiveness of interventions, as observed in Kenya for *T. parva* (Walker *et al.*, 2014a). Acaricides also require regular application, and there is a possibility of contaminating food supplies and the environment, and disrupting endemic stability (George *et al.*, 2004). Extensive reliance on acaricides and inappropriate use (Byaruhanga *et al.*, 2015; Vudriko *et al.*, 2018a) has resulted in the development of acaricide resistant tick populations (Vudriko *et al.*, 2016). The one-host ticks *Rhipicephalus* (*Boophilus*) species are more prone to acaricide resistance, because they complete their life cycle on the same host and therefore exposed to more acaricide treatments (Ntondini *et al.*, 2008). Moreover, the short life cycle of *Rhipicephalus* (*Boophilus*) species (only 21 days) allows the ticks to complete 3-5 generations in a year, which subjects them to intense selection pressure (George *et al.*, 2004; Abbas *et al.*, 2014). . Resistance in *R. microplus* and *R. decoloratus* to most classes of old acaricides was reported in southern and eastern Africa (George *et al.*, 2004; Abbas *et al.*, 2014; Ntondini *et al.*, 2008), but very few resistant populations of the three-host ticks (*Amblyomma*, *Hyalomma* and *Rhipicephalus*) were found (George *et al.*, 2004; Abbas *et al.*, 2014). Resistant *R. microplus* populations have also been reported in Benin in West Africa, using pyrethroids and formamidines, after introduction of the invasive species (Adehan *et al.*, 2016). In South Africa, high levels of amitraz and pyrethroid resistance and genetic diversity within *R. microplus* populations have been reported (Robbertse *et al.*, 2016), and in Uganda, multi-acaricide resistance and super resistance (0% tick mortality) were observed in the usage of synthetic pyrethroids, organophosphates and co-formulations among *R. appendiculatus* and *R. decoloratus* populations (Vudriko *et al.*, 2016). The cost of managing resistant tick populations makes acaricide treatments more expensive and raises the costs of tick control programmes (George *et al.*, 2004). In the absence of a management strategy, acaricide resistance will be detrimental to the national herd, and livelihoods of resource-poor farmers who depend on cattle production for their mainstay. An Evidence-Based Acaricide Tick Control (EBATIC) strategy to manage acaricide resistance was suggested (Vudriko *et al.*, 2018b), and this includes continuous Identification, Testing, Intervention and Eradication (IT-IE). This strategy needs to be facilitated by a surveillance programme, and requires specialized laboratories to identify and test (IT) acaricide resistant ticks for timely intervention and eradication (IE) (Vudriko *et al.*, 2018b). The results from the laboratory can then be used to guide farmers and extension workers in appropriate management strategies, and in the production and dissemination of advisory material. For effective intervention, this requires working with all stakeholders. After nearly a century of acaricides utilization, it is widely believed that acaricides alone do not provide a sustainable solution to tick and TBDs control.

Anti-tick vaccines can be cost-effective, reduce environmental contamination, and significantly contribute to the reduction in the use of acaricides and therapeutics, therefore preventing the emergence of acaricide resistance. Anti-tick vaccines have been developed against *R. microplus* (Suarez *et al.*, 2016), but these may not provide broad protection against other important tick species. Research is needed to identify novel antigens from the most important tick species circulating among cattle populations in Africa, and then liaise with private sector for vaccine development and distribution.

Immunization of cattle by the infection and treatment method (ITM) offers a valuable option for control of ECF. The challenges faced include disease outbreak due to transmission of vaccine components from vaccinated carriers to non-immunized co-grazing animals, and failure to protect against buffalo-derived *T. parva* parasites (Sitt *et al.*, 2015). There are also variations in challenge intensity associated with vector dynamics, and therefore controlled spraying or dipping against ticks is necessary to maintain immunity and control other TBDs. Widespread use of the ECF vaccine in Africa is also limited by high cost of the vaccine and vaccination, limited cold-chain facilities in rural areas and inadequate expertise (. However, more than 1.6 million cattle have been immunized against ECF using the Muguga cocktail vaccine in eastern, central and southern Africa (Perry, 2016). Large scale trials are underway in Uganda and these are likely to influence regulatory approval in the country (Bishop *et al.*, 2020). Livestock keepers can be organized into cooperative associations to facilitate supply and utilization of the vaccine, which individual herd owners cannot afford. A study in northern Tanzania showed that ITM was adapted by cross-border pastoral communities, which led to reduced incidence of ECF (Lynen et al., 2006). Other studies in southern Kenya and northern Tanzania have shown that the method is profitable and results in increased production and household income that enhances wellbeing, including girl education, nutrition and health (Marsh *et al.*, 2016). Vaccination also reduced acaricide and antibiotic use in animals and decreased annual calf mortality by 80-95% (Marsh *et al.*, 2016). Vaccination is an effective means of controlling bovine anaplasmosis, however, up to date, there is no safe and efficacious vaccine for this disease (Hove *et al.*, 2018). The use of surface proteins of *A. marginale* is a prospect for vaccine development, however, due to high pathogen diversity, developing a vaccine that is cross-protective remains a challenge (Hove et al., 2018).

Live vaccines against *B. bovis* and *B. bigemina* are currently commercialized around the world (Florin-Christensen *et al.*, 2014), and have a desirable level of efficacy (over 95% from a single vaccination) (Bock *et al.*, 2004). However, *B. bovis* and *B. bigemina* live vaccines do not eliminate the parasites but rather produce disease-resistant carrier animals that may act as reservoirs for tick-transmission (Bock *et al.*, 2004). It is possible to develop alternative subunit vaccines, however, relatively few blood stage antigens have been tested as vaccine candidates. For *E. ruminantium*, antigenic strain differences and high diversity limit the protection conferred by Ball3 genotype against some common virulent genotypes, because cross protection between immunotypes of *E. ruminantium* may be minimal (Barbet *et al.*, 2009). In order to develop *E. ruminantium* vaccines, it is important to find genotypes which can confer cross-immunity to as wide a range of *E. ruminantium* field strains as possible (Nakao et al., 2012).

In situations where prevention and control fail, treatment is the secondary strategy in the control of TBDs. However, chemotherapy is expensive, effective only for a short period, and often may be difficult to apply effectively in extensively managed cattle. Moreover, accurate differential diagnosis in the field is difficult and may lead to wrong or delayed treatment. In the rural and remote areas, accessibility and affordability of the drugs can be a challenge (SNV, 2013a). Inappropriate application of drugs leads to development of carriers, residues in the food chain, proliferation of tick-borne infections, and development of drug resistance (SNV, 2013a. Some communities have used ethnoveterinary knowledge in the control of ticks in rural livestock keeping communities. Although ethnoveterinary plants have been evaluated for acaricidal activities (Fouche *et al.*, 2016), further *in vivo* studies are needed to establish the feasibility, effectiveness and safety of the plant extracts on animals.

Given aforementioned shortcomings, an integrated approach for control of ticks and TBDs should be considered, taking into account local epidemiological situations. An effective control strategy can be achieved by integrating the strategic use of acaricides, exploitation of endemic stability, application of vaccines, and the use of tick-resistant cattle breeds (Norval *et al.*, 1992; Minjauw and McLeod, 2003; George *et al.*, 2004). In a population where a large proportion of cattle are susceptible, and vector ticks are absent or of low density, and therefore not a reliable indicator of occurrence of infection, it is logically and financially difficult to sustain immunization. Moreover, there would be little or no challenge from naturally infected ticks which are required to sustain the immunity (Gachohi *et al.*, 2012). In such areas, the most cost-effective control methods seem to be strategic tick control during periods when the vector is found, and effective treatment in clinical cases of successful transmission (Gachohi *et al.*, 2012).

Given the limited number and capacity of extension services for farmers, as observed in South Africa (Raidimi and Kabiti, 2019) and Uganda (Byaruhanga *et al.*, 2015), increased deployment and re-training of the existing lower level personnel is necessary to increase the quality and availability of advice to farmers on the choice of appropriate control methods. More creative and ‘farmer-friendly’ information networks are needed. Farmers will need knowledge on the biology and local ecology of tick species, as well as the clinical picture of TBDs, so as to optimize both the use of acaricides and management of infection (Minjauw and McLeod, 2003).

Control approaches will not be uniform, even when integrated, but will take into account production systems, epidemiological contexts, socio-economic aspects, and level of farmer organization. For example, interventions regarding resistance will be more required in semi-intensive and intensive systems, because these have better access and are likely to apply acaricides more frequently, compared to extension and nomadic systems. Other practices such as footbaths may not be easily accepted or incorporated into usual practices among nomadic farmers due to their transhumance nature (De Meneghi *et al.*, 2016). However, communal cattle dips can be operationalized in traditional systems by organizing the farmers into cooperative societies for a sustainable community-based management of ticks. In more intensive commercial farms that have adequate land for grazing, integration with rotational grazing, pasture spelling, alternative feeding of hay and silage can reduce exposure of cattle to ticks.

FUTURE RESEARCH DIRECTIONS

The current trends in climate change, human settlements and animal movements have a significant effect on the abundance and diversity of pathogens and vectors in Africa. For example, *R. microplus* is spreading to many areas on the continent, requiring further investigation into the patterns of spread, underlying factors and implications, given its competence to transmit *B. bovis*. *Theileria parva* has been detected in Cameroon, even in the absence of *R. appendiculatus*, requiring further surveys in the region with respect to cross-border cattle movements. High antigenic and genetic diversity of tick-borne pathogens is evident, and this has consequences on the effectiveness of potential vaccines. Perhaps an effective vaccine against TBDs will involve a cocktail of several antigens derived from different parasite developmental stages, so contributing to lifelong protective and robust immunity against a spectrum of economically devastating diseases. It is also critically important that the new generation of recombinant antigen vaccines permit the artificial establishment and maintenance of endemic stability.

More information on impact of the TBDs from different production systems and geographical regions is required, for evidence-based data to prioritize control of ticks and TBDs. Attention should focus on identifying the significance of tick-borne infections in the specific production systems and then integrate strategic control measures based on the epidemiology and economics of TBDs. Currently, in most countries in Africa, control of ticks and TBDs is liberalized, which complicates control because poor farmers, or those that have limited access to drugs, may not effectively control the ticks and diseases, leading to persistent/carrier states. While for farmers that have easy access to the drugs, inadequate veterinary services force them to try local solutions to mitigate problems of efficacy of the available acaricides, but this potentially may worsen the negative consequences on future use of chemicals in tick control, and can be detrimental to animal welfare and public health. Given the emergence and spread of acaricide resistance, consistent surveillance for resistant populations and a centralized control in the usage of acaricides are required.

Research is also needed to understand the molecular basis of this widespread acaricide resistance. Further research on sustainable integrated control strategies for ticks and TBDs should include multiple host species and multiple pathogen strains, given high pathogen diversity and the complex dynamics of vector and disease transmission and maintenance at the wildlife-livestock interfaces.

CONCLUSION

This chapter has highlighted the current dynamics in the epidemiology of ticks and TBDs of cattle in Africa. With the current global climate change, their distribution has shifted from the previously expected trends, and therefore epidemiological modelling of the future trends would be essential for effective control. Furthermore, the establishment of the current economic burden of ticks and TBDs of cattle in Africa is paramount.

REFERENCES

Abbas, R. Z., Zaman, M. A., Colwell, D. D., Gilleard, J., & Iqbal, Z. (2014). Acaricide resistance in cattle ticks and approaches to its management: The state of play. *Veterinary Parasitology*, *203*(1-2), 6–20. doi:10.1016/j.vetpar.2014.03.006 PMID:24709006

Adehan, S. B., Adakal, H., Gbinwoua, D., Yokossi, D., Zoungrana, S., Toé, P., Ouedraogo, M., Gbaguidi, A. M., Adoligbé, C., Fandohan, A. B., Hounmanou, G., Kakaï, R. G., Farougou, S., & De Clercq, E. M. (2018). West African cattle farmers' perception of tick-borne diseases. *EcoHealth*, *15*(2), 437–449. doi:10.100710393-018-1323-8 PMID:29536206

Adehan, S. B., Biguezoton, A., Adakal, H., Assogba, M. N., Zoungrana, S., Gbaguidi, A. M., Tonouhewa, A., Kandé, S., Achi, L., Kagone, H., Adehan, R., Mensah, G. A., De Deken, R., Madder, M., & Farougou, S. (2016). Acaricide resistance of *Rhipicephalus microplus* ticks in Benin. *African Journal of Agricultural Research*, *11*(14), 1199–1208. doi:10.5897/AJAR2015.10619

Allsopp, B. A. (2010). Natural history of *Ehrlichia ruminantium. Veterinary Parasitology*, *167*(2-4), 123–135. doi:10.1016/j.vetpar.2009.09.014 PMID:19836892

Allsopp, B. A., Bezuidenhout, J. D., & Prozesky, L. (2004). Heartwater. In J. A. W. Coetzer & R. C. Tustin (Eds.), *Infectious Diseases of Livestock* (2nd ed., pp. 507–535). Oxford University Press Southern Africa.

Allsopp, M. T. E. P., Theron, J., Coetzee, M. L., Dunsterville, M. T., & Allsopp, B. A. (1999). The occurrence of *Theileria* and *Cowdria* parasites in African buffalo (*Syncerus caffer*) and their associated *Amblyomma hebraeum* ticks. *The Onderstepoort Journal of Veterinary Research*, *66*, 245–249. PMID:10631711

Aubry, P. & Geale, D. W. (2010). *Transboundary and Emerging Diseases. A Review of Bovine Anaplasmosis*. Academic Press.

Aubry, P., & Geale, D. W. (2011). A review of bovine anaplasmosis. *Transboundary and Emerging Diseases*, *58*(1), 1–30. doi:10.1111/j.1865-1682.2010.01173.x PMID:21040509

Barbet, A. F., Byrom, B., & Mahan, S. M. (2009). Diversity of *Ehrlichia ruminantium* major antigenic protein 1-2 in field isolates and infected sheep. *Infection and Immunity*, *77*(6), 2304–2310. doi:10.1128/IAI.01409-08 PMID:19307215

Barré, N. (1989). *Biology and ecology of the tick Amblyomma variegatum (Acarina: Ixodina) and Guadeloupe (French West Indies)* (Doctoral thesis in science). University of Paris-Sud-Orsay.

Barré, N., & Uilenberg, G. (2010). Spread of parasites transported with their hosts: Case study of two species of cattle tick. *Revue Scientifique et Technique (International Office of Epizootics)*, *29*(1), 149–160. PMID:20617654

Bishop, R. P., Odongo, D., Ahmed, J., Mwamuye, M., Fry, L. M., Knowles, D. P., Nanteza, A., Lubega, G., Gwakisa, P., Clausen, P. H., & Obara, I. (2020). A review of recent research on *Theileria parva*: Implications for the infection and treatment vaccination method for control of East Coast fever. *Transboundary and Emerging Diseases*, *67*(S1), 56–67. doi:10.1111/tbed.13325 PMID:32174044

Biswas, S. (2003). Role of veterinarian in the care and management during harvest of skin in livestock species. *Proceedings of the National Seminar Leather Industry in Today's perspective*.

Bock, R., Jackson, L., De Vos, A., & Jorgensen, W. (2004). Babesiosis of cattle. *Parasitology*, *129*(S1), S247–S269. doi:10.1017/S0031182004005190 PMID:15938514

Byaruhanga, C., Oosthuizen, M. C., Collins, N. E., & Knobel, D. (2015b). Using participatory epidemiology to investigate management options and relative importance of tick-borne diseases amongst transhumant zebu cattle in Karamoja Region, Uganda. *Preventive Veterinary Medicine*, *122*(3), 287–297. doi:10.1016/j.prevetmed.2015.10.011 PMID:26527312

CABI. (2020). *Theileria annulata infections. Invasive Species Compendium*. CAB International. https://www.cabi.org/isc/datasheet/62107#tooverview

Chaisi, M. E., Sibeko, K. P., Collins, N. E., Potgieter, F. T., & Oosthuizen, M. C. (2011). Identification of *Theileria parva* and *Theileria* sp. (buffalo) 18S rRNA gene sequence variants in the African buffalo (*Syncerus caffer*) in southern Africa. *Veterinary Parasitology*, *182*(2-4), 150–162. doi:10.1016/j.vetpar.2011.05.041 PMID:21700394

Cumming, G. S. (2002). Comparing climate and vegetation as limiting factors for species ranges of African ticks. *Ecology*, *83*(1), 255–268. doi:10.1890/0012-9658(2002)083[0255:CCAVAL]2.0.CO;2

De Deken, R., Martin, V., Saido, A., Madder, M., Brandt, J., & Geysen, D. (2007). An outbreak of East Coast Fever on the Comoros: A consequence of the import of immunized cattle from Tanzania? *Veterinary Parasitology*, *143*(3-4), 245–253. doi:10.1016/j.vetpar.2006.08.018 PMID:16996692

De Meneghi, D., Stachurski, F., & Adakal, H. (2016). Experiences in tick control by acaricide in the traditional cattle sector in Zambia and Burkina Faso: Possible environmental and public health implications. *Frontiers in Public Health*, *4*, 239. doi:10.3389/fpubh.2016.00239 PMID:27882313

De Vos, A. J., De Waal, D. T., & Jackson, L. A. (2004). Bovine babesiosis. In J. A. W. Coetzer & R. C. Tustin (Eds.), *Infectious Diseases of Livestock* (pp. 406–424). Oxford University Press Southern Africa.

Deem, S. L., Perry, B. D., Katende, J. M., McDermott, J. J., Mahan, S. M., Maloo, S. H., Morzaria, S. P., Musoke, A. J., & Rowlands, G. J. (1993). Variations in prevalence rates of tick-borne diseases in Zebu cattle by agro-ecological zone: Implications for the East Coast fever immunization. *Preventive Veterinary Medicine*, *16*(3), 171–187. doi:10.1016/0167-5877(93)90064-Z

Estrada-Peña, A. (2003). Climate change decreases habitat suitability for some tick species (Acari: Ixodidae) in South Africa. *The Onderstepoort Journal of Veterinary Research*, *70*(2), 79–93. PMID:12967169

Estrada-Peña, A., Horak, I. G., & Petney, T. N. (2008). Changes in climate and their suitability for the ticks *Amblyomma hebraeum* and *Amblyomma variegatum* (*Ixodidae*) in Zimbabwe (1974–1999). *Veterinary Parasitology*, *151*(2-4), 256–267. doi:10.1016/j.vetpar.2007.11.014 PMID:18083309

Eygelaar, D., Jori, F., Mokopasetso, M., Sibeko, K. P., Collins, N. E., Vorster, I., Troskie, M., & Oosthuizen, M. C. (2015). Tick-borne haemoparasites in African buffalo (*Syncerus caffer*) from two wildlife areas in Northern Botswana. *Parasites & Vectors*, *8*(1), 26. doi:10.118613071-014-0627-y PMID:25589064

Florin-Christensen, M., Suarez, C. E., Rodriguez, A. E., Flores, D. A., & Schnittger, L. (2014). Vaccines against bovine babesiosis: Where we are now and possible roads ahead. *Parasitology*, *141*(12), 1563–1592. doi:10.1017/S0031182014000961 PMID:25068315

Fouche, G., Ramafuthula, M., Maselela, V., Mokoena, M., Senabe, J., Leboho, T., Sakong, B. M., Adenubi, O. T., Eloff, J. N., & Wellington, K. W. (2016). Acaricidal activity of the organic extracts of thirteen South African plants against *Rhipicephalus* (*Boophilus*) *decoloratus* (Acari: Ixodidae). *Veterinary Parasitology*, *224*, 39–43. doi:10.1016/j.vetpar.2016.05.011 PMID:27270388

Gachohi, J., Skilton, R., Hansen, F., Ngumi, P., & Kitala, P. (2012). Epidemiology of East Coast fever (*Theileria parva* infection) in Kenya: Past, present and the future. *Parasites & Vectors*, *5*(1), 194. doi:10.1186/1756-3305-5-194 PMID:22958352

George, J. E., Pound, J. M., & Davey, R. B. (2004). Chemical control of ticks on cattle and the resistance of these parasites to acaricides. *Parasitology*, *129*(S1), S353–S366. doi:10.1017/S0031182003004682 PMID:15938518

Gomes, A. F., & Neves, L. (2018). *Rhipicephalus microplus* (Acarina, Ixodidae) in Angola: Evidence of its establishment and expansion. *Experimental & Applied Acarology*, *74*(1), 117–122. doi:10.100710493-017-0207-x PMID:29380168

Hemmink, J. D., Sitt, T., Pelle, R., Klerk-Lorist, L. M., Shiels, B., Toye, P. G., Morrison, W. I., & Weir, W. (2018). Ancient diversity and geographical sub-structuring in African buffalo *Theileria parva* populations revealed through metagenetic analysis of antigen-encoding loci. *International Journal for Parasitology*, *48*(3-4), 287–296. doi:10.1016/j.ijpara.2017.10.006 PMID:29408266

Homewood, K., Trench, P., Randall, S., Lynen, G., & Bishop, B. (2006). Livestock health and socio-economic impacts of a veterinary intervention in Maasai land: Infection-and-treatment vaccine against East Coast fever. *Agricultural Systems*, *89*(2-3), 248–271. doi:10.1016/j.agsy.2005.09.004

Hove, P., Chaisi, M. E., Brayton, K. A., Ganesan, H., Catanese, H. N., Mtshali, M. S., Mutshembele, A. M., Oosthuizen, M. C., & Collins, N. E. (2018). Co-infections with multiple genotypes of *Anaplasma marginale* in cattle indicate pathogen diversity. *Parasites & Vectors*, *11*(1), 5. doi:10.118613071-017-2595-5 PMID:29298712

Hurtado, O. J. B., & Giraldo-Ríos, C. (2018). Economic and Health Impact of the Ticks in Production Animals. In *Ticks and Tick-Borne Pathogens*. https://www.intechopen.com/books/ticks-and-tick-borne-pathogens/economic-and-health-impact-of-the-ticks-in-production-animals

Jongejan, F., & Uilenberg, G. (2004). The global importance of ticks. *Parasitology*, *129*(S1), S3–S14. doi:10.1017/S0031182004005967 PMID:15938502

Jonsson, N. N. (2006). The productivity effects of cattle tick (*Boophilus microplus*) infestation on cattle, with particular reference to *Bos indicus* cattle and their crosses. *Veterinary Parasitology*, *137*(1-2), 1–10. doi:10.1016/j.vetpar.2006.01.010 PMID:16472920

Kamani, J., Apanaskevich, D. A., Gutiérrez, R., Nachum-Biala, Y., Baneth, G., & Harrus, S. (2017). Morphological and molecular identification of *Rhipicephalus* (*Boophilus*) *microplus* in Nigeria, West Africa: A threat to livestock health. *Experimental & Applied Acarology*, *73*(2), 283–296. doi:10.100710493-017-0177-z PMID:28887701

Kanduma, E. G., Emery, D., Githaka, N. W., Nguu, E. K., Bishop, R. P., & Šlapeta, J. (2020). Molecular evidence confirms occurrence of *Rhipicephalus microplus* Clade A in Kenya and sub-Saharan Africa. *Parasites & Vectors*, *13*(1), 432. doi:10.118613071-020-04266-0 PMID:32854747

Kivaria, F.M. (2006). Estimated direct economic costs associated with tick-borne diseases on cattle in Tanzania. *Tropical Animal Health and Production, 38*, 291-299.

Kivaria, F. M., Kapanga, A. M., Mbassa, G. K., Mtui, P. F., & Wani, R. J. (2012). Epidemiological perspectives of ticks and tick-borne diseases in South Sudan: Cross-sectional survey results. *The Onderstepoort Journal of Veterinary Research*, *79*(1), E1–E10. doi:10.4102/ojvr.v79i1.400 PMID:23327317

Kocan, K. M., de la Fuente, J., Blouin, E. F., Coetzee, J. F., & Ewing, S. A. (2010). The natural history of *Anaplasma marginale*. *Veterinary Parasitology*, *167*(2-4), 95–107. doi:10.1016/j.vetpar.2009.09.012 PMID:19811876

Kocan, K. M., de la Fuente, J., Guglielmone, A. A., & Meléndez, R. D. (2003). Antigens and Alternatives for Control of *Anaplasma marginale* Infection in Cattle. *Clinical Microbiology Reviews*, *16*(4), 698–712. doi:10.1128/CMR.16.4.698-712.2003 PMID:14557295

Latif, A. A., Rowlands, G. J., Punyua, D. K., Hassan, S. M., & Capstick, P. B. (1994). An epidemiological study of tick-borne diseases and their effects on productivity of zebu cattle under traditional management on Rusinga Island, western Kenya. *Preventive Veterinary Medicine*, *22*(3), 169–181. doi:10.1016/0167-5877(94)00408-B

Latif, A. A., & Walker, A. R. (2004). *An introduction to the biology and control of ticks in Africa*. www.alanrwalker.com/assets/PDF/tick-biology-africa.pdf

Lawrence, J. A., Perry, B. D., & Williamson, S. M. (1992). Infectious Diseases of Livestock. Oxford University Press.

Lawrence, J. A., Perry, B. D., & Williamson, S. M. (2004a). Corridor disease. In J. A. W. Coetzer & R. C. Tustin (Eds.), *Infectious Diseases of Livestock* (2nd ed., pp. 468–471). Oxford University Press Southern Africa.

Lawrence, J. A., Perry, B. D., & Williamson, S. M. (2004b). East Coast fever. In J. A. W. Coetzer & R. C. Tustin (Eds.), *Infectious Diseases of Livestock* (2nd ed., pp. 448–467). Oxford University Press Southern Africa.

Lawrence, J. A., Perry, B. D., & Williamson, S. M. (2004c). Zimbabwe theileriosis. In J. A. W. Coetzer & R. C. Tustin (Eds.), *Infectious Diseases of Livestock* (2nd ed., pp. 472–474). Oxford University Press Southern Africa.

Lubembe, D. M., Odongo, D. O., Salih, D. A., & Sibeko-Matjila, K. P. (2020). Microsatellite and mini-satellite genotyping of *Theileria parva* population from southern Africa reveals possible discriminatory allele profiles with parasites from eastern Africa. *Ticks and Tick-Borne Diseases*, *11*(6), 101539. doi:10.1016/j.ttbdis.2020.101539 PMID:32993948

Lynen, G., Zeman, P., Bakuname, C., Di Giulio, G., Mtui, P., Sanka, P., & Jongejan, F. (2008). Shifts in the distribution ranges of *Boophilus* ticks in Tanzania: Evidence that a parapatric boundary between *Boophilus microplus* and *B. decoloratus* follows climate gradients. *Experimental & Applied Acarology*, *44*(2), 147–164. doi:10.100710493-008-9134-1 PMID:18266058

Mahan, S. M., Peter, T. F., Semu, S. M., Simbi, B. H., Norval, R. A. I., & Barbet, A. F. (1995). Laboratory reared *Amblyomma hebraeum* and *Amblyomma variegatum* ticks differ in their susceptibility to infection with *Cowdria ruminantium*. *Epidemiology and Infection*, *115*(2), 345–353. doi:10.1017/S0950268800058465 PMID:7589273

Marcellino, W. L. (2004). Current situation of East Coast fever in Bahr el Jebel State. *East Coast Fever in southern Sudan. Proceedings of a workshop*, 8-9.

Marsh, T. L., Yoder, J., Deboch, T., McElwain, T. F., & Palmer, G. H. (2016). Livestock vaccinations translate into human capital and school attendance by girls. *Science Advances*, *2*(12), e1601410. doi:10.1126ciadv.1601410 PMID:27990491

Martin, V., Chevalier, V., Ceccato, P., Anyamba, A., De Simone, L., Lubroth, J., de La Rocque, S., & Domenech, J. (2008). The impact of climate change on the epidemiology and control of Rift Valley fever. *Revue Scientifique et Technique (International Office of Epizootics)*, *27*(2), 413–426. doi:10.20506/rst.27.2.1802 PMID:18819669

Marufu, M. C., Chimonyo, M., Mans, B. J., & Dzama, K. (2013). Cutaneous hypersensitivity responses to *Rhipicephalus* tick larval antigens in pre-sensitized cattle. *Ticks and Tick-Borne Diseases*, *4*(4), 311–316. doi:10.1016/j.ttbdis.2012.12.001 PMID:23453577

Mattioli, R. C., Pandey, V. S., Murray, M., & Fitzpatrick, J. L. (2000). Immunogenetic influences on tick resistance in African cattle with particular reference to trypano-tolerant N'Dama (*Bos taurus*) and trypano-susceptible Gobra zebu (*Bos indicus*) cattle. *Acta Tropica*, *75*(3), 263–277. doi:10.1016/S0001-706X(00)00063-2 PMID:10838210

Meltzer, M. I., Norval, R. A. I., & Donachie, P. L. (1995). Effects of tick infestation and tick-borne disease infections (heartwater, anaplasmosis and babesiosis) on the lactation and weight-gain of Mashona cattle in southeastern Zimbabwe. *Tropical Animal Health and Production*, *27*(3), 129–144. doi:10.1007/BF02248956 PMID:7502343

Minjauw, B., & McLeod, A. (2003). *Tick-borne diseases and poverty. The impact of ticks and tick-borne diseases on the livelihood of small-scale and marginal livestock owners in India and eastern and southern Africa. Research report, DFID Animal Health Programme*. Centre for Tropical Veterinary Medicine, University of Edinburgh.

Muhanguzi, D., Byaruhanga, J., Amanyire, W., Ndekezi, C., Ochwo, S., Nkamwesiga, J., Mwiine, F. N., Tweyongyere, R., Fourie, J., Madder, M., Schetters, T., Horak, I., Juleff, N., & Jongejan, F. (2020). Invasive cattle ticks in East Africa: Morphological and molecular confirmation of the presence of *Rhipicephalus microplus* in south-eastern Uganda. *Parasites & Vectors*, *13*(1), 165. doi:10.118613071-020-04043-z PMID:32245511

Mukhebi, A. W., Chamboko, T., O'collaghan, C. J., Peter, T. F., Kruska, R. L., Medley, G. F., Mahan, S. M., & Perry, B. D. (1999). An assessment of the economic impact of heartwater (*Cowdria ruminantium* infection) and its control in Zimbabwe. *Preventive Veterinary Medicine*, *39*(3), 173–189. doi:10.1016/S0167-5877(98)00143-3 PMID:10327437

Mukhebi, A. W., Perry, B., & Kruska, R. (1992). Estimated economics of theileriosis control in Africa. *Preventive Veterinary Medicine*, *12*(1-2), 73–85. doi:10.1016/0167-5877(92)90070-V

Nah, K., Bede-Fazekas, A., Trájer, A. J., & Wu, J. (2020). The potential impact of climate change on the transmission risk of tick-borne encephalitis in Hungary. *BMC Infectious Diseases*, *20*(1), 34. doi:10.118612879-019-4734-4 PMID:31931734

Nakao, R., Morrison, L. J., Zhou, L., Magona, J. W., Jongejan, F., & Sugimoto, C. (2012). Development of multiple-locus variable-number tandem-repeat analysis for rapid genotyping of *Ehrlichia ruminantium* and its application to infected *Amblyomma variegatum* collected in heartwater endemic areas in Uganda. *Parasitology*, *139*(1), 69–82. doi:10.1017/S003118201100165X PMID:22008706

Norval, R. A., Fivaz, B. H., Lawrence, J. A., & Daillecourt, T. (1983). Epidemiology of tick-borne diseases of cattle in Zimbabwe. I. Babesiosis. *Tropical Animal Health and Production*, *15*(2), 87–94. doi:10.1007/BF02239802 PMID:6868134

Norval, R. A. I., Perry, B. D., Young, A. S., Lawrence, J. A., Mukhebi, A. W., Bishop, R., & McKeever, D. (1992). *The Epidemiology of Theileriosis in Africa*. Academic press.

Nyangiwe, N., Goni, S., Hervé-Claude, L. P., Ruddat, I., & Horak, I. G. (2011). Ticks on pastures and on two breeds of cattle in the Eastern Cape Province, South Africa. *The Onderstepoort Journal of Veterinary Research*, *78*(1), 1–9. doi:10.4102/ojvr.v78i1.320 PMID:23327215

Nyangiwe, N., Haarrison, A., & Horak, I. G. (2013). Displacement of *Rhipicephalus decoloratus* by *Rhipicephalus microplus* (Acari: Ixodidae) in the Eastern Cape Province, South Africa. *Experimental & Applied Acarology*, *61*(3), 371–382. doi:10.100710493-013-9705-7 PMID:23771510

Nyangiwe, N., Yawa, M., & Muchenje, V. (2018). Driving forces for changes in geographic range of cattle ticks (Acari: Ixodidae) in Africa: A review. *South African Journal of Animal Science*, *48*(5), 829. doi:10.4314ajas.v48i5.4

Obara, I., Githaka, N., Nijhof, A., Krücken, J., Nanteza, A., Odongo, D., Lubembe, D., Atimnedi, P., Mijele, D., Njeri, A., Mwaura, S., Owido, G., Ahmed, J., Clausen, P. H., & Bishop, R. P. (2020). The *Rhipicephalus appendiculatus* tick vector of *Theileria parva* is absent from Cape buffalo (*Syncerus caffer*) populations and associated ecosystems in northern Uganda. *Parasitology Research*, *119*(7), 2363–2367. doi:10.100700436-020-06728-x PMID:32500369

Ocaido, M., Muwazi, R. T., & Opuda, J. A. (2009). Economic impact of ticks and tick-borne diseases on cattle production systems around Lake Mburo National Park in South Western Uganda. *Tropical Animal Health and Production*, *41*(5), 731–739. doi:10.100711250-008-9245-z PMID:18958579

Okello-Onen, J., Tukahirwa, E. M., Perry, B. D., Rowlands, G. J., Nagda, S. M., Musisi, G., Bode, E., Heinonen, R., Mwayi, W., & Opuda-Asibo, J. (1999). Population dynamics of ticks on indigenous cattle in a pastoral dry to semi-arid rangeland zone of Uganda. *Experimental & Applied Acarology*, *23*(1), 79–88. doi:10.1023/A:1006058317111 PMID:10353189

Olwoch, J. M., Reyers, B., Engelbrecht, F. A., & Erasmus, B. F. N. (2008). Climate change and the tick-borne disease, theileriosis (East Coast fever) in sub-Saharan Africa. *Journal of Arid Environments*, *72*(2), 108–120. doi:10.1016/j.jaridenv.2007.04.003

Onono, J. O., Wieland, B., & Rushton, J. (2013). Constraints to cattle production in a semiarid pastoral system in Kenya. *Tropical Animal Health and Production*, *45*(6), 1415–1422. doi:10.100711250-013-0379-2 PMID:23417827

Oura, C. A. L., Tait, A., Asiimwe, B., Lubega, G. W., & Weir, W. (2011). Haemoparasite prevalence and *Theileria parva* strain diversity in Cape buffalo (*Syncerus caffer*) in Uganda. *Veterinary Parasitology*, *175*(3-4), 212–219. doi:10.1016/j.vetpar.2010.10.032 PMID:21074945

Pegram, R. G., & Banda, D. S. (1990). Ecology and phenology of cattle ticks in Zambia: Development and survival of free-living stages. *Experimental & Applied Acarology*, *8*(4), 291–301. doi:10.1007/BF01202139 PMID:2350995

Perry, B. D. (2016). The control of East Coast fever of cattle by live parasite vaccination: A science-to-impact narrative. *One Health*, *2*, 103–114. doi:10.1016/j.onehlt.2016.07.002 PMID:28616483

Perry, B. D., Chamboko, T., Mahan, S. M., Medley, G. F., Minjauw, B., O'callaghan, C. J., & Peter, T. F. (1998). The economics of integrated tick and tick-borne disease control on commercial farms in Zimbabwe. *Zimbabwe Veterinary Journal*, *29*, 21–29.

Raidimi, E. N., & Kabiti, H. M. (2019). A review of the role of agricultural extension and training in achieving sustainable food security: A case of South Africa. *South African Journal of Agricultural Extension*, *47*(3). Advance online publication. doi:10.17159/2413-3221/2019/v47n3a520

Regitano, L., & Prayaga, K. (2011). Ticks and tick-borne diseases in cattle. In S. Bishop, R. Axford, F. Nicholas, & J. Owen (Eds.), *Breeding for Disease Resistance in Farm Animals* (3rd ed., pp. 295–314). CAB International.

Robbertse, L., Baron, S., Van der Merwe, N. A., Madder, M., Stoltsz, W. H., & Maritz-Olivier, C. (2016). Genetic diversity, acaricide resistance status and evolutionary potential of a *Rhipicephalus microplus* population from a disease-controlled cattle farming area in South Africa. *Ticks and Tick-Borne Diseases*, *7*(4), 595–603. doi:10.1016/j.ttbdis.2016.02.018 PMID:26965787

Rubaire-Akiiki, C. M., Okello-Onen, J., Nasinyama, G. W., Vaarst, M., Kabagambe, E. K., Mwayi, W., Musunga, D., & Wandukwa, W. (2004). The prevalence of serum antibodies to tick-borne infections in Mbale District, Uganda: The effect of agro-ecological zone, grazing management and age of cattle. *Journal of Insect Science*, *4*(8), 1–8. doi:10.1673/031.004.0801 PMID:15861224

Salih, A. D., Marcellino, W. L., & Elhussein, J. A. M. (2011). Economic impact of East Coast fever in central equatorial state of South Sudan. *International Research Journal of Agricultural Science and Soil Science*, *1*(6), 218–220.

Salih, D. A., Mwacharo, J. M., Pelle, R., Njahira, M. N., Odongo, D. O., Mbole-Kariuki, M. N., Marcellino, W. L., Malak, A. K., Kiara, H., El Hussein, A. R. M., Bishop, R. P., & Skilton, R. A. (2018). Genetic diversity and population structure of Theileria parva in South Sudan. *Ticks and Tick-Borne Diseases*, *9*(4), 806–813. doi:10.1016/j.ttbdis.2018.03.002 PMID:29534988

Siddig, A., El Malik, K., & Hameida, T. A. (2003). Estimation of the financial losses during an outbreak of tropical Theilerosis in dairy farm. *Sudan J. Vet. Sci. Anim. Husb.*, *42*, 346.

Silatsa, B. A., Kuiate, J. R., Njiokou, F., Simo, G., Feussom, J. M. K., Tunrayo, A., Amzati, G. S., Bett, B., Bishop, R., Githaka, N., Opiyo, S. O., Djikeng, A., & Pelle, R. (2019). A countrywide molecular survey leads to a seminal identification of the invasive cattle tick *Rhipicephalus* (*Boophilus*) *microplus* in Cameroon, a decade after it was reported in Cote d'Ivoire. *Ticks and Tick-Borne Diseases*, *10*(3), 585–593. doi:10.1016/j.ttbdis.2019.02.002 PMID:30765191

Silatsa, B. A., Simo, G., Githaka, N., Kamga, R., Oumarou, F., Tiambo, C. K., Machuka, E., Domelevo, J.-B., Odongo, D., Bishop, R., Kuiate, J.-R., Njiokou, F., Djikeng, A., & Pelle, R. (2020). First detection of *Theileria parva* in cattle from Cameroon in the absence of the main tick vector *Rhipicephalus appendiculatus*. *Transboundary and Emerging Diseases*, *67*(S1), 68–78. doi:10.1111/tbed.13425 PMID:32174039

Sitt, T., Henson, S., Morrison, W. I., & Toye, P. (2019). Similar levels of diversity in the gene encoding the p67 sporozoite surface protein of *Theileria parva* are observed in blood samples from buffalo and cattle naturally infected from buffalo. *Veterinary Parasitology*, *269*, 21–27. doi:10.1016/j.vetpar.2019.04.006 PMID:31079824

Sitt, T., Poole, E. J., Ndambuki, G., Mwaura, S., Njoroge, T., Omondi, G. P., Mutinda, M., Mathenge, J., Prettejohn, G., Morrison, W. I., & Toye, P. (2015). Exposure of vaccinated and naive cattle to natural challenge from buffalo-derived Theileria parva. *International Journal for Parasitology. Parasites and Wildlife*, *4*(2), 244–251. doi:10.1016/j.ijppaw.2015.04.006 PMID:26005635

SNV. (2013a). *Management of Livestock Drug Systems. Dairy Platform*. Policy Brief 03. SNV Netherlands Development Organization. http://www.snvworld.org/download/publications/ecf_brief_3_management_pdf

Stachurski, F. (2000). Invasion of West African cattle by the tick *Amblyomma variegatum*. *Medical and Veterinary Entomology*, *14*(4), 391–399. doi:10.1046/j.1365-2915.2000.00246.x PMID:11129703

Suarez, M., Rubi, J., Pérez, D., Cordova, V., Salazar, Y., Vielma, A., Barrios, F., Gil, C. A., Segura, N., Carrillo, Y., Cartaya, R., Palacios, M., Rubio, E., Escalona, C., Ramirez, R. C., Baker, R. B., Machado, H., Sordo, Y., Bermudes, J., ... Estrada, M. P. (2016). High impact and effectiveness of Gavac™ vaccine in the national program for control of bovine ticks *Rhipicephalus microplus* in Venezuela. *Livestock Science*, *187*, 48–52. doi:10.1016/j.livsci.2016.02.005

Taylor, M., Coop, R., & Wall, R. (2016). Veterinary entomology. In M. Taylor, R. Coop, & R. Wall (Eds.), *Veterinary Parasitology* (4th ed., pp. 161–258). Wiley-Blackwell.

Tønnesen, M. H., Penzhorn, B. L., Bryson, N. R., Stoltsz, W. H., & Masibigiri, T. (2004). Displacement of *Boophilus decoloratus* by *Boophilus microplus* in the Soutpansberg region, Limpopo Province, South Africa. *Experimental & Applied Acarology*, *32*(3), 199–208. doi:10.1023/B:APPA.0000021789.44411.b5 PMID:15139085

Vudriko, P., Okwee-Acai, J., Byaruhanga, J., Tayebwa, D. S., Okech, S. G., Tweyongyere, R., Wampande, E. M., Okurut, A. R. A., Mugabi, K., Muhindo, J. B., Nakavuma, J. L., Umemiya-Shirafuji, R., Xuan, X., & Suzuki, H. (2018a). Chemical tick control practices in southwestern and northwestern Uganda. *Ticks and Tick-Borne Diseases*, *9*(4), 945–955. doi:10.1016/j.ttbdis.2018.03.009 PMID:29606621

Vudriko, P., Okwee-Acai, J., Byaruhanga, J., Tayebwa, D. S., Omara, R., Muhindo, J. B., Lagu, C., Umemiya-Shirafuji, R., Xuan, X., & Suzuki, H. (2018b). Evidence-based tick acaricide resistance intervention strategy in Uganda: Concept and feedback of farmers and stakeholders. *Ticks and Tick-Borne Diseases*, *9*(2), 254–265. doi:10.1016/j.ttbdis.2017.09.011 PMID:28978456

Vudriko, P., Okwee-Acai, J., Tayebwa, D. S., Byaruhanga, J., Kakooza, S., Wampande, E., Omara, R., Muhindo, J. B., Tweyongyere, R., Owiny, D. O., Hatta, T., Tsuji, N., Umemiya-Shirafuji, R., Xuan, X., Kanameda, M., Fujisaki, K., & Suzuki, H. (2016). Emergence of multi-acaricide resistant *Rhipicephalus* ticks and its implication on chemical control in Uganda. *Parasites & Vectors*, *9*(1), 4. doi:10.118613071-015-1278-3 PMID:26727991

Walker, A. R., Latif, A. A., Hans, N., Nijhof, A., & Jongejan, F. (2014b). *Electronic Tick Guide for Africa. Developed by the International Consortium on Ticks and Tick-borne Diseases*. ICTTD.

Walker, J. G., Klein, E. Y., & Levin, S. A. (2014a). Disease at the wildlife-livestock interface: Acaricide use on domestic cattle does not prevent transmission of a tick-borne pathogen with multiple hosts. *Veterinary Parasitology*, *199*(3-4), 206–214. doi:10.1016/j.vetpar.2013.11.008 PMID:24315187

ADDITIONAL READING

Perry, B. D. (2009). Economic impacts of tick-borne diseases in Africa: Tick-borne diseases. *The Onderstepoort Journal of Veterinary Research*, *76*(1). Advance online publication. doi:10.4102/ojvr.v76i1.63

Young, A., Groocock, C., & Kariuki, D. (1988). Integrated control of ticks and tick-borne diseases of cattle in Africa. *Parasitology*, *96*(2), 403–432. doi:10.1017/S0031182000058388 PMID:3287285

KEY TERMS AND DEFINITIONS

Acaricide: Chemicals used for controlling ticks.

Antigenic/Genetic Diversity: Variation in the genetic composition or within the antigen-encoding genes.

Carrier: An apparently healthy animal which is infected with a pathogenic organism. The carrier can transmit the organism to other animals which may then develop disease.

Endemic: The occurrence of a disease in a population or region at 'normally' expected levels.

Epidemiology: Is the study of the distribution and determinants of health-related states or events in specified populations.

Host: Is an animal that harbours an infectious agent.

Incidence: Is a measure of disease frequency. It quantifies the number of new cases (animals newly acquiring a disease) in a population at risk of developing the disease over a given period of time.

One-Host Tick: Tick that completes all its molts on the same individual host animal.

Prevalence: Is a measure of disease frequency. It quantifies the number of existing cases (animals with a disease) in a population at a point in time or during a period of time.

Seroprevalence: Frequency of animals with antibodies to a specific parasite.

Three-Host Tick: Tick that seeks a new host animal for a blood meal after each of its three molts.

Two-Host Tick: Tick that remains in one host for the larval and nymphal stages, drops off to molt an adult, and seeks a second host on which to take another blood meal and copulate.

Chapter 8
An Overview of Hard and Soft Ticks and Their Control Methods

Charles Muleke Inyagwa
Department of Veterinary Pathology, Microbiology and Parasitology, Faculty of Veterinary Medicine and Surgery, Egerton University, Kenya

Frederick Ojiambo Obonyo
Department of Animal Science, School of Agriculture and Food Science, Meru University of Science and Technology, Kenya

Donald Lubembe Mukolwe
Department of Veterinary Pathology, Microbiology and Parasitology, Faculty of Veterinary Medicine and Surgery, Egerton University, Kenya

Eric O. Mungube
Kenya Agricultural and Livestock Research Organization, Veterinary Science Research Institute, Kenya

ABSTRACT

Ticks are important ectoparasites of livestock in tropical and sub-tropical countries. They are responsible for the enormous economic losses both through the direct effects of blood sucking and indirectly as vectors of disease pathogens and toxins. Morphologically, ticks are classified into two families known as Ixodidae (hard ticks) and Argasidae (soft ticks). In this chapter, the authors provide a description of the soft and hard ticks. Emphasis is laid on the genera morphology, species, mammalian hosts, and distribution and the diseases transmitted by the ticks species. Various methods for controlling ticks and tick-borne diseases using chemicals are described. A broad range of chemical acaricides used for the control of ticks on livestock are described. Conventional methods and unconventional methods for applying acaricides to protect livestock against ticks are described. Alternative methods for tick control are described. Only the most important hard and soft ticks and associated diseases are discussed within the limits of this review.

DOI: 10.4018/978-1-7998-6433-2.ch008

AN OVERVIEW OF HARD AND SOFT TICKS AND THEIR CONTROL METHODS

Introduction

Ticks are obligate blood sucking ectoparasites of vertebrates, particularly mammals, birds, terrestrial reptiles. Ticksare arachnids in the subclass Acari and are closely related to mites. Two well-defined families of ticks exist, the Ixodidae or hard ticks and the Argasidae or soft ticks, which differ from each other in morphological features, habits, host spectrum, and life cycle, (Shivairo *et al.*, 2018). Over 80% of the world's 1,281 million cattle are infested with ticks and 186 million cattle are at risk of tick infestation and the accompanying tick-borne diseases (TBDs) (Getahun *et al.*, 2016). Besides the direct losses, ticks have the greatest impact due to their large numbers and variety of microbial disease pathogens that they transmit among domestic animals. Most of the tick-borne infections of humans, farm and companion animals are essentially associated with wildlife animal reservoirs (Baneth, 2014). Ticks and the diseases they transmit are widely distributed throughout the world, (Shivairo *et al.*, 2018), particularly in tropical and subtropical countries (Getahun *et al.*, 2016). Ticksare responsible for severe losses caused either by the direct effect of the tick through: blood loss, damage to hides and skins and the injection of toxins or through mortality or debility caused by the diseases they transmit (Getahun *et al.*, 2016). Globally, ticks transmit a wide spectrum of pathogenic microorganisms such as protozoa, rickettsia, spirochetes and viruses to humans, farm and companion animals essentially harbored by wildlife animal reservoirs (Baneth, 2014). In Africa, TBDssuch as theileriosis, babesiosis, and rickettsia diseases like anaplasmosis and heartwater (cowdriosis) are devastating to domestic animals (Shivairo *et al.*,2018). It is estimated that the annual global costs associated with ticks and TBDs in cattle lies between US $ 13.9 and 18.7 billion, globally, (DeCastro, 1997). In Africa, TBDs are considered to be the most important impediment to the development of the livestock subsector (Young *et al.*, 1988). Although it is difficult to quantity the economic losses due to ticks, foregoing figures provide a rough estimate on the impact and effects of ticks to the livestock industry globally. This review provides a description of the taxonomy, morphology and disease transmitted by various tick species in the Ixodidae and Argasidae families and their control methods. Only the most important hard and soft ticks and associated diseases are briefly discussed within the limits of this review.

Back ground

Ticks are obligated blood feeding ectoparasite of vertebrates, surviving for several years. During this time, they feed periodically taking large blood meals, often interspersed with long intervals each meal (Wall and Shearer, 1997). Ticks and the diseases they transmit are widely distributed throughout the world (Nicholson *et al.,* 2010), particularly in tropical and subtropical countries. Ticks are responsible for severe losses caused either by the direct effect of the tick through: blood loss, damage to hides and udders and the injection of toxins or through mortality or debility caused by the disease transmitted (Rajput *et al.*, 2006). In Africa, tick-borne diseases are considered to be the most important animal disease problem (Young *et al.*, 1988).

Classification of Hard Ticks

Ticks are classified under the phylum Arthropoda; subphylum Chelicerata; Class Arachnida; Order Acarina (Urquhart *et al.*, 1996). Ticks are classified (together with mites) in the order Acari. The present tick numbers worldwide are estimated to be 889 with Ixodid being predominant (702) followed by Argasid (186) and 1-Nuttalliella tick (sub-) species), (Nijhof *et al.*, 2016). Family names Ixodidaeor "hard ticks" and Argasidae or "soft ticks" is by the virtue of their hard-dorsal shield and their flexible leathery cuticle respectively (Getahun *et al.*,2016). Ticks are dioecious having separate sex, with hard ticks showing sexual dimorphism relative to soft ticks (Getahun *et al.*, 2016).

Taxonomy and Biology Hard Ticks

All ticks are obligate ectoparasites of vertebrates and are widely distributed among host taxa, which include marsupial and placental mammals, birds, reptiles (snakes, iguanas, and lizards) and amphibians (Walker and Olwage, 1987). Ticks have four pairs of legs as nymphs and adults, and the body is divided into the capitulum (which bears the mouthparts) or cephalo-thorax and the opisthosoma or abdomen. Antennae in ticks are absent and the mouth parts lack mandibles while the palps, chelicerae and hypostome are borne by the false head or capitulum. Ticks evolved during the Cretaceous period, the most common form of fossilization being amber immersion. Tick species are widely distributed around the world wherever their host species occur, (Getahun *et al.*, 2016), but they tend to flourish more in countries with warm, humid climates, because they require a certain amount of moisture in the air to undergo metamorphosis, and because low temperatures inhibit their development from eggs to larvae (Getahun *et al.*, 2016). Migrating birds carry ticks with them on their journeys; a study of migratory birds passing through Egypt found more than half of the bird species examined were carrying ticks. The tick species often differed between the autumn and spring migrations, probably because of the seasonal periodicities of different species (Walker and Olwage, 1987). For a detailed descriptionof thetaxonomy and tick biology of the Ixodidae, the reader is referred to the recent Veterinary Parasitology book by (Taylor *et al.*, 2015).

Life-Cycle of Ixodidae Ticks

For the detailed account of the life cycle of hard ticks, readers are referred to the recent book publication by, (Shivairo *et al.*, 2018) and the OIE publication (Jongejan and Ulineberg, 1994). In brief Ixodidae, exhibit three types of life cycles that can be distinguished based on the tropisms revealed by ticks at different instars. These are the one-host, two-host and the three-host ticks according to the species, (Walker and Olwage, 1987). In one-host ticks, the larvae emerge from the eggs three to four weeks after deposition attach themselves to host animals where they complete their entire development. On the host they develop to nymph then to adult and then copulate. Afterwards, they drop off and deposit their eggs on the ground. The entire development cycle takes mostly 19-21 days as a rule, with minimum of 15 and maximum of 40 days, each stage taking one week (Jongejan and Ulineberg, 1994). In these ticks a stricter adaptation eliminates the need to drop to the ground for metamorphosis. All the instars occur on a single vertebrate, attacked by the larva. The larval and nymphal metamorphoses take place on the host, at the point of attachment of the larva and nymph. There is only one parasitic phase (Walker and Olwage, 1987). The two-host ticks attach on the host at the larval stage, then feed on blood and develop into the nymphal stage. After a maximum of 14 days, it drops off on to the ground where it reaches the imago stage in 20-30 days' time. Male as well as female ticks then search for another host, feed on blood and copulate. After 6-11 days, the female drops to the ground and deposits its eggs. The entire cycle from the time the larva emerges until the engorged female deposits the eggs mainly depends on the time the adult spent on the ground to find a new host. According to the species the nymph may survive on the ground for several weeks (Jongejan and Ulineberg, 1994). In these ticks the three stages develop on two different individuals that may or may not belong to the same species. In the first phase, the engorged larva molts on the host and the nymph reattaches close by. At the end of the blood meal, the nymph detaches and metamorphoses on the ground. Engorgement of adults occurs during the second parasitic phase. There are only two searches for a host, which eliminates the risks linked with the need for nymphal attachment, (Jongejan and Ulineberg, 1994). The three-host ticks look for a new host during each stage of development in order to feed. The larva emerges from the egg on the ground, looks for a host, feeds on it for three to seven days, drops off and molts after three to four weeks on the ground. The nymph attacks a second host for three to seven days feed on it and drops and develops into an adult on the ground after two to eight weeks. After that, the adult tick looks for a third host to feed on and for copulation that takes one to three weeks. Finally, it drops off and completes the cycle with oviposition on the soil. Because of the different time spent in each stage on the ground, the entire development cycle may last up to one year (Jongejan and Ulineberg, 1994). These ticks require three hosts for development, irrespective of the host species. There are three parasitic phases, separated by two phases on the ground, when metamorphosis occurs (Walker and Olwage, 1987). The life cycle of Ixodidae ticks is influenced by availability and access to suitable hosts in the habitant, predilection sites on the hosts, prevailing climatic factors (temperature and relative humidity conditions), and vegetation (Getahun *et al.*, 2016).

FAMILY: IXODIDAE

The family Ixodidae is distinguished from soft-ticks by the presence of a rigid chitinousscutum which covers the entire dorsal surface of adult males. In the adult female, nymph and larva, the scutum extends for only a small area permitting the abdomen to swell after feeding. The capitulum projects anteriorly and is easily visible dorsally. For a detailed account of the distinguishing morphological features of the Ixodidae and Argasidae ticks, the reader is referred to a book recently published by (Shivairo *et al.*, 2018).

Main Genera of Ixodid Ticks

The description of adult tick species under the genera Ixodid is summarized in Table 1.

Major characteristics of the seven (7) genera of ixodid ticks (*Amblyomma, Rhipicephalus* (*Boophilus*), *Dermacentor*, *Haemophysalis*, *Hyalomma*, *Ixodes*, *Rhipicephalus*are described with emphasis on their morphology, species, mammalian hosts and distribution including the disease pathogens they transmitted.

Table 1. Key to the adult Ixodid genera

<table>
<tr><th colspan="6">Description</th><th>Tick Species</th></tr>
<tr><td colspan="6">Anal grooves curves around in-front of the anus, inornate, eyes absent</td><td>Ixodes</td></tr>
<tr><td rowspan="6">Anal grooves located behind the anus</td><td>Eyes absent</td><td colspan="4">Palps, conical short, article 2 projects beyond lateral margin of the basis capitulum</td><td>Haemophysalis</td></tr>
<tr><td rowspan="5">Eyes present</td><td rowspan="3">Palps wider than long; or slightly longer than wide</td><td colspan="3">Large, ornate, scutum, basis capitulum rectangular dorsally</td><td>Dermacentor</td></tr>
<tr><td>Small moderate: mostly inornate: basis capitulum not rectangular dorsally</td><td>Festoons absent; spiracular plates round or oval; anal grooves faint or absent, inornate males, very small</td><td>Palps very short and ridged dorsally and laterally; males with normal legs</td><td>Rhipicephalus (Boophilus)</td></tr>
<tr><td></td><td>Festoons present; spiracular plates with tail like projections, anal grooves distinct, males of moderate size</td><td>Adanal shields in males, accessory coxa IV of male normal size; no spur on palpal article-1 of the male and female, usually inornate</td><td>Rhipicephalus</td></tr>
<tr><td rowspan="2">Palps much longer than wide</td><td colspan="3">Ventral plate is absent in males, palpal article 2 at least twice as long as article 3, festoons well developed and regular, scutum usually ornamented</td><td>Amblyomma</td></tr>
<tr><td colspan="3">Ventral plate present in males; palpal article 2 less than twice long as article 3, festoons less well developed and irregular, scutum inornate.</td><td>Hyalomma</td></tr>
</table>

Amblyomma **Species**

Morphology of Amblyomma Ticks

Amblyomma species are large, ornated ticks whose legs have bands of coloration. The eyes and festoons are present. The palps and hypostome are long and ventral plates are absent in the males (Fig: 1). *Amblyomma americanum*– nicknamed the "one-star tick" possesses a single white spot on the female scutum.

Figure 1. a. Amblyomma variegatum adult female (dorsal view); b. Amblyomma variegatum adult female (ventral view); c. Amblyomma variegatum adult male (dorsal view); d. Amblyomma variegatum adult male (ventral view). (Courtesy Walker et al., 2004)

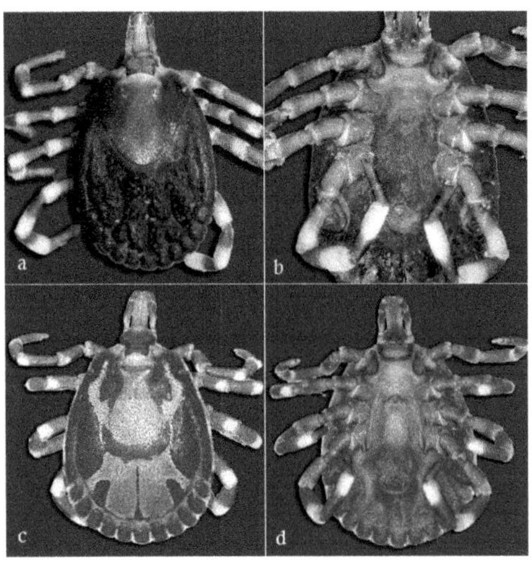

Species, Mammalian Host and Distribution of Amblyomma Ticks

The genus *Amblyomma* is a three-host ticks comprise of over 100 species. These are widespread in tropical and sub-tropical zones, where they parasitize a wide variety of mammalian, reptiles and amphibians. The mature stages of these species infest birds, where they play an important role in spreading the tick. *Amblyomma variegatum* (Figure 1), is the most important species found on the African continent, being well adapted to domestic livestock and having the widest distribution: throughout tropical sub-Saharan Africa, Madagascar, Reunion Island, Mauritius and other islands near Africa (Walker and Olwage,1987). *Amblyomma hebraeum* inhabits the south-eastern part of the African continent. *Amblyomma lepidum* and *A. gemma*- infests domestic ruminants and is found in parts of eastern and north-eastern sub-Saharan Africa. *Amblyomma pomposum* - infests domestic ruminants mainly in Angola, while *Amblyomma cohaerens*- which has adapted from buffalo to cattle, mainly in Ethiopia infests domestic ruminants. In addition to widespread distribution in Africa, *A. variegatum* has been introduced into the Caribbean region, presumably in the 18th or 19th century with cattle from West Africa. To date, *A. variegatum* is the only African *Amblyomma* tick to which has successfully established itself outside Africa (Walker and Olwage, 1987). *A. americanum, A. cajennense* and *A. maculatum)* are of economic significance in the USA, since the adultticks prefer to feed on cattle (Walker and Olwage, 1987).

Disease Pathogens Transmitted by Amblyomma Ticks

Amblyomma ticks are vectors of cowdriosis (heartwater) of cattle and small ruminants, important TBDs in Africa (Uilenberg and Camus, 1993). Heartwater is caused by the rickettsia organism known as *Ehrlichia (Cowdria) ruminantium* characterized by high fever, respiratory distress, nervous symptoms and hydro pericardium. Apart from Africa, the disease has also been confirmed in the Islands of Guadeloupe, Marie- Galante and Antigua (Jongejan and Ulineberg, 1994). The actual distribution of heartwater and *A. variegatum* in the Caribbean was extensively reviewed by (Casmus *et al.*, 1993). Thirteen species of *Amblyomma* are known to be able to transmit cowdriosis. Ten of these are of African origin, and three are American species, *A. maculatum, A. cajennense* and *A. dissimile,* are capable of transmitting the disease under experimental conditions (Jongejan and Ulineberg, 1994). Not all of these vectors are of equal significance in the natural transmission of cowdriosis. For instance, reptiles have been shown to be infected with cowdriosis in South Africa (Bezuidenhout, 1990). *A. maculatum,* the Gulf Coast tick, and *A. cajennense* are widespread and common on cattle. However, while *A. maculatum* has been demonstrated to be an efficient experimental vector, *Amblyomma cajennense*is a poor vector. A number of American *Amblyomma* species have been investigated as potential vectors of cowdriosis, with negative results. These include *A. rotundatum,* which feeds on amphibians and reptiles in Central and South America, and which has established itself in South Florida (Jongejan and Ulineberg, 1994).

Rhipicephalus (*Boophilus*)Species

Morphology of Rhipicephalus (Boophilus) Ticks

Ticks in this genus have no scutum, ornamentation and festoons. They possess eyes which are sometimes difficult to discern. *Rhipicephalus (Boophilus)* species possess short palps which are ridged dorsally and laterally and the hypostomes are short. The ticks have short mouthparts which are unable to penetrate

Figure 2. a. Rhipicephalus (Boophilus) decoloratus adult female (dorsal view); b. R. (Boophilus) decoloratus adult female (ventral view); c. R. (Boophilus) decoloratus adult male (dorsal view); d. R. (Boophilus) Decoloratus Adult Male (Ventral View). (Courtesy: Walker Et Al., 2004)

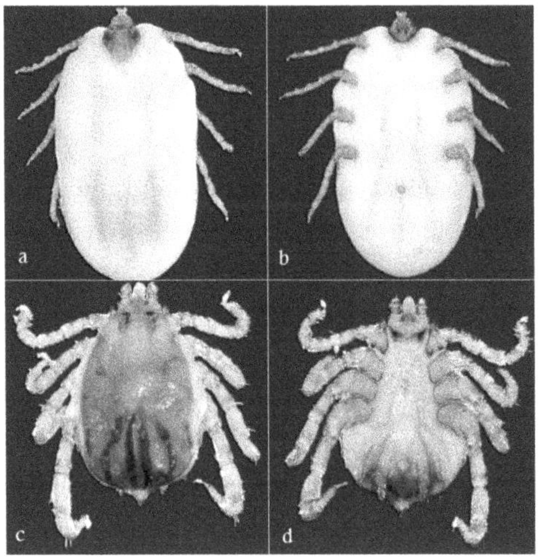

deeply into the animal skin (Fig. 2). The males have seven non-salient ventral plates (Jongejan and Ulineberg, 1994; Shivairo *et al.*, 2018).

Species, Mammalian Host and Distribution of Rhipicephalus (Boophilus) Ticks

The genus has only five species of small ticks, all of which exhibit one host tick cycle. *R. (Boophilus)* take three weeks to complete their blood meal on the host, preferably on cattle (except for *R. (Boophilus) kohlsi*, the Middle Eastern species with a predilection for shoats.*R. (Boophilus) microplus* is the most important species whose origin is in South East Asia. The species has since moved to the tropical countries including Australia and America. The *R. (Boophilus) annulatus* is found in the Mediterranean region, Ukraine and Russia Middle East, and Africa from West Africa to Sudan, (Jongejan and Ulineberg, 1994). *R. (Boophilus) annulatus* is found in Mexico, and southern parts of the USA where it was eradicated alongside the *B. microplus* tick. Another species, known as R. (*Boophilus) geigyi* is common on cattle and small ruminants in West Africa, and parts in the southern Sudan (Jongejan *et al.*, 1987). To date, *R. (Boophilus) decoloratus*, has remained confined to Africa where it has successfully replaced the *R. (Boophilus) microplus*, (Jongejan and Ulineberg, 1994).

Disease Transmitted by Rhipicephalus (Boophilus) Ticks

Boophilus ticks are the main vectors of bovine babesiosis caused by the protozoan parasites *Babesia bigemina* and *Babesia bovis*. *B. bigemina* is only moderately pathogenic for adult cattle, but the widely distributed of the *Babesia* spp., as it is transmitted by both *B. microplus* and *B. annulatus,* as well as *B. decoloratus* and probably *B. geigyi*. *Babesia bovis* is highly pathogenic in exotic European breeds of cattle. The parasite is found in all continents, but its distribution in Africa is restricted, as *B. bovis* cannot be transmitted by *B. decoloratus*. *Rhipicephalus (Boophilus)* ticks, together with other numerous tick species, transmit *Anaplasmamarginale,* the rickettsia which causes anaplasmosis in cattle on all continents including temperate regions. Since *Rhipicephalus (Boophilus)* spp. are one-host ticks, they become very numerous in cattle herds, especially in herds with a low degree of resistance to infestation, and cause considerable direct damage due to the continual loss of blood and effects on the host metabolism, (Jongejan and Ulineberg, 1994).

Hyalomma Species

Morphological Features of Hyalomma Ticks

Hyalomma ticks are also referred to as the "Bont-legged ticks" due to banded legs (Fig. 3). Theyare medium to large sized ticks with long palps and eyes typically positioned in the sockets. They have non-ornamented scutum and may or may possess festoons. The males possess adanal shields.

Figure 3. a. Hyalomma truncatum adult female (dorsal view); b. Hyalomma truncatum adult female (ventral view); c. Hyalomma truncatum adult male (dorsal view); d. Hyalomma truncatum adult male (ventral view). (Courtesy: Walker et al., 2004)

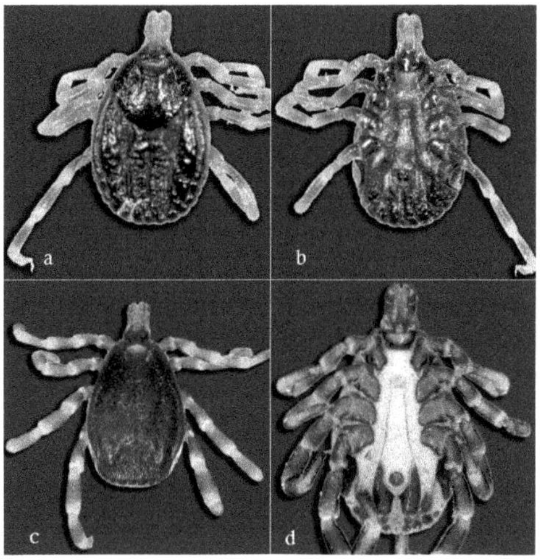

Species, Mammalian Host and Distribution of Hyalomma Ticks

The genus *Hyalomma* consists of about 30 species, most of which follow the three-host life cycle. However, some species undergo either a two-host or a three-host life cycle, depending on the host species affected. *Hyalomma scupense*is unique as it parasitizes cattle in Europe and Russia and has a one host cycle. Adult *Hyalomma* ticks actively run out from their resting sites when a host approaches, unlike most other Ixodidae ticks, which wait on vegetation, (Jongejan and Ulineberg, 1994). *Hyalomma* species parasitize both domestic and wild mammals and birds, and are abundant in semi-arid zones, more so, in Africa, Asia and South Europe (Shivairo *et al.*, 2018).

Disease Pathogens Transmitted by Hyalomma Ticks

The long mouth parts of *Hyalomma* ticks cause large feeding lesions, which are often exposed to secondary bacterial infections. Lameness is frequently associated with these ticks. Some populations of *H. truncatum* in Africa cause a toxic syndrome known as sweating sickness, the aetiology of which remains unclear. Numerous *Hyalomma* ticks transmit *Theileriaannulata,* the causative protozoan pathogen of tropical theileriosis. The distribution of this disease ranges from China in the east to Spain and Mauritania in the west. *Hyalomma anatolicum* and *Hyalomma detritum* are the most important vectors of *T. annidata,* and occur in Central Asia, and Southern Russia and Europe, Middle East, and Northern Africa. All stages of these ticks have adapted to cattle, explaining their role as vectors, since theileriosis is transmitted only trans-stadially (Jongejan and Ulineberg, 1994). *Hyalomma dromedarii,* the camel tick, is found both in India and Africa. *Hyalomma marginatum* is the main vector for theileriosis in parts of Spain (Jongejan and Ulineberg., 1994). Crimean-Congo hemorrhagic fever virus is an important zoonotic disease agent in humans in Europe, Asia and Africa, transmitted by *Hyalomma* ticks, (Jongejan and Ulineberg, 1994; Shivairo *et al.*, 2018).

Genus: *Rhipicephalus*

Morphological Features of Rhipicephalus Ticks

These are small to medium sized ticks with short, broad palps and hypostome; most species have an inornate scutum. When viewed from the dorsal aspect, the basis capituli has a distinct hexagonal shape with protruding lateral margins (Fig. 4). Males have got ventral plates near the anus, adanal plates and accessory shields. Both sexes have got distinct eyes and festoons (Shivairo *et al.*, 2018).

Figure 4. a. Rhipicephalus appendiculatus adult female (dorsal view); b. Rhipicephalus appendiculatus adult female (ventral view); c. Rhipicephalus appendiculatus adult male (dorsal view); d. Rhipicephalus appendiculatus adult male (ventral view). (Courtesy: Walker et al., 2004)

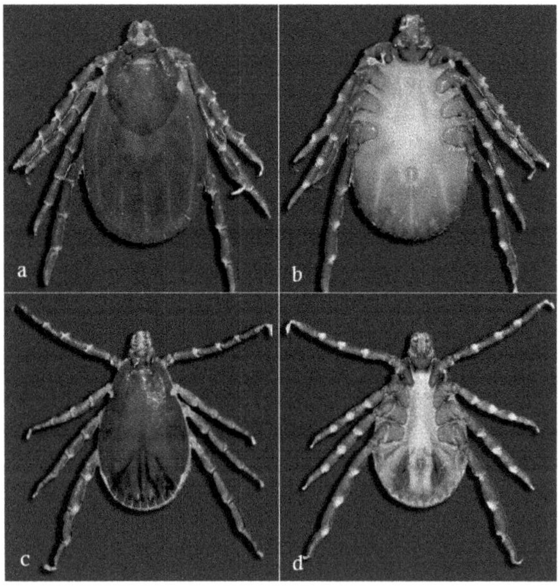

Species, Mammalian Host and Distribution of Rhipicephalus Ticks

Ticks in the genus *Rhipicephalus* comprise over 70 species ad most of these are found on mammals in Africa. They are usually three host ticks, although some have two host cycle (e.g., *R. evertsi).* The taxonomic identification of Rhipicephalid ticks has several taxa of *Rhipicephalus* ticks for example, *R. sanguineus sensustricto* exhibits host specificity and is exclusively found on dogs and some wild carnivores (Jongejan and Ulineberg, 1994). *R. sanguineus* of ticks have been collected from domestic ruminants (Jongejan *et al*., 1987). Such ticks could easily be misidentified as *R. sanguineus,* but in fact belong to other species, which could be far less host-specific (Jongejan and Ulineberg, 1994). The later phenomenon signifies the importance for precise and accurate identification of prevalent tick species in a given locality to ascertain their potential capacity as vector and their capacity to harbor and transmit an infectious agent. *Rhipicephalus appendiculatus,* the brown ear tick, is the most important rhipicephalid tick in East and Southern Africa, where it occurs on a wide variety of domestic and wild ruminants (Shivairo *et al*., 2018).

Diseases Pathogens Transmitted by Rhipicephalus spp.

Rhipicephalus appendiculatus and the closely related *Rhipicephalus zambeziensis* are vectors of *Theileriaparva,* the causative agent of East Coast fever. A comprehensive review of the epidemiology of theileriosis in Africa was published in 1992 (Norval *et al*., 1992). *Rhipicephalus* species are also vectors of *Babesia ovis,* transmitted by the two-host tick *R. bursa* and *Erlichiaovis,* of which both *R. bursa* and *R. evertsi*are vectors). The two etiological agents cause ovine babesiosis and ehrlichiosis, respectively. *Rhipicephalus bursa* is probably also one of the main vectors of anaplasmosis of small ruminants. *Rhipicephalus sanguineus sensustricto* is a vector of *B. canis* in dogs and also of *E. canis,* the cause of tropical pancytopaenia in dogs, a severe rickettsia disease which occurs world-wide (Shivairo *et al*., 2018). This species is also a vector of *Rickettsia conorïi* the etiological agent of classical tick-bite fever in humans

Genus: *Haemophysalis*

Morphological Features of Haemophysalis Ticks

All *Haemaphysaliss* pecies are eyeless and easy to differentiate from other genera due to the characteristic lateral projection of palpal article two beyond the margins of the basis capituli. The ticks possess festoons, and have short sensory palps (Shivairo *et al*., 2018). The males have no ventral shields and anal groove contours posteriorly (Fig. 5).

Species, Mammalian Host and Distribution of Haemophysalis Ticks

The genus *Haemaphysalis* contains about 155 species with only a few of these are adapted to domestic livestock. *Haemaphysalis bispinosa* is adapted to cattle in the Indian subcontinent. *Haemaphysalis longicornis* is found on cattle and other domestic animals in the East Asian, Australia, New Zealand and New Caledonia. In Australia, *H. longicornis* maintains itself by parthenogenesis (reproducing without males), whereas normal bisexual populations of the same tick species are found in Asia, side-by-side with the parthenogenetic populations; the reason for this difference is not yet known (Jongejan and

Figure 5. a. Haemophysalis leachi adult female (dorsal view) ; b. Haemophysalis leachi adult female (ventral view) ; c. Haemophysalis leachi adult male (dorsal view) ; d. Haemophysalis leachi adult male (ventral view) (Courtesy: Walkeret al., 2004)

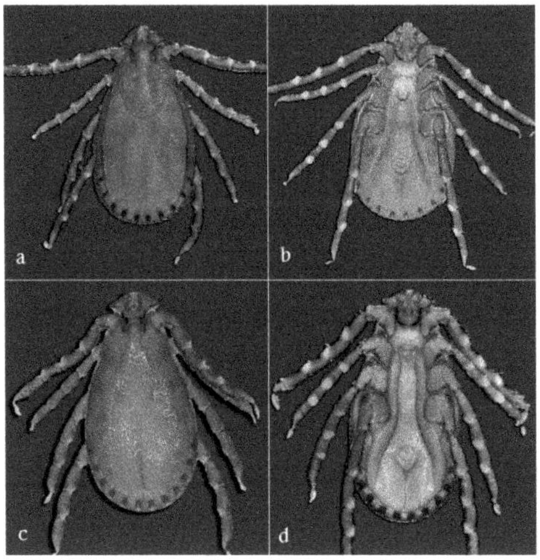

Ulineberg, 1994). *Haemaphysalis leachi* the "yellow dog tick" is common in Africa, Asia and Australia and transmits *B. canis* in dogs. *Haemaphysalis punctata* that is found inEurope transmits *Babesia* spp. in cattle, sheep and carries non-pathogenic *Theileria* spp., and also *Anaplasmamarginale* in cattle. This tick has been reported to cause tick paralysis.

Disease Pathogens Transmitted by Haemophysalis Ticks

Haemaphysalis species are of importance to livestock in Europe and Asia. *Haemaphysalis punctata* is the tick vector for *Babesia major* in cattle and *Babesia motasi* in small ruminants in Europe (Jongejan and Ulineberg,1994). *Haemaphysalis* spp. Isalso involved in the transmission of human rickettsioses (Shivairo *et al.*, 2018).

Genus: *Dermacentor*

Morphological Features of Dermacentor Ticks

Dermacentor ticks are large, ornated, eyed and have a basis capitulum which is rectangular dorsally (Fig. 6). The ticks usually follow a three-host cycle (exceptions include the American *Dermacentor pictus* [the so-called 'winter tick'] and *D. nitens*, which are both one-host ticks (Jongejan and Ulineberg,1994; Shivairo*et al.*, 2018).

Figure 6. a. Dermacentor rhinocerinus adult female (dorsal view); b. Dermacentor rhinocerinus adult female (ventral view); c. Dermacentor rhinocerinus adult male (dorsal view); d. Dermacentor rhinocerinus adult male (ventral view). (Courtesy: Walker et al., 2004)

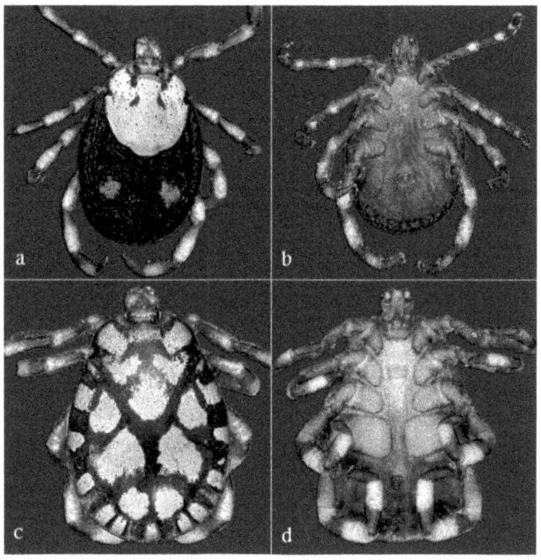

Species, Mammalian Host and Distribution of Dermacentorticks

The genus *Dermacentor* comprises of 30 species which are found on all continents, except Australia. In Eurasia, many species (e.g., *Dermacentor marginatus* and *Dermacentor reticulatus)* infest domestic animals. In North America *Dermacentor variabilis* and *Dermacentor andersoni* are important ectoparasites of livestock. *Dermacentor* (or *Anocentor) nitens,* the tropical horse tick, is present in Central and South America. *Dermacentornitens* is a one-host tick, originally found on deer in South America, and has a predilection for the ears of the host. In Africa, *Dermacentor* ticks do not play a significant role as disease vectors for livestock.

Disease Pathogens Transmitted by Dermacentor Ticks

In temperate regions, *Dermacentor* spp. are important vectors in bovine and small ruminant anaplasmosis and equine babesiosis. Several species are important in human medicine as vectors of rickettsia diseases. For example, *Dermacentor andersoni* is the vector of *Rickettsia rickettsia* which causes rock mountain spotted fever of humans in North America and, also causes tick paralysis in livestock. *Dermacentor marginatus* is the reservoir of *Coxiella burnettii* which causes Q fever in Europe whereas, *D. reticulatus* is the main vector of *Babesia canis* which causes European canine babesiosis. *Dermacentor nitens* causes loss of condition when large numbers feed on horses, as well as predisposing to attack by myiasis flies. It is the vector of *Babesia caballi* of horses in the western hemisphere.

Genus: *Ixodes*

Morphology Features of Ixodes Ticks

All tick species in the genus *Ixodes* are characterizedby lack of eyes and presence of the anal groove curving anteriorly to the anus (Fig.7). This easily differentiates *Ixodes* from the other genera of hard ticks (Metastriata), in which this groove curves posteriorly around the anus. The scutum is inornate and lacks festoons (Jongejan and Ulineberg,1994; Shivairo *et al.*, 2018).

Figure7. a. Ixodes rubicundus adult female (dorsal view); b. Ixodesrubicundus adult female (ventral view); c. Ixodes rubicundus adult male (dorsal view); d. Ixodes rubicundus adult male (ventral view). (Walker et al., 2004)

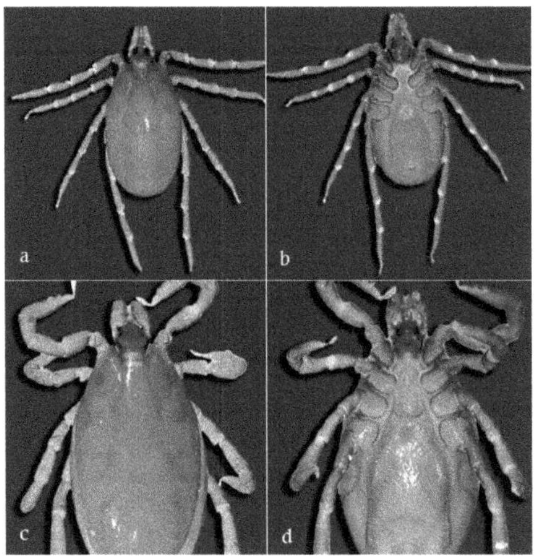

Species, Mammalian Host and Distribution of Ixodesticks

Ixodes spp. comprises the largest genus within the hard ticks with over 250 species that inhabit nests or burrows, (Jongejan and Ulineberg, 1994). They belong to a separate branch of the Ixodidae, the Prostriata. Only a few *Ixodes* species parasitize large mammals, and the genus is widely distributed in the wooded or grassy environments, (Jongejan and Ulineberg, 1994). The two most important species in North America are *I. scapularis* (for which *I. dammini* is now considered to be a synonym (Jongejan and Ulineberg, 1994). The most common species in Europe and Asia are *I. Ricinus* and *I. persulcatus. Ixodes* spp. Are not of any economic importance in the tropics, except *I. rubicundus* which is on sheep and goats in South Africa. In Australia, *I. holocyclus* usually occur on bandicoots, but can also feed on domestic animals, tropical bats *(Pteropus* spp.) or humans (Jongejan and Ulineberg, 1994; Shivairo *et al.*, 2018).

Disease Transmitted by Ixodes Ticks

Both *Ixodes rubicundus* in South Africa and *Ixodes holocyclus* in Australia can induce paralysis when they feed on animals or humans, (Jongejan and Ulineberg,1994; Shivairo *et al.*, 2018). The indiscriminate feeding behavior of *Ixodes ricinus* means that this species is found on many different hosts and is an important vector of many disease pathogens in Europe. These include *Babesia divergens* in cattle, *Ehrlichiaphagocytophila* (tickborne fever) in domestic ruminants, and louping ill virus in sheep, (Jongejan and Ulineberg,1994; Shivairo *et al.*, 2018). *Ixodes* ticks are of great importance as vectors of human diseases. Lyme disease (Borelliosis) in Europe is transmitted by *Ixodesricinus,* in Europe and in Asia by *Ixodes persulcatus* and in America by *Ixodes scapularis.* The species *Ixodes Ricinus* and *Ixodes persulcatus* are also vectors of tick-borne encephalitis virus in humans (Rosenberg *et al.*, 2018).

CONTROL OF HARD TICKS

Within the limits of this review, it is possible to outline only briefly various aspects of conventional tick control and emerging alternatives approaches. Eradication of ticks is not possible, except on isolated small islands, where successful campaigns have rigorously been implemented (Jongejan and Ulineberg, 1994. On large islands and in the continental landscapes, eradication of ticks is not economically realistic. Tick control methods including chemical, ecological, biological, genetic, and the newer vaccination regimes are discussed as follows.

Chemical Control

There are several methods for controllinghard ticks and TBDs using chemicals. Use of chemical acaricides has been successful; however, it has several setbacks including accumulation of toxic residues in the human food chain and environmental pollution (Shivairo *et al.*, 2018). Continuous application of acaricides leads to selection of resistant tick populations (Getahum *et al.*, 2016). The development of new safe acaricides is a long and expensive process, which justifies the need for alternative approaches to control tick infestations. A brief overview of acaricides is provided as follows:

Acaricides

A wide range of chemical acaricides such as arsenical, chlorinated hydrocarbons, organophosphates, carbamates and synthetic pyrethroids have been used for the control of ticks on livestock, Table 2. The performance of each acaricide depends on both the activity of a product, the quality and quantity of the active ingredient deposited on the animal and/or delivered internally (Getahum *et al.*, 2016).

Arsenic

Arsenic (As_2O_3) was first used for tick control in 1893 in South Africa since it was inexpensive, stable, and water soluble, and had an accurate vat-side test, (Getahum *et al.*, 2016). Organic arsenicals pesticides used include cacodylic acid (CA), HDMO (hydroxydimethylarsine oxide), MSMA (monosodium methanearsonate) and DSMA (disodium methanearsonate) (Amrit *et al.*, 2011). The exact mechanism

Table 2. Chemicals used for tick control

Group Name	Chemical Name
Organophosphates	Bromocyclene, Coumaphos, Crotoxyphos, Dichloruos, Diazinon, Malathion, Propoxur, Phoxin, Stirofos, Trichlofo (Metrifonol)
Carbamates	Carbaryl, Lindane
Chlorinated Hydrocarbons	Methoxychlor, Flumetrin, Fenualerate
Pyrethrins and Pyrethroids	Permethrin, Pyrethrin
Avermectin	Ivermectin

(Source: Getahum *et al.*, 2016)

of action of arsenic is not known, but several hypotheses have been proposed. In the trivalent state, inorganic and organic (methylated) arsenic may react with critical thiols in proteins and inhibit their activity, (Amrit *et al.*, 2011). Arsenic metabolism is important as it exerts its toxicity by inhibiting over 200 enzymes involved in cellular energy pathways and DNA synthesis and repair, (Amrit *et al.*, 2011). Arsenic dips were used successfully to eradicate *Rhipicephalus (Boophilus)* ticks from the southern United States. Unfortunately, arsenic has a very short residual effectiveness (less than one to two days), and the *Rhipicephalus (Boophilus)* ticks have become resistant to arsenic (Getahum *et al.*, 2016).

Synthetic Chlorinated Hydrocarbons

Due to resistance to arsenicals in many tick species, Arsenicwas rapidly replaced by chlorinated hydrocarbons (CHCs) (Getahum *et al.*, 2016). These CHCs were developed in the 1940s after the discovery of the insecticidal properties of Dichloro-Diphenyl-Trichloroethane (DDT), (Narahashi, 2013). CHCs are diverse synthetic chemicals with varying chemical structure. DDT is the best known of these CHCs. First synthesized in 1874, DDT remained obscure until its insecticidal properties became known in 1939, a discovery that earned a Nobel Prize in 1948 (Narahashi, 2013). The insecticidal properties of cyclodienes, (which include aldrin, dieldrin, chlordane, heptachlor, and endrin) were discovered about 1945 (Narahashi, 2013). CHCs became widely used in the United States following World War II. Their primary uses included broad spectrum applications for agricultural crops and forestry and, to a lesser extent, human health protection by spraying to destroy mosquitoes and other potential disease carriers. These compounds also became widely used to combat insect carriers of domestic animal diseases. CHCs mode of action is by interfering with nerve conduction of ticks, (Getahum *et al.*, 2016). CHCs had long residual effects in the environment and were used extensively for controlling ticks. Due to their high toxicity and long life span these compounds were withdrawn from the market.

Organophosphates

Organophosphates (OPs) replaced CHCs after hard ticks developed resistance, (Getahum *et al.*, 2016). OPs are esters of phosphoric acid and have a wide range of activities against ticks at very low concentration. Common OPs used as pesticides included; parathion, malathion, methyl parathion, chlorpyrifos, diazinon, dichlorvos, and fenitrothion. The mechanism of action for organophosphate pesticides (OPPs), is by disrupting the enzyme acetylcholinesterase. In this process,acetylcholinesterase breaks down the acetylcholine neurotransmitter, which sends out signals to other nerve endings in the body, (Peter *et al.*, 2014). OPs affect the enzyme in varied ways;one way is through irreversible covalent inhibition, (Peter *et al.*, 2014). The brain sends out neurotransmitters to the nerve endings in the body; organophosphates disrupt this process from occurring. The residual effectiveness of OPs was shorter than that of chlorinated hydrocarbons, and the risk of acute toxicity in livestock was even much greater and soon ticks developed resistance to the organophosphates (Wharton, 1967).

Carbamates

Carbamates are esters of carbamic acids. They closely resemble OPs and are more toxic and expensive compared to the OPs (Getahum *et al.*, 2016; Shivairo *et al.*, 2018). Carbamate insecticides are toxic to insects and mammals by virtue of their ability to inhibit the enzyme acetylcholinesterase.The insecticides kill insects by reversibly inactivating the enzyme acetylcholinesterase. This is achieved by carbamylation of a serine hydroxyl residue at the active site of the enzyme, a process that involves cleavage of the carbamate molecule, which in effect is treated by the enzyme as if it were acetylcholine (Getahum *et al.*, 2016). Examples of the carbamate insecticides are aldicarb (temik), carbofuran (furadan), carbaryl (sevin), ethienocarb, fenobucarb, and methomyl, oxamyl (vydate), thiodicarb (larvin), methomyl (lannate).

Synthetic Pyrethroids

Pyrethrum extract, a natural product of the flowers of the plants, *Chrysanthemum cinerariaefolium* and *C. cineum*, has been used as a natural insecticide for a long period of time. Initially the extracts of pyrethrum were used and later, the specific synthetic chemical pyrethroids analogs were produced. The six active insecticidal compounds of pyrethrum are popularly known as pyrethrins. Synthetic pyrethroids (SPs) are synthetic analogs and derivatives of the original pyrethrins and include a variety group of about 1,000 insecticides (Mader *et al.*, 2015). Though they are analogs of pyrethrins, their production has involved extensive chemical modifications which have made them highly toxic and less degradable in the environment (Mader *et al.*, 2015). Pyrethroids are grouped into two classes namely: Type 1: (allethrinpermethrin, phenothrin, resmethrin, teramethrin, tefluthrin) and Type11: (cyfluthrin, cypermethrin, deltamethrin, fenvalerate, flumethrirn, tralomethrin, cyhalothrin), based on the toxicological and physical properties. Type I pyrethroids are derivatives of pyrethrin that do not have a cyano group and elicit tumors while type II pyrethroids have cyoano group and cause chloreoathetosis and salivation. Due to complex chemical structure, pyrethroids are composed of two, four or eight isomers, and a range of products may contain a mixture of these various isomers. The production of individual pyrethroids with varying isomeric ratios may be the reason for the variations in the toxicities of the same compound. For increasing the efficiency of the insecticides, the pyrethroids are formulated with compounds like piperonylbutoxide, piperonylsulfoxide and sesamex, which act as synergists. Formulated commercial pyrethroids have a high percentage of other inert ingredients, which are highly toxic, (Mader *et al.*, 2015). Pyrethrins delay the closure of voltage-gated sodium channels in the nerve cells of insects, resulting in repeated and extended nerve firings. This hyperexcitation causes the death of the insect due to paralysis and loss of motor coordination, (Mader *et al.*, 2015). Resistance to pyrethrin has been bypassed by pairing the insecticide with synthetic synergists such as piperonylbutoxide. Together, these two compounds prevent detoxification in the insect, ensuring insect death (Mader *et al.*, 2015). Synergists make pyrethrin more effective even at lower doses. Pyrethrins are effective insecticides because they selectively target insects rather than mammals due to higher insect nerve sensitivity, smaller insect body size, lower mammalian skin absorption, and more efficient mammalian hepatic metabolism (Mader *et al.*, 2015). Because of their insecticide and insect repellent effect, pyrethrins were very successful in reducing insect pest populations that affect humans, crops, livestock, and pets, such as ants, spiders, and lice, as well as potentially disease-carrying mosquitoes, fleas, and ticks. With the reduction in the use of organochlorine pesticides in the late 1960s, organophosphorous compounds and carbamates, synthetic pyrethroids came into the market. Although pyrethroids do not accumulate in to the environment compared to the organochlorines they resulted in to increased mortality of fish and wildlife (Mader *et al.*, 2015).

Avermectins

Avermectins or amitraz & macrocyclic lactonesare products of the fungus *Actinomyces* species that are divided into two major components A and B depending on the structures. Avermectins compound is active against intestinal parasites, lungworms, warble flies, lice, midges and different genera of ticks. The active substance inhibits the transmission of stimuli between the interneurons of the ventral nerves and the motor neurons of the parasites. The compound is deposited for a short time in the liver, fat and only in small amounts in muscle and kidney. About 50% is excreted unchanged with the faces. Therefore, a withdrawal time of 38 days is required for milk and beef. Because of this, it also does not make sense to utilize the compound in tropical animal production (Getahum *et al.*, 2016).

Application of Chemical Acaricides

Conventional methods for application of chemical acaricidesto protect livestock against ticks include dipping, spray races, spot treatment and hand dressing (Getahum *et al.*, 2016). Unconventional methods for applying acaricides include use of ear tags, neck bands, tail bands and pour-on. The direct application of acaricides to animals is the most popular method (Shivairo *et al.*, 2018).

Dipping

In this method, animals are totally immersed in a dipping bathtub containing solution of chemicals. By 1893 in Australia, Africa, and the United States the use of "dipping-vats" to immerse tick-infested cattle was heavily used for the control of ticks and TBD affecting cattle (Getahum *et al.*, 2016). In general, dipping vats provide an effective method of applying acaricides for tick control. However, due to their immobility, high initial investment cost of dip vat construction, and the cost of acaricides make it impractical for small animal ranching operations. Dip vats must be well managed and maintained at the proper concentration and animals must be immersed properly (Getahum *et al.*, 2016).

Spray Races

Spray races have successfully been used for the control of ticks in livestock. Spray race equipment are portable, and only small amounts of acaricides need to be mixed for a single application. However, all the full animal body parts coverage is less efficient in spray races compared to dip vat immersion (Getahum *et al.*, 2016).

Spot Treatment and Hand Dressing

Certain tick species have particular predilections sites which may not effectively be treated by spray races or dip vats. For example, the inner parts of the ear, under part of the tail, the tail brush and areas between the teats and legs in cattle with large udder, are likely to escape acaricide treatment. Acaricides can be applied to these sites by hand dressing or spot treatment (Getahum *et al.*, 2016). The application of insecticides with aerosols and in oils, smears, and dusts by hand to limited body areas is time-consuming and laborious, but it may be more effective and economical (in terms of cost of acaricide) than treating the entire animal body (Getahum *et al.*, 2016).

UNCONVENTIONAL METHODS FOR APPLYING ACARICIDES

Unconventional methods for applying acaricides include use of ear tags, neck bands, tail bands and pour-on, especially so for pyrethroids which have long residual activity. In Kenya, an intra-ruminal ivermectin slow-release device has been demonstrated to provide 90days protection against tick damage (Getahum *et al.*, 2016). Using alternative methods for applying chemical acaricides has revealed variable and inconsistent results and needs further validation before widespread adoption.

Alternativetick Control Methods

Prolonged use of chemic alacaricides produce resistant tick populations, accumulation of toxic residues in meat and milk and result into environmental pollution (Jongejan and Ulineberg, 1994). These shortcomings have necessitated the need for alternative methods for tick control such as biological, ecological, genetic and use of tick vaccines.

Biological Control

In biological control microorganisms (entomopathogens) that attack and kill arthropods (ticks and insects) are utilized. Biological control using entomopathogenic fungi is a promising alternative to chemical acaricides being used for their control. It has been demonstrated that isolates of Metarhiziumanisopliae and of Beauveria bassiana fungal isolates are pathogenic to both amitraz-susceptible and amitraz-resistant strains of R. decoloratus (Murigu *et al.*, 2020). Effective entomopathogenic activities of tick predators especially ants and birds (Buphagus spp. Oroxpeckers spp, Crotophagus spp, and various magpies, village fowl) have been tested, (Getahun *et al.*, 2016). Depending on the conditions, these predators are able to kill a large number of ticks (Getahun *et al.*, 2016). Some hyperparasites such as the chalcidflies and Hunterellus are also thought to be entomopathogenic (Getahun *et al.*, 2016).

Ecological Control

In ecological approach information on the ecosystem of different instars is used for habitat and host associated tick control. The environment and vegetation is modified in the habitant by removal of plants that shelter the ticks. Plants are sporadically removed by burning. The influence of bush fire on tick population varies markedly with the time of year, intensity of burn, and the tick species present (Baars,

1999). Replacement of natural foliage, cropping, and soil cultivation are integrated approaches that allow pasture improvement and enhance tick eradication. Under this approach acaricide may be used tosanitize localized habitats such as stables, sheds, poultry houses, kennels markets and areas around wells. Wild ungulates and carnivores are eliminated in the environment. These wild ungulates are alternative hosts and can maintain tick populations and their removal is an effective control measure. Pasture rotation can be used exclusively, or together with the use of acaricides to temporarily withdraw domestic hosts, (Jongejan and Ulineberg, 1994).

Genetic Control

The ability to resist ticks is an acquiredcharacteristic; however, it develops according to genetic factors. If it is possible to breed cattle for tick resistance, it would provide the much-needed solution that will allow the costly and unsafe use of acaricides to be discarded, with the related risk of the emergence of non-sensitive tick strains. Each animal develops its own level of resistance in response to tick challenge; the level may be high (as in most local zebu cattle) or low (as in most exotic European breeds of cattle), but a wide range of resistance occurs in all breeds of cattle (Jongejan and Ulineberg, 1994). Tick resistance is a heritable, and selection and breeding for this trait is possible both in local zebu and exotic European breeds. Breeding of tick resistance cattle in combination with pasture rotation, pasture burning and planting certain special grasses have been considered for tick control (Jongejan and Ulineberg,1994). Reports elsewhere have revealed that *Bos indicus* had innate resistance toinfectionwith *babesiabovis*, *B. bigemina* and *Anaplasmamarginale* compared to *Bostaurus* breed, (Jongejan and Ulineberg, 1994). Host-parasite relationships that evolved in nature reveal that cattle survived in Asia and Africa despite *Babesia*, *Theileria* and their natural *Rhipicephalus (Boophilus)* and *Rhipicephalus* vectors (Getahun *et al.*, 2016). Host resistance, expressed by an animal's ability to prevent the maturing of large numbers of ticks, and disease immunity, are survival mechanisms for the host and for external and internal parasites. The problem is not only to utilize these attributes, but also to increase productivity. Considerable progress has been made in evolving resistant *Bosindicus Bostaurus* beef and dairy cattle that limit the effects of ticks while retaining high productivity (Jongejan and Ulineberg, 1994).

Anti-Tick Vaccines

The first attempts to develop anti tick vaccinesincluded the use of complex tick extracts (Willadsen *et al.*, 1988). Previous anti-tickvaccine development has focused on the economically important TBD of cattle, such as Babesiosis (*B. bovis*, *B. bigemina*), Theileriosis (*T. parva*, *T. annulata*), Anaplasmosis (*A. marginale*) and Cowdriosis (*C. ruminantium*). The three strain *Theileria parva* stabilize vaccine, known as the Muguga cocktail was developed for integrated tick control in Eastern, Central and Southern Africa, (FAO, 1984). The Muguga cocktail vaccine was used in combination with an antibiotic treatment, known as the infection-and-treatment method. Later the mild Boleni strain was isolated for mass vaccine production in Zimbabwe and was used without the antibiotic treatment component (Getahun *et al.*, 2016).These anti tick vaccineshave several side effects, and are very expensive to produce (Willadsen, 2001).The new Recombinant DNA technology has provided a breakthrough in the production of protein vaccines with less side effects. A membrane bound tick gut glycoprotein was isolated from*R. (Boophilus) microplus,* and the gene coding for this protein was cloned and expressed in *Escherichia coli*. The recombinant protein was capable of inducing high degree of immune protection in cattle against homologous infestation with *R. (Boophilus) microplus.* (Rand *et al.*, 1989). This recombinant protein holds great promise for future use and has been released on the market (Willadsen,1990). The development of anti-tick vaccines represents a promising alternative to chemical control and has advantages such as environmental safety, low cost of production, and relative ease of product registration. Alternative approaches based on the use of vaccines containing recombinant midgut antigen from *Rhipicephalus (Boophilus) microplus* (Bm86) have been used for control of *R. microplus* infestations in the field in different regions of Central and South America and the Caribbean (Willadsen and Jongejan, 1999). The effectiveness of Bm86-based vaccines includes both reduction of adult tick engorgement weights and female fecundity, as assessed in laboratory vaccination trials (Rodriguez *et al.*, 1994). Presently, Bm-gut antigen-based commercial vaccines are marketed under the trade names TickGARD® (Hoechst Australia Ltd, Melbourne, VIC, Australia) and Gavac® (HeberBiotec S.A., Havana, Cuba) and have been shown to be effective in the integrated control of *B. microplus* populations and other *Boophilus*species (De Vos *et al.*, 2001). The potential of using TickGARD Plus to control *Amblyomma variagatum*and *R. appendiculatus* has been assessed; however, no significant protection was observed despite the description of a homolog of Bm86 in *R. appendiculatus,* (De Vos *et al.*, 2001). Confirmation of the success and economic viability of this vaccine would considerably raise the hopes of releasing similar vaccines for use in future integrated programmes against other species of ticks. The adaption of anti-tick vaccines should reduce the over reliance on chemical acaricides and allow more widespread exploitation of tick-susceptible exotic cattle breeds.

FAMILY: *ARGASIDAE*

General Morphology of Soft Ticks

Soft ticks have a characteristic leathery integument, and most unfed ticks (adults and nymphs) have marked grooves on the dorsum. They lack scutum and the mouth parts (gnathosoma) are invisible from the dorsal aspect in nymphs and adults (Fig. 8A and 9.8B), but visible in some larval stages. They have four pairs of legs, similar in morphology to the Ixodidae (hard ticks), and the pulvilli are usually present in most larval stages but rudimentary or absent in the nymphs and adults, Fig. 8A and8B. Soft ticks are

Figure 8. General morphology of soft ticks from the dorsal and ventral views
A. Dorsal view (1- Marked grooves, 2- leathery integument, 3-terminal claws). B. Ventral view (1-gnathosoma, 2- gnathosoma with a central toothed hypostome, a pair of palps, and post-palpal seta, 3-postpalpal seta, 4-genital aperture, 5-spiracle, 6-basis capitulum without post-palpal seta)(Walker et al., 2004)

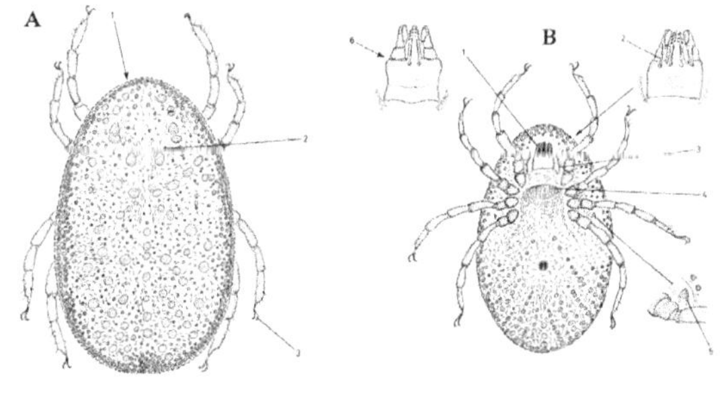

mainly found in dry areas and are tolerant to drought conditions, where some species can live for several years even in the absence of a suitable host, but majority occur near their hosts in the hosts' housing. The three main genera of veterinary importance are; Argas, Ornithodoros and Otobius.

General Life Cycle of Argasidae

Soft ticks have multi-host developmental lifecycles with the larvae feeding once before moulting to first nymphal stage. There are between two and seven nymphal stages with each stage feeding on a separate host before moulting to the subsequent nymphal stage. Adults mate away from the host, and females lay eggs in batches (400-500) after every blood meal. Unlike the Ixodidae, the Argasidae are quick feeders, feeding moderately and often, and the females do not engorge as much as the hard ticks.

Table 3. Argas species of veterinary importance

Species	Common name	Host	Site	Pathogenic significance
Argus persicus	Fowl tick, chicken tick, adobe tick, blue bug	Chicken, turkey and wild birds	Skin	Transmits *Borreliaanserina*, *Aegyptianellapullorum.*
Argasreflexus	Pigeon tick	Pigeon	Skin	Transmits *Borrelia anserina*
Argaswalkerae	Chicken tick, southern fowl tampan	Chicken	Skin	Transmits *Borrelia anserina*
Argasminiatus		Wild birds	Skin	-
Argas radiatus		Wild birds	Skin	-
Argasrobertsi		Wild birds	Skin	-
Argassnachezi		Wild birds	Skin	-

Adapted from Taylor *et al.* (2015)

Genus: *Argas*

Species within the genus, *Argas*, are dorsoventrally flattened with a lateral suture (distinct margin separating the dorsal and ventral surfaces), well developed mouth parts and most of themare nocturnal. The average size for adult ticks is approximately 6mm with the integument texture comprising of convoluted patterns, but in some species, the integument is smooth or has a mamillated texture (Fig.9). The scutum (in females) and conscutum (in males) is absent, the legs are slender and without pale rings and pulvilli. Basis capituli is characterized by straight lateral margins,with small palps. Eyes and anal groove are absent. Festoons are absent in males, and the spiracular plates are small and occur postero-laterally between the 3rdand 4th coxae. The female genital aperture is formed by a broad horizontal slit, whereas in males, it occurs in the same position as females, but its smaller. Currently, 61 species in the genus *Argas* have been described, where *Argas persicus*, *Argas walkerae*, and *Argas reflexus* are the common species that parasitize on domesticated birds. Other species comprising of *Argas miniatus*, *Argas radiatus*, *Argas robertsi* and *Argas snachezi* parasitize on wild birds (Table 3). In Africa, *A. persicus* and *A. walkerae* are the common species, with *A. persicus* occurring in most parts of western, eastern, northern

Figure 9. Morphological features of Argas species (Walker et al., 2004)

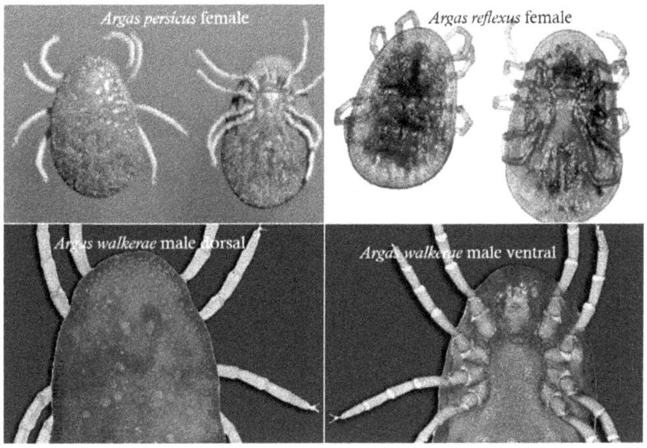

Figure 10. The distribution of Argas and Ornithodoros species in Africa (Walker et al., 2004). Countries with reported presence of these species are highlighted in yellow. The distribution of the ticks in each country is indicated by grey dots or shading.

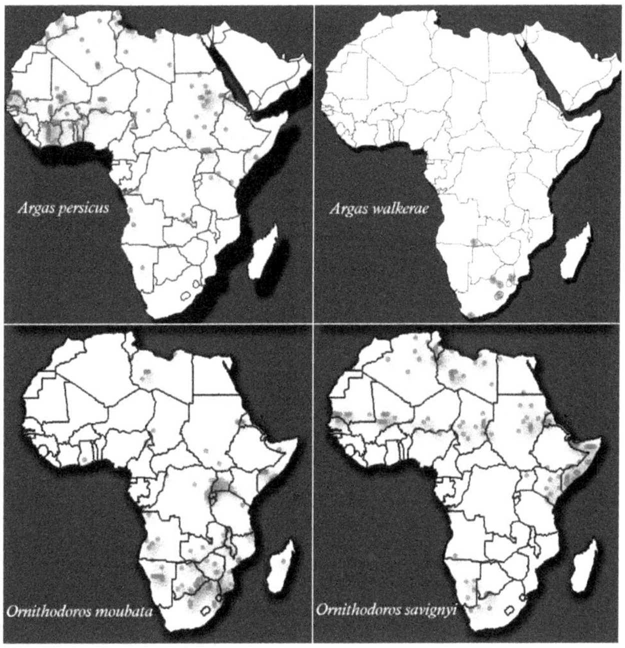

and southern Africa, while *A. walkerae* is common only in South Africa (Fig.10). All stages of the life cycle are parasitic, where heavy infestations may cause death of the host due to severe anemia. Some species are important vectors of fowl spirochaetosis and the rickettsia parasite of poultry, *Aegyptianella pullorum* (Table 9.3).The larval stages excrete a toxin that causes paralysis in chickens.

Genus: *Ornithodoros*

Adults are grey-brown in colour and approximately 8mm in size, with well-developed mouth parts and are nocturnal. Lateral suture is absent, havea rounded body margin, and the integument texture comprises of mamillated patterns, where the mammillae are without stalks (Fig.11). The integument also has some areas that are sunken and of rough texture. Scutum and conscutum are absent, the legs have no pale rings, are slender and without pulvilli. Eyes are absent in some species (e.g., *Ornithodoros moubata*, *Ornithodoros erraticus*) and present in some (*Ornithodoros savignyi*), where they occur as a pair of distinctly convex, shiny-blackstructureson each lateral margin of the body. Similar to *Argas*, the basis capitulumhas straight lateral margins with small palps. Festoons are absent in males, and the spiracular plates are small and occur between the 3rdand 4th coxae. Similar to females, the genital pore in males is situated between the 2nd coxae, but it is smaller and circular. About 113 species of this genus have been reported, where the nymphs and adults are parasitic with the host range including domesticated and wild animals, and human. The common species are *O. moubata*and *O. savignyi* which occur in most parts of Africa including South Africa, Botswana, Namibia, Zimbabwe, Tanzania, Kenya, Uganda, Ethiopia, Somalia and Sudan, and in Central and someparts of West Africa (Fig. 10). The parasitic nymphs and adults cause irritation and blood loss, and some species are vectors of pathogens responsible for various infections in animals and human (Table 4).

Table 4. Ornithodoros species of veterinary importance

Species	Common name	Hosts	Site	Pathogenic significance
Ornithodoros savignyi	Sand tampan, eyed tampan	Camels, cattle, poultry, human	Skin	Transmits *CandidatusBorrelia kalaharica*
Ornithodoros moubata/porcinus	Eyeless tampan, hut tampan	Warthog, bushpig, porcupine, pig.	Skin	Transmits African swine fever virus, *Borrelia anserina, Aegyptianella pullorum.*
Ornithodoros hermsi		Rodents	Skin	Transmits*Borrelia duttoni*
Ornithodoros coriaceus		Cattle, deer, human	Skin	Transmits *Borrelia duttoni and* African swine fever virus
Ornithodoros perkeri		Rodents	Skin	Transmits *Borelliaduttoni*
Ornithodoros turicata		Rodents	Skin	Transmits *Borelliaduttoni*
Ornithodoros tholozani		Birds and reptiles	Skin	-
Ornithodoros erraticus		Livestock, human	Skin	-
Ornithodoros rudis		Rodents, human	Skin	-
Ornithodoros lahorensis		Sheep and goats	Skin	-

Adapted from Taylor *et al.* (2015)

Ormithodoros moubata transmits *Borrelia duttoni*, that causes African relapsing fever in humans while, *O. moubata* complex transmits *Borrelia anserina* that causes fowl spirochaetosis, and*Aegyptianella pullorum* to chickens. The warthog species, *O. porcinus*is is important in the transmission of African swine fever virus from the warthog (reservoir host) to domesticated pigs.

Figure 11. Morphological features of Ornithodoros species (Walker et al., 2004).

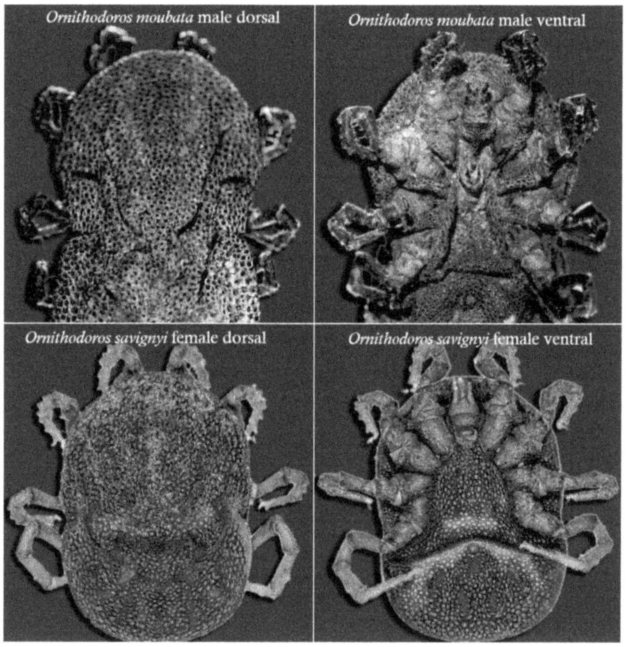

Figure 12. Morphological features of Otobius megnini and its distribution in Africa (Walker et al., 2004). Countries with reported presence of this species are highlighted in yellow. The distribution of the tick in each country is indicated by grey dots or shading.

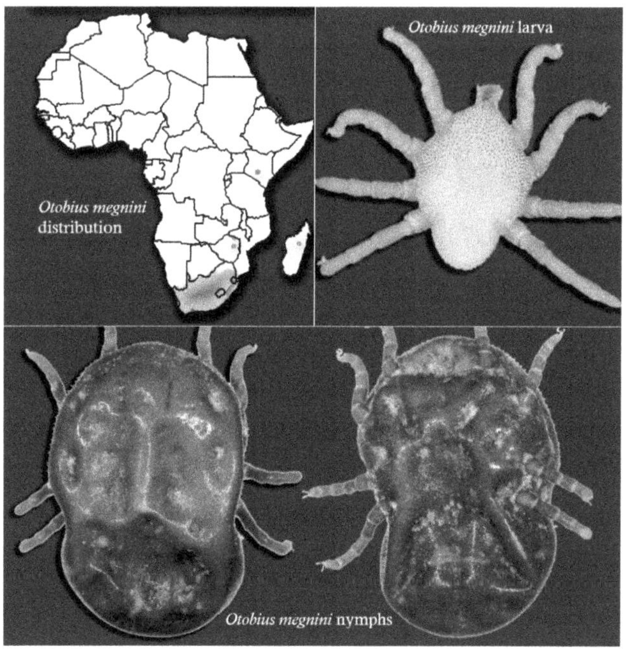

Table 5. Otobius species of veterinary importance

Species	Common name	Hosts	Site	Pathogenic significance
Otobius megnini	Spinose ear tick	Cattle, horses, sheep, goats, dogs, human, wild ruminants	Ear canal	Causes otitis externa
Otobius lagophilus	Rabbit ear tick	Rabbits	Ear canal	Causes otitis externa

Adapted from Taylor *et al.* (2015)

Genus: *Otobius*

Adults are dark, violin-shaped measuring approximately 6mm, without the lateral suture, and the integument texture comprises of spines (making it appear granular). However, in between the spines, the integument is smooth. Nymphs are diamond-shaped, while larvae are pear-shaped with projecting mouth parts clearly visible anteriorly (Fig. 12). Scutum, conscutum and eyes are absent, the legs have no pale rings, are slender, and without pulvilli (Fig. 12). Similar to *Argas* and *Ornithodoros*, the basis capitulumhas straight lateral margins, with small palps. Festoons are absent in males, and the spiracular plates are small and located above the area between the 3rdand 4th coxae. Genital aperture is absent in the 1st and 2nd nymphal stages. Two species in this genus have been reported; *Otobiusmegnini* and *Otobiuslagophilus*, of which *Otobiusmegnini* is the common species occurring in South Africa, Zimbabwe, Madagascar and Kenya (Fig. 12). The two species are one-host ticks where only the larval and nymphal stages are parasitic on a wide range of hosts causing severe irritation leading otitis externa (Table 5).

FUTURE RESEARCH DIRECTIONS

As we move into the future, identification of tick species need to be carried out for a particular production system in a specific area. The tick surveillance by serological methods and control through vaccination could be possible and this opens new avenues for the development and advance of new tests and further research on other tick species. Further research on the detailed epidemiology of ticks needs to be undertaken as well. The recent characterization of non-specific viruses in argasid vectors and all the argasid-associated pathogens mentioned in this review suggest the great potential of argasids for viral and bacterial disease transmission requires further research. The possibility that ticks might serve as vectors for many more pathogens than expected requires a greater effort in implementing control measures, such as the search for new protective antigens to be included in a broad-spectrum anti-tick vaccine as well as specific coordinated and urgent epidemiological and parasite-surveillance programs.

CONCLUSION

Hard and soft ticks are distributed worldwide and global climatic changes, along with social factors, influence tick habitats and their hosts. These factors hinder the prediction of the tick-borne diseases distribution patterns as several factors could influence the tick vector competence. Ticks are the most important ectoparasites of livestock in tropical and sub-tropical areas, and are responsible for severe economic losses both through the direct effects of blood sucking and indirectly as vectors of pathogens and toxins. The major losses caused by ticks are due to the ability to transmit protozoan, rickettsia and viral diseases of livestock, which are of great economic importance world-wide. Since there is no single ideal solution for the control of ticks, an integrated control approach is probably the most effective. Vector and reservoir surveillance are an important component of such a strategy. Designing an economical, integrated tick control strategy for a particular production system in a specific area comprising of chemical control, ecological control biological control, genetic control, and immunization is imperative.

REFERENCES

Amrit, P. S., Rajesh, K. G., & Tajpreet, K. (2011). Mechanisms pertaining to arsenic toxicity. *Toxicology International Journal*, *18*(2), 87–93. doi:10.4103/0971-6580.84258 PMID:21976811

Baars, R. M. T. (1999). The effect of rangeland fires on cattle tick infestation in Western Zambia. *Tropical Animal Health and Production*, *31*(5), 275–278. doi:10.1023/A:1005299605959 PMID:10509420

Baneth G. (2014). Tick-borne infections of animals and humans: a common ground. *International Journal. Parasitol*ogy, *44*(9), 591-6. doi: .011 doi:10.1016/j.ijpara.2014.03

Bezuidenhout, J. D. (1990). Recent research findings on cowdriosis. In C. Williams & I. Kakoma (Eds.), *Ehrlichiosis* (Vol. J, pp. 125–135). Kluwer Academic Press. doi:10.1007/978-94-009-1998-3_11

Camus, E., Martinez, D., Beauperthuy, L., Benderdouche, A., Coisne, S., Corbette, C., Denormandie, N., Garrís, G., Harris, E., King, T., Lannoy, L., Louvet, A., Lunel, E., Montrose, M., Nisbett, B., Nyack, B., Robinson, I., Rouchosse, P., Swanston, G. B., ... Thye, G. D. (1993). Heartwaterin Guadeloupe and the Caribbean.InProceedings of the 2nd Biennial Meeting of the American Society of Tropical Veterinary Medicine.2-6 February. *Revue d'Élevage et de Médecine Vétérinaire des Pays Tropicaux*, *46*, 109–114. PMID:8134620

Dantas-Torres, F., Oliveira-Filho, E. F., Soares, F. A. M., Souza, B. O. F., Valença, R. B. P., & Sá, F. B. (2008). Ticks infesting amphibians and reptiles in pernambuco. *Revista Brasileira de Parasitologia Veterinária*, *17*(4), 218–221. doi:10.1590/S1984-29612008000400009 PMID:19265581

De Vos, S., Zeinstra, L., Taoufik, O., Willadsen, P., & Jongejan, F. (2001). Evidence for the utility of the Bm86 antigen from *Boophilus microplus* in vaccination against other tick species. *Experimental & Applied Acarology*, *25*(3), 245–261. doi:10.1023/A:1010609007009 PMID:11523920

DeCastro, J. J. (1997). Suitable tick and tick-borne disease control in livestock improvement. *Veterinary Parasitology*, *71*, 77–97. doi:10.1016/S0304-4017(97)00033-2 PMID:9261972

FAO. (1977). *Review of recent progress in the control of ticks and tick-borne diseases in East*. Africa. FAO.

FAO. (1984). *Ticks and TBDs control; in a practical field manual* (Vol. 1). FAO.

Getahun, A., Yacob, H., & Basu, A. (2016). Overview of the Biology, Epidemiology and Control Methods against Hard Ticks. *Global Journal of Science Frontier Research, 16*(2).

Habela, M., Van, H. A. I., Rol, J. A., Navarrete, I., & Jongejan, F. (1999). Transmisión experimental.de *Theileriaannulata. Acta Parasitologica PortuguesaActa., 1*(2), 8.

Hornok, S., Kontschán, J., Estrada-Peña, A., de Mera, I. G. F., Tomanović, S., & de la Fuente, J. (2015). Contributions to the morphology and phylogeny of the newly discovered bat tick species, *Ixodesariadnae* in comparison with *I. vespertilionis* and *I. simplex. Parasites & Vectors, 8*(1), 47. doi:10.118613071-015-0665-0 PMID:25616683

Jongejan, E. & Ulineberg, G. (1994). Ticks and control methods. *Revue Scientifiqueet Technique International Office of Epizootics, 13*(4), 1201-1226.

Jongejan, F., Zivkovic, D., Pegram, R. G., Tatchell, R. J., Fison, T., Latif, A. A., & Paine, P. (1987). Ticks of the blue and White Nile ecosystems in the Sudan with particular reference to the *Rhipicephalus sanguineus* group. *Experimental & Applied Acarology, 3*, 331–346. doi:10.1007/BF01193169 PMID:3331134

Mercy, M. (2016). Laboratory and field evaluation of entomopathogenic fungi for the control of amitraz-resistant and susceptible strains of *Rhipicephalus decoloratus. Veterinary Parasitology, 225*, 12–18. doi:10.1016/j.vetpar.2016.05.026 PMID:27369570

Metcalf, R. L. (2000). *Insect Control.* Ullmann's Encyclopedia of Industrial Chemistry. doi:10.1002/14356007.a14_263

Narahashi, T. (2013). *Toxicology of Halogenated Hydrocarbons Health and Ecological Effects*. doi:10.1016/B978-0-08-027530-7.50023-6

Nicholson, W. L., Allen, K., McQuiston, J. H., Breitschwerdt, E. B., & Little, S. E. (2010). The increasing recognition of rickettsial pathogens in dogs and people. *Trends in Parasitology, 26*(4), 205–212. doi:10.1016/j.pt.2010.01.007 PMID:20207197

NijhofA. M.GuglielmoneA. A.HorakI. G. (2016). www.catalogueoflife.org/col.ISSN

Norval, R. A. I., Perry, B. D., & Young, A. S. (1992). *The epidemiology of theileriosis in Africa.* Academic Press.

Pegram, R. G., Keirans, J. E., Clifford, C. M., & Walker, J. B. (1987). Acari, Ixodoidae. *Systematic Parasitology, 10*, 3–44. doi:10.1007/BF00009099

Pegram, R. G., Perry, B. D., Musisi, E. L., & Mwanaumo, B. (1986). Ecology and phenology of ticks in Zambia: Seasonal dynamics on cattle. *Experimental & Applied Acarology, 2*(1), 25–45. doi:10.1007/BF01193353 PMID:3451861

Rajput, Z. I., Hu, S., Chen, W., Arijo, A. G., & Xiao, C. (2006). Importance of ticks and their chemical and immunological control in livestock. *J Zhejiang Univ Science B, 7*(11), 912-921. www.springerlink.com

Rand, K. N., Moore, T., Sriskantha, A., Spring, K., Tellam, R., Willadsen, P., & Cobon, G. S. (1989). Cloning and expression of a protective antigen from the cattle tick *Rhipicephalus (Boophilus) microplus. Proceedings of the National Academy of Sciences of the United States of America, 86*, 9657 – 9661.

Rodriguez, M., Rubiera, R., & Montesinos, R. (1994). High level expression of the *B. microplus*Bm86 antigen in the yeast *P. pastoris* forming highly immunogenic particles for cattle. *Journal of Biotechnology*, *33*(2), 135–146. doi:10.1016/0168-1656(94)90106-6 PMID:7764729

Rosenberg, R., Lindsey, N. P., Fischer, M., Gregory, C. J., Hinckley, A. F., Mead, P. S., Paz-Bailey, G., Waterman, S. H., Drexler, N. A., Kersh, G. J., Hooks, H., Partridge, S. K., Visser, S. N., Beard, C. B., & Petersen, L. R. (2018). Vital Signs: Trends in Reported Vector borne Disease Cases - United States and Territories, 2004-2016.Morbidity and Mortality Weekly Report. *Rep*, *67*, 496. PMID:29723166

Saimo, M., & Odongo, D. O. (2011). Recombinant *Rhipicephalus appendiculatus* gut (Ra86) and salivary gland cement (Trp64) proteins as candidate antigens for inclusion in tick vaccines: Protective effects of Ra86 on infestation with adult *R. appendiculatus*. *Vaccine: Development and Therapy*, *1*, 15–23. doi:10.2147/VDT.S20827

Shivairo, R. S., Muleke, C. I., & Lubembe, D. M. (2018). *Students Guide Book* (1st ed.). Veterinary Parasitology and Parasitic Diseases.

Soulsby, E. J. I. (1986). *Helminths, Arthropods and Protozoa of Domesticated Animals by E. J. I. Soulsby* (7th ed.). Lea and Febiger.

Taylor, M. A., Coop, R. L., & Wall, R. L. (2015). Veterinary Entomology. In M. Taylor, R. Coop, & R. Wall (Eds.), *Veterinary Parasitology* (pp. 161–258). doi:10.1002/9781119073680.ch3

Uilenberg, G., & Camus, E. (1993). Heartwater (cowdriosis). In W. M. Ristic (Ed.), *Rickettsial and chlamydial diseases of domestic animals* (Vol. Z, pp. 293–332). Pergamon Press.

Urquhart, G. M., Armour, J., Ducan, J. L., Dunn, A. M., & Jennings, F. W. (1996). *Veterinary Parasitology (2nd ed.)*. Blackwell Science.

Walker, A. R., Latif, A. A., Hans, N., Nijhof, A., & Jongejan, F. (2004). *Electronic Tick Guide for Africa. Developed by the International Consortium on Ticks and Tick-borne Diseases*. ICTTD.

Walker, J. B., & Olwage, A. (1987). The tick vectors of *Cowdriaruminantium*(Ixodoidea, Ixodidae, genus *Amblyomma*) and their distribution. *The Onderstepoort Journal of Veterinary Research*, *54*, 353–379. PMID:3329325

Walker & Bouattour. (2003). Ticks of domestic animals in Africa: A guide to identification of species. *Bioscience Reports*.

Wall, R., & Shearer, D. (1997). *Veterinary Entomology*. Chapman and Hall Publishers. doi:10.1007/978-94-011-5852-7

Wall, R., & Shearer, D. (2001). Veterinary Ectoparasites: Biology, Pathology and Control (2nd ed.). Blackwell Science.

Wharton, R. H. (1976). Tick-borne livestock diseases and their vectors. Acaricide resistance and alternative methods of tick control. *World Animal Review*, *2*, 8–15.

Willadsen, P. (1990). Perspectives for subunit vaccines for the control of ticks. InProc. FAO expert consultation on revision of strategies for the control of ticks and tick-borne diseases. *Parassitologia*, *32*, 195–200. PMID:2284131

Willadsen, P., & Jongejan, F. (1999). Immunology of tick-host interaction and the control of ticks and tick-borne diseases. *Parasitology Today (Personal Ed.)*, *15*(7), 258–262. doi:10.1016/S0169-4758(99)01472-6 PMID:10377526

Willadsen, P., Mckenna, R. V., & Riding, G. A. (1988). Isolation from the cattle tick, *Rhipicephalus (Boophilus) microplus*), of antigenic material capable of eliciting a protective immunological response in the bovine host. *International Journal for Parasitology*, *18*(2), 183–189. doi:10.1016/0020-7519(88)90059-8 PMID:3372125

Young, A. S., Groocock, C. M., & Kariuki, D. P. (1988). Integrated control of ticks and tick-borne diseases of cattle in Africa. *Parasitology*, *96*(2), 403–432. doi:10.1017/S0031182000058388 PMID:3287285

ADDITIONAL READING

Shivairo, R. S., Muleke, C. I., & Lubembe, D. M. (2018). *Students Guide Book* (1st ed.). Veterinary Parasitology and Parasitic Diseases.

Soulsby, E. J. I. (1986). *Helminths, Arthropods and Protozoa of Domesticated Animals by E. J. I. Soulsby* (7th ed.). Lea and Febiger.

Chapter 9
East Coast Fever:
Theileria Parva Infection of Cattle

Sammy Gichuhi Ndungu
Kenya Agricultural and Livestock Research Organisation, Kenya

Sebastian K. Waruri
Kenya Agricultural and Livestock Research Organisation, Kenya

James M. Wanjohi
Kenya Agricultural and Livestock Research Organisation, Kenya

ABSTRACT

East coast fever, a disease of cattle caused by the protozoan parasite Theileria parva and transmitted by the three-host tick Rhipicephalus appendiculatus (the brown ear tick), is a major constraint to cattle production in Eastern, Central, and Southern Africa. In Kenya it is the most important tick-borne disease and a major constraint in cattle productivity. This is due to the high morbidity and mortality it causes in susceptible herds, the cost of control of the vector ticks, and the cost of treatment of clinical cases. Animals that recover from the disease also suffer from reduced productivity which can be long term. The limited distribution of the tick and the disease to only East, Central and Southern Africa also means that the market for therapeutic drugs and acaricides is small. Therefore, drug companies are not keen on funding research and development of new drug and acaricide molecules when resistance occurs.

INTRODUCTION

East Coast fever disease is caused by the obligate intracellular protozoan parasite *Theileria parva* and is transmitted by the three-host brown ear tick *Rhipicephalus appendiculatus.* The disease is of major economic importance in the endemic areas of east, central and southern Africa where it accounts for an estimated 1.1 million cattle deaths and $168 million loss in productivity annually (Mukhebi et al, 1992). In susceptible cattle, the morbidity and mortality can reach 100% and over 90% respectively. Local zebu cattle show a degree of resistance to the disease, but the more productive taurine breeds and grade

DOI: 10.4018/978-1-7998-6433-2.ch009

cattle are highly susceptible (Ndungu et al, 2005) requiring strict control of the disease for successful introduction of these breeds and upgrading of the local zebus in these areas.

Control of the disease is through eradication of the vector by use of acaricides Acaricides are expensive (Minjauw and McLeod, 2003), ticks develop resistance to them (Keating, 1983) and they are toxic to both humans and animals. They contaminate the environment and may end up in the food chain. Use of acaricides also interferes with endemic stability and creates susceptible populations (Lawrence et al, 1980). Curative drugs are expensive and are beyond the majority of resource-poor smallscale livestock keepers. The vaccines available are based on the infection and treatment method. They are effective (Perry, 2016), but they are not very affordable by small-holder livestock keepers as the packaging is in many doses. Currently, it is not possible to determine the number of doses per package at the production stage. The dose package is determined by titration of the vaccine stabilate in cattle to determine the user dilution. Research on development of sub-unit vaccines for East Coast fever is on-going (Nene et al, 2016) and once they become available, the cost of the vaccines and their packaging may no longer be a challenge in the uptake of the vaccine.

AETIOLOGY

The aetiological agent, *Theileria parva*, and its intralymphocytic stages, the schizonts in lymphoid cells, was first described by Koch (1903).

The taxonomic classification of *Theileria parva* is as follows: (Irvin, 1987).

Sub Kingdom - Protozoa
Phylum - Apicomplexa
Class - Sporozoea
Subclass - Piroplasma
Order - Piroplasmida
Family - Theileriidae
Genus - Theileria
Species - *Theileria parva*

T. parva is a parasite of the Cape buffalo (*Syncerus caffer*) that has adapted to infect and undergo transmission in cattle (Uilenberg, 1981, Morrison, *et al,* 2020) and is highly pathogenic in cattle. Three subspecies of *T. parva* were recognized, namely *Theileria parva parva, Theileria parva lawrence*i and *Theileria parva bovis*. These parasites cause East Coast fever (ECF), Corridor disease and January disease respectively in cattle. They were distinguished on clinical and epidemiological parameters as they are morphologically and serologically indistinguishable (Burridge et al, 1973). The parasite(s) are also transmitted by the same tick vector *Rhipicephalus appendiculatus*. In classical ECF caused by *T. parva parva*, macroschizonts and piroplasms are plentiful. The clinical and epidemiological differences between the subspecies are that *T. parva lawrencei* is a buffalo parasite which produces relatively few, small schizonts and very few, if any piroplasms in cattle thus making inter–cattle transmission by the tick inefficient (Young et al., 1973, Morrison *et al,* 2020). *T. parva bovis* differs from *T. parva lawrencei* in that plentiful piroplasms are produced in cattle and it can therefore, be transmitted between cattle (Norval et al., 1992). *T. parva bovis* was distinguished from *T. parva parva* because it produced a carrier state in recovered animals and the number of parasites in *T. parva bovis* infections are normally lower than in *T. parva parva* infections.

That the three subspecies of the parasite are actually the same parasite has been shown by passaging *T. parva lawrencei* through ticks and cattle (Maritim *et al.*, 1992). *T. parva lawrencei* changes so that it resembles *T. parva bovis* and later becomes indistinguishable from *T. parva parva*. There is varying cross immunity between the three subspecies which is not observed between other *Theileria* species (Irvin, 1987) and carrier state is now known to occur in *T. parva* infections (Maritim *et al.*, 1992). Monoclonal antibody profiles (Conrad *et al.*, 1987a), DNA hybridization and oligonucleotide probes studies (Allsopp *et al.*, 1993), have also shown that the three subspecies are one species. It has therefore, been recommended that name *T. parva* should encompass all the three subspecies (Anon, 1989).

TRANSMISSION

The Target Cell Types in *T. parva* Infection

Following inoculation of a susceptible bovine host with *T. parva* sporozoites, the first evidence of infection is the appearance, after several days, of schizont infected lymphoblastoid cells in the lymph node draining the site of inoculation. The infected cells lack both surface and intracellular immunoglobulins (Emery, 1981a). This characteristic, together with the lymphoblastoid appearance of the cells, prompted the suggestion that the infected cells are T. lymphocytes. Lineage-specific monoclonal antibodies have confirmed that infected cell lines can be established from B cells, T cells and null cells but not from monocytes or granuloctyes (Baldwin *et al.*, 1988).

Life Cycle of *Theileria parva* in the Vector Tick

The life cycle of *T. parva* (Melhorn and Shein, 1984) is diagrammatically illustrated in Figure 1. Ticks acquire infection when they are feeding on blood from cattle or buffalo containing *T. parva piroplasms* in the erythrocytes. Once in the gut lumen of the tick, the erythrocytes lyse releasing the piroplasms. The piroplasms differentiate into male and female gametes. These undergo syngamy to form zygotes which enter the lumen epithelial cells. During moulting the zygotes develop into motile kinetes and these enter the haemolymph from where they invade the newly-formed salivary gland acini. They invade type III acini and enter type e cells to form a sporont. Sporogony by meiosis occurs when the tick is feeding on a mammalian host and sporozoites are formed. The sporozoites that are produced enter the saliva of the tick and eventually the mammalian host when the newly moulted tick commences feeding. A single acinar cell contains approximatelly 40,000 – 50,000 sporozoites (Fawcett *et al.*, 1985).

Life Cycle in the Mammalian Host

Once the sporozoites are released at the site of inoculation, entry into the host target cell is very rapid (Shaw *et al.*, 1991). The nature of the target cell at the site of inoculation is not known. In *in vitro* systems, *T. parva* sporozoites infect BoT4+, BoT8+ T cells, B cells and null cells (Baldwin *et al.*, 1988). The sporozoite membrane becomes tightly attached to the host cell membrane, the host cell surface invaginates and the rims of the invagination close over the parasite. The parasite is thus internalized

Figure 1. The lifecycle of T. parva in the vector Rhipicephalus appendiculatus tick and mammalian host. Photo credits – KALRO - VSRI

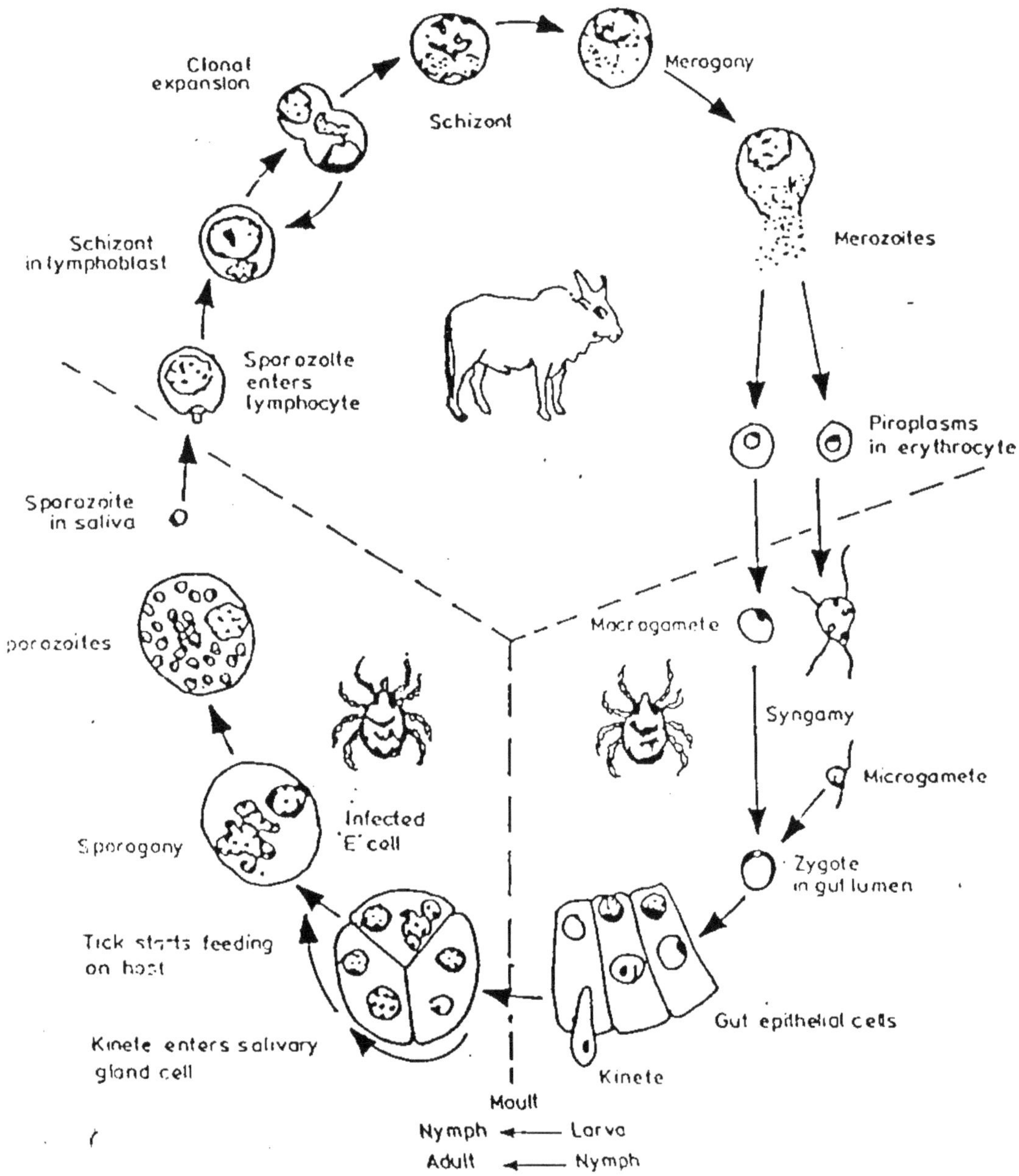

while still surrounded by the closely apposed host membrane. The host membrane then dissolves and the parasite enters the host cell cytoplasm. .

In the bovine host cells, the sporozoites undergo nuclear division resulting in multinuclear schizonts. The picture below shows a cell undergoing division on the left and a cell with a schizont on the right

Figure 2. (a) A cell undergoing division (b) Schizonts. Photo credits – KALRO - VSRI

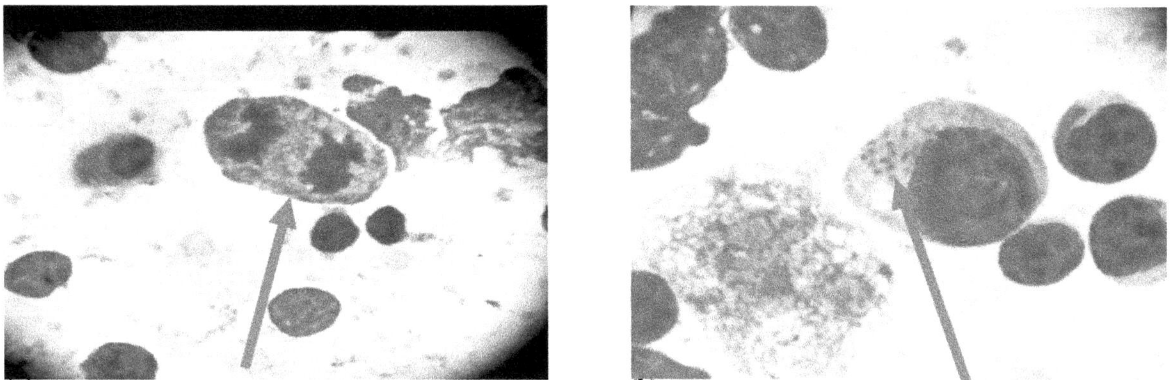

(a) **(b)**

At the same time the host cell undergoes blast transformation and starts to proliferate. Division of the host cell occurs synchronously with the division of the schizonts and this results in the distribution of the parasite into each daughter cell. This process results in clonal expansion of the population that was initially infected. The clonal expansion induced by the *Theileria* parasite is continuous and can go on indefinitely ((Stagg *et al.*, 1981), but can be reversed with theilericidal drugs. As the infection progresses, the number of nuclei per schizont increases and some of the schizonts undergo merogony to

Figure 3. Piroplasms in cattle erythrocytes. Photo credits – KALRO - VSRI

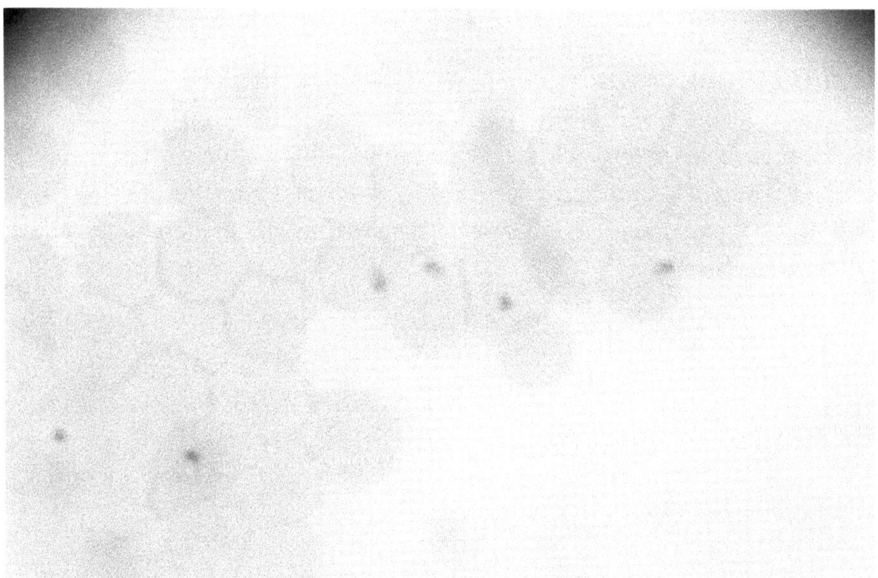

form merozoites. The merozoites are released from the cell and enter erythrocytes to form piroplasms which in turn infect ticks.

The Vector of *Theileria parva*

T. parva is transmitted by the three-host tick *Rhipicephalus appendiculatus,* commonly called the brown ear tick. The picture below shows a male and engorged female brown ear tick.

Figure 4. (a) A male brown ear tick (b) An engorged female brown ear tick. Photo credits – KALRO - VSRI

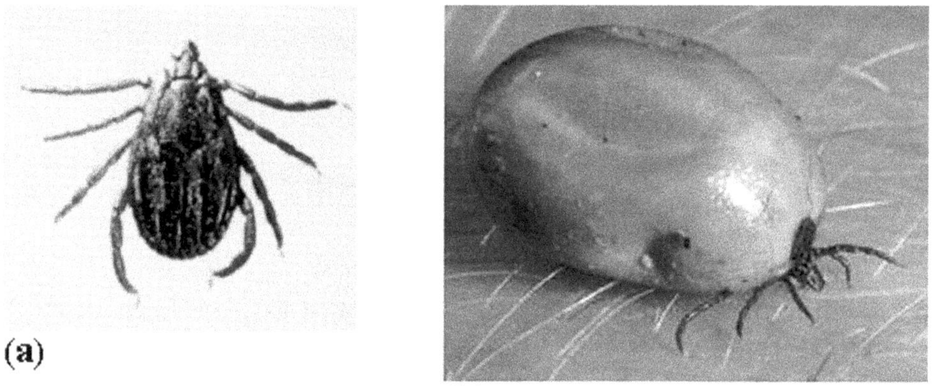

(a) (b)

The transmission is transtadial with one instar acquiring the infection and the subsequent instar transmitting it after moulting i.e. larvae will transmit as nymphs, and nymphs as adults. *Rhipicephalus evertsi* have been shown to be capable of transmitting *T. parva* in the laboratory but, ticks that have been implicated in the transmission of *T. parva* in the field include *Rhipicephalus duttoni* (Walker, 1991) and *Rhipicephalus zambeziensis* (Lawrence *et al.,* 1983).

Geographical Distribution

The distribution of East Coast fever is closely linked to the distribution of *R. appendiculatus* (Figure 5). The tick is found at altitudes from sea level to over 7,000 feet where there is suitable vegetation and adequate rainfall. It is not found in areas where annual rainfall is less that 20 inches and where the altitude is over 8,000 (Walker, 1974).

Figure 5. Map showing the distribution of R. appendiculatus in Africa (Lessard et al., 1990)

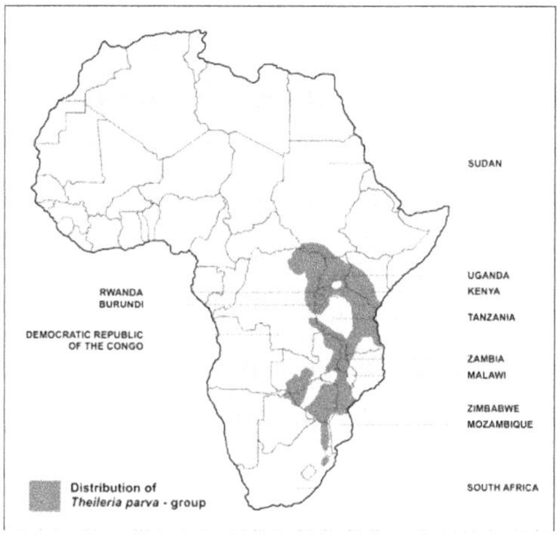

Figure 6. Map showing the distribution of Rhipicephalus appendiculatus in Kenya. Photo credits – KALRO - VSRI

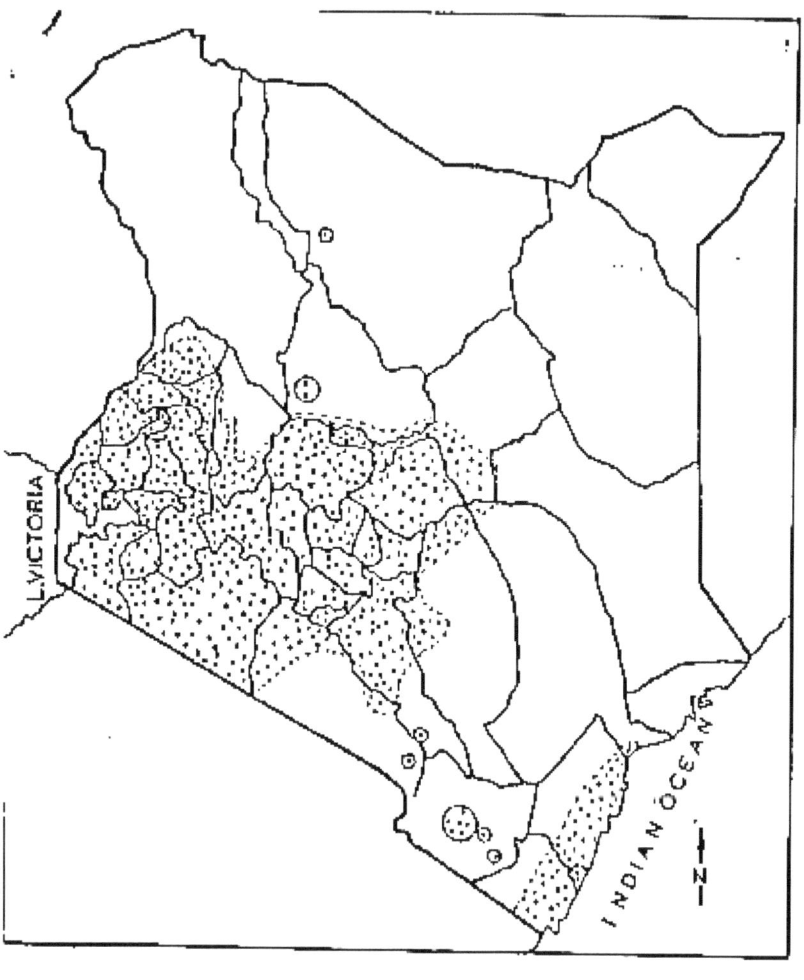

EPIDEMIOLOGY AND RISK FACTORS

Factors Affecting the Distribution of *R. Appendiculatus*

Factors affecting the distribution of the vector ticks and hence the disease include climate, vegetation and host availability. Climate has two major components: temperature which determines development rates and to a lesser extent survival of the ticks, and moisture which has a direct influence on survival of the ticks. In the case of species that undergo diapause, day length may also play a significant part in the distribution of the tick. Suitable vegetation is important for the survival of *R. appendiculatus*. The tick tends to be absent from open plains and dense forests and is mainly found in savannah and woodland savannah habitats (Walker, 1974). Grass and limited tree cover form the best microhabitat for the tick (Minshull and Norval, 1982).

R. appendiculatus becomes very abundant in the presence of hosts that have a low level of tick resistance such as susceptible *Bos taurus* cattle (Norval et al, 1988). There are some suitable areas where the tick is not found such as Central Tanzania, large areas of Central and Western Africa and the Ethiopian Highlands. It is thought that this may be due to low densities of host herbivores for various reasons such as presence of trypanosomiasis or dense forests (Perry *et al.*, 1990).

Suitable environment for the survival of the tick may be extending through the effects of climate change. A prediction model by Olwoch et al (2008) has shown that some areas of South Africa which were unsuitable for the survival of the tick are now habitable.

Factors Affecting Survival of *T. parva* in the Tick

A number of factors affect the survival of *T. parva* in the tick vector. These include factors within the tick's body, climatic factors and the immunological status of the mammalian host that the tick feeds on.

After the zygotes are formed, the moulting process seems to have a profound effect on the differentiation of the zygotes to kinetes. The moulting determines the timing of kinete formation so they can enter the newly formed salivary gland cells (Fawcett et al., 1985).

Temperature has a major influence on the survival of the parasite in the tick. Young and Leitch (1981) have demonstrated in the laboratory that the time over which gametes are produced and infection rates in the tick are temperature dependent. Kinete migration after moulting of the tick is also temperature dependent with kinetes migrating in ticks held at 28°C and not at 18°C or lower (Bell, 1983).

In field studies it has been shown that adult *R. appendiculatus* remain infective to cattle for up to 15 months (Young *et al.*, 1987).

The immunological status of the host on which the tick feeds indirectly affects the survival of the parasite. Tick-naïve hosts produce higher infection rates in the ticks than tick-resistant hosts. Ticks feeding on tick resistant hosts have reduced blood meal weights and have low mean survival time as compared to ticks on naïve cattle. The tick may ingest very low numbers of erythrocytes as most of the meal of the tick is tissue fluid. Ticks may die on resistant hosts and those that persist may be dislodged by enhanced grooming that is evident in resistant cattle (De Castro *et al*, 1985).

Animals that recover from ECF become immune, but may become carriers (Maritim *et al.*, 1992). Such animals have low levels of piroplasms and ticks feeding on them have low levels of infections (Walker *et al.*, 1981).

Factors Affecting Transmission of *T. parva* From Ticks to Cattle

When a kinete enters an 'e' cell of the tick salivary gland, it undergoes sporogony for some time then stops. It remains dormant until the tick feeds on a suitable host. The sporoblast then develops into sporozoites that can be transmitted to cattle or other hosts. This process takes 3-5 days (Purnell *et al.*, 1973). Other factors not inherent in the tick may also be involved as unfed ticks have been shown to be capable of transmitting *T. parva* (Young *et al.*, 1987). Temperature has been shown to play a major role in the maturation of the sporoblast to sporozoites. In field studies, it was shown that heat stimulated ticks were capable of transmitting *T. parva* within 24 hours of application on cattle (Ochanda 1988). Female ticks have more type III acini and more cells than male ticks and consequently have on average higher infection rates than male ticks (Fawcett et al., 1985).

In southern Africa, there is a seasonal occurrence of *R. appendiculatus* and *R. zambeziensis* adult ticks (Short *et al*, 1989). The ticks are found active on vegetation during the wet and warm summer months when day length is long. For the rest of the year, the ticks are found inactive at or near the soil surface. This seasonal pattern of occurrence is caused by climatic variation in temperature and humidity which affect development and survival rates (Kaiser *et al.*, 1988) as well as the diurnal behavior of host seeking (Punyua *et al.,* 1985). The seasonal occurrence of adult ticks results in only one generation per year in the southern regions of Africa. Since there is little or no overlap between the adult larval and nymphal stages, onward transmission of *T. parva* is dependent on ticks picking infection from carrier cattle (Norval *et al.,* 1991). In tropical Africa where there is no seasonal variation in daylight and rainfall is spread more or less uniformly throughout the year, all stages of the tick are found on the host throughout and pass through more than one generation per year. Onward transmission of *T. parva* is rapid and continuous and occurs from both carrier cattle and those cattle undergoing clinical disease.

Factors Affecting Transmission of *T. parva* From Cattle to Ticks

Infection rates in ticks are mainly affected by the level of piroplasm parasitaemia in the host animal. Primary infections have higher piroplasm parasitaemias than carrier infections and therefore ticks feeding on cattle with primary infections will generally have higher infection rates than those feeding on carrier animals. Epidemiologically, primary infections are less important than carrier infections since the period of high parasitaemia is brief, resulting in only a few ticks picking up the infection. In an endemically stable situation, it is the calves that suffer primary infections while the majority of adults are immune carriers. Calves normally carry a lower tick burden than adult cattle (Moll *et al.*, 1986) and therefore, are less significant in the epidemiology of theileriosis than adult cattle.

Continuous tick infestation on cattle induces resistance to ticks. Ticks feeding on tick-resistant cattle have lower infection rates than those feeding on tick-naïve cattle. Such ticks have lower engorgement weights and have lower survival time than those engorging on tick naïve cattle (Newson and Chiera, 1989).

PATHOGENESIS AND HISTOPATHOLOGY

Prior to detection of schizonts in the local drainage lymph node, the events that take place after inoculation of sporozoites into cattle are not known. Cowdry and Danks (1933) were unable to find the parasite in the skin by histological methods, but the sporozoites are quickly taken up by the host target cells

(Morrison *et al.* 1981) and rapidly become associated with the host lymphocytes (Shaw *et al.* (1991). There is a dramatic increase in cellularity of the local drainage lymph node with plenty of blast cells and parasitized cells and in efferent lymph. The hyperplasia and expansion of the infected cell population is followed by a phase of lymphoid disorganization and depletion, characterized by gradual involution of the germinal centres and lymphocytolysis, first observed as small foci of necrosis which gradually become more extensive.

The pattern of cellular changes in the general body lymphoid tissues distant to the site of inoculation, including the thymus and spleen, are similar to those in the local drainage lymph node, except that the parasites appear later and hyperplasia and tissue destruction are less pronounced and occur more gradually except in the mediastinal and mesenteric lymph nodes where the cellular changes are more severe and mildest respectively (Barnett, 1960). In the thymus the initial appearance of parasitized lymphocytes is associated with marked atrophy of the organ. Infiltration of the bone marrow with parasitized cells results in degeneration, destruction and ultimately replacement of the normal bone marrow tissue with lymphoid cells. The resident bone marrow cells are themselves not parasitized. The destruction of the bone marrow, the thymus and the spleen leads to a depression of leucopoiesis. The leucopaenia evident in ECF is probably due to the depression of leucopoiesis and the extensive lymphocytolysis in the lymphoid organs. Other organs affected include the kidney, liver, lungs, adrenal glands, the gastro-intestinal tract and in some cases the brain. In the kidneys, liver, lungs and adrenal glands, there is extensive infiltration of the organs with both parasitized and non-parasitised lymphocytes. In the lungs, presence of these cells seem to cause increased permiability of the blood vessels resulting in effusions of fluids into the air passages, hence the pulmonary oedema. In the gastro-intestinal tract, there is extensive infiltration of the lamina propria with both parasitized and non-parasitised lymphocytes together with necrosis and desquamation of the epithelium. In the brain the changes include infiltration of blood vessels with parasitized and non-parasitised lymphocytes. Occasional demyelination, perivascular cuffing and haemorrhages are seen.

CLINICAL MANIFESTATIONS

Clinical signs of ECF have been described by Brown (1990). The severity of the disease is dependent on the dose of parasite inoculated (Dolan et al., 1984) and the immunological status of the host. The symptoms may be acute, subacute, mild or inapparent. There is also the nervous syndrome referred to as cerebral theileriosis or 'turning sickness'. The incubation period varies from 8 to 16 days when cattle are exposed to tick challenge.

The acute form of disease is the most common and is characterized by fever, a cessation of rumination, serous nasal discharge, lachrymation, ocular opacity due to infiltration with parasitized and unparazitised lymphoctes and sometimes blindness. In some cases, there is swelling of the eyelids and bottle-jaw. A characteristic sign is the swelling of superficial lymph nodes especially those draining the site of inoculation. Later there is dyspnea, tachypnea, diarrhea and the animal becomes emaciated. There is a moist cough and when the animal becomes recumbent, large amounts of froth may escape from the nostrils. Petechiation of the mucous membranes is common. The course of disease varies from 8-25 days. Death is invariably the final outcome.

The subacute form of disease is mainly encountered in calves and, to a lesser degree, in adults in enzootic regions and in partially immune animals. The symptoms are similar to those of the acute form but are less pronounced. Animals usually recover from this form of disease. The mild form of disease is less severe than the subacute form and the animals may show mild fever for a few days. There is swelling of the superficial lymph nodes in which schizonts are present. The inapparent form of disease is seen in experimental animals that are solidly immune to the strain of parasite used for challenge. There are no overt clinical signs manifested by such animals.

Cerebral theileriosis was described by Mettam and Carmichael (1936) in animals that had recovered from ECF in Uganda. The disease affects both sexes and all breeds, but is most common in young stock between 6 months and 2 years of age. The course of disease can either be acute, mild or chronic. In the acute form the animal develops nervous signs such as turning or spinning in one direction, hence the name 'turning sickness'. In mild cases the animal may stand with its head pushed against immovable objects. In the chronic form the animal may show muscular incoordination and blindness.

The animals have no fever but invariably die of the disease. Clinical signs appear within a variable period after the animal has been exposed to ECF, usually between 2 months and five years.

Haematology

There is a marked drop in the white blood cell count and a pronounced thrombocytopenia. The red blood cell count remains within normal ranges in acute cases, but in protracted cases there may be variable degrees of anaemia (Maxie *et al.*, 1982). Both haematocrit and haemoglobin levels follow the red blood cell pattern. The leucopenia may be due to retardation of myelopoiesis and extramedullary lympholysis may play a significant role (Allison, 1981).

Gross Pathology

The gross pathology varies with the course of disease. In acute cases, the general nutritional state of the carcass is good but in protracted cases these is loss of condition. There may be variable amounts of froth escaping from the nostrils of the dead animal. The lymph nodes are swollen and oedematous. Hydrothorax, hydropericardium and hydroperitoneum may be present in some cases. The myocardium is flabby and there are marked splash haemorrhages of the epicardium and endocardium. The visceral and parietal pleura may be spotted with petechial haemorrhages which may also be present in the mucous membranes of the trachea and bronchi. The lungs are congested and oedematous, and on section fluid oozes from the cut surface. Froth is found in the trachea and bronchi. The liver may be increased in size and in some cases may have a yellowish tinge on the cut surface. The spleen lesions depend on the duration of infection and also the strain of the parasite. With the South African strains of *T. parva*, the spleen is always enlarged with prominent malpighian corpuscles irrespective of the duration of the illness. With the East African strains of *T. parva*, acute cases have enlarged spleens but in protracted cases the spleen is small and fibrotic. The kidneys may be pale and congested and have a variegated appearance due to the presence of greyish-white spots called 'infracts'. In the majority of cases petechiae may also be present in the kidney (Barnett, 1960). There may be occasional haemorrhages in the serosa of the rumen. In the majority of cases there are multiple superficial erosions and ulcers in the mucosal surface of the abomasum. They are mainly found on the pylorus and occasionally on the fundus. The ulcers may have a haemorrhagic zone around them. The intestines, particularly the large intestine, have mucosal haemorrhages and the Peyer's patches are swollen and prominent. Ulcers may occasionally be observed.

In cerebral theileriosis (turning sickness) the pathological lesions are exclusively in the brain (Mattem and Carmichael, 1936). In acute cases, there are haemorrhages in the brain and meninges. In chronic cases necrosis, atrophy, fibrosis and cavity formation, when haemorrhages are absorbed, are the main lesions. The lesions mainly found in the white matter of the cerebral hemispheres. A constant feature of chronic cases is a yellow discolouration of the white matter which may be localized or diffuse. The cavities contain a yellow fluid.

DIAGNOSIS

Diagnosis of ECF is based on clinical signs, haematoloy which is characterized by a marked leucopoenia and demonstration of schizonts in Giemsa stained lymph node biopsy smears. Presence of piroplasms in blood smears is not diagnostic as theilerial piroplasms may be present in clinically normal carrier animals. Its reliability is, however, limited by the fact that it is difficult to differentiate schizonts of *T. parva* from those of *T. taurotragi, T. mutans* and *T. annulata* by their morphological appearance alone. Other methods of differential diagnosis have been developed. They include serological tests, DNA probes and polymerase chain reaction (PCR).

Among the serological tests is the indirect fluorescent antibody test (IFAT) using tissue culture schizont antigen (Burridge and Kimber, 1972). IFAT is cumbersome to carry out and relies on subjective observation of the degree of fluorescence. A fluorescent microscope is required and the test is not very specific as there are cross-reactions between *T. taurotragi, T. annulata* and *T. parva.* The antibodies detected by IFAT occur late in the course of disease, and cannot therefore, be used for clinical diagnosis. It is useful in epidemiological studies as the antibodies are detectable six months post-infection in the absence of re-challenge (Burridge and Kimber, 1973c). Complement fixation test, the capillary agglutination test and the indirect haemagglutination test have been used, but their value has been limited by lack of specificity. Other serological tests that have been developed for detection of *T. parva* antibodies and antigens include the enzyme-linked immunosorbent assay (ELISA) test. The test is specific for *T. parva.*

Other tools for diagnosis of ECF include DNA probes (Conrad *et al.*, 1987b), polymerase chain reaction (PCR) (Saiki et al., 1988) and oligonucleotide probes that use species-specific synthetic oligonucleotides designed to probe small sub-unit ribosomal RNA (Allsopp *et al* 1993). It probes a naturally amplified target, the RNA, it is very sensitive without the contamination problems of the diagnostic PCR and cheaper.

IMMUNE RESPONSES AGAINST *T. PARVA*

The immune responses are directed against the various *T. parva* life-cycle stages in the bovine host, the sporozoite, the schizont, the merozoite and the piroplasm. The merozoite and piroplasm stages occur relatively late in the course of infection and the immune responses directed against them are of little value in protective immunity.

Antibody-mediated immune responses play a part in the immunity of cattle to *T. parva* (Musoke *et al.* 1982). The feeding of infected ticks on cattle provokes production of antibodies which neutralize the infectivity of sporozoites. The antibodies are directed against a surface antigen on the sporozoite and binding to this antigen prevents the sporozoite from entering the target cell (Dobbelaere *et al.,* 1985).

The role of antibodies against sporozoites in the control of primary infection may not be significant, but in immune animals, presence of neutralizing antibodies in sufficient quantities at the inoculation site may reduce the dose of sporozoites and therefore, the severity of the disease, as severity of disease is dose dependent.

Cattle that recover from *T. parva* infection, either naturally, through treatment with anti-theilerial drugs, or which are immunized by infection and treatment, are solidly immune to challenge with the homologous parasite stock and show varying resistance to other parasite stocks (Radley, 1981). These animals produce antibodies against the schizont and piroplasm stages of the parasite. However such antibodies do not seem to play a significant role in immunity (Mohammed et al., 1975). Neither do colostoral antibodies (Mining, 1992). In contrast transfer of thoracic duct lymphocytes from immune co-twins to susceptible chimaeric twins protected them from lethal ECF challenge (Emery *et al,* 1981). These findings suggest immunity to *T. parva* infection is mainly cell mediated. The cytotoxic cells are a subpopulation of T. lymphocytes bearing the BoT2+ BoT4- BoT8+ phenotype and their activity is restricted by determinants on class I MHC molecules (Morrison *et al.* 1987a,b).

Other Immune Responses

Macrophages/monocytes have also been implicated in the regulation of *T. parva* infection in cattle. Peripheral blood mononuclear cells from animals undergoing spontaneous recovery from infection with *T. parva* have been shown to have a cytostatic effect on the growth of an allogeneic cell line infected with *T. parva* sporozoites. The cytostatic effect was mediated by macrophages (Preston, 1981). Further work by Preston and Brown (1988), with *T. annulata* immunized calves, showed that macrophages isolated during immunization and after challenge have the capacity to inhibit lymphoproliferation of infected cells *in vitro* and may therefore, contribute to recovery and protective immunity.

IMMUNE RESPONSES AGAINST THE TICK

Johnson and Bancroft (1918), in Australia, observed that ticks were more abundant on European breeds of cattle (*Bos taurus*) than Zebu cattle (*Bos indicus*). Cattle develop resistance to tick infestation with *Bos indicus* cattle developing resistance more readily than *Bos taurus* cattle (Ali and De Castro, 1993). Resistance is expressed by reduced numbers of engorged ticks, reduced blood meal weights, reduced numbers and viability of ova, prolonged duration of engorgement and death of ticks on the host. Grooming has also been reported to be an expression of resistance caused by release of histamine and other mediators of the hypersensitivity response (De Castro *et al.*, 1985). Such grooming results in dislodging of attached ticks.

A variety of immune responses are involved in the rejection and/or destruction of the tick. They include humoral and cellular immunity and immediate type hypersensitivity which is manifested by a variety of physical responses such as licking, rubbing or scratching and swelling of the skin at the site of tick attachment (De Castro and Newson, 1993).

There are reports that the resistance to tick infestation in *Bos indicus* cattle is innate (De Castro *et al.*, 1985, Norval *et al.*, 1988), but this is not universally agreed. *Bos indicus* cattle however acquire resistance to tick infestation more readily than *Bos taurus* cattle (De Castro, 1990) and the resistance is heritable (Hewetson, 1972).

Treatment and Control

Following the development of techniques for growing *T. parva* schizonts *in vitro* by Malmquist *et al.* (1970), a new tool for drug screening *in vitro* for antitheilerial activity was established. Theilericidal activity of menoctone, a naphthoquinone compound, against *T. parva* was demonstrated (McHardy, *et al.*, 1976, McHardy, 1978). Its anti-theilerial activity was also confirmed *in vivo* (Dolan and McHardy 1978). Menoctone is active against both the sporozoite and schizont stages of the parasite (Dolan, 1981). This drug was not developed further as it was too expensive to synthesise (McHardy and Rae, 1981). Subsequently search for cheaper analogues yielded a number of compounds with an almost equal activity. Parvaquone (2-cyclohexyl-3-hydroxy-l, 4-naphthoquinone), was chosen for development. It was very effective when administered intramuscularly against tick or sporozoite stabilate-induced clinical theileriosis (Dolan, 1981). Some animals treated before the appearance of schizonts did not seroconvert and were fully susceptible to re-challenge and some of the recovered animals became carriers. Field trials conducted in Kenya showed the drug to be effective against naturally induced theileriosis (Chema *et al.,* 1986). Parvaquone kills both schizonts and piroplasms without affecting the host cells (McHardy, 1984). Another naphthoquinone analogue, buparvaquone, has been developed. It is 20 times and 8 times more effective than parvaquone *in vitro* and *in vivo* respectively against *T. parva* infection (McHardy *et al.,* 1985).

A combination of parvaquone with the diuretic frusemide has been shown to improve the response of cattle treated for clinical ECF (Musoke et al, 2004)

Halofuginone, a quinozoline compound, has been shown to be effective against clinical theileriosis (Schein and Voigt, 1979; Chema *et al.,* 1987). It is used as a lactic salt which is water soluble. The drug is administered orally at 1.2 mg/kg body weight. Doses above 3.0mg/kg are toxic to cattle. The mode of action is unknown, but appears to act against the schizont, not the piroplasm.

CONTROL MEASURES

Control methods against ECF can be divided into (1) measures against the vector tick and (2) measures against the parasite in the bovine host. Control measures against the tick include the use of acaricides, exploitation of host resistance to ticks and the use of tick vaccines. Control measures aimed at the parasite in the bovine host include immunization against the parasite and chemotherapy. Other less commonly used method for control of ECF include pasture spelling and use of tick predators.

Acaricides

The application of acaricides to cattle is the most widely used method of tick control (De Castro and Newson, 1993). There are several methods for acaricide application. They include dipping, spraying and topical application. Dipping involves complete immersion of the animal in a dip tank containing an aqueous emulsion, suspension or solution of acaricide. This ensures adequate exposure of the ticks to acaricide. Spraying, either as an individual hand-spray or a spray-race, involves spraying the animal with the acaricide. For the smallholder, hand-spraying is more economical. This method, however rarely achieves complete wetting of the animal and there is a lot of wastage as the dripping acaricide is not recycled. Topical application involves the use of 'pour on' formulations that contain solvent/propellants that spread rapidly over the surface of skin and hair after being applied to limited areas of the body (Stendel, 1985).

Other methods that have been used include slow-release devices such as acaricide-impregnated ear tags for control of ear ticks (Young *et al.*, 1985).

A main disadvantage of the chemical control of ticks include development of resistance in ticks to these compounds (Keating 1983). Development of acaricide resistance occurs first in the one-host ticks, followed by two-host ticks and eventually the three-host ticks, the reason being that one-host ticks have a shorter generation time than the two and three-host ticks and therefore have more frequent exposure to acaricides. The application of acaricides to cattle is potentially dangerous to the human population and also to the cattle themselves. Direct contact, particularly from hand-spraying, is hazardous as these chemicals are toxic. The chemicals may also contaminate animal products. Death of animals through acaricide application is common. Another serious disadvantage is that intensive use of acaricide destroys endemic stability for tick-borne diseases and a highly susceptible cattle population is created. Failure of tick control programmes can spell disaster for the livestock industry (Lawrence *et al.*, 1980). Acaricides are expensive to develop and purchase.

Infection and Treatment Immunization

The infection and treatment method (ITM) of immunization is the cheapest effective control tool for ECF (Perry, 2016, Babo Martins, 2010). It involves inoculation of cattle with a sporozoite stabilate and simultaneous administration of a blocking agent (oxytetracline). Oxytetrcyclines slow down the multiplication of the parasite (Spooner, 1990) thus allowing the animal to mount an effective immune response.

Production of cryopreserved sporozoite stabilate by Cunningham *et al* (1973b), allowed cattle to be infected with a predetermined and reproducible dose, and to reduce the drug regimen required for immunization. Two intramuscular doses (10mg/kg bodyweight 1/m) of short-acting tetracyclines given on days 0 and 4 gave adequate results with *T. parva* (Muguga) (Radley, 1978). The development of a long acting formulation of oxytetracycline (Terra, LA. Pfizer Corporation) reduced the number of doses that were required. One dose of this formulation at 20mg/kg given on day 0 at the same time as the *T. parva* sporozoite stabilate resulted in the production of homologous immunity with minimal reactions occurring in the process (Radley, 1978). Development of the 30% oxytetracycline formulation (Norbrook) has eliminated the need for monitoring as the rate of reactors is almost non-existent.

The regimens of 4 doses at 5mg/kg from day 0-3 (Brown et al., 1977) or 2 doses at 10mg/kg on day 0 and 4 of short acting oxytetracycline formulations, or one dose on day 0 for long-acting formulation are not adequate for the buffalo-derived *T. parva* isolates as the latter breaks through the chemoprophylactic cover (Radley, 1978, 1981). This breakthrough can be overcome by increasing the duration of chemoprophylactic cover to 6 days (day 0-5 inclusive) of short-acting tetracycline at 5mg/kg or long-acting tetracycline at 20mg/kg on day 0 and 4 (Mutugi *et al.*, 1991).

Parvaquone, if administered during the incubation period, inhibits the development of immunity, but if administered after the establishment of the parasite (Day 8) it does not interfere with immunity (Dolan *et al.,* 1988). Buparvaquone does not seem to interfere with immunity irrespective of the time at which it is administered (Mutugi *et al.* 1991).

Other Methods of Control

Other methods that have been tried in the control of tick-borne diseases include the exploitation of the development of immunity to tick infestation by the host

Cattle have been immunized against *Rhipicephalus (Boophilus) microplus* using 'concealed' tick antigens, non-salivary tick antigens to which the host immune system is not exposed during normal tick feeding, prepared from homogenized whole ticks (Kemp *et al.,* 1986). Adult ticks feeding on immunized animals died and survivors showed obvious signs of damage to the gut epithelium. There was also marked reduction of weight and egg production of surviving engorged adult females.

Recombinant vaccines have now been produced based on this information and are commercially available for control of *R. (B) microplus* (de la Fuente et al, 2007). Recombinant vaccines so far tried have not been successful in the control of *R. appendiculatus* (Odongo et al, 2007).

Other control measures that have been successfully employed in the control of arthropod pests such as predators, parasites and pathogens have been found to be of little practical value in tick control.

Control of animal movement and quarantine of farms where there is an ECF outbreak have been used to contain the spread of disease, but strict enforcement of these measures is required for them to be effective.

FUTURE RESEARCH DIRECTION

Immunisation against ECF using the infection and treatment method remains the cheapest and most effective control method. The cost however remains out of reach for most small-scale livestock keepers. Currently, dose numbers depend on the titration of the stabilate which is unpredictable. Development of recombinant vaccines against the disease and the vector tick would go a long way in eliminating these challenges.

CONCLUSION

East Coast fever poses a major challenge to cattle production in both the pastoral and dairy systems due to the high cost of control, mortalities, morbidities and the reduction of productivity in animals that recover. A lot has been achieved so far, but there is still a lot to be done to effectively control and hopefully eliminate the disease. This will go a long way in financially empowering farmers and reducing poverty.

REFERENCES

Ali, M., & De Castro, J. J. (1993). Host resistance to ticks (Acari: Ixodidae) in different breeds of cattle at Bako, Ethiopia. *Tropical Animal Health and Production*, *25*(4), 215–222. doi:10.1007/BF02250871 PMID:8109054

Allison, A. C. (1981). Cellular immune responses in theileriosis. In A. D. Irvin, M. P. Cunnigham, & A. S. Young (Eds.), *Advances in the control of theileriosis* (pp. 273–288). Martinus Nijhoff Publishers. doi:10.1007/978-94-009-8346-5_49

Allsopp, B. A., Baylis, H. A., Allsopp, M. T. E. P., Cavalier-Smith, T., Bishop, R. P., Carrington, D. M., Sohanpal, B., & Spooner, P. (1993). Discrimination between six species of Theileria using oligonucleotide probes which detect small ribosomal RNA sequencies. *Parasitology*, *107*(2), 157–165. doi:10.1017/S0031182000067263 PMID:8414670

Allsopp, B. A., Carrington, M., Baylis, H., Sohal, S., Dolan, T. T., & Iams, K. (1989). Improved characterization of Theileria parva isolates using the polymerase chain reaction and oligonucleotide probes. *Molecular and Biochemical Parasitology*, *35*(2), 137–148. doi:10.1016/0166-6851(89)90116-3 PMID:2570351

Babo Martins, S., Di Giulio, G., Lynen, G., Peters, A., & Rushton, J. (2010). Assessing the impact of East Coast fever immunization in Tanzanian pastoralists systems. *Preventive Veterinary Medicine*, *97*, 175–182. doi:10.1016/j.prevetmed.2010.09.018 PMID:20974501

Bailey, K. P. (1960). Notes on the rearing of *Rhipicephalus appendiculatus* and their infection with *Theileria parva* for experimental transmission. *Bulletin of Epizootiological Diseases of Africa*, *8*, 33–43.

Baldwin, C. L., Black, S. J., Brown, W. C., Conrad, P. A., Goddeeris, B. M., Kinuthia, S. W., Lalor, P. A., MacHugh, N. D., Morrison, W. I., Morzaria, S. P., Naessens, J., & Newson, J. (1988). Bovine T cells, B cells and null cells are transformed by the protozoan parasite *Theileria parva. Infection and Immunity*, *56*(2), 462–467. doi:10.1128/IAI.56.2.462-467.1988 PMID:3123392

Barnett, S. F. (1960). Connective tissue reactions in acute fatal East Coast fever (*Theileria parva* of cattle). *The Journal of Infectious Diseases*, *107*(3), 253–282. doi:10.1093/infdis/107.3.253 PMID:13687137

Barnett, S. F., & Brocklesby, D. W. (1966). The passage of *Theileria lawrencei* (Kenya) through cattle. *The British Veterinary Journal*, *122*(9), 396–409. doi:10.1016/S0007-1935(17)40406-4

Bell, L. J. (1983). *Development of Theileria in tick tissue culture* (MPhil Thesis). University of Edinburgh.

Brown, C. G. D. (1990). Theileriosis. In M. M. H. Sewell (Ed.), Handbook on animal Diseases in the Tropics (4th ed.). Academic Press.

Burridge, M. J., & Kimber, C. D. (1972). The indirect fluorescent antibody test for experimental East Coast fever (Theileria parva infection of cattle). Evaluation of a cell culture schizont antigen. *Research in Veterinary Science*, *13*(5), 451–455. doi:10.1016/S0034-5288(18)33992-4 PMID:4564334

Burridge, M. J., & Kimber, C. D. (1973). Duration of serological response to the indirect fluorescent antibody test of cattle recovered from Theileria parva infection. *Research in Veterinary Science*, *14*(2), 270–271. doi:10.1016/S0034-5288(18)33929-8 PMID:4217920

Burridge, M. J., Kimber, C. D., & Young, A. S. (1973). The use of indirect fluorescent antibody technique in serological studies of Theileria lawrencei infections in cattle. *American Journal of Veterinary Research*, *13*, 897. PMID:4198343

Chema, S. (1984). Evaluation of new acaricides in Kenya and views on selection for acaricide resistance in past and future. In *Report of the Third FAO Expert Consultation on Research on Tick-borne Diseases and Their Vectors* (p. 29). FAO.

Chema, S., Chumo, R. S., Dolan, T. T., Gathuma, J. M., James, A. D., Irvin, A. D., & Young, A. S. (1987). Clinical trials of halofunone lactate for treatment of East Coast fever in Kenya. *The Veterinary Record, 120*(24), 575–577. doi:10.1136/vr.120.24.575 PMID:3303642

Chema, S., Waghala, S., James, A. D., Dolan, T. T., Young, A. S., Masiga, W. N., Irvin, A. D., Mulela, G. H. M., & Wekesa, L. M. (1986). Clinical trials of parvaquone for the treatment of East Coast fever in Kenya. *The Veterinary Record, 118*(21), 588–589. doi:10.1136/vr.118.21.588-a PMID:3526698

Conrad, P. A., Iams, K., Brown, W. C., Sohanpal, B., & Ole-Moi Yoi, O. (1987b). DNA probes detect genomic diversity in *Theileria parva* stocks. *Molecular and Biochemical Parasitology, 25*(3), 213–226. doi:10.1016/0166-6851(87)90085-5 PMID:2892129

Conrad, P. A., Ole-Moi Yoi, O., Baldwin, C. L., Dolan, T. T., O'Callaghan, C. J., Njamungeh, R. E. G., Grootenhuis, J. G., Stagg, D. A., Leitch, B. L., & Young, A. S. (1989). Characterisation of buffalo derived theilerial parasites with monoclonal antibodies and DNA probes. *Parasitology, 98*(2), 179–183. doi:10.1017/S0031182000062089 PMID:2569709

Conrad, P. A., Stagg, D. A., Grootenhuis, J. G., Irvin, A. D., Newson, J., Njamungeh, R. E. G., Rossiter, P. B., & Young, A. S. (1987a). Isolation of *Theileria* parasites from the African buffalo (*Syncerus caffer*) and characterization with anti-schizont monoclonal antibodies. *Parasitology, 94*(3), 413–423. doi:10.1017/S0031182000055761 PMID:3112700

Cowdry, E. V., & Danks, W. B. C. (1933). Studies on East Coast fever. II Behaviour of the parasite and the development of distinctive lesions in susceptible animals. *Parasitology, 25*(1), 1–63. doi:10.1017/S003118200001920X

Cunningham, M. P. (1977). Immunisation of cattle against Theileria parva. In L. H. Miller & J. J. McKelvey Jr., (Eds.), *Immunity to Blood Parasites of Animal and Man* (pp. 1–132). Planum Press. doi:10.1007/978-1-4615-8855-9_11

Cunningham, M. P. (1981). Biological control of ticks with particular reference to Rhipicephalus appendiculatus. In Advances in the Control of Theileriosis. Martinus Nijhoff Publishers.

Cunningham, M. P., Brown, C. D. G., Burridge, M. J., & Purnell, R. E. (1973b). Cryopreservation of infective particles of Theileria parva. *International Journal for Parasitology, 3*(5), 583–587. doi:10.1016/0020-7519(73)90082-9 PMID:4200405

Cunningham, M. P., Brown, C. G. D., Burridge, M. J., Musoke, A. J., Purnell, R. E., Radley, D. E., & Sempebwa, C. (1974). East Coast fever: Titration in cattle of suspension of *Theileria parva* derived from ticks. *The British Veterinary Journal, 130*(4), 336–345. doi:10.1016/S0007-1935(17)35836-0 PMID:4214585

Cunningham, M. P., Brown, C. G. D., Joyner, L. P., Purnell, R. E., & Bailey, K. P. (1973a). Infection of cattle with East Coast fever by inoculation of the infective stage of *Theileria parva* harvested from the tick vector *Rhipicephalus appendiculatus*. *Bulletin of Epizootic Diseases of Africa, 21*, 235–238.

De Castro, J. J. (1991). Resistance to ixodid ticks in cattle with an assessment and its role in tick control in Africa. In J. B. Owen & R. F. E. Axford (Eds.), *Breeding for Disease Resistance in farm Animals* (pp. 244–262). C.A.B. International.

De Castro, J. J., & Newson, R. M. (1993). Host resistance in cattle tick control. *Parasitology Today (Personal Ed.), 9*(1), 13–17. doi:10.1016/0169-4758(93)90154-8 PMID:15463656

De Castro, J. J., Young, A. S., Dransfield, R. D., Cunnighm, M. P., & Dolan, T. T. (1985). Effects of tick infestation on Boran (Bos indicus) cattle immunized against theileriosis in an endemic area of Kenya. *Research in Veterinary Science, 39*(3), 279–288. doi:10.1016/S0034-5288(18)31714-4 PMID:4081331

de la Fuente, J., Almazan, C., Canales, M., de la Lastra, J. M. P., Kocan, K. M., & Willadsen, P. (2007). A ten-year review of commercial vaccine performance for control of tick infestations on cattle. *Animal Health Research Reviews, 8*(1), 23–28. doi:10.1017/S1466252307001193 PMID:17692140

Dobbelaere, D. A. E., Shapiro, S. Z., & Webster, P. (1985). Identification of a surface antigen on *Theileria parva* sporozoites by monoclonal antibodies. *Proceedings of the National Academy of Sciences of the United States of America, 82*(6), 1771–1775. doi:10.1073/pnas.82.6.1771 PMID:3920654

Dolan, T. T. (1981). Progress in the chemotherapy of theileriosis. In A. D. Irvin, M. P. Cunnigham, & A. S. Young (Eds.), *Advances in the control of theileriosis* (pp. 186–208). Martinus Nijhoff Publishers. doi:10.1007/978-94-009-8346-5_34

Dolan, T. T. (1989). Theileriasis: A comprehensive review. *Rev. Sci. Tech. Off. Int. Epiz, 8*(1), 11–36. doi:10.20506/rst.8.1.398 PMID:32731664

Dolan, T. T., Linyonyi, A., McHardy, N., Bond, A. L., & Clampit, R. B. (1988). Chemotherapy of East Coast fever: Parvaquone treatment of *Theileria parva parva* at intervals after infection. *Research in Veterinary Science, 44*(1), 15–20. doi:10.1016/0034-5288(88)90005-7 PMID:3375580

Dolan, T. T., & McHardy, N. (1978). The chemotherapy of experimental *T. parva* infection. In J. K. H. Wilde (Ed.), *Tick-borne Diseases and Their Vectors* (pp. 318–323). University of Edinburgh.

Dolan, T. T., Young, A. S., Losos, G. J., McMillan, T., Minder, Ch. E., & Soulsby, K. (1984). Dose-dependent responses of cattle to *Theileria parva* stabilate. *International Journal for Parasitology, 14*(1), 89–95. doi:10.1016/0020-7519(84)90017-1 PMID:6423557

Emery, D. L., MacHugh, N. D., & Morrison, W. I. (1988). *Theileria parva* (Muguga) infects bovine T lymphocytes *in vivo* and induces coexpression of BoT4 and BoT8. *Parasite Immunology, 10*(4), 379–391. doi:10.1111/j.1365-3024.1988.tb00228.x PMID:3262855

Emery, D. L., Morrison, W. I., Nelson, R. T., & Murray, M. (1981). The induction of cell mediated immunity in cattle inoculated with cell lines parasised with *Theileria parva*. In A. D. Irvin, M. P. Cunnigham, & A. S. Young (Eds.), *Advances in the control of theileriosis* (pp. 295–310). Martinus Nijhoff Publishers. doi:10.1007/978-94-009-8346-5_51

Fawcet, D. W., Young, A. J., & Leitch, B. L. (1985). Sporogony in Theileria (Ampicomplexa: Piroplasmida). A comparative ultrastructural study. *Journal of Submicroscopic Cytology, 17*, 299–314.

Hewetson, R. W. (1972). The inheritance of resistance by cattle to cattle ticks. *Australian Veterinary Journal*, *48*(5), 299–303. doi:10.1111/j.1751-0813.1972.tb05161.x PMID:5068812

Irvin, A. D. (1987). Characterisation of species and strains of *Theileria. Advances in Parasitology*, *26*, 145–197. doi:10.1016/S0065-308X(08)60296-1 PMID:3118658

Irvin, A. D., & Morrison, W. I. (1987). Immunopathology, immunology and immunoprophylaxis of Theileria infections. In E. J. L. Soulsby (Ed.), *Immune Responses in Parasite Infections: Immunopathology, Immunology and Immunoprophylaxis* (Vol. 3, pp. 223–274). CRC Press.

Kaiser, M. N., Sutherst, R. W., Bourne, A. S., Gorissen, L., & Flyod, R. B. (1988). Population dynamics of ticks on Ankole cattle in five ecological zones in Burundi and strategies for their control. *Preventive Veterinary Medicine*, *6*(3), 199–122. doi:10.1016/0167-5877(88)90031-1

Keating, M. I. (1983). Tick control by chemical ixodicides in Kenya: A review 1912-1981. *Tropical Animal Health and Production*, *15*(1), 1–6. doi:10.1007/BF02250752 PMID:6342241

Kemp, D. H., Agbede, R. I. S., Johnson, L. A. Y., & Gough, J. M. (1986). Immunisation of cattle against Boophilus microplus using extracts derived from adult female ticks: Feeding and survival of the parasite on vaccinated cattle. *International Journal for Parasitology*, *16*(2), 115–120. doi:10.1016/0020-7519(86)90096-2 PMID:3721702

Koch, R. (1903). *Interim report on Rhodesia Redwater or 'African Coast fever'.* Argus Printing and Publishing Co. Ltd.

Lawrence, J. A., Norval, R. A. I., & Uilenberg, G. (1983). *R. zambeziensis* as a vector of bovine *Theileria. Tropical Animal Health and Production*, *15*(1), 39–42. doi:10.1007/BF02250760 PMID:6405516

Lessard, P., Le Plattenier, R., Norval, R. A. I., Kundert, K., Dolan, T. T., Croze, H., Walker, J. B., Irvin, A. D., & Perry, B. D. (1990). Geographical information systems for studying the epidemiology of cattle diseases caused by *Theileria parva. The Veterinary Record*, *126*, 255–262. PMID:2327044

Malmquist, W. A., Nyindo, M. B. A., & Brown, C. G. D. (1970). East Coast fever: Cultivation in vitro of bovine spleen cell lines infected and transformed by *Theileria parva. Tropical Animal Health and Production*, *2*(3), 139–150. doi:10.1007/BF02359680

Maritim, A. C., Kariuki, D. P., Young, A. S., & Mutugi, J. J. (1989b). The importance of carrier state of *Theileria parva* in the epidemiology of theileriosis and its control by immunization. In T. T. Dolan (Ed.), *Theileriosis in Eastern, Central and Southern Africa* (pp. 121–128). ILRAD Nairobi.

Maritim, A. C., Young, A. S., Lesan, A. C., Ndungu, S. G., Stagg, D. A., & Ngumi, P. N. (1992). Transformation of Theileria parva derived from the African buffalo *(Syncerus caffer)* by tick passage in cattle and its use in infection and treatment immunization. *Veterinary Parasitology*, *43*(1-2), 1–14. doi:10.1016/0304-4017(92)90043-9 PMID:1496792

Martin, H. M., Barnett, S. F., & Vidler, B. O. (1964). Cyclic development and longevity of *Theileria parva* in the tick *Rhipicephalus appendiculatus. Experimental Parasitology*, *15*(6), 527–555. doi:10.1016/0014-4894(64)90045-1 PMID:14242265

Maxie, M. G., Dolan, T. T., Jura, W. G. Z., Tabel, H., & Flowers, M. J. (1982). A comparative study of the disease in cattle caused by *Theileria parva* or *Theileria lawrencei*: II haematology, clinical chemistry, coagulation studies and complement. *Veterinary Parasitology*, *10*(1), 1–19. doi:10.1016/0304-4017(82)90002-4 PMID:7201709

McHardy, N. (1978). *In vitro* studies on the action of menoctone and other compounds on *Theileria parva* and *Theileria annulata. Annals of Tropical Medicine and Parasitology*, *72*(6), 501–511. doi:10.1080/00034983.1978.11719354 PMID:104669

McHardy, N. (1984). Recent advances in the chemotherapy of theileriosis. In H. P. Riemann & M. J. Burridge (Eds.), *Impact of Diseases on Livestock Production in the Tropics* (pp. 179–192). Elsevier. doi:10.1016/0167-5877(84)90061-8

McHardy, N., Haigh, A. J. B., & Dolan, T. T. (1976). Chemotherapy of *Theileria parva* infection. *Nature*, *261*(5562), 698–699. doi:10.1038/261698a0 PMID:934316

McHardy, N., & Wekesa, L. S. (1985). Buparvaquone (BW 720), a new anti-theilerial naphthoquinone, its role in therapy and prophylaxis. In A. D. Irvin (Ed.), *Immunisation Against Theileriosis in Africa* (p. 88). ILRAD.

Melhorn, H., & Schein, E. (1984). The piroplasms: Lifecycle and sexual stages. *Advances in Parasitology*, *23*, 37–103. doi:10.1016/S0065-308X(08)60285-7 PMID:6442536

Mettam, R. W. M., & Carmichael, J. (1936). Turning disease, a protozoan encephalitis of cattle in Uganda. Its relationship with East Coast fever. *Parasitology*, *28*(2), 254–283. doi:10.1017/S0031182000022447

Minami, T., Spooner, P. R., Irvin, A. D., Ocama, J. G. R., Dobbelaere, D. A. E., & Fujinaga, T. (1983). Characterisation of stocks of Theileria parva by monoclonal antibody profiles. *Research in Veterinary Science*, *35*(3), 334–340. doi:10.1016/S0034-5288(18)32029-0 PMID:6420853

Mining, S. K. (1992). *The investigation of maternally derived immunity in calves* (PhD Thesis). University of Liverpool.

Minjauw, B., & McLeode, A. (2003). Tick-borne diseases and poverty. The impact of ticks and tick-borne diseases on the livelihood of small scale and marginal livestock owners in India and eastern and southern Africa. Research Report, DFID Animal Health Programme, Centre for Tropical Veterinary Medicine, University of Edinburgh, UK.

Minshull, J. I., & Norval, R. A. I. (1982). Factors influencing the spatial distribution of Rhipicephalus appendiculatus in Kyle Reacreational Park in south-eastern Zimbabwe. *South African Journal of Wildlife Research*, *12*, 118–123.

Moll, G., Lodhing, A., Young, A. S., & Leitch, B. L. (1986). Epidemiology of theileriosis in an endemic area in Kenya. *Veterinary Parasitology*, *19*(3-4), 255–273. doi:10.1016/0304-4017(86)90073-7 PMID:3085323

Morrison, W. I., Buscher, G., Emery, D. L., Nelson, R. I., & Murray, M. (1981). The kinetics of infection with Theileria parva in cattle and the relevance of the development of immunity. In A. D. Irvin, M. P. Cunningham, & A. S. Young (Eds.), *Advances in the Control of Theileriosis* (pp. 311–327). Martinus Nijhoff Publishers. doi:10.1007/978-94-009-8346-5_52

Morrison, W. I., Goddeeris, B. M., Brown, W. C., Baldwin, C. L., & Teale, A. J. (1989). *Theileria parva* in cattle: Characterisation of infected lymphocytes and immune responses they provoke. *Veterinary Immunology and Immunopathology*, *20*(3), 213–237. doi:10.1016/0165-2427(89)90003-2 PMID:2497579

Morrison, W. I., Goddeeris, B. M., & Teale, A. J. (1987b). Bovine cytotoxic T cell clones which recognize lymphoblasts infected with two antigenically different stocks of the protozoan parasite *Theileria parva*. *European Journal of Immunology*, *17*(12), 1703–1709. doi:10.1002/eji.1830171205 PMID:3121356

Morrison, W. I., Goddeeris, B. M., Teale, A. J., Groocock, C. M., Kemp, S. J., & Stagg, D. A. (1987a). Cytotoxic cells elicited in cattle challenged with *Theileria parva* (Muguga): Evidence for restriction by class 1 MHC determinants and parasite strain specificity. *Parasite Immunology*, *9*(5), 563–578. doi:10.1111/j.1365-3024.1987.tb00530.x PMID:3120135

Morrison, W. I., Hemmink, J. D., & Toye, P. G. (2020). *Theileria parva*: A parasite of African buffalo, which adapted to infect and undergo transmission in cattle. *Parasitology*, *50*, 403–412. PMID:32032592

Muhammed, S.T., & Luerman, L.H. (1975). Effect of humoral antibodies on the course of *Theileria parva* (East Coast fever) of cattle). *American Journal of Veterinary Research*, *36*, 389–402. PMID:47727

Mukhebi, A. W., Perry, B. D., & Kruska, R. (1992). Estimated economics of theileriosis control in Africa. *Preventive Veterinary Medicine*, *13*(1-2), 73–85. doi:10.1016/0167-5877(92)90070-V

Musoke, A. J., Nantulya, V. M., Buscher, G., Masake, R. A., & Otim, B. (1982). Bovine immune response to Theileria parva: Neutralizing antibodies to sporozoites. *Immunology*, *45*, 663–668. PMID:6802745

Musoke, R. A., Twenyongyere, R., Bizimenyera, E., Waiswa, C., Mugisha, A., Biryomumaisho, S., & McHardy, N. (2004). Treatment of East Coast fever of cattle with a combination of parvaquone and frusemide. *Tropical Animal Health and Production*, *36*(3), 233–245. doi:10.1023/B:TROP.0000016832.69136.68 PMID:15080540

Mutugi, J. J., Ndungu, S. G., Linyonyi, A., Maritim, A. C., Mining, S. K., Ngumi, P. N., Kariuki, D. P., Leitch, B. L., D'Souza, D., Maloo, S., & Lohr, K. F. (1991). Responses to vaccine trial for East Coast fever in two cattle herds at the Kenya Coast. *Preventive Veterinary Medicine*, *10*(3), 173–183. doi:10.1016/0167-5877(91)90002-J

Mutugi, J. J., Young, A. S., Martim, A. C., Linyonyi, A., Mbogo, S. K., & Leitch, B. L. (1988). Immunisation of cattle using varying doses of *Theileria parva lawrencei* sporozoites derived from an African buffalo (*Syncerus caffer*) and treatment with buparvaquone. *Parasitology*, *96*(2), 391–402. doi:10.1017/S0031182000058376 PMID:3131722

Ndungu, S. G., Brown, C. G. D., & Dolan, T. T. (2005). *In vivo* comparison of susceptibility and disease resistance between *Bos indicus* and *Bos taurus* cattle types to *Theileria parva* Muguga infection. *The Onderstepoort Journal of Veterinary Research*, *72*, 13–22. doi:10.4102/ojvr.v72i1.220 PMID:15991702

Neitz, W. O. (1950). The specific action of pamaquin on the haemotropic parasites of *Theileria parva. South African Journal of Science*, *46*, 218–219.

Neitz, W. O. (1953). Aureomycin in *Theileria parva* infection. *Nature*, *171*(4340), 34–35. doi:10.1038/171034a0 PMID:13025470

Nene, V., Kiara, H., Lacasta, A., Pelle, R., Svitek, N., & Steina, L. (2016). The biology of Theileria parva and control of East Coast fever- current status and future trends. *Ticks and Tick-Borne Diseases*, *7*(4), 549–564. doi:10.1016/j.ttbdis.2016.02.001 PMID:26972687

Newson, R. M., & Chiera, J. W. (1989). Development of resistance in calves to nymphs of *Rhipicephalus appendiculatus* (Acarina:Ixodidae) during test feeds. *Experimental & Applied Acarology*, *6*(1), 19–27. doi:10.1007/BF01193230 PMID:2707106

Newson, R. M., Chiera, J. W., Young, A. S., Dolan, T. T., Cunningham, M. P., & Radley, D. E. (1984). Survival of Rhipicephalus appendiculatus (Acarina:Ixodidae) and persistence of Theileria parva (Ampicomplexa:Theileriidae) in the field. *International Journal for Parasitology*, *14*(5), 483–489. doi:10.1016/0020-7519(84)90029-8 PMID:6439668

Norval, R. A. I. (1977). Tick problems in relation to land utilization in Rhodesia. *Rhodesia Veterinary Journal*, *8*, 33–38.

Norval, R. A. I., Lawrence, J. A., Young, A. S., Perry, B. D., Dolan, T. T., & Scott, J. (1991). *Theileria parva*: Influence of vectors, parasite and host relationships on the epidemiology of theileriosis in southern Africa. *Parasitiology*, *102*(3), 347–356. doi:10.1017/S0031182000064295 PMID:1907728

Norval, R. A. I., Perry, B. D., & Young, A. S. (1992). Development of Theileria. In The Epidemiology of Theileriosis in Africa. Academic Press.

Norval, R. A. I., Sutherst, R. W., Kurki, K., Gibson, J. D., & Kerr, J. D. (1988). The effect of the brown ear tick Rhipicephalus appendiculatus on the growth of Sanga and European breed of cattle. *Veterinary Parasitology*, *30*(2), 149–164. doi:10.1016/0304-4017(88)90162-8 PMID:3245108

Norval, R. A. I., Walker, J. B., & Colborne, J. (1982). The ecology of *Rhipicephalus zambeziensis* and *Rhipicephalus appendiculatus* (Acarina: Ixodidae) wiyh particular reference to Zimbabwe. *The Onderstepoort Journal of Veterinary Research*, *49*, 181–190. PMID:7185034

Ochanda, H., Young, A. S., Mutugi, J. J., Mumo, J. M., & Omwoyo, P. L. (1988). The effect of temperature on the rate of transmission of *Theileria parva* infection to cattle by its tick vector *Rhipicephalus appendiculatus. Parasitology*, *97*, 239–245. PMID:3143989

Odongo, D., Kamau, L., Skilton, R., Mwaura, S., Nitsch, C., Musoke, A., Taracha, E., Daubenberger, C., & Bishop, R. (2007). Vaccination of cattle with TickGARD induces cross-reactive antibodies binding to conserved linear peptides of Bm86 homologues in *Boophilus decoloratus. Vaccine*, *25*(7), 1287–1296. doi:10.1016/j.vaccine.2006.09.085 PMID:17070625

Olwoch, J. M., Reyers, B., Engelbrecht, F. A., & Erasmus, B. F. N. (2008). Climate change and the tick-borne diseases, Theileriosis (East Coast fever) in sub-Saharan Africa. *Journal of Arid Environments*, *72*(2), 108–120. doi:10.1016/j.jaridenv.2007.04.003

Perry, B.D. (2016). The control of East Coast fever of cattle by live parasite vaccination. *One Health*, *2*, 103-114.

Perry, B. D., Lessard, P., Norval, R. A. I., Kundert, K., & Kruska, R. (1990). Climate, vegetation and distribution of *Rhipicephalus appendiculatus* in Africa. *Parasitology Today (Personal Ed.)*, *6*(4), 100–110. doi:10.1016/0169-4758(90)90224-R PMID:15463309

Preston, P. M. (1981). The role of macrophages in protective immunity and immunosuppression in bovine theileriosis. In A. D. Irvin, M. P. Cunnigham, & A. S. Young (Eds.), *Advances in the Control of Theileriosis* (pp. 354–356). Martinus Nijhoff Publishers. doi:10.1007/978-94-009-8346-5_57

Preston, P. M., & Brown, C. G. D. (1988). Macrophage-mediated cytostasis and lymphocyte cytotoxicity in cattle immunized with *Theileria annulata* sporozoites or macroschizont-infected cell lines. *Parasite Immunology*, *10*(6), 631–647. doi:10.1111/j.1365-3024.1988.tb00250.x PMID:3146048

Punyua, D. K., Newson, R. M., & Mutinga, M. J. (1985). Diurnal and seasonal activity of unfed adult *Rhipicephalus appendiculatus* (Acarina: Ixodidae) in relation to some intrinsic factors. I. Factors regulating activity. *Insect Science and Its Application*, *6*, 63–70.

Purnell, R. E., Brown, C. G. D., Cunningham, M. P., Burridge, M. J., Kirimi, T. M., & Ledger, M. A. (1973). East Coast fever: Correlation between the morphology and infectivity of *Theileria parva* developing in its tick vector. *Veterinary Parasitology*, *66*(3), 539–544. doi:10.1017/S0031182000046084 PMID:4198658

Radley, D. E. (1978). *Immunisation against East Coast fever by chemoprophylaxis*. FAO Technical Report 1, AG:DP, RAF/67/077, FAO, Rome.

Radley, D. E. (1981). Infection and treatment method against theileriosis. In A. D. Irvin, M. P. Cunnigham, & A. S. Young (Eds.), *Advances in the Control of Theileriosis* (pp. 227–237). Martinus Nijhoff Publishers. doi:10.1007/978-94-009-8346-5_41

Radley, D. E., Brown, C. G. D., Burridge, M. J., Cunnigham, M. P., Pierce, M. A., & Purnell, R. E. (1974). East Coast fever: Quantitative studies of *Theileria parva* in cattle. *Experimental Parasitology*, *36*(2), 278–287. doi:10.1016/0014-4894(74)90067-8 PMID:4213046

Radley, D. E., Brown, C. G. D., Burridge, M. J., Cunningham, M. P., Kirimi, I. M., Purnell, R. E., & Young, A. S. (1975). East Coast fever, I. chemophylactic immunization of cattle against *Theileria parva* (Muguga) and five *Theileria* strains. *Veterinary Parasitology*, *1*(1), 35–41. doi:10.1016/0304-4017(75)90005-9

Saiki, R. K., Gelford, D. H., & Stoffel, S. (1988). Primer directed enzymatic amplification of DNA with thermostable DNA polymerase. *Science*, *239*(4839), 487–491. doi:10.1126cience.239.4839.487 PMID:2448875

Schein, E., & Voigt, W. P. (1979). Chemotherapy of bovine theileriosis with halofuginone. *Acta Tropica*, *36*, 391–394. PMID:44104

Shaw, M. K., Tilney, L. G., & Musoke, A. J. (1991). The entry of *Theileria parva* sporozoites in bovine lymphocytes. Evidence of MHC class 1 involvement. *The Journal of Cell Biology*, *113*(1), 87–101. doi:10.1083/jcb.113.1.87 PMID:1901066

Short, N. J., Floyd, R. B., Norval, R. A. I., & Sutherst, R. W. (1989). Development rates, fecundity and survival of development stages of the tick *Rhipicephalus appendiculatus, Boophilus decoloratus* and *Boophilus microplus* under field conditions in Zimbabwe. *Experimental & Applied Acarology*, *6*(2), 123–141. doi:10.1007/BF01201643 PMID:2707110

Spooner, P. B. (1990). The effect of oxytetracycline on *Theileria parva* in vitro. *Parasitology*, *100*(1), 11–17. doi:10.1017/S0031182000060066 PMID:2107502

Stagg, D. A., Dolan, T. T., Leitch, B. L., & Young, A. S. (1981). The initial stages of infection of cattle cells with *Theileria parva* sporozoites in vitro. *Parasitology*, *86*, 243–254. doi:10.1017/S0031182000050411 PMID:6791118

Stendel, W. (1985). Experimental studies on the tickcidal effect of BayticolR pour-on. *Veterinary Medical Review*, 132–145.

Uilenberg, G. (1981). *Theileria* species of domestic livestock. In A. D. Irvin, M. P. Cunningham, & A. S. Young (Eds.), *Advances in the Control of Theileriosis* (pp. 4–37). Martinus Nijhoff Publishers. doi:10.1007/978-94-009-8346-5_2

Walker, A. R. (1990). Parasitic adaptions in the transmission of Theileria by ticks. A review. *Tropical Animal Health and Production*, *22*(1), 23–33. doi:10.1007/BF02243494 PMID:2108514

Walker, A.R., Young, A.S. & Leitch, B.L. (1981). Assessment of *Theileria* infections of *R. appendiculatus* ticks collected in the field. *Z. Parasitenkunde*, *65*, 63-69.

Walker, J. B. (1974). *The Ixodid ticks of Kenya. A review of Present Knowledge of Their Hosts and Distribution*. Commonwealth Institute of Entomology.

Walker, J. B. (1991). A review of the Ixodid ticks (Acari: Ixodidae) occurring in Southern Africa. *The Onderstepoort Journal of Veterinary Research*, *58*, 81–105. PMID:1881661

Wilde, J. K. H., Brown, C. G. D., Hulliger, L., Gall, D., & MacLeod, W. G. (1968). Experiments with tissues of infected ticks. *The British Veterinary Journal*, *124*(5), 196–208. doi:10.1016/S0007-1935(17)39402-2 PMID:5649029

Yeoman, G. H., & Walker, J. B. (1967). *The ixodid ticks of Tanzania. A study of the zoogeography of the ixodidae of an East African country*. Commonwealth Institute of Entomology.

Young, A. S., De Castro, J. J., & Kiza-aura, P. P. (1985). Control of African ticks by application of acaricide impregnated ear tags to cattle. *Bulletin of Entomological Research*, *75*, 609–619. doi:10.1017/S0007485300015868

Young, A. S., Groocock, C. M., & Kariuki, D. P. (1988). Integrated control of ticks and tick-borne diseases of cattle in Africa. *Parasitology*, *96*(2), 403–432. doi:10.1017/S0031182000058388 PMID:3287285

Young, A. S., & Grotenhuis, J. G. (1985). Influence of wildlife on immunization of cattle against theileriosis in East Africa. In A. D. Irvin (Ed.), *Immunisation against Theileriosis in Africa* (pp. 104–109). ILRAD.

Young, A. S., & Leitch, B. L. (1981). Epidemiology of East Coast fever: Some effects of temperature on the development of *Theileria parva* in the tick vector, *Rhipicephalus appendiculatus. Parasitology, 83*(1), 199–211. doi:10.1017/S0031182000050162 PMID:6791119

Young, A. S., Leitch, B. L., Morzaria, S. P., Irvin, A. D., Omwoyo, P. L., & De Castro, J. J. (1987). Development and survival of *Theileria parva parva* in *Rhipicephalus appendiculatus* exposed in the Trans Mara, Kenya. *Parasitology, 94*(3), 433–441. doi:10.1017/S0031182000055785 PMID:3112702

Young, A. S., Leitch, B. L., & Mutugi, J. J. (1984). Some factors controlling the stimulation of sporogony of *Theileria parva* in its vector *Rhipicephalus appendiculatus* infecting cattle. *Parasitology, 90*, 391–399. doi:10.1017/S0031182000051088

Young, A. S., Leitch, B. L., Newson, R. M., & Cunningham, M. P. (1986). Maintenance of *Theileria parva* infection in an endemic area of Kenya. *Parasitology, 93*(1), 9–16. doi:10.1017/S0031182000049787 PMID:3092172

Young, A. S., Leitch, B. L., & Omwoyo, P. L. (1979). The induction of infective stages of Theileria parva by exposure of host tick to high temperature. *The Veterinary Record, 105*, 531–533. PMID:119344

Chapter 10
Bovine Anaplasmosis and Control

Sebastian Kironji Waruri
Kenya Agricultural and Livestock Research Organisation, Kenya

James Muriuki Wanjohi
Kenya Agricultural and Livestock Research Organisation, Kenya

Leonard Khaluhi
Kenya Agricultural and Livestock Research Organisation, Kenya

Sam Ndungu Gichuhi
Kenya Agricultural and Livestock Research Organisation, Kenya

Erick O. Mungube
Kenya Agricultural and Livestock Research Organisation, Kenya

ABSTRACT

Bovine anaplasmosis is one of the most important tick borne diseases of ruminants worldwide causing significant economic losses in the livestock industries due to the high morbidity and mortality in susceptible cattle herds. Bovine anaplasmosis, caused by Anaplasma marginale, is an infectious but non-contagious disease. The mode of transmission of bovine anaplasmosis includes mechanical (blood contaminated fomites (needles, ear tagging, dehorning and castration equipment), biological (tick bites) and transplacental (mother to fetus). Bovine Anaplasmosis occurs in tropical and subtropical regions worldwide. Cattle of all ages are susceptible to infection with A. marginale, but the severity of disease increases with age. The common clinical sign of bovine anaplsmosis includes; fever, anorexia, rapid loss of body condition, severe decrease in milk production, pale and icteric mucous membranes, increased heart and respiratory rates, muscle weakness and depression. Diagnosis of bovine anaplasmosis can be made by demonstration of A. marginale on stained blood smears from clinically infected animals during the acute phase of the disease, but it is not reliable for detecting infection in pre-symptomatic or carrier animals. Instead, serological demonstration of antibodies and confirmation of antigen with molecular detection tools are used for diagnosis. Anaplasmosis can be treated by administration of oxytetracyclines,

DOI: 10.4018/978-1-7998-6433-2.ch010

however oxytetracyclines do not clear the parasite from carrier animals. Control measures for bovine anaplasmosis vary with geographical location and include maintenance of Anaplasma free herds, vector control, administration of antibiotics and vaccination. Intensive acaridae application to control ticks has a number of limitations, therefore, immunization together with strategic tick control is recommended for exotic and crossbred cattle. Further studies on epidemiology of bovine anaplasmosis is needed

INTRODUCTION

Anaplasmosis is an infectious, non-contagious haemotropic disease of cattle, sheep and goats caused by obligate intra-erythrocytic bacterium *Anaplasma* spp. In cattle severe debility, emaciation, anaemia and jaundice are the major clinical signs. In sheep and goats the disease is usually sub-clinical. The main mode of transmission of Bovine anaplasmosis are tick bites but can also be transmiited mechanically (blood contaminated fomites (needles, ear tagging, dehorning and castration equipment) and transplcental (mother to fetus). Bovine anaplasmosis occurs in tropical and subtropical regions worldwide. Cattle of all ages are susceptible to infection with anaplasmosis but the severity of disease increases with age. The disease which was formerly known as gall sickness, traditionally refers to a disease of ruminants which is not only notable in both dairy and beef cattle but also in other ruminants including water buffalo, bison, African antelopes, and some species of deer which can become persistently infected (Aubry & Geale, 2011). Infected wildlife may have significance as reservoirs of the parasite. The disease occurs worldwide in tropical and subtropical areas including the North and South Americas, Africa, the Caribbean, Russia, European countries bordering the Mediterranean, and the Middle and Far East.

When anaplasma infects a previously uninfected herd, the producer can expect a 3.6% reduction in successful calving, 30% increase in the cull rate, and 30% mortality in adults showing signs (Aubry & Geale, 2011). Animals surviving an acute attack often make a slow recovery, resulting in losses in milk or meat production. Tetracycline antibiotics and imidocarb are currently used for treatment. Cattle treated with these drugs remain immune to severe anaplasmosis for at least 8 months. Control measures for bovine anaplasmosis may vary with geographical location, but they have not varied markedly during the past 50 years (Kocan *et al.*, 2000). They include arthropod control, chemoprophylaxsis, vaccination, and maintenance of an anaplasma-free herd.

AETIOLOGY

Anaplasma marginale is the causative agent in cattle and wild animals and *Anaplasma ovis (A. ovis)* in sheep and goats. *Anaplasma centrale* does not cause clinical disease but is used as a live vaccine for cattle in Israel, South Africa, South America and Australia (de la Fuente *et al.*, 2005). The *Anaplasma* spp belong to the family Anaplasmataceae in the order Rickettsiales. *A. marginale* does not cause clinical disease in domestic small ruminants, *A.ovis*, the agent of ovine anaplasmosis, may cause mild to severe disease in sheep, deer and goats but is not infectious for cattle (Aubry & Geale, 2011).

The *Anaplasma* genus has recently been expanded to include three species transferred from the genus(Table1) *Ehrlichia*: (i) *Anaplasma phagocytophilum*, which was transferred from species previously known as *Ehrlichia phagocytophila*, *Ehrlichia equi* and human granulocytic ehrlichiosis agent (now human granulocytic anaplasmosis or HGA), (ii) *Anaplasma bovis* (previously *Ehrlichia bovis*) and (iii)

Table 1. Reclassification of genus anaplasma organisms (Rymaszewska & Grenda, 2008)

Aetiological agent		Disease	Vector	Infected organism or host
Before 2001	**After 2001**			
Anaplasma marginale	*Anaplasma marginale*	Bovine anaplasmosis	*Ixodes* spps; *dermancentor* spps	Ruminants
Anaplasma central	*Anaplasma central*	Bovine anaplasmosis	*Haemaphysalis* spps, *Ixodes* spps	Ruminants
E.equi *E.phagocytophila*	*Anaplasma phagocytophilum*	Human and animal granulocytic anaplsmosis	*Ixodes* spps; *Dermacentor* spps	Small ruminants forming and wild,horses,dogs, Human

Anaplasma platys (previously *Ehrlichia platys*) (Dumler *et al* 2001). These three species invade blood cells other than erythrocytes of their respective mammalian hosts. The amino acid sequences of the major surface proteins (MSPs) of *A. phagocytophilum* are similar to those of *A. marginale.*

Anaplasma phagocytophilum is present worldwide (including in the USA) but seems to cause the disease called tick-borne fever in domestic and wild ruminants in Europe only (Hoar *et al.*,2008)

Several strains of *A. marginale* have been identified from various geographical areas. They differ in morphology, protein sequence, antigenic characteristics and their ability to be transmitted by ticks (Smith *et al.*, 1986; Allred *et al.*, 1990; Rodriguez *et al.*, 2000; de la Fuente *et al*; 2001c.cd, 2003c., Palmer, 2001). The strains are sufficiently similar to avoid the need for the separate isolates to be included in a vaccine (Kuttler *et al* 1984)

Major surface proteins play a crucial role in the interaction of *A. marginale* with host cells and thus their ability to cause infection. Six MSPs have been identified on *A. marginale* derived from bovine erythrocytes (Palmer *et al.*, 1999) and were found to be conserved on tick and cell culture-derived organisms (Kocan and de la Fuente, 2003). Three of these MSPs, namely MSP1a, MSP4 and MSP5, are from single genes and do not vary antigenically within strains, while the other three, MSP1b, MSP2 and MSP3, are from multigene families and may vary antigenically, most notably in persistently infected cattle (Kocan *et al.*, 2003). Because of the variation in the repeated portion of the *msp1α* gene, it has been used as a stable genetic marker for the identification of *A. marginale* geographical strains (Barbet *et al.*, 1987, 1999; Allred *et al.*, 1990; de la Fuente *et al.*, 2001a). The gene, *msp1α*, that encodes MSP1a is conserved during the multiplication of the rickettsia in cattle and ticks (Palmer *et al.*, 2001; Bowie *et al.*, 2002) and has been shown to be involved in adhesion to bovine erythrocytes and tick cells (McGarey and Allred, 1994; McGarey *et al.*, 1994; de la Fuente *et al.*, 2001b). Major surface protein 5 is also a highly conserved surface protein that has been proven effective as a diagnostic antigen and is used in a commercially available competitive enzyme-linked immunosorbent assay (cELISA) (Torioni de Echaide *et al.*, 1998).

Disease Transmission

The parasite *Anaplasma marginale* whch causes the disease is transmitted by one host ticks in the genus *Rhipicephalus* (*Boophilus*). The ticks have been in the genus *Boophilus* until 2003. The change from *Boophilus* to *Rhipicephalus* was first published by Horak, Camecas & Keirans (2002) and confirmed by Murell & Becker (2003). The ticks *Rhipicephalus* (*Boophilus*) *microplus* and *Rhipicephalus* (*Boophilus*) *decorates* are the main ticks which transmits *A. marginale*. At least 20 other different ixodidae tick species among them *Boophilus* spp., *Dermacentor* spp., *Ixodes ricinus* and *Rhipicephalus* spp. may be involved in the transmission of A. *marginale* (Davidson& Goff. 2001; Kocan *et al*., 2004). The only exception is *Amblyomma* spp. as there is no evidence to suggest its role in the transmission of this parasite from infected to susceptible hosts (Kocan *et al*., 2004).

Male ticks have an intermittent feeding behaviour: they usually remain on their hosts, feed repeatedly and inseminate several females (Aubry & Geale, 2011). They can also feed, detach, wander about looking for females, seek a new host and then feed and mate thus playing a role in the transmission of *A. marginale*. *Anaplasma marginale* is also transmitted mechanically through blood sucking flies in fresh erythrocytes during a blood meal. The mechanical transmitters include *Stomoxys calcitrans* (stable flies), Tabanids (horse flies) and Cullicidae (midges) (Aubry and Geale, 2011). In these flies, *A. marginale* does not undergo cyclic development as is the case in the tick vectors.

Vertical transmission (transplacental) of *A. marginale* has also been experimentally demonstrated in one of two cows exposed during the second or third trimester but not during the first (Zaugg, 1985). In South Africa, incidence of in-utero transmission involving *A. central* or *A. marginale* of 15.6% infections was reported among 77calves born to splenectomized and intact cows, chronically infected or undergoing primary reactions during the first, second or third trimester of gestation (Potgieter & VanRensburg, 1987).

Anaplasma marginale also gets transmitted from infected to non-infected animals within a herd through iatrogenic means related to operations involving blood-contaminated fomites like needles, dehorning equipment, nose tongs, tattooing instruments, ear-tagging devices and castration or spaying equipment (Kocan *et al*., 2000).

Tick Vectors of Anaplasma

Rhipicephalus Decoratus- African Blue Tick

The common name of this tick is rather confusing as it is derived from the color of the engorged female ticks, whereas the engorged females of several other tick species are also a blue color. It is found in Africa south of Sahara in Namibia, Botswana, Mozambique, Zimbabwe, Angola, Zambia, Malawi, parts of Tanzania, Burundi, Uganda, western Kenya and in the wetter highlands of Ethiopia(Figure 1). It is also found in parts of West Africa, such as Mali and Senegal (Mason & Norval, 1980; FAO, 1984; Matthysse & Colbo, 1987; Norval, 1994).

In South Africa it is being replaced by *R. microplus*. There are several reasons for the displacement of *R. decoloratus* by *R. microplus*. The total length of the life cycle of *R. decoloratus* is longer than that of *R. microplus* and female *R. decoloratus* produce slightly less eggs than those of *R. microplus*. Part of the increase in distribution of *R. microplus* may thus be coupled to its shorter life cycle and higher egg production. Cross mating between the two species results in the production of sterile eggs, and although the male ticks prefer to mate with ticks of the same species, they will also mate with females of the other species. Males of *R. microplus* (Figure1) are sexually mature a few days sooner than those of *R. decoloratus* (Figure2) and consequently in mixed infestations they stand a better chance of mating with *R. microplus* females. Should the sex ratio of male to female *R. microplus* be 2:1, as it is for *R. decoloratus* on field infested animals, the *R. microplus* males that do not find a *R. microplus* female could mate with *R. decoloratus* females, instead of the converse. If the females mate once only, the cross-mated females would produce sterile eggs and *R. microplus* would consequently comprise increasing proportions of future mixed populations of the two ticks.

Figure 1. Rhipicephalus decoloratus male
Photo credits – KALRO - VSRI

Rhipicephalus decoloratus is also suspected of being a vector of *Babesia trautmanni*, the cause of porcine babesiosis. However under field conditions, the small numbers encountered on warthog, even in regions where large numbers of this tick are present, make this seem unlikely. Cattle, impalas (*Aepycero smelampus*), eland, bushbuck (*Tragelaphus scriptus*), greater kudu (*Tragelaphus strepsiceros*) and also horses and zebras are hosts of *R. decoloratus*. In addition to *A. marginale* it also transmits *A. mesaeterum, A. ovis, Babesia bigemina, Babesia bovis, Babesia trautimanni*, Crimean-Congo Haemorrhagic fever virus *Ehrichia ruminatum, Theileria velifera and Trypanosoma vivax.* The preferred predilection sites are the sides of the body, shoulders, neck and dewlap. The immature stages may be found on the tips and upper edges of the ears and on the legs.

Rhipicephalus microplus

Originally from Asia, it is found in all continents in South and Central America, including Mexico, and is a major problem in Brazil (Evans, 1992). It occurs in much of southern Asia, China, South Africa, Mozambique, Kenya and Tanzania. It is the predominant tick in Australia (Angus, 1996). Recently, after 2000, *R. (B.) microplus* has also been found in West Africa, mainly in the Ivory Coast, Benin, Burkina Faso and Mali (Figure 3). The route of introduction is not known but is probably related to the importation of cattle, the principal host for this tick species, from Brazil. Cattle have been imported into the Ivory Coast and Benin between 2000 and 2006 on several occasions and genetic analyses of *R. (B.) microplus* ticks collected in the Ivory Coast suggest a Latin American origin. There is a serious concern that this tick could spread to the entire West African region because of the favorable environment, high cattle densities, and cattle movements throughout the entire region.

Rhipicephalus microplus (Figure 2) has a one-host life cycle, which in total takes about two months to complete. Several generations are possible, especially in those areas with a prolonged rainy season. Because of its one-host life cycle all stages of development occur on the host at the same time and the presence on a single host of approximately 70 engorged females can represent a total parasitic population exceeding 10 000 ticks.

Figure 2. Rhipicephalus microplus male
Photo credit KALRO VSRI

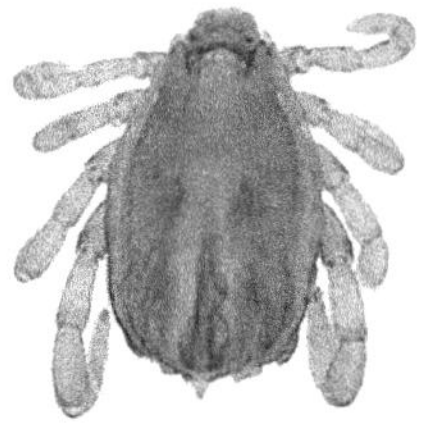

Domestic cattle are probably the only effective hosts of this tick, although several collections, including engorged females have been made from goats. There are very few records from wildlife and then only when domestic cattle were also present.In addition to *A. marginale* it also transmits *A. mesaeterum*, *A. ovis, Babesia bigemina*, *Babesia bovis*, *Babesia trautimanni*, Crimean-Congohaemorrhagic fever virus, *Ehrichia ruminatum*, *Theileria velifera*,*Theileria annulata*, *Theileria ovis*, *Theileria orientalis*, Rift Valley fever virus and Bovine leukemia virus, *Trypanosoma vivax*. *Trypanosoma brucei brucei*.

Geographical Distribution

Anaplasmosis is distributed across the tropical and sub-tropical regions worldwide in areas situated between latitudes 40°N and 32°S, including South and Central America, the United States (US), southern Europe, Africa, Asia and Australia (The Merck Veterinary Manual, 2008). In the temperate regions, anaplasmosis occurs sporadically Radostitis *et al*., 2010.

Figure 3. Distribution Rhipicephalus decoratus
Source: *https://www.cabi.org/isc/datasheet/91606#toDistributionMaps (accessed 27.10.2020)*

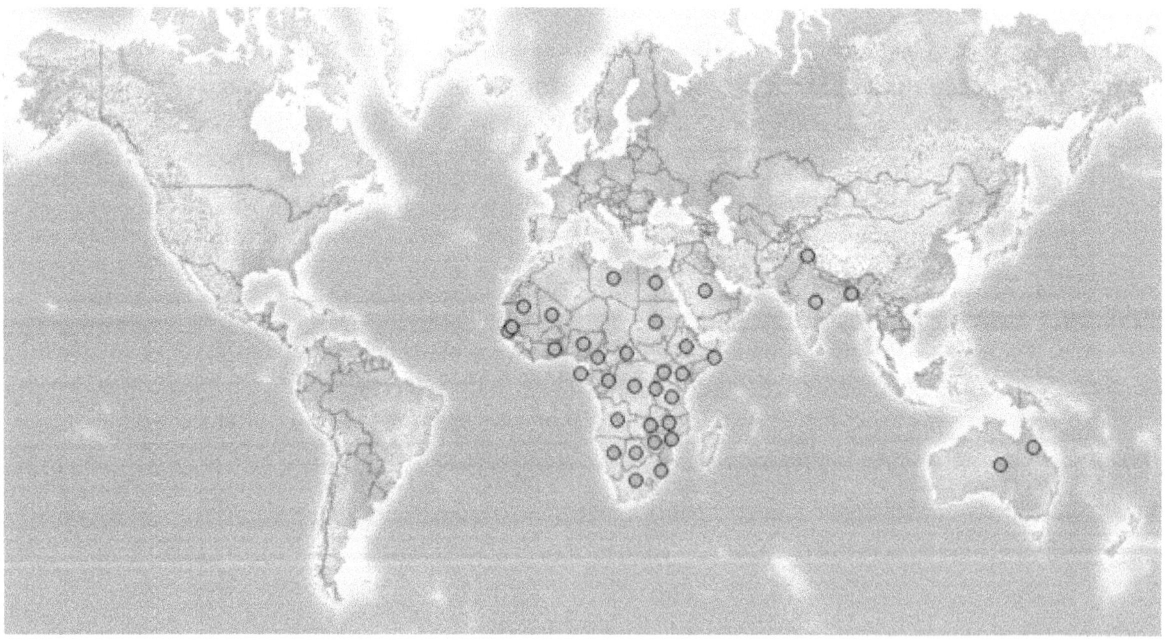

Figure 4. Distribution of Rhipicephalus microplus
Source: *https://www.cabi.org/isc/datasheet/91606#toDistributionMaps (accessed 27.10.2020)*

The distribution of the disease follows the vector distribution as shown in figures 3 and 4

It is also possible that anaplasmosis may occur in the temperate areas. It is worthy stating that the epidemiology of anaplasmosis is complicated by the life-long carrier state which occurs in animals that have recovered from the clinical disease. The distribution is expected to keep changing in relation to climate change and global warming .Climate change and other enviromanatal changes will modify the prevalence, activity,and the range of a variety of ticks and the pathogens they carry. This will lead to an increased risk of anaplasmosis and other tickborne diseases. (Bouchard *et al* 2019). The current distribution may be expected to remain the same due to the stability of the tick niche suitability in currently suitable regions. However an increase in the distribution is expected as an increase in niche suitability increases in various parts of the world (Marques *et al* 2020)

Epidemiology and Risk Factors

Anaplasmosis is a disease of cattle and their close relatives in the wild. *Anaplasma marginale* is host specific, with only a few reported cases found in sheep and goats (Zabel & Agusto, 2018).*Bos taurus* and *Bos indicus* and their crosses have equal susceptibility and show the same age susceptibility However, under field conditions, *Bos indicus* are not commonly affected probably since they have ability to tolerate heavy tick infestation (Nazar *et al.*, 2018). Further, cattle breeds with black or red skin colour have a higher risk of getting infected with *Anaplasma* especially where mechanical transmission through biting flies is concerned (Hungerford & Smith, 1997). *Bos indicus* breeds of cattle appear to possess a greater resistance to *A marginale* infection than *B taurus* breeds, but variation of resistance of individuals within breeds of both species occurs. Differences in virulence between *Anaplasma* strains and the level and duration of the rickettsemia also play a role in severity of clinical manifestations.

Anaplasmosis presents with high morbidity but mortality is generally low to moderate as it ranges from 5 and 40 percent, but may reach 70 percent during a severe outbreak. Prevalence of infection in cattle in endemic areas is very high with sero-positivity rates exceeding 60% and often approaching 90% (Radostitis *et al.*, 2010). However, seropositivity is much lower in regions that interface between endemic and non-endemic zones (Hungerford and & Smith, 1997).

The severity of the disease is age-dependent since it increases with increasing age (Nazar *et al.*, 2018; Radostits *et al.*, 2007). Generally, the disease is rare in calves under 6 months of age but the severity increases with age with cattle aged between 6 months and 1 year developing the mild form of the disease. The acute form of anaplasmosis is notable in cattle aged between 1 and2 years while those over 2 years of age will develop acute disease which may often be fatal with mortality risks ranging between 29% and 49% (Richey, 1991; Kocan *et al.*, 2003).Often older animals will be found dead. Newly infected adult cattle over 3 years will show the most severe clinical signs, and 30 percent to 50 percent will die if they are not treated early. Once cattle become infected with *A. marginale they* remain persistently infected carriers for life, whether or not they develop clinical disease and regardless of the age at the time of infection (Richey, 1991).

Pathogenesis

In the mammalian host, after an incubation period of 7-60 days (with an average of 28 da*ys) A.marginale* moves to the erythrocytes as shown in figure5 in a membrane bound inclusion called the initial body with 4-8 ricketssias. It then undergoes a series of replication and is subsequently phagocytozed by the host's reticulo-endothelial system to further re-invade other erythrocytes. As many as 109 erythrocytes per milliliter of blood (70% of all erythrocytes) can be infected (Anziani *et al* 1982, Arulkanthan *et al* 1999). The phagocytosis of erythrocytes results in mild to severe haemolytic anaemia without hemoglobinaemia or hemoglobinuria (Bowie *et al* 2002, Bradway *et al* 2001). Between 10% and 90% of the erythrocytes may be parasitized in the acute stage of the infection depending on the strain of infecting *Anaplasma* and also on the host susceptibility (Aubry and Geale, 2011). At least 15% of the erythrocytes have to be parasitized before there is clinical disease (Constable *et al.*, 2016).

Anaplasmosis has an incubation period (the time infection is introduced into a susceptible animal to the time clinical signs become manifest) of between 7 and 60 days with an average of 28 days (Kocan *et al.*, 2003) Ticks acquire .*A. marginale* (Figure5) while feeding on carrier hosts.in the ticks *A. marginale* invades the midguts and undergoes a first cycle of replication to form dense (infective) forms which moves to other tissues. After a series of further replications the dense forms move to the salivary glands from where the *A.marginale* is transmitted to a new mammalian host.

Clinical Signs of Anaplasmosis (Constable *et al.*, 2016)

§ In herds where cattle are not observed closely, unexplained dead animals are frequently the first signs noticed in an anaplasmosis outbreak.

§ Anaplsomosis infected cattle show marked weakness as the first clinical sign as such tend to fall behind the rest of the herd

§ Infected cattle will not eat or drink.

Figure 5. Life cycle of A. marginale (source: Kocan et al 2003)

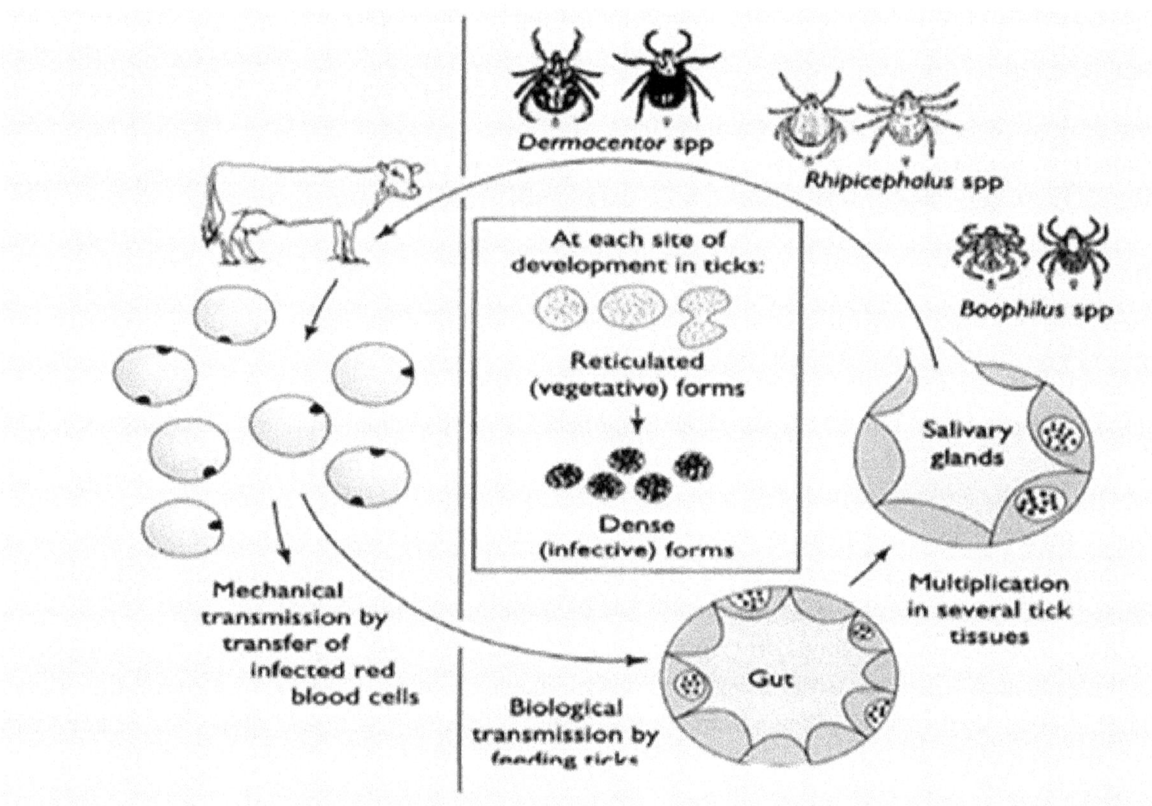

§ Cows with light skin will initially look pale around the eyes and muzzle, but later this can change to a yellowish color (jaundice). This jaundice is due to the destruction of the blood cells and their contents being released into the blood stream.
§ Weight loss is rapid.
§ Cattle can become extremely aggressive because of oxygen deprivation occasioned by severe anemia.
§ Oxygen deprivation can also result in abortions in pregnant cows.
§ Constipation, high fever, and labored breathing can also be seen.
§ The most critical period is the first 4 to 9 days after clinical signs appear with any infected animal having an increased chance of survival.

In animals <1 year old anaplasmosis is usually subclinical, in yearlings and 2-year-olds it is moderately severe, and in older cattle it is severe and often fatal. Anaplasmosis is characterized by progressive anemia due to extravascular destruction of infected and uninfected erythrocytes. The prepatent period of *A. marginale* is directly related to the infective dose and typically ranges from 15–36 days (although it may be as long as 100 days). After the prepatent period, peracute (most severe but rare), acute, or chronic anaplasmosis may follow. Rickettsemia approximately doubles every 24 hr during the exponential growth phase. Generally, 10%–30% of erythrocytes are infected at peak rickettsemia, although this figure may be as high as 65%. RBC count, PCV, and hemoglobin values are all severely reduced. Macrocytic anemia with circulating reticulocytes may be present late in the disease.

Animals with peracute infections succumb within a few hours of the onset of clinical signs. Acutely infected animals lose condition rapidly. Milk production falls. Inappetence, loss of coordination, breathlessness when exerted, and a rapid bounding pulse are usually evident in the late stages. The urine may be brown but, in contrast to babesiosis, hemoglobinuria does not occur. A transient febrile response, with the body temperature rarely exceeding 106°F (41°C) occurs at about the time of peak rickettsemia. Mucous membranes appear pale and then yellow. Pregnant cows may abort. Surviving cattle convalesce over several weeks, during which hematologic parameters gradually return to normal.

Diagnosis

Identification of *Anaplasma marginale*

Blood smears (a blood sample is taken in anticoagulant from the jugular vein) and organ smears can be obtained from suspected clinical cases and stained in 10% Giemsa stain for approximately 30 minutes after fixation in absolute methanol for 1 minute (OIE, 2008). After staining, the smears are rinsed three or four times with tap water to remove adhering stain, and are then air-dried. Smears are examined under oil immersion at a magnification of ×700–1000 to confirm presence of *A. marginale* and of the disease. The parasites become visible microscopically 2-6 weeks following transmission.

In Giemsa-stained thin blood films, *Anaplasma* spp appear as dense, homogeneously staining blue-purple inclusions 0.3–1 µm in diameter. *A marginale* inclusions are usually located toward the margin of the infected erythrocyte as shown in figure 6, whereas *A. centrale* inclusion bodies are located more

Figure 6. Anaplasma marginale, bovine blood smear
Photo credits – KALRO - VSRI

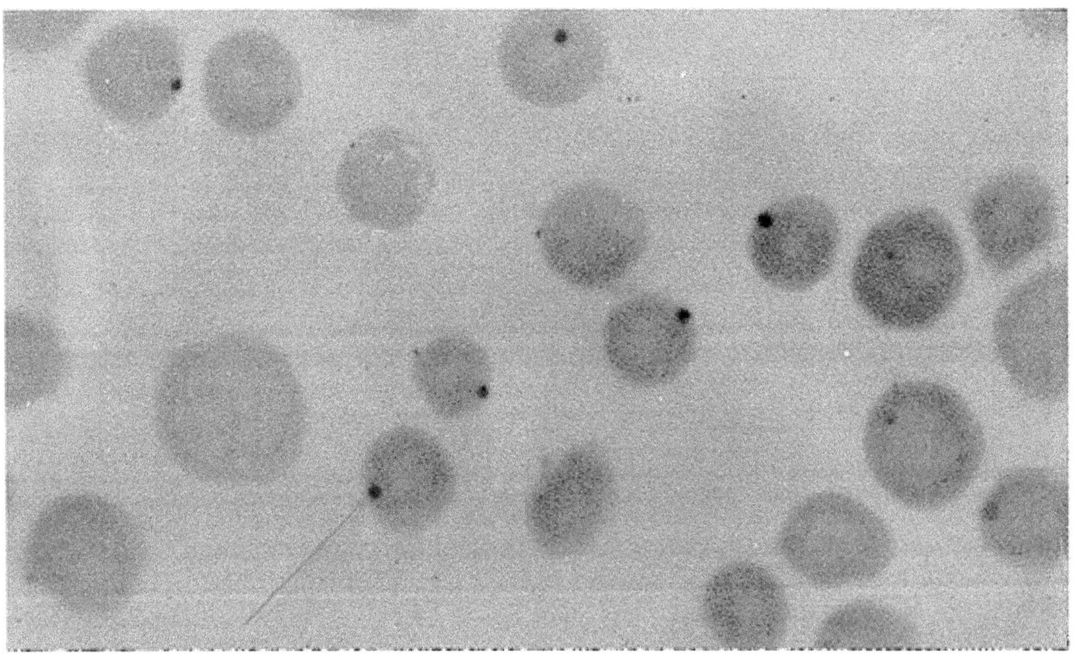

centrally. *A caudatum* cannot be distinguished from *A. marginale* using Giemsa-stained blood films. Special staining techniques are used to identify this species based on observation of characteristic appendages associated with the bacteria. *A caudatum* has been reported only in North America and could possibly be a morphologic form of *A. marginale* and not a separate species. Inclusion bodies contain 1–8 initial bodies 0.3–0.4 µm in diameter, which are the individual rickettsiae. The percentage of infected erythrocytes varies with the stage and severity of disease; maximum rickettsemias in excess of 50% can occur with *A. marginale*. Microscopically, the parasites becomes visible 2–6 weeks after transmission. During the course of infection, the rickettsemia can double each day for up to 10 days and then decreases. Severe anemia can persist for weeks after parasites cannot be detected in blood smears. However, this method is not only reliable especially for the pre-symptomatic or carrier animals but also has low sensitivity of between 20 to 30% (Aubry & Geale, 2011).

Serological Techniques

Serological tests are appropriate for pre-symptomatic or carrier cattle and the tests detect antibodies which is confirmation of exposure to the *A. marginale*. Extensive epidemiological studies can employ a number of serological tests including complement fixation (CF) test, capillary agglutination assay, card agglutination test (CAT), indirect fluorescent antibody (IFA) test, cELISA, indirect ELISA and dot ELISA (OIE, 2008).

The CAT is quite sensitive, may be undertaken either in the laboratory or in the field, and gives a result within a few minutes (OIE, 2008). However, nonspecific reactions may be a problem, and subjectivity in interpreting assay reactions can result in variability in test interpretation. In addition, the CAT antigen, which is a suspension of *A. marginale* particles, can be difficult to prepare and can vary from batch to batch and laboratory to laboratory.

The antigen is obtained from splenectomised calves. The calves are infected by intravenous inoculation with blood containing *Anaplasma*-infected erythrocytes. When the parasitaemia exceeds 50%, the animal is exsanguinated, the infected erythrocytes are washed, lysed, and the erythrocyte ghosts and *Anaplasma* particles are pelleted. The pellets are sonicated, washed, and then resuspended in a stain solution to produce the antigen suspension.

Competitive enzyme-linked immunosorbent Assay(c-ELISA) using a recombinant antigen termed rMSP5 and MSP5-specific monoclonal antibody (MAb) has proven very sensitive and specific for detection of *Anaplasma*-infected animals (Torini de Eschades *et al.*, 1998). All *A. marginale* strains tested, *A. ovis* and *A. centrale*, express the MSP5 antigen and induce antibodies against the immune dominant epitope recognised by the MSP5-specific MAb.

Indirect enzyme-linked immunosorbent assay (I-ELISA) based on the use of a normal red blood cell antigen (negative antigen) and an *A. marginale*-infected red blood cell antigen (positive antigen) has been found to be reliable for the detection of *A. marginale* positive sera (Duzgun *et al.*, 1988). Although more cumbersome than tests using only one antigen, this test eliminates those sera that have high levels of nonspecific activity due to iso-antibodies to normal red blood cell components.

Dot enzyme-linked immunosorbent assay (dot ELISA) has also been described. Compared with the I-ELISA, the dot ELISA has the potential advantages of being rapid, inexpensive and simple to perform. The dot ELISA has been reported to have a sensitivity of 93% and a specificity of 96% (Montenegro-James *et al.*, 1990).

The complement fixation test (CFT) although extensively used for many years shows variable sensitivity (ranging from 20 to 60%), possibly reflecting differences in techniques for antigen production, and poor reproducibility (OIE, 2008). CFT assays may also fail to detect a significant proportion of carrier cattle (5). It is also uncertain as to whether or not the CF test can identify antibodies in acutely infected animals prior to other assays (6, 24). Therefore, the CF test is no longer recommended as a reliable assay for detecting infected animals.

Indirect fluorescent antibody test (IFA) Because of the limitations on the number of indirect fluorescent antibody (IFA) tests that can be performed daily by one operator, other serological tests are generally preferred to the IFA test. A serious problem encountered with the test is nonspecific fluorescence. Antigen made from blood collected as soon as adequate parasitaemia (5–10%) occurs is most likely to be suitable. Nonspecific fluorescence due to antibodies adhering to infected erythrocytes can be reduced by washing the erythrocytes in an acidic glycine buffer before antigen smears are prepared (Montenegro-James *et al.*, 1985). Infected erythrocytes are washed twice in 0.1 M glycine buffer (pH 3.0, centrifuged at 1000 g for 15 minutes at 4°C) and then once in PBS, pH 7.4

Polymerase chain reaction (PCR)–based methods have been developed that are capable of detecting low levels of infection in carrier animals but can still misdiagnose a proportion of carrier cattle that would test negative. The detection limit (analytical Se) of the PCR was reported to be 24 infected erythrocytes per microlitre blood (equivalent to 0.0001% infected erythrocytes) (Figueroa *et al.*, 1993; Gale *et al.*, 1996). A more sensitive nested PCR (nPCR) was capable of identifying as few as 30 infected erythrocytes per millilitre blood (equivalent to a0.0000001% parasitemia) under experimental conditions, which would be well below the lowest levels in carrier animals (Torioni de Echaide *et al.*, 1998).

Real-time PCR (RT-PCR) can successfully be applied in the detection and quantification of *A. marginale*DNA (msp1b gene) in the blood of naturally infected cattle (Carelli *et al.*, 2007). The test is highly specific with no cross-reactions with other haemoparasites of ruminants (*A. centrale*, *A. bovis*,*A. phagocytophilum*, *B. bigemina*, *Theileriabuffeli*) as determined by reverse line blot (RLB) hybridization.

The most accurate serological test currently available for identifying *Anaplasma*-infected cattle is a cELISA that uses a monoclonal antibody (MAb) specific for MSP5 [*Anaplasma* Antibody Test Kit (cELISA); VMRD Inc., Pullman, WA, USA]. This assay specifically detects the presence of serum antibodies that target a surface protein, MSP5 of *Anaplasma* spp. It has proven very sensitive and specific for the detection of *Anaplasma* spp.-infected animals (Knowles *et al.*, 1996, Toroni de Echaide *et al.*, 1998, Strik *et al.*, 2007) However, the test cannot differentiate between *A. marginale* and some of the other *Anaplasma* species, because they all express the MSP5 antigen(Visser et al 1992)and induce antibodies recognized by the MSP5-specific MAb. cELISA and CAT are the two OIE preferred serological tests currently used for identifying infected animals (OIE, 2008).

Differential Diagnosis

1. Cattle

Anaplasmosis causes haemolytic anaemia. There are other causes of haemolytic anaemia which may be confused with anaplasmosis which includes babesiosis, leptospirosis, bacillary haemaglobinuria, rapeseed poisoning, eperythrozoonosis and chronic copper poisoning

Anaplasmosis is commonly confused with babesiosis, as many of the epidemiological features are similar. Haemoglobinuria is a distinguishing feature of babesiosis, but is not always present in the early stages of *Babesia bovis* infection.

Anaemia and progressive weight loss are also clinical signs associated with *Trypanosoma* spp. infections. Trypanosomiasis in Africa (nagana), generally occurs within well defined, tsetse-infested areas (Potgieter and Stoltsz, 1994).

Some forms of theileriosis (*T. mutans* transmitted *Amblyoma variegatum*) should be differentiated from anaplasmosis, and are usually associated with enlarged lymph nodes, the microscopic presence of schizonts in lymph node biopsy smears and presence of vector tick (Potgieter & Stoltsz, 1994).

Anaemia and icterus may be encountered in cattle suffering from leptospirosis, post-parturient haemoglobinuria, chronic copper poisoning and intoxications with *Brassica* and *Allium* spp. In these conditions haemoglobinaemia and haemoglobinuria are present. Causes of hepatotoxicity, such as *Lantana camara* and *Microcystis aeruginosa* poisoning in which icterus is one of the principal signs, should also be considered (Potgieter & Stoltsz, 1994).

Changes in behaviour, such as aggression are occasionally manifested in anaplasmosis and need to be differentiated from diseases like heart water, cerebral babesiosis and cerebral theileriosis, rabies as well as conditions such as lead poisoning.

2. Sheep and Goats

Many of the differential diagnoses listed for cattle also need to be considered for sheep and goats.

Ehlichiosis, eperythrozoonosis and verminosis (eg haemonchosis) need also to be considered and may occur concurrently.

Postmortem Findings

Lesions are typical of those found in animals with anemia due to erythrophagocytosis. The carcasses of cattle that die from anaplasmosis are generally markedly anemic and jaundiced. Blood is thin and watery. The spleen is characteristically enlarged and soft, with prominent follicles. The liver may be mottled and yellow-orange. The gallbladder is often distended and contains thick brown or green bile. Hepatic and mediastinal lymph nodes appear brown. There are serous effusions in body cavities, pulmonary edema, petechial hemorrhages in the epi- and endocardium, and often evidence of severe gastro-intestinal stasis. Widespread phagocytosis of erythrocytes is evident on microscopic examination of the reticuloendothelial organs. A significant proportion of erythrocytes are usually found to be parasitized after death due to acute infection.

Microscopic findings include bile duct hyperplasia, bile retention, increase in liver capsule size, fatty changes in liver, lungs emphysema and myocardial degeneration.

Treatment

Tetracyclines and imidocarb dihydrochloride or dipropionate are currently used for treatment. Cattle may be sterilized by treatment with these drugs and remain immune to severe anaplasmosis subsequently for at least 8 months.

Prompt administration of tetracyclines (tetracycline, chlortetracycline, oxytetracycline, rolitetracycline, doxycycline, minocycline) in the early stages of acute disease (eg, PCV >15%) usually ensures survival. A commonly used treatment consists of a single IM injection of long-acting oxytetracycline at a dosage of 20 mg/kg. Blood transfusion to partially restore the PCV greatly improves the survival rate of more severely affected cattle. The carrier state may be eliminated by administration of a long-acting oxytetracycline preparation (20 mg/kg, Intramuscular, at least two injections with a 1-week interval). Withholding periods for tetracyclines apply in most countries. Injection into the neck muscle rather than the rump is preferred.

Imidocarb is also highly efficacious against *A marginale* as a single injection (as the dihydrochloride salt at 1.5 mg/kg, SC, or as imidocarb dipropionate at 3 mg/kg). Elimination of the carrier state requires the use of higher repeated doses of imidocarb (e.g., 5 mg/kg, IM or SC, two injections of the dihydrochloride salt 2 weeks apart). Imidocarb is a suspected carcinogen with long withholding periods and is not approved for use in the USA or Europe.

Control of Anaplasmosis

Control measures for bovine anaplasmosis may vary with geographical location, but they have not varied markedly during the past 50 years (Kocan *et al.*, 2000). They include arthropod control, chemoprophylaxsis, vaccination, and maintenance of an anaplasmosis-free herd. Control measures may be different depending on geographic location, availability, cost, and the feasibility of application.

Maintenance of anaplasmosis-free herds works where the prevalence of infection is low. All the animals in the herd and any new additions are tested for presence of *Anaplasma* parasites with those testing positive getting eliminated from the herd as they are sources of infection for other animals. Alternatively, some carriers could be treated and tested to confirm chemosterilization (treating carrier animals to completely rid them of *A. maginale*) using long acting tetracycline (Reinbold *et al.*, 2010). Chemosterilization is costly as it involves a long period of administration of long acting tetracycline with long withdrawal periods and also increasing the risk of antimicrobial resistant *Anaplasma marginale* strains.

Since diagnostic tests sometimes lead to false negative results, it is advisable to source cattle from low-risk herds if possible or from areas can also help in preventing the accidental introduction of *A. marginale* into a herd.

Vector control is a popular method of preventing and controlling tick-borne diseases in Africa (Aubry and Geale, 2011). Vectors involved in the transmission of *A. marginale* including ticks and other haematogenous flies must be controlled to interrupt the transmission pathway for an effective anaplasmosis control programme

Vector control can only partially prevent against *A. marginale* transmission if iatrogenic transmission through blood contaminated fomites is also present (Kocan *et al.*, 2003).Despite this, it is worth noting that controlling ticks and biting flies is not only labour-intensive and expensive but has associated negative impacts on the environment. Additionally, the repeated use of acaricides is likely to select for acaricide resistant tick and fly populations (Kocan *et al.*, 2000). The other disadvantage is the possibility of creating a susceptible population of cattle in endemic areas since the endemic stability gets destroyed.

In areas where both ticks and biting flies are involved in the spread of *A. marginale*, products that are effective against both are necessary (or combination of products).

Keeping animals in environments that are not conducive to the survival and completion of tick life cycle can help in a tick control programme (Aubry & Geale, 2011). This is appropriate for dairy cattle reared under zero grazing conditions. A buffer zone should probably be used between the area where the cattle are housed and any vegetation or other groups of cattle that have access to vegetation (Jonsson *et al.*, 1998).

Flies are best managed in farms when producers coordinate several methods that offer good control while reducing costs and dependence on chemicals. Integrated pest management combines sanitation practices, biological control and chemical agents to reduce fly populations (Watson *et al.*, 1994; Texas AgriLife Extension Service, 1997; Alberta Agriculture and Rural Development, 2008).Sanitation includes cleaning out hutches, stalls, pens, feeding lanes/bunks on a regular basis, manure management and covering and protecting feed. Biological control by natural enemies of flies (beetles, mites and parasitic wasps, or parasitoids) can be very effective if care is taken by using insecticides (chemical control) that are compatible with their survival (space sprays and bait but not residual premise spray). Parasitoids can also be bought and released to 'jump start' their population growth in the spring–early summer

Iatrogenic transmission through blood-contaminated fomites can contribute to the spread of*A. marginale* (and other blood-borne pathogens) within a herd (Aubry &Geale, 2011). Avoid using the same needle for administration treatment for more than one animal. Where possible, needle-free administration systems should be preferred for parenteral administration of vaccines or antimicrobials (Reinbold *et al.*, 2010).

Surgical castration, dehorning, ear tagging among others should be performed using properly cleaned and disinfected surgical instruments. Other implements like nose tongs, prolapse and implant needles should also be properly cleaned and disinfected between animals.

Antimicrobial therapy is useful for the management of active infection (typically used on apparently healthy animals during the vector season to limit the clinical effects of infection) and treatment of clinical anaplasmosis (in sick animals).

Tetracyclines and imidocarb are used as antimicrobials (Richey, 1981). Other chemotherapeutic agents include arsenicals, anti-malarials, antimony derivatives and dyes which have little to no effect on acute anaplasmosis (Potgieter & Stoltsz, 2004). Administration of antimcrobials does not reliably eliminate persistent infections, and there is no evidence that it prevents cattle from becoming infected with *A. marginale*.

Vaccination is an economical and relatively effective way to control bovine anaplasmosis worldwide (Aubry & Geale, 2011). Both killed and live vaccines have relied on erythrocyte-derived antigen sources to induce protective immunity or to prevent clinical disease. However, neither one prevents cattle from becoming persistently infected with *A. marginale* or becoming reservoirs of infections (Kocan *et al.*, 2003). Development of effective vaccines to control anaplasmosis is hampered by the increasing number of field *A. maginale* strains which occur in a given geographical area (de la Fuente *et al.*, 2005) The extensive purification, dependence upon live animals as an antigen source, requirement of booster immunizations and difficulty of standardization have contributed to the expense of producing and using killed vaccines for bovine anaplasmosis (Kocan *et al.*, 2000).

Live vaccines involve inoculating cattle with erythrocytes infected with less pathogenic (attenuated) strains of*A. marginale* or*A. centrale*. The immune response is similar to natural infection with vaccinated animals developing mild and inapparent infections and becoming persistently infected with the vaccine strain (Aubry & Geale, 2011). *Anaplasma centrale* is used as a vaccine in Africa, Australia, Israel and Latin American countries. However, it does not provide effective cross-protection in widely separated geographical areas, as was demonstrated in Paraguay (Brizuela *et al.*, 1998). As with *A. centrale*, premunizing isolates of *A. marginale* may not be cross-protective in widely separated geographic areas. In addition, attenuated *A. marginale* vaccines could become virulent after successive passage through cattle or by ticks (Aubry & Geale, 2011).as with any whole- blood vaccine there is risk of transmitting unsuspected viruses such as that of bovine leucosis (Rodgers *et al* 1994).

Economic Importance of Anaplasmosis

When *Anaplasma* infects a previously uninfected herd, the producer can expect a 3.6% reduction in successful calving, 30% increase in the cull rate, and 30% mortality in adults showing signs (Aubry and Geale, 2011). Animals surviving an acute attack often make a slow recovery, resulting in losses in milk or meat production. Generally, the effects of tick-borne diseases on animals are often synergistic, where animals are infected with more than one pathogen at a time (De Waal, 2000). Although studies to quantify the losses that are specifically attributable to bovine anaplasmosis are yet to be carried out in South Africa and other African countries, the disease causes significant losses (De Waal, 2000; Potgieter & Stoltsz, 2004). It has been reported that costs arising from bovine anaplasmosis have been estimated from US $300–800 million (Kocan *et al.*, 2003).

Table 2. Breakdown of the economic losses caused by the cattle ticks

	Percentage of total loss
Increased labour costs	36
Loss of beef	20
Loss in dairy production	16
Cost of acaricides	11
Loss by death	7
Hide damage	5
Increased drought loss	5

Apart from the diseases they transmit, both *R. (B.) decoloratus* and *R. (B.) microplus* can have a significant impact on cattle production (Table2). On a mixed bushveld farm in Mpumalanga Province, South Africa, the weaning weights of calves of which the dams were predominantly infested with *R (B). decoloratus*, were reduced by approximately 8 g for each engorged female present on the cows. In Australia mid-lactation Holstein-Friesian cows were infested at weekly intervals with gradually increasing numbers of R *(B.) microplus* larvae for a period of 15 weeks. By week 15 uninfested control cows produced 2.86 litres more milk and 0.14 kg more butterfat per day and had gained 10.6 kg more live-weight than the infested cows. It was estimated that each engorging female tick was responsible for a loss of 8.9 ml in daily milk production and 1,0g in body weight gain over the trial period.

FUTURE RESEARCH DIRECTIONS

1. Redefine the role of wildlife reservoirs of *A. marginale* especially the role of deer (white tailed and mule) which may impact the epidemiology of anaplasmosis using molecular diagnostic tools.

2. Constant mapping of ticks population as ticks distribution is changing with climate change and this may cause tickborne diseases in new geographical areas.
3. Development ofvaccines that areprotective against a wide range of isolates owing the genetic diversity of *A. marginale.*

CONCLUSION

Anaplasmosis and other tick-borne diseases will continue to be of major economic importance in the world. The control methods for all of them revolves around the vector control and vaccination/immunization. The major vector control method is acaricide application which faces the danger of acaricide resistance which will lead to an escalation of the vector and hence the diseases. The situation is further complicated in East and Central Africa where the three host ticks which causes east coast fever and heart water co-exists with the anaplasma vector .here the vector control involves the use of acaricide weekly as opposed to once every three weeks where analpasmosis exists on its own. There is therefore a need for an integrated approach which involves the use of save vaccines for all the tick borne diseases and vector control. For anaplasmosis the risk for spreading other diseases through the vaccine needs to be addressed. There is also an increase in the vector niche suitability which requires continued surveillance to avoid spread of disease to otherwise clean areas and unexpected outbreaks.

REFERENCES

Abba, Y., Jesse, F. F. A., Sadiq, M. A., Ibrahim, H. H., Chung, E. L. T., Bitrus, A. A., Lila, M. A. M., & Hambali, I. U. (2016). Clinical management and gross pathology findings of a severe anaplamosis in a dairy cow. *Journal of Advanced Veterinary and Animal Research*, *3*(2), 195–199. doi:10.5455/javar.2016.c150

Abba, Y., Jesse, F. F. A., Sadiq, M. A., Ibrahim, H. H., Chung, E. L. T., Bitrus, A. A., Lila, M. A. M., & Hambali, I. U. (2016). Clinical management and gross pathology findings of a severe anaplamosis in a dairy cow. *Journal of Advanced Veterinary and Animal Research*, *3*(2), 195–199. doi:10.5455/javar.2016.c150

Abdisa, T. (2019). Epidemiology of Bovine Anaplasmosis. *SOJ Vet Sci*, *5*(1), 1–6. doi:10.15226/2381-2907/5/1/00165

Alberta Agriculture and Rural Development. (2008). *A guide for the control of flies in Alberta confined feeding operations*. Author.

Allred, D. R., McGuire, T. C., Palmer, G. H., Leib, S. R., Harkins, T. M., McElwain, T. F., & Barbet, A. F. (1990). Molecular basis for surface antigen size polymorphisms and conservation of a neutralization-sensitive epitope in *Anaplasma marginale*. *Proceedings of the National Academy of Sciences of the United States of America*, *87*(8), 3220–3224. doi:10.1073/pnas.87.8.3220 PMID:1691504

Angus, B. M. (1996). The history of the cattle tick *Boophilus microplus* in Australia and achievement in its control. *International Journal for Parasitology*, *26*(12), 1341–1355. doi:10.1016/S0020-7519(96)00112-9 PMID:9024884

Arulkanthan, A., Brown, W. C., McGuire, T. C., & Knowles, D. P. (1999). Biased immunoglobulin G1 isotype responses induced in cattle with DNA expressing *msp*1a of *Anaplasma marginale*. *Infection and Immunity*, *67*(7), 3481–3487. doi:10.1128/IAI.67.7.3481-3487.1999 PMID:10377129

Aubry, P., & Geale, D. W. (2011). A review of bovine anaplasmosis. *Transboundary and Emerging Diseases*, *58*(1), 1–30. doi:10.1111/j.1865-1682.2010.01173.x PMID:21040509

Bouchard, C., Dibernardo, A., Koffi, J., Wood, H., Leighton, P. A., & Lindsay, L. R. (2019). N Increased risk of tick-borne diseases with climate and environmental changes. *Canada communicable disease report = Releve des maladies transmissibles au Canada, 45*(4), 83–89. doi:10.14745/ccdr.v45i04a02

Bowie, M. V., de la Fuente, J., Kocan, K. M., Blouin, E. F., & Barbet, A. F. (2002). Conservation of major surface protein 1 genes of *Anaplasma marginale* during cyclic transmission between ticks and cattle. *Gene*, *282*(1-2), 95–102. doi:10.1016/S0378-1119(01)00845-9 PMID:11814681

Bradway, D. S., Torioni de Eschaide, S., Knowles, D. P., Hennager, S. G., & McElwain, T. F. (2001). Sensitivity and specificity of the complement fixation test for detection of cattle persistently infected with *Anaplasma marginale*. Journal of. Veterinary. Diagnostic atic. *Investigation.*, *13*, 79–81. PMID:11243370

Brizuela, C. M., Ortellado, C. A., Sanabria, E., Torres, O., & Ortigosa, D. (1998). The safety and efficacy of Australian tick-borne disease vaccine strains in cattle in Paraguay. *Veterinary Parasitology*, *76*(1-2), 27–41. doi:10.1016/S0304-4017(97)00047-2 PMID:9653988

CABI Data Mining. (2001). CAB Abstracts Data Mining. CABI, current year. *Fallopia japonica*. In *Invasive Species Compendium*. Wallingford, UK: CAB International. www.cabi.org/isc

Carelli, G., Decaro, N., Lorusso, A., Elia, G., Lorusso, E., Mari, V., Ceci, L., & Buonavoglia, C. (2007). Detection and quantification of *Anaplasma marginale*DNA in blood samples of cattle by real-time PCR. *Veterinary Microbiology*, *124*(1-2), 107–114. doi:10.1016/j.vetmic.2007.03.022 PMID:17466470

Constable, P., Hinchcliff, K., Done, S., & Gruenberg, W. (2016). *Veterinary Medicine: A Textbook of the Diseases of Cattle*. Horses, Sheep, Pigs and Goats.

De la Fuente, J., Lew, A., Lutz, H., Meli, M. L., Hofmann-Lehmann, R., Shkap, V., Molad, T., Mangold, A. J., Alamaza'n, C., Naranjo, V., Gorta'zar, C., Torina, A., Caracappa, S., Garcı'a-Pe'rez, A. L., Barral, M., Oporto, B., Ceci, L., Carelli, G., Blouin, E. F., & Kocan, K. M. (2005). Genetic diversity of Anaplasma species major surface proteins and implications for anaplasmosis sero-diagnosis and vaccine development. *Animal Health Research Reviews*, *6*(1), 75–89. doi:10.1079/AHR2005104 PMID:16164010

De Waal, D. T. (2000, December). Anaplasmosis control and diagnosis in South Africa. *Annals of the New York Academy of Sciences*, *916*(1), 474–483. doi:10.1111/j.1749-6632.2000.tb05327.x PMID:11193662

Dumler, J. S., Barbet, A. F., Bekker, C. P. J., Dasch, G. A., Palmer, G. H., Ray, S., Rikihisa, C. Y., & Rurangirwa, F. R. (2001). Reorganization of the genera in the families Rickettsiaceae and Anaplasmataceae in the order Rickettsiales: Unification of some species of *Ehrlichia* with *Anaplasma,Cowdria* with *Ehrlichia* and *Ehrlichia* with *Neorickettsia*, descriptions of six new species combinations and designation of *Ehrlichia equi* and "HGE agent" as subjective synonyms of *Ehrlichia phagocytophila. International Journal of Systematic and Evolutionary Microbiology*, *51*(6), 2145–2165. doi:10.1099/00207713-51-6-2145 PMID:11760958

Duzgun, A., Schunter, C. A., Wright, I. G., Leatch, G., & Waltisbuhl, D. J. (1988). A sensitive ELISA technique for the diagnosis of *Anaplasma marginale* infections. *Veterinary Parasitology*, *29*(1), 1–7. doi:10.1016/0304-4017(88)90002-7 PMID:3176298

Evans, D. E. (1992). Tick infestsation of livestockand tick control methods in Brazil: A situation report. *Insect Science and Its Application*, *13*(4), 629–643.

Figueroa, J. V., Chieves, L. P., Johnson, G. S., & Buening, G. M. (1993). Multiplex polymerase chain reaction based assay for the detection of *Babesia bigemina, Babesia bovis* and *Anaplasma marginale* DNA in bovine blood. *Veterinary Parasitology*, *50*(1-2), 69–81. doi:10.1016/0304-4017(93)90008-B PMID:8291198

Gale, K. R., Dimmock, C. M., Gartside, M., & Leatch, G. (1996). *Anaplasma marginale*: Detection of carrier cattle by PCR-ELISA. *International Journal for Parasitology*, *26*(10), 1103–1109. doi:10.1016/S0020-7519(96)80009-9 PMID:8982791

Handistatus, O. I. E. (2002). *World Animal Health Publication and Handistatus II (dataset for 2001).* Office International des Epizooties.

Handistatus, O. I. E. (2003). *World Animal Health Publication and Handistatus II (dataset for 2002).* Office International des Epizooties.

Handistatus, O. I. E. (2004). *World Animal Health Publication and Handistatus II (data set for 2003).* Office International des Epizooties.

Handistatus, O. I. E. (2005). *World Animal Health Publication and Handistatus II (data set for 2004).* Office International des Epizooties.

Horak, I. G., Camicas, J.-L., & Keirans, J. E. (2002). The Ixodidae,Amblyommidae and Nuttalliellidae (Acari: Ixodida): a world list of valid tick names. *Experimental & Applied Acarology*, *28*(1-4), 7–54. doi:10.1023/A:1025381712339

Hungerford, L. L., & Smith, R. D. (1997). Variations in seroprevalence and host factors for bovine anaplasmosis in Illinois. *Veterinary Research Communications*, *21*(1), 9–18. doi:10.1023/B:VERC.0000009696.68224.a8 PMID:9060138

Jonsson, N. N., Mayer, D. G., Matschoss, A. L., Green, P. E., & Ansell, J. (1998). Production effects of cattle tick (*Boophilus microplus*) infestation of high yielding dairy cows. *Veterinary Parasitology*, *78*(1), 65–77. doi:10.1016/S0304-4017(98)00118-6 PMID:9703620

Kocan, K. M., Blouin, E. F., & Barbet, A. F. (2000). Anaplasmosis control. Past, present, and future. *Annals of the New York Academy of Sciences*, *916*(1), 501–509. doi:10.1111/j.1749-6632.2000.tb05329.x PMID:11193665

Kocan, K. M., de la Fuente, J., Blouin, E. F., & Garcia-Garcia, J. C. (2004). *Anaplasmamarginale* (Rickettsiales: Anaplasmataceae):recent advances in defining host-pathogen adaptations of a tick-borne rickettsia. *Parasitology*, *129*(S1), S285–S300. doi:10.1017/S0031182003004700 PMID:15938516

Kocan, K. M., de la Fuente, J., Guglielmone, A. A., & Melendez, R. D. (2003). Antigens and alternatives for control of *Anaplasma marginale* infection in cattle. *Clinical Microbiology Reviews*, *16*(4), 698–712. doi:10.1128/CMR.16.4.698-712.2003 PMID:14557295

Kocan, K. M., Halbur, T., Blouin, E. F., Onet, V., Fuente, J. L., Garcia-Garcia, J. C., & Saliki, J. T. (2001). Immunization of cattle with *Anaplasma marginale* derived from tick cell culture. *Veterinary Parasitoloby, 102*, 151-161.

Marques, R., Krüger, R. F., Peterson, A. T., de Melo, L. F., Vicenzi, N., & Jiménez-García, D. (2020). Climate change implications for the distribution of the babesiosis and anaplasmosis tick vector, *Rhipicephalus* (*Boophilus*) *microplus. Veterinary Research (Faisalabad)*, *51*(1), 81. doi:10.118613567-020-00802-z PMID:32546223

Mason, C. A., & Norval, R. A. I. (1980). The ticks for Zimbabwe. 1. The genus Boophilus. *Zimbabwe Veterinary Journal*, *11*, 36–43.

Matthysse, J. G., & Colbo, M. H. (1987). *The ixodid ticks of Uganda*. Entomological Society of America.

Montenegro-James, S., & Guillen, A.T., Ma, S.-J., Tapang, P., Abdel-Gawad, A., Tori, M., & Ristic, M. (1990). Use of the dot enzyme-linked immunosorbent assay with isolated *Anaplasma marginale* initial bodies for serodiagnosis of anaplasmosis in cattle. *American Journal of Veterinary Research*, *51*, 1518–1521. PMID:2240769

Montenegro-James, S., James, M. A., & Ristic, M. (1985). Modified indirect fluorescent antibody test for the serodiagnosis of *Anaplasma marginale* infections in cattle. *American Journal of Veterinary Research*, *46*, 2401–2403. PMID:3907438

Murrell, A., & Barker, S. C. (2003). Synonomy of *Boophilus*Curtice, 1981 with *Rhipicephalus* Koch, 1844 (Acari: Ixo didae). *Systematic Parasitology*, *56*(3), 169–172. doi:10.1023/B:SYPA.0000003802.36517.a0 PMID:14707501

Nazar, M., Khan, M. A., Shah, A. A., Rahman, S. U., Khan, I., Ullah, A., & Khan, I. U. (2018). Occurrence and transplacental transmission of Anaplasma marginale in dairy cattle. *Slovenian Veterinary Research*, *55*(3), 183–191. https://www.cabi.org/isc/datasheet/91734. doi:10.26873/SVR-499-2018

OIE. (2009). *World Animal Health Information Database - Version: 1.4. World Animal Health Information Database*. Paris, France: World Organisation for Animal Health. http://www.oie.int

OIE (World Organization for Animal Health). (2008). Chapter 2.4.1 Bovine Anaplasmosis. In *Manual of Diagnostic Tests and Vaccines for Terrestrial Animals* (6th ed., pp. 599-610). World Organization for Animal Health.

Palmer, G. H., Rurangirwa, F. R., & McElwain, T. F. (2001). Strain composition of the ehrlichia *Anaplasma marginale* within persistently infected cattle, a mammalian reservoir for tick transmission. *Journal of Clinical Microbiology*, *39*(2), 631–635. doi:10.1128/JCM.39.2.631-635.2001 PMID:11158120

Potgieter, F. T., & Stoltsz, W. H. (2004). Bovine anaplasmosis. In J. A. W. Coetzer & R. C. Tustin (Eds.), *Infectious Diseases of Livestock, 3* (2nd ed., pp. 594–616). Oxford University Press, Southern Africa.

Potgieter, F. T., & Van Rensburg, L. (1987). The persistence of colostral *Anaplasma* antibodies in incidence of in uterotransmission of *Anaplasma* infections in calves under laboratory conditions. *The Onderstepoort Journal of Veterinary Research*, *54*, 557–560. PMID:3444609

Reinbold, J. B., Coetzee, J. F., Hollis, L. C., Nickell, J. S., Riegel, C., Christopher, J. A., & Ganta, R. R. (2010). Comparison ofiatrogenic *Anaplasma marginale* transmission in Holstein steers by needle and needle-free injection techniques. *American Journal of Veterinary Research*, *71*(10), 1178–1188. doi:10.2460/ajvr.71.10.1178 PMID:20919904

Richey, E. J. (1981). Bovine anaplasmosis. In R. S. Howard (Ed.), *Current Veterinary Therapy Food Animal Practice* (pp. 767–772). W. B. Saunders.

Richey, E. J. (1991). Bovine anaplasmosis. *Proceedings of the 24th Annual Conference of the American Association of Bovine Practitioners*, 3-11.

Rodgers, S. J., Welsh, R. D., & Stebbins, M. E. (1994). Seroprevalence of bovine anaplasmosis in Oklahoma from 1977 to 1991. *Journal of Veterinary Diagnostic Investigation*, *6*(2), 200–206. doi:10.1177/104063879400600211 PMID:8068752

Rodriquez, M., Massard, C., Henrique da Fonseca, A., Fonseca, R. N., Machado, H., Labarta, V., & De La Fuente, J. (1995). Effect of vaccination with a recombinant Bm86 antigen preparation on natural infestations of Boophilus microplus in grazing dairy and cross-bred cattle in Brazil. *Vaccine*, *13*(18), 1804–1808. doi:10.1016/0264-410X(95)00119-L PMID:8701597

Smith, R. D., Levy, M. G., Kuhlenschmidt, M. S., Adams, J. H., Rzechula, D. L., Hardt, T. A., & Kocan, K. M. (1986). Isolate of Anaplasma marginale not transmitted by ticks. *American Journal of Veterinary Research*, *47*, 127–129. PMID:3946891

Texas AgriLife Extension Service. (1997). *Integrated pest management of flies in Texas dairies*. animalscience.tamu.edu/images/pdf/dairy/dairy-integrated-pestmgmt

Torioni De Echaide, S., Knowles, D. P., Mcguire, T. C., Palmer, G. H., Suarez, C. E., & Mcelwain, T. F. (1998). Detection of cattle naturally infected with *Anaplasma marginale* in a region of endemicity by nested PCR and a competitive enzyme-linked immunosorbent assay using recombinant major surface protein 5. *Journal of Clinical Microbiology*, *36*(3), 777–782. doi:10.1128/JCM.36.3.777-782.1998 PMID:9508311

Watson, D. W., Waldron, J. K., & Rutz, D. A. (1994). *Integrated management of flies in and around dairy and livestock barns*. Academic Press.

Wickwire, K. B., Kocan, K. M., Barron, S. J., Ewing, S. A., & Smith, R. D. (1987). Infectivity of three *Anaplasma marginale* isolates for *Dermacentor andersoni. American Journal of Veterinary Research, 48*, 96–99. PMID:3826850

Zable, T.A. & Agusto, F.B. (2018). *Transmission dynamics of Bovine Anaplasmosis in a cattle herd interdisciplinary perspectives in infectious diseases*. Academic Press.

Chapter 11
Heartwater and Control

James Wanjohi
Kenya Agricultural and Livestock Research Organisation, Kenya

Sebastian K. Waruri
Kenya Agricultural and Livestock Research Organisation, Kenya

Sammy Gichuhi Ndungu
Kenya Agricultural and Livestock Research Organisation, Kenya

Leonard Muchenditsi Khaluhi
Kenya Agricultural and Livestock Research Organisation, Kenya

Erick M. Mungube
Kenya Agricultural and Livestock Research Organisation, Kenya

Moses Okoth Olum
Kenya Agricultural and Livestock Research Organisation, Kenya

ABSTRACT

Heartwater (HW) is an acute, febrile, tick-borne disease of cattle, sheep, goats, and wild ruminants characterized by nervous signs and high mortality. The disease is caused by a rickettsia agent, Erlichia ruminantium, formally classified as Cowdria ruminantium. The disease is transmitted by several ixodid ticks of the genus Ambylomma. Chemoprophylaxis has been used as a method to facilitate the movement of heartwater susceptible stock into heartwater endemic areas while allowing them to acquire immunity by limited tick exposure.

INTRODUCTION

Heartwater (HW) is an acute, febrile, tick-borne disease of cattle, sheep, goats and wild ruminants characterized by nervous signs and high mortality (Camus et al. 1996). The disease is caused by a rickettsia agent, Eh*rlichia ruminantium* (Dumler et al. 2001). The disease was formerly classified as *Cowdria ru-*

DOI: 10.4018/978-1-7998-6433-2.ch011

minantium (Cowdry, 1925). The disease is transmitted by several 3-host ticks of the genus *Ambylomma* (Henning, 1956). Although therapeutic drugs are available, treatment often fails because the disease usually runs a very rapid course.

Heartwater was first diagnosed in South Africa by Lois Trichardt in 1838 in his sheep 3 weeks after they had a heavy tick infestation. In Kenya the disease was first reported By Daubney (1939-40). Documentation of heartwater in Kenya like most other African countries is negligible and comprises annual Veterinary reports and from authors who have attempted to carry out some studies on heartwater (Kocan et al. 1987).

Heartwater is not a notifiable disease in most countries and is overshadowed by the threat from East Coast fever, which is the major tick-borne disease limiting livestock production in many countries in East, Central and Southern Africa. Few actual cases of heartwater are reported occasionally by field Veterinarians and other workers as a by- the way even in areas where *Ambylomma* ticks are endemic.

Since the tick vector is extremely sensitive to acaricides, strict tick control has caused the vector to all but disappear, together with the disease from farmlands where a majority of ruminants are kept.

In *Ambylomma*-endemic areas, the disease incidence in indigenous animals is low and most farmers are not encouraged to keep exotic animals, creating the impression that heartwater is unimportant. Coupled with the difficulty of diagnosing the disease clinically and at post mortem, most cases go unreported even with experienced staff.

Background

With the advent of ECF immunization, it is anticipated that farmer's response to strict regular acaricidal control of ticks will be reduced in favor of strategic control (Purnell, 1984). *Ambylyoma* ticks which had been eradicated in high potential areas are therefore likely to re-establish and spread heartwater in susceptible populations and cause massive livestock deaths. The A*bylomma* ticks are also likely to spread in other areas suitable for the tick survival. There is therefore the need to develop more effective vaccines for heartwater bearing in mind that there may be different antigenic variations of *Ehrlichia* parasites in different areas.

AETIOLOGY AND LIFE CYCLE

The causative agent of heartwater is a rickettsia with a trans-stadial transmission cycle in the *Amblyomma* ticks and a wide ruminant host population. It has been shown that the rickettsial organism is transmitted between ruminant and vector host while undergoing a simple development cycle in distinct reticulate bodies (Rbs), elementary bodies (EBs) and intermediate bodies (IBs), (Jongejan et al. 1991a). It is thought that single elementary bodies are picked up by the tick while feeding. These invade the gut epithelium where they multiply and invade other organelles via the haemolymph. They find their way to the salivary glands and from there, they are injected into the ruminant as single organisms that eventually find their way to the endothelium of blood vessels. In endothelium tissue culture, the infectious cycle lasts 5-6 days (Jongejan et al. 1991). They multiply greatly in the host endothelium and are released by disruption into circulation from where they are picked up by the tick as EBs colonies of the organism have so far been demonstrated in gut, salivary gland cells, haemocytes and malpighian tubules of infected Amblyomma ticks (Kocan and Bezuidenhout, 1987). With the ELISA test, the parasite was revealed in various tissues and haemolymph of adult *A. hebraeum.* The gut, salivary gl ands, hypodermis and synganglion showed the highest concentration of the parasites in females while in males it was in the salivary glands and gut (Viljoen et al. 1988).

The predominant type seen by Kocan and Bezuidenhout, (1987) in both nymphs and adult stages was the reticulated form that appeared to divide by binary fusion, electron dense forms were also present. Colonies of *Cowdria (Ehrlichia) ruminantium* in salivary glands and in haemolymph of feeding ticks had different morphologic forms of the organism. Forms similar to those documented in endothelial cells of the vertebrate host were seen (Kocan & Bezuidenhout, 1987). Colonies of*Cowdria (Ehrlichia) ruminantium* were studied in midgut epithelial cells and salivary gland acini of nymphal *Amblyomma hebraeum* that were infected experimentally as larvae. Colonies observed within gut cells frequently contained 2 types of the organism: electron-dense and reticulated forms (Kocan et al. 1987). The morphology of colonies from salivary glands, as seen with light microscopy, varied from compact, densely staining, small colonies to large ones in which individual organisms were apparent. Females of *A. hebraeum*, experimentally infected as larvae with the Ball 3 stock of *Cowdria (Ehrlichia) ruminantium,* had predominantly reticulate morphologic form of Cowdria bodies in gut epithelial cells after 1-3 days of feeding (Hart et al.1991). Single intermediate bodies and no elementary bodies were observed. Organisms were found within a membrane-bound vacuole and each organism had a double-unit membrane (Hart et al. 1991). Infrequently, colonies contained homogenous electron-dense inclusions. Groups of *Cowdria* organisms within a haemocyte suggested a possible dissemination of organisms from the gut to various other tissues by haemocytes (Hart et al. 1991).

Ultrastructural studies of *Cowdria (Ehrlichia) ruminantium* cultured *in vitro* concurred to a large extent with that in previous *in vivo* studies. Two distinct forms, elementary and reticulate bodies, and a 3rd group of intermediate organisms were identified (Prozesky et al. 1986). Organisms within a particular vacuole were generally a specific form, but in cells containing many colonies different forms were present in the same colony. Most organisms were enveloped by two membranes and a few were surrounded by a 'capsule' (Prozesky et al. 1986). It was concluded that *Cowdria (Ehrlichia) ruminantum* multiples mainly by binary fusion, but that multiplication can also take place by means of budding (Prozesky et al. 1986). After density-gradient centrifugation of crude suspensions of infected sheep brain and nymphs of *Ambylomma hebraeum* ticks, *Cowdria* was seen by electron microscopy in all density fractions (Neitz et al. 1986). Different populations of *Cowdria (Ehrlichia)* were presemt at different densities, with larger organisms occurring at lower densities. There was no significant ultrastructural difference between *Cowdria* from brain and ticks, apart from the presence of a single defined electron-transparent region in brain isolates; organisms in ticks had several electron transparent areas (Neitz et al. 1986). Colonies were not observed in gut tissues obtained from nymphs during the feeding period in *Amblyomma variegatum* that had become infected by feeding as nymphs on cattle (Kocan. Et al. 1987), colonies were small (1-10aem) initially, but as tick development progressed their diameter increased to as much as 60aem. Many colonies contained a large, electron-base inclusion that was morphologically similar to haemoglobin deposits found in the cytoplasm of midgut epithelial cells of recently fed ticks. *Cowdria (Ehrlichia) ruminantium* was often observed adhered to these inclusions (Kocan et al. 1987).

Although there is a lack of information on the development of *Ehrlichia ruminantium* there is some evidence that the parasite undergoes a sequential development in both the vertebrate and invertebrate hosts (Plessis, 1982). The organism replicates mainly by binary fission, and possibly by endosporulation (Pienaar, 1970). It appears that the reticulate bodies are predominately proliferative, while the elementary bodies represent the infective Stage (Jongejan et al. 1990). Transmission electron microscope studies of in vitro-cultivated organisms demonstrated the presence of intracellular reticulate bodies 2 to 4 days after Infection and intermediate bodies 4 to 5 days after infection. Large numbers of elementary bodies are seen after rupture of endothelial cells 5 to 6 days after infection (Jongejan et al. 1991).

The host *Amblyomma* spp. ticks become infected during the larval and nymphal stages when they feed on infected domestic and wild ruminants, and possibly also on certain game birds and reptiles while *E. ruminantium* is circulating in the blood of these hosts. Nymphae and adult ticks transmit *E. ruminantium* to susceptible hosts without losing the infection. Intrastadial transmission has been demonstrated (Andrew &. Norval, 1989), but transovarial transmission has only been reported once under laboratory conditions and probably does not occur in nature (Bezuidenhout & Jacobsz, 1986). The development cycle of the organism in the tick and the infectivity of successive stages of the tick are poorly understood. It is thought that after an infected blood meal, initial replication of the organism takes place in the intestinal epithelium of the tick and that the salivary glands eventually become parasitized (Kocan et al .1987b). Transmission of the parasite to the vertebrate host probably takes place either by regurgitation of their gut contents or through the saliva of the tick while feeding. The minimum period required for transmission of the parasite after tick attachment is between 27-38 hours in nymphs and 21-75 h in adults (Bezuidenhout, 1988).

Disease Transmission

Heartwater occurs when an infected *Amblyomma* tick feeds on a susceptible ruminant. The immatures of *Amblyonma* species feed on a wide range of hosts while adults are more host specific (Petney et al., Rechav, 1987). A number of species require a specific intermediate host making them less efficient transmitters of heartwater than others. There is one report of progeny of *A.hebraeum* females that had become infected as larva which transmitted the infection as larvae, nymphs and adults (Bezuidenhout & Jacobsz, 1986).Except for this report of transovarial transmission, transmission is transstadial (Bezuidenhout, 1987). *A.dissimile* larvae transmitted *C. ruminantium* as nymphs to a goat in the laboratory transovarialy (Jongejan, 1992; Barre & Camus 1987). *C.ruminantium* isolated from a naturally infected steer in Zambia was transmitted by *Amblyomma variegatum* nymphs previously fed on an infected goat as larvae (Musisi & Hussein, 1984). Engorged *A. hebraeum* nymphs allowed to moult to adults were fed in small pools of heartwater-susceptible sheep, some of which became infected. (Norval et al. 1990).

In transmission experiments with Friesian calves and Dutch goats *Amblyomma* spp. from various tropical countries, *A. astrion* from West Africa infected in the larval stage with *Cowdria (Ehrlichia) ruminantium* were able to transmit the rickettsia trans-stadially. *A. cohaerens* from East Africa transmitted *Cowdria (Ehrlichia) ruminantium* trans-stadially from larva to nymph, nymph to adult and larva to adult. A single individual of *A. cajennense* (F.) from Mexico infected in the larval stage with *Cowdria (Ehrlichia) ruminantium* transmitted it to a goat in the nymphal stage; this species appears to be a poor vector. Experiments with *A. lepidum* Don. From Tanzania (infected as larvae) and *A. hebraeum* .Koch from South Africa (infected as nymphs) showed that these ticks were able to transmit *T.velifera* transtadially (Uilenberg, 1983). Males transferred from live or dead hosts (sheep or goats) to live hosts (sheep or cattle) were shown to transmit *Cowdria (Ehrlichia).ruminantium* repeatedly (Andrew & Norval, 1989). This shows that they are able to transmit disease both interstadially and intrastadially (Garris, 1984; Andrew & Norval, 1989).

Ticks may obtain the infection while feeding on reacting animals, subclinically infected hosts or on immune animals after reinfection (Bezuidenhout, 1987). Infectivity is increased in infected ticks during feeding (Bezuidenhout, 1987). The demonstration of *Cowdria(Ehrlichia) ruminantium* in salivary glands of ticks suggests that transmission takes place via the saliva and that regurgitation from the gut may not be as important as previously thought (Bezuidenhout, 1987). The infection rate of ticks in the field is difficult to estimate because *Cowdria (Ehrlichia)* is not an easy organism to demonstrate, but it is thought to be generally low depending on the endemic status in an area. Engorged *A.hebraeum* nymphs collected from the field and allowed to moult to adults were fed in small pools on heartwater-susceptible sheep, some became infected. (Norval et al. 1990). The infection rates of these ticks estimated statistically were 0.0-44.9% for males, 20.0-36.1 for females and 0.0-13.4% for nymphs. These rates were considerably higher than those previously believed to occur (Norval et al. 1990).

That heartwater may be transmitted by vectors other than *Amblyomma* ticks is contested by an *Ehrlichia*-like agent isolated from an adult *Hyalomma truncatum* female that was confirmed to have a close relationship to *Cowdria* using the indirect fluorescent antibody test. After being passed through 3 generations of *A. hebraeum*. It elicited a disease in sheep indistinguishable from heartwater (Du Plessis, 1990). Also *Cowdria* persisted in the haemolymph of *Ornithodoros coriaceus* for at least 2 years (Endris et al., Birnie & Hess 1989) and *Melophagus ovinus* were suspected to be the vectors in one outbreak of sheep where they were present and no ticks could be found (Logan et al. 1989)

Epidemiology

Factors relating to the tick vector, causative organism, and vertebrate host are important in the epidemiology of heartwater. These include possible immunological strain differences of *E. ruminantium*, availability of wild animal reservoir hosts or vectors for the organisms, infection rate in ticks, age and genetic resistance of domestic ruminant populations, seasonal changes influencing tick abundance and activity, and the intensity of tick control (Uilenberg, 1983).

Figure 1. Ambylomma tick species

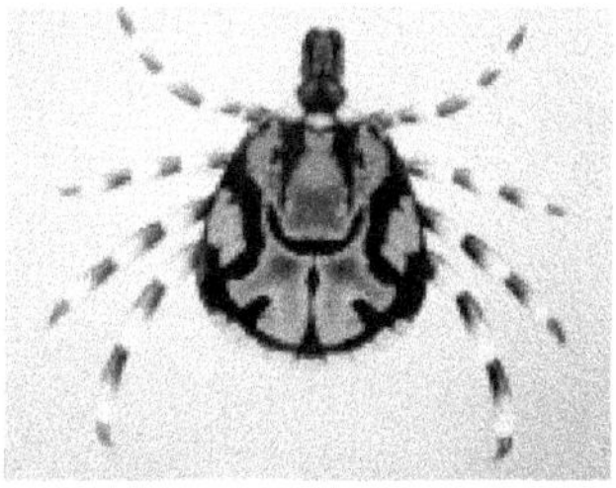

. The main vectors of heartwater are *Amblyomma variegatum* and *A. hebraeum,* although a number of other *Amblyomma* sp have been shown experimentally to be able to transmit the organism. *A. maculatum*, occurring in the USA is also capable of transmitting the disease (Uilenberg, 1982). Not all are equally good vectors,and their importance in the transmission of heartwater depends not only on their vector competence, but also on their distribution and association with domestic stock (Uilenberg, 1983). Furthermore, the activity and abundance of the ticks is influenced by temperature and humidity (Petney et al., 1987). Ticks can acquire the infection from the host from about the time of the febrile reaction for up to 361 days, or even longer (Andrew and Norval, 1989) and probably retain their infectivity for life (Neitz, 1968; Ilemobade, 1976). Infection rates in ticks vary, from 0-44.9% for males, 20-36.1% for females and 0-13.4% for nymphs, depending on the season and the locality in which they are collected (Plessis, 1985; Plessis& Malan 1987; Norval et al., 1990).

The existence of antigenically different stocks of *E. ruminantium* with varying virulence has been demonstrated. There is also variable cross-protection between these different varieties (Jongejan et al., 1988; Plessis et al., 1989). The introduction of animals which are immune to a particular variant of *E. ruminantium* into an endemic area where a different variety occurs may therefore result in cases of heartwater.

Various factors such as species, breed, age, degree of natural resistance and immune status play a role in determining whether asymptomatic or overt disease will develop in a susceptible host after infection. Young calves, lambs and kids possess a non-specific resistance which is independent of the immune status of the dam and is of short duration: the first 4-6 weeks of life in calves and only the first week in lambs and kids (Neitz & Alexander, 1941; Alexander et al. 1946; Uilenberg, 1981; Plessis et al., 1987). The susceptibilities of different breeds of cattle and sheep vary. Some sheep breeds, such as the Blackhead Persian, possess a certain degree of natural resistance (Alexander, 1931; Uilenberg, 1983). Angora goats are highly susceptible to heartwater and their immunity is of short duration (Plessis et al. 1983). Genetic resistance, which is due to a recessive sex-linked gene, has been demonstrated in Creole goats in Guadeloupe (Matheron et al. 1987).

Wild animals have been shown to be involved in the maintenance and transmission of heartwater (Oberem & Bezuidenhout,1987).They may have been the original host of *cowdria(Ehrlichia)* and most no longer suffer from clinical heartwater (Uilenberg,1981). The introduction of domestic ruminants in Africa introduced new susceptible hosts leading to epidemic outbreaks of heartwater and natural selection of the less susceptible cattle, sheep and goats. The next introduction of susceptible hosts must have been during the colonization period when settlers brought their exotic animals with them. *Cowdria (Ehrlichia) ruminantium* has been demonstrated in a wide range of hosts over the years from mammalian to avian and reptiles (Horak et al. 1987., Petney & Horak 1988).

An old African elephant that died was diagnosed with *Cowdria (Ehrlichia) ruminantium* in stained impression smears from the elephant brain. The elephant was thought to have died of Anthrax and heartwater (Okewole et al. 1993).

Albino mice have long been known and used as laboratory animals for virulent and latent strains of *Cowdria (Ehrlichia)* (Haig, 1952; Du Plessis, 1985., MacKenzie & McHardy, 1987). Other species suspected to be susceptible to heartwater but lacking in definitive proof are nilgai (Boselaphus tragocamelus), fallow deer (Dama dama), Himalayan tahr (Hemitragus jemlahicus), barbary sheep (Ammotragus lervia), mouflon (Ovis aries), blackbuck (Antilope cervicapra), and white and black rhinoceros. Other animals known to harbor Cowdria are the leopard tortoise and the helmeted guinea fowl (Oberem & Benzuidenhout, 1987).

Of ruminants in the western hemisphere deer*(Odocoileus virginianus)* experimentaly infected with *Cowdria(Ehrlichia) ruminantium* developed clinical signs some dying of heartwater while some responded to treatment(Dardiri et al. 1987).

It was once believed that the guinea fowl and leopard tortoises were the non-ruminant hosts of *E. ruminantium*, but recent data have confirmed that these species are not susceptible and do not transmit to vector ticks that usually feed on them.

RISK FACTORS AND IMMUNE MECHANISMS

Immunity to heartwater has been shown to be dependent of cell mediated immunity. This was demonstrated by the inability of athymic nude mice to make a drug-aided recovery from infection with either the Kumm or the Welgevonden stocks of *Cowdria (Ehrlichia) ruminantium* and their inability to mount an immune response (Du Plessis, Berche and Van Gas, 1991). Spleen cells transferred from Welgevonden immune mice protected the nude mice suggesting that the immunity in heartwater is cell mediated. The lyt-2+ cells were shown to be a necessary component whereas L3T4 T cells did not have a significant effect on the protection conferred (Du Plessis, Berche & Van Gas, 1991) Further supportive evidence was obtained when the number of Lyt-2+ T cells increased 3-fold in the circulating blood of mice challenged with the Welgevonden stock of *Cowdria (Ehrlichia).ruminantium* they had been infected with (Du Plessis et al. 1992). Lyt-2+ cells increased only after reinfection of the mice indicating that the development of immunity to heartwater is dependent on the unhindered and adequate replication of *Cowdria (Ehrlichia) ruminantium* (Du Plessis et al 1992).

The resistance of cattle to heartwater from calf to adult-hood was shown to have both specific and non- specific components (serum K), which takes effect from 5 months and persists for life (Du Plessis, 1985).

Protective and Acquired Immunity

The role of acquired immunity and the value of exposing calves to *Cowdria (Ehrlichia)* at an early age was demonstrated when a group of young calves that had been exposed early under different husbandry management was compared with calves artificially infected at an average

Age of 12 months. Whereas no calves died in the first group more than 50% died of heartwater in the second group. The resistance of cattle to heartwater from calf to adulthood was seen to have both specific and non-specific components (serum K) which takes effect from 5 months and persists for life (Du Plessis, (1985). There was no parallel between the prevalence of heartwater on the farms and the immune status of the animals. The seropositivity of the animals was determined largely by the tick loads to which they were subjected and that the influence of the tick infection rate was less evident as higher tick counts were recorded in herds where strategic tick control was practiced than on farms where total tick control programme was practiced(Du Plessis & Malan 1987)

Innate Resistance

Young cattle, sheep and goats have partial innate resistance not related to the immunological status of the dam, Neits & Alexander 1941; Neitz & Adelaar 1946). Calves born to cows fully susceptible to heartwater were no different from calves bred in heartwater endemic areas, they resisted artificial infection with *Cowdria (Ehrlichia) ruminantum* irrespective of the natural challenge they had been exposed to (Du Plessis & Malan 1988). Blood collected from these calves 14-26 days after infection and sub-inoculated into mice revealed *Cowdria* (*Ehrlichia)* was in the blood of 8-10 calves that were displaying only mild to moderate fever (Du Plessis & Malan 1988).

The combination of a mouse model and the indirect fluorescent antibody test revealed considerable variation in the degrees to which calves became infected and reacted to artificial infection (Du Plessis & Malan 1988).

Certain breeds of cattle, sheep and goats are more susceptible than others. The Creole goat in the Caribbean are relatively susceptible while the Creole Zebu are highly resistant. Black face Persian sheep in South Africa are also more resistant than other sheep. Indigenous animals raise in endemic areas have acquired some resistance through natural selection while exotic animals being exposed for the first time, are fully susceptible (Matheron et al. 1987).

Antigenic and Immunogenic Variation

The expression of antigenic variation in different populations is a fact of many parasite species. In the early history of Cowdriosis (Ehrlichiosis), it was believed that the Ball 3 strain could immunize and protect against all other strains of *Cowdria (Ehrlichia)*. Emerging facts in cross immunity and antigenic studies point to the contrary being true of *Cowdria (Ehlichia)*. The first signs that there might be important strain differences in *Cowdria (Ehrlichia)* was when the Du Plessis & Kumm (1971) found a mouse infective isolate that was very pathogenic to sheep but did not infect cattle (Du Plessis, 1981).Other workers got similar results (Uilenberg et al. 1983) were of the opinion that the differences in reaction of goats in cross immunity studies were significant between African isolates of different geographical origin. The Ball 3 stock was first noted not to be as immunogenic as was previously thought when it failed to protect against the Senegalese strain although it was fully cross protective against the Zambian strain (Jongejan et al. 1988)

Again partial protection was observed between different strains used in the laboratory, (Stewart et al. 1989) conferred by the Ball 3 strain against two strains, the Caribbean (Gardel) isolate and west African (Mali) isolate in goats but a large proportion of the animals challenged with the Mali strain were not protected (Brown et al. 1989).

Du Plessis, and Van Gas, (1989) found only 74% protection in seropositive sheep and goats infected naturally in heartwater endemic areas of South Africa when challenged with the Ball 3 and Welgevonden stocks of *C.ruminantum*. They ascribed this to the poor cross-protection between stocks of *C.rminantium* as all seronegative animals in the indirect fluorescent antibody test used reacted when they were challenged.

A comparison of ten stocks of *Cowdria(Ehrlichia) ruminantium* (Ball 3, Breed, Comoro, Germishuys, Kumm, Kwanyanga, Mali, Mara, Nonile and Welgevonden) for cross-immunity, serological and pathogenicty in mice indicated there was varying degrees of differences between them Du Plessis et al.1989). The Ball 3, Comoro and Germishuys stocks were found to be similar but not identical. There was no pattern in the antigenic diversity to decipher the differences between the 10 stocks. The Welgevonden stock conferred the broadest spectrum of protection against the other South African stocks and by far exceded the reference Ball 3 stock (Du Plessis et al. 1989). The antigenic diversity of the 10 stocks could not be correlated with the antibody test, using the Kumm stock antigen since the sera against all 10 stocks reacted positively to and the titre variation was not stock-related (Du Plessis et al. 1989). Further, differences in pathogenicity and cross-immunity observed amongst 9 stocks could not be explained by examining identical or unique antigenic properties of two antigenically conserved proteins of *Cowdria (Ehrlichia) ruminantium* reacting with sheep or bovine antiserum (Rossouw et al. 1990)

Elsewhere, various degrees of cross-protection was found between 5 stocks of *Cowdria(Ehrlichia),* Ball 3, Kumm, Kwanyanga, Senegal, and Welgevoden), in goats and sheep indicating there was immunological and antigenic differences between them, (Jongejan et al. 1991). Complete lack of cross protection was observed between two of them, the Kumm and Kwanyanga stocks (Jongejan et al 1991). Within the confines on one farm 3 isolates of C.ruminantium that differed in pathogenicity and immunogenicity were recovered from ticks. One of these isolates was almost non-pathogenic to cattle (Du Plessis et al 1992). Sheep immune to the Ball 3, the Kwanyanga or the Mara stocks of *Cowdria (Ehrlichia) ruminantium* or a combination of these 3 stocks were protected against challenge with the Welgevonden stock, but not the Mali stock (Du Plessis et al. 1990). Complete absence of cross-immunity was found between a Dutch isolate of *E.phagocytophila* and antigenically distinct stocks of *Cowdria (Ehrlichia) ruminantium* from Senegal and South Africa, in goats, despite previous observations that both rickettsial organisms have certain antigenic determinants in common (Jongejan & Wassink, 1991).

Clinical Disease

Heartwater can present as a mild fever or as an acute lethal disease depending on the age, breed, species and immune status of the affected animal. Clinical signs include sudden pyrexia, which may be missed if not taken, later on depressed appetite, respiratory distress, nervous signs such as incoordination, muscular tremor, lip- licking and death may occur very rapidly (Van de Pypekemp & Prozesky,1982). The presence of the vector is usually an indicator of the likely indicator that the disease may be there. A postmortem having characteristic, though not pathognomonic hydropericardiam, hydrothorax and Ascites are usually present.

Infected domestic ruminants exhibit a wide range of clinical signs varying from a peracute to mild (clinically inapparent) form. The incubation period in naturally infected cattle ranges 9-29 days (average 18 days) and that of sheep and goats 7-35 days (average 14 days). Peracutely affected animals die within a few hours after the initial fever, either with or without any clinical signs (Alexander, 1931; Neitz, 1968; Uilenberg, 1983). Acute heartwater is the most common form of the disease in endemic areas. Fever of 40°C or higher, which usually persists for 3-6 days is followed by a drop to subnormal levels shortly before death. Animals gradually show inappetance and eventually stop feeding. Cessation of rumination and difficult breathing follow. Petechiae are visible on the mucous membranes of the conjunctiva (mainly cattle). Tachypnoea, abducted front limbs due to hydrothorax during the latter stage of acute heartwater, the majority of animals manifest nervous symptoms ranging from a mild incoordination to pronounced convulsions (Alexander, 1931). They are hypersensitive when handled or startled. The gait of affected animals becomes progressively more unsteady, whereas some animals show hypermetria, especially of the forelegs (mainly cattle). They eventually become prostrate, assume a position of lateral recumbency and show intermittent leg-paddling, chewing movements and licking of the lips. A large amount of froth is usually present at the mouth and nostrils. Diarrhoea is occasionally seen in cattle. Less severe cases (subacute and mild) occur with clinical signs ranging from slightly less intense than the acute form to little or no signs at all.

Figure 2. Sheep suffering from Heartwater

ECONOMIC IMPORTANCE OF HEARTWATER

Most authorities regard heartwater in southern Africa as an economically important disease (Uilenberg, 1983), ranked it second only to East Coast fever and tsetse-transmitted trypanosomosis. (Neitz, 1968) stated that in endemic areas, mortalities due to heartwater were three times as great as those due to babesiosis and anaplasmosis. However, there have been no definitive studies designed to quantify this importance. A study was conducted under the auspices of the UF/USAID/SADC heartwater research project to evaluate the economic impact of heartwater in Zimbabwe. The total annual losses were estimated at US $5.6 million (Mukhebi et al., 1999). Annual economic losses per animal in the commercial production system in Zimbabwe were 25 times greater than losses in the communal system. The greatest components of economic loss were acaricide costs (76%), followed by milk loss (18%) and treatment cost (5%). However, no other reliable figures are available on the economic impact of heartwater in the region.

Gross Pathology of Heartwater

Lesions in cattle, sheep and goats are similar, although quite variable in extent and some changes are commoner in certain species than in others. Effusion of body cavities, (hydropericardium, hydrothorax, and, in some cases a degree of ascitis) is a very common change in most fatal cases of heartwater. The transudate is usually transparent or slightly turbid, light yellow fluid that often coagulates on exposure to air. The volume of fluid ranges from 20 ml in goats, about 0.5 litres in sheep to several litres in cattle (Steck, 1928). A hydropericardium, as indicated by the name 'heartwater' is a striking change in most animals that die of the disease and is usually more pronounced in sheep and goats than in cattle (Henning, 1956). Oedema of the lungs is a regular finding and appears to be more severe in most animals that die peracutely from the disease (Pypekamp & Prozesky, 1987). The interlobular septa of the lungs, mediastinum and associated lymph nodes are oedematous and serous frothy fluid oozes from the cut surface of the lung. The trachea and bronchi often contain serofibrinous exudates, and their mucosae are congested, with petechiae and ecchymoses. Splenomegaly is present although less strikingly in sheep and goats. The cut surface is dark red in colour and has a pulpy consistency. In animals that die peracutely it is often impossible to make a diagnosis on macroscopical lesions alone; splenomegaly, epi- and endocardial haemorrhages are sometimes the only significant changes (Alexander, 1931). Hepatic lesions are less striking with only a mild hepatomegaly.

Figure 3. Oedemotous lungs in sheep suffering from heartwater

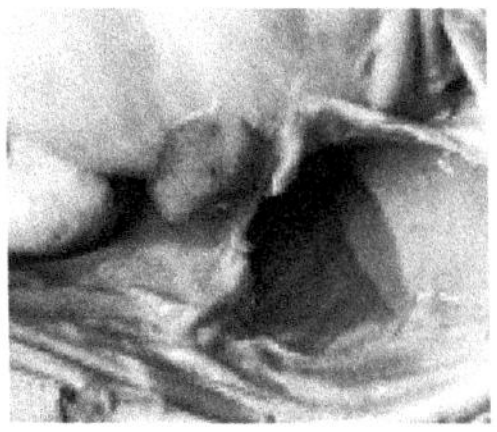

Laboratory Diagnosis

The confirmation of a diagnosis based on clinical signs and postmortem lesions requires the demonstration of the organisms in the cytoplasm of endothelial cells of blood vessels. The easiest, most efficient and quickest way of doing this is to visualize them in stained smears of the brain (Purchase, 1945) although they may also be found in histological sections such as the brain and kidneys. The examination of brain biopsies in live animals for the confirmation of a diagnosis of heartwater is useful in experimental animals, but is not practical under field conditions (Synge, 1978; Camus & Barré, 1982; Amstel, 1987).

Figure 4. Ehrlichia Ruminantium colonies in the brain of an infected sheep

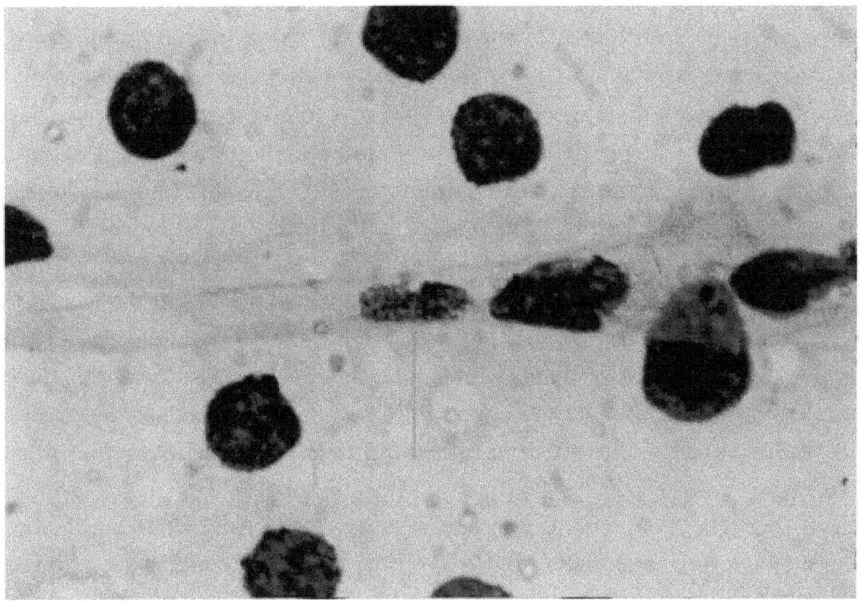

Smears should be air-dried before staining, and stains such as Giemsa or the CAM's Quick stain usually give the best results.

Serology

The indirect immunofluorescence test (IFA test) is not used anymore (Plessis and Malan, 1987a, b). Many attempts at developing a diagnostic serological test for heartwater have failed due to the high degree of cross-reaction occurring between antigens from different stocks of *Ehrlichia* and antibodies against Cytoecetes phagocytophila, and some *Ehrlichia* spp. *(Ehrlichia equi, E. canis, E. ovina* and *E. bovis*(Camus, 1987; Holland et al.; Plessis & Malan,). To minimize the degree of cross-reaction, two ELISA were developed using recombinant MAP-1. The first one is an indirect ELISA, ELISA MAP1-B using an immunogenic fraction of MAP-1, the recombinant antigen MAP1-B (Van Vliet et al, 1985). The second one is a competitive ELISA using MAP-1 gene cloned in baculovirus and monoclonal antibodies raised against MAP1 (Katz et al., 1997). Both tests improved the specificity of the tests. Low sero-positivity of cattle (even cattle that had previously also been vaccinated) occurs in a heartwater endemic areas. The antibody detection is possible 2 weeks after natural infection and lasts few months in naturally infected domestic ruminants moreover the period is shorter for bovine than for small ruminants. Serology as a diagnostic tool for detecting individual animals exposed specifically to *E. ruminantium* is unreliable. Serological analysis should be considered at herd level taking into account the epidemiological environment and should be complemented by molecular diagnosis.

Molecular Diagnosis

There have been significant improvements in the development of molecular tools for the diagnosis of heartwater and the genetic typing of the different strains of *E. ruminantium*. Concerning the molecular diagnosis, two primers, AB128 and AB129 have been designed to target a fragment of a unique and specific gene of *E. ruminantium*, *pCS20*. These primers amplify a 280 pb *pCS20* fragment which is revealed by a labelled *pCS20* probe (Peter et al., 1995). The PCR/hybridization allows increasing the sensitivity of the method with an experimental detection threshold of 1 to 10 organisms per sample. However the sensitivity of the PCR assay is lower and drops to 61% and 28% with tick samples containing 10^3 and 10^2 organisms, respectively (Peter et al., 2000). A hemi-nested PCR targeting *pCS20* fragment of gene was developed using as external primers AB128 and a new primer AB130 followed by a second run on first PCR product using AB128 and AB129 (Martinez et al., 2004). The detection limit (6 organisms per sample) is similar to the PCR/hybridization method described above but the nested PCR method is easier and less time consuming. The diagnosis of heartwater based on examination of brain smears from dead ruminants is much less sensitive than by molecular diagnosis. As an example, the comparison of methods, brain smear observations and *pCS20* nested PCR on the same brain samples, demonstrated the improvement of detection threshold with a percentage of heartwater positive cases after brain smears observations of 75% compare to 97% by *pCS20* nested PCR (Adakal et al., 2010). The main disadvantage with nested PCR is the higher contamination risk. The range of strain detection was increased by the use of primers including universal nucleotides AB128' AB130' and AB129' and this method is used routinely for *E. ruminantium* detection in field samples, especially in ticks (Molia et al., 2008, Adakal et al 2010). The *pCS20* hemi-nested PCR allowed detection in organs (lung and brain) from infected dead animals, blood from infected animals during hyperthermia and ticks fresh, frozen or preserved in 70% ethanol. The detection of *E. ruminantium* by nested PCR is possible in the blood of animals 1 or 2 days before hyperthermia and during the hyperthermia period but not on asymptomatic animals. PCR based methods appear to be more reliable in detecting infection in ticks and this could have epidemiological value in determining the *E. ruminantium* prevalence in ticks and the geographical distribution of *E. ruminantium*.

Several quantitative real time PCRs have been developed for the detection of *E. ruminantium* targeting *map-1*, *map1-1* and *pCS20* (Peixoto et al., 2005, Postigo et al., 2002, Steyn et al., 2008). These methods allow the quantification of the pathogen with a similar sensitivity to nested PCR.*pCS20* real time PCR can be used for diagnosis due to its ability to detect up to 18 different *E. ruminantium* strains.

The genetic characterization and structure of *E. ruminantium* population at regional scale is essential in order to select potential vaccinal strains. The genetic typing of strains was previously done using RFLP on the polymorphic gene *map-1* after PCR amplification (Faburay et al., 2007b; Adakal et al., 2010). Based on the genome analysis of 2 different strains Gardel and Welgevonden, truncated and unique coding sequences specific of strains have been identified. This analysis allows the development of a differential strain-specific diagnosis using nested PCRs targeting 6 unique and 4 truncated

CDS (Vachiéry et al., 2008a). New multi-locus methods adapted to *E. ruminantium* recently have been validated such as multi-locus sequence typing (Adakal et al., 2009) and multi-locus variable number of tandem repeated sequence analysis (Pilet et al., 2012). These tools are recently used on field samples for molecular epidemiological studies.

Differential Diagnosis

The peracute form of heartwater can be confused with anthrax. The acute form may resemble rabies, tetanus, bacterial meningitis or encephalitis,cerebral babesiosis, anaplasmosis, cerebral trypanosomiasis, or theileriosis. It must also be differentiated from poisoning with strychnine, lead, ionophores and other myocardial toxins, organophosphates, arsenic, chlorinated hydrocarbons, or some poisonous plants. Accumulations of fluid similar to heartwater are also sometimes seen in heavy helminth infestations (haemonchosis).

Treatment of Hertwater

The first chemotherapeutic agent that was found to be effective against the disease was a sulfonamide drug, "Uleron", (Neitz, 1939). Later tetracyclines were found to be effective (Weiss,Hig & Alexander, 1952), and have currently been adopted for regular use. For treatment 1 ml/10 kg bwt of a long acting tetracycline if given early will cure the disease. Under field conditions, heartwater progresses too rapidly making chemotherapy frequently too late to save the animals. This is because of the difficulties encountered in early diagnosis when therapy is likely to be effective, (Van Amstrel & Oberem, 1987). However, the best use of chemotherapeutic agents has been made in immunization where infection is induced and treated predictably to enable immunity to develop. Chemoprophylaxis has been used as a method to facilitate the movement of heartwater susceptible stock into heartwater endemic areas while allowing them to acquire immunity by limited tick exposure. This is done by injecting a long acting oxytetracycline preparation once a week for 4 weeks, as demonstrated in Cape Province,South Africa(Purnell & Schroder,1984).

Tick Control

This method of disease control is fraught with many difficulties and does not provide a long term solution (Norval et al. 1991). Though effective, tick control is affected by cost of acarides and organizational problems, both political and economic. Regular acaricide application allows tick resistance to develop through selection pressure on the ticks. Also leaves the animals highly susceptible to the disease and also upsetting endemic stability for heartwater and other tick-borne diseases. Acaricides also cause environmental pollution which does not auger well for the practice (Norval et al. 1991). Not, withstanding these drawbacks, intensive application of acaricides remains the most dependable method of controlling tick infestations of livestock and the disease pathogens transmitted by them (Benzuidenhout & Bigalke, 1987). This is often combined with drugs, pasture management such as pasture spelling and zero grazing and breeding of cattle for tick resistance ((Walker, 1984). The efficacy of ivermectin, administered in a sustained release for mulation by intraruminal pumps at approximate daily dose rates of 20, 40 and 60 aeg/kg body weight was evaluated in 16 attle against induced infestations of 3 strains of adult Ambylomma hebraem. Engorged female ticks were mass- measured and incubated, and reproductive data recorded.

There was an increase in mortality of both male and female ticks compared to that of controls with increasing daily dose of ivermectin, and a decrease in the number of ticks engorging (Soll,Carmichael & Grosss,1987).Ticks fed on ivermectin-treated cattle had a smaller mass when engorged and laid smaller egg masses, both absolutely and as a propotion of engorged mass(Soll,Carmichael & Grosss,1987). Index of reproduction was reduced 100% at 60 aeg/kg/day, >99% at 40aeg/kg/day and 96% at 20 aeg/kg/day ((Soll,Carmichael & Grosss,1987). Differences occurred between the 3 strains of *A.hebraeum* used in the study, especially with regard to engorged mass and reproductive variables. This had practical implications on the application of sustained release of ivermectin on *A.hebraeum* with reference to the control of heartwater. In Botswana heartwater is a major problem of livestock in cattle sheep and goats and *Ambylomma hebraeum* is the main vector. It is considered that in addition to using the current methods of heartwater control, future methods may be in use of tick-derived vaccines (Windsor, 1987).

Immunization Against Heartwater

The only commercial vaccine available consists of the blood of sheep infected with live virulent *E. ruminantium* organisms (BALL3-strain) (Bezuidenhout, 1989). This vaccine is only used on a large scale in South Africa and to a very limited extent in other African countries.

Vaccination with heartwater-infected sheep blood will complement natural tick infection of young animals and also ensure immunity in those animals which escape natural infection. Since the vaccine contains live heartwater organisms, vaccinated, susceptible animals will develop variable degrees of heartwater, which may or may not require appropriate treatment. A number of different vaccination methods are used.

Calves vaccinated before the age of 4-6 weeks usually show no clinical signs but will nevertheless develop an immunity. Although young calves generally do not require any treatment, approximately 10-20% of calves may show a temperature reaction (fever), and a few of these may even develop clinical signs and therefore require treatment. The immunity which thus develops in calves is usually of shorter duration and it is important that the calves be exposed to infected ticks within 3 months. Newborn lambs and kids must be vaccinated within the first week of their lives to avoid clinical reactions and therefore the need for treatment.

It is important that animals be kept under daily observation for at least one month after vaccination, so that additional treatment can be given should any of the animals develop clinical signs. Vaccinated animals should be exposed to *Ambylomma*t ticks as soon as possible to boost their immunity. Pregnant animals should not be vaccinated. Susceptible pregnant animals introduced into heartwater endemic areas should be kept free from ticks until they have lambed/calved and thereafter they should be immunized. The vaccine is administered intravenously into the jugular vein at a dosage of 3 ml for cattle, sheep and goats, irrespective of their size or age (Combrink et al., 1997). The vaccine reaction usually occurs 1-3 weeks after vaccination. The peak of the reaction is normally accompanied by fever (40°C or higher, measured early in the morning), and maybe symptoms such as listlessness, poor appetite, and decreased milk yield, and even nervous symptoms. Animals which develop vaccine reactions must immediately receive appropriate treatment. Immunity to heartwater develops within approximately 4-6 weeks after immunization and immunized Animals will be protected against serious disease caused by most (but not all) naturally occurring strains of heartwater.

This method is risky (possible transmission of other pathogens and possible loss of animals), expensive (cold chain required for storage, close monitoring of animals during one month) and thus inadequate in low-input farming systems. Other experimental vaccines have been developed such as inactivated, attenuated vaccine and recombinant vaccines. The first one is the entire killed bacteria emulsified in oily adjuvant, ISA50. Inactivated vaccines have several advantages as they contain killed bacteria and their storage conditions are compatible with field use (-20°C or refrigerated). Two injections are necessary with one month delay and animals should be protected from tick infestation during at least these 2 months. After natural or experimental challenges, animals develop hyperthermia and are clinically affected. Inactivated vaccine emulsified in ISA50 had been tested both in experimental and field conditions and demonstrated its efficiency (Martinez et al., 1994; Mahan et al., 1995, 2001; Adakal et al., 2010).

Lack of vaccine efficiency is related to the diversity of strains within a restricted area. A vaccine including a second local strain improved significantly the protection in a field trial in Burkina Faso (Adakal et al., 2010). Improvement of the production of E. ruminantium antigen at industrial scale and evaluation of the minimal efficient dose of vaccine (35µg) gives the opportunity to produce this vaccine at low cost (0.11 euros per dose) (Marcelino et al., 2006; Vachiéry et al., 2006; Marcelino et al., 2007). As soon as regional isolates are available in culture after isolation, it could be possible to produce an inactivated vaccine including a cocktail of regional strains. The main difficulty is to choose the strains which could protect against other circulating strains. The choice will depend on genetic characteristics and markers which are not yet defined.

The second experimental vaccine is the in vitro attenuated vaccine. Senegal, Welgevonden and Gardel strains were attenuated by successive in vitro passages and demonstrated their efficiency in experimental conditions (Jongejan et al., 1991; Faburay et al., 2007). Senegal attenuated vaccine confers good protection against homologous strains but poor protection against heterologous strains. Welgevonden attenuated vaccine confers good protection in controlled conditions against homologous and four different strains but is not yet tested in field conditions. The main disadvantage of attenuated vaccines is the possible reversion to virulence. Moreover as any live vaccine, it should be stored in liquid nitrogen during cold chain storage.

The third vaccine has been developed based on prime DNA/boost recombinant protein vaccine (Pretorius et al., 2008). There was an efficient protective effect of a cocktail of 4 ORFs against homologous challenge but it did not give satisfactory results during field tick challenge. Moreover, simple intramuscular immunisation is not sufficient to induce protection, and the use of gene gun is necessary for prime DNA injection which is not suitable for large vaccination campaign.

FUTURE RESEARCH DIRECTIONS

One polymorphic gene has been identified as an efficient component of a recombinant vaccine against heartwater using prime/boost method (Pretorius et al., 2010). However, as this gene is polymorphic, a recombinant vaccine should include almost 3 different genotypes.

For any kind of vaccine, live, inactivated, attenuated or recombinant vaccines, the main problem is the presence of numerous strains in the field with high genetic diversity and the choice of vaccine strains genotypes depending on the region. The future research should involve identification of a broad based vaccine, which most likely would be a recombinant vaccine. In Kenya we are currently evaluating a local isolate of *Ehrlichia ruminantium* on its safety and efficacy.

CONCLUSION

There is a lot that is known about heartwater from its aetiology to its biology and its tick vectors, and in Africa most of this work in the past was done in South Africa. This is mainly because in South Africa, it is the most important livestock disease. However, following widespread vaccination against East Coast fever and relaxation of tick control, heartwater may become of great economic importance in many countries. Therefore all efforts should go towards development of a noval vaccine as the present methods of intensive tick control are not sustainable and with time ticks develop resistance to the available acaricides.

REFERENCES

Alexander, R., Neitz, W. O., & Adelaar, T. F. (1946). Heartwater. *Farming in South Africa*, *21*, 548–552.

Alexander, R. A. (1931). Heartwater - the present state of our knowledge of the disease. 17th Report of the Director of Veterinary Services and Animal Industry. *Union of South Africa*, *17*(1), 89–150.

Andrew, H. R., & Norval, R. A. I. (1989). The carrier status of sheep, cattle and African buffalo recovered from heartwater. *Veterinary Parasitology*, *34*(3), 261–266. doi:10.1016/0304-4017(89)90056-3 PMID:2617830

Benzuidenhout, J. D., Peterson, C. L., & Bernard, B. J. H. (1985). Article. *The Onderstepoort Journal of Veterinary Research*, *52*, 113–120. PMID:3900854

Bezuidenhout, J. D. (1988). *Certain aspects of the transmission of heartwater, the occurrence of the organism in ticks and in vitro culture* (DVSc Thesis). University of Pretoria.

Bezuidenhout, J. D., & Bigalke, R. D. (1987). The control of heartwater by means of tick control. *The Onderstepoort Journal of Veterinary Research*, *54*(3), 525–528. PMID:3329336

Bezuidenhout, J. D., & Jacobsz, C. J. (1986). Proof of transovarial transmission of Cowdria ruminantium by Amblyomma hebraeum. *The Onderstepoort Journal of Veterinary Research*, *53*(1), 31–34. PMID:3960489

Bezuidenhout, J. D., & Prozesky, L. (1994). Heartwater. In Infectious diseases of Livestock with special reference to Southern Africa. Cape Town, South Africa: Oxford University Press.

Carasco-Lacombe, C., Lefrançois, T., Meyer, D. F., Martinez, D., Boulouis, H. J., & Haddad, N. (2012). A new typing technique for the Rickettsiales Ehrlichia ruminantium: Multiple-locus variable number tandem repeat analysis. *Journal of Microbiological Methods*, *88*(2), 205–211. doi:10.1016/j.mimet.2011.11.011 PMID:22143037

Combrink, Waal, & Troskie. (1997). Evaluation of a 3 ml heartwater (cowdriosis) infective blood vaccine dose. *The Onderstepoort Journal of Veterinary Research*, *64*(4), 309–311. PMID:9551483

Conceiçao. (1949). *A heartwater (Rickettsia ruminantium) in Angola*. Academic Press.

Cowdry, E. V. (1925a). Studies on the etiology of heartwater I. Observation of a rickettsia, Rickettsia ruminantium (n. sp.), in the tissues of infected animals. *The Journal of Experimental Medicine*, *42*(2), 231–252. doi:10.1084/jem.42.2.231 PMID:19869049

Daubney, R. (1939). Heartwater (Rickettsia ruminantium). Department of Agriculture Kenya.

du Plessis, J. L. (1970). Pathogenesis of heartwater: I. Cowdria ruminantium in the lymph nodes of domestic ruminants. *The Onderstepoort Journal of Veterinary Research*, *37*(2), 89–95. PMID:4115310

du Plessis, J. L. (1982). *Mice infected with a Cowdria ruminantium-like agent as a model in the study of heartwater* (DVSc thesis). University of Pretoria.

du Plessis, J. L. (1985). A method for determining the Cowdria ruminantium infection rate of Amblyomma hebraeum: Effects in mice injected with tick homogenates. *The Onderstepoort Journal of Veterinary Research*, *52*(2), 55–61. PMID:3900855

du Plessis, J. L. (1990). Increased pathogenicity of an Ehrlichia-like agent after passage through Amblyomma hebraeum: A preliminary report. *The Onderstepoort Journal of Veterinary Research*, *57*(4), 233–237. PMID:2293132

du Plessis, J. L., Jansen, B. C., & Prozesky, L. (1983). Heartwater in Angora goats. I. Immunity subsequent to artificial infection and treatment. *The Onderstepoort Journal of Veterinary Research*, *50*(2), 137–143. PMID:6634087

du Plessis, J. L., & Malan, L. (1987). The application of the indirect fluorescent antibody test in research on heartwater. *The Onderstepoort Journal of Veterinary Research*, *54*(3), 319–325. PMID:3329322

Faburay, B., Geysen, D., Munstermann, S., Taoufik, A., Postigo, M., & Jongejan, F. (2007). Molecular detection of Ehrlichia ruminantium infection in Amblyomma variegatum ticks in The Gambia. *Experimental & Applied Acarology*, *42*(1), 61–74. doi:10.100710493-007-9073-2 PMID:17476576

Henning, M. W. (1956). Heartwater. In *Animal Diseases in South Africa* (3rd ed., pp. 1155–1178). Central News Agency Ltd.

Ilemobade, A. A. (1976). *Study on heartwater and the causative agent Cowdria ruminantium (Cowdry, 1925) in Nigeria* (PhD thesis). Ahmadu Bello University.

Jongejan, & Zandbergen, Wiel, Groot, & Uilenberg. (1991). The tick borne rickettsia Cowdria ruminantium has a Chlamydia-like developmental cycle. *The Onderstepoort Journal of Veterinary Research*, *58*(4), 227–237. PMID:1780122

Jongejan, F. (2007). Immunisation of sheep against heartwater in The Gambia using inactivated and attenuated Ehrlichia ruminantium vaccines. *Vaccine*, *25*(46), 7939–7947. doi:10.1016/j.vaccine.2007.09.002 PMID:17920167

Jongejan, F., & Zandbergen, T. A. (1990). *The tick-borne rickettsia Cowdria ruminantium has a Chlamydia-like developmental cycle. In: Tick host interaction and disease transmission with special reference to Cowdria ruminantium (Rickettsiales)* (PhD thesis). Utrecht University.

Kocan, K. M., & Bezuidenhout, J. D. (1987). Morphology and development of Cowdria ruminantium in Amblyo mma ticks. *The Onderstepoort Journal of Veterinary Research*, *54*(3), 177–182. PMID:3448559

Mahan, Andrew, Tebele, Burridge, & Barbet. (1995). Immunisation of sheep against heartwater with inactivated Cowdria ruminantium. *Research in Veterinary Science, 58*(1), 46-49.

Marcelino, I., Sousa, M. F. Q., Veríssimo, C., Cunha, A. E., Carrondo, M. J. T., & Alves, P. M. (2006). Process development for the mass production of Ehrlichia ruminantium. *Vaccine*, *24*(10), 1716–1725. doi:10.1016/j.vaccine.2005.08.109 PMID:16257480

Marcelino, I., Vachiéry, N., Amaral, A. I., Roldão, A., Lefrançois, T., Carrondo, M. J. T., Alves, P. M., & Martinez, D. (2007). Effect of the purification process and the storage conditions on the efficacy of an inactivated vaccine against heartwater. *Vaccine*, *25*(26), 4903–4913. doi:10.1016/j.vaccine.2007.04.055 PMID:17531356

Martinez, D., Lefrançois, T., Vachiery, N., & Frutos, R. (2009). MLST scheme of Ehrlichia ruminantium: Genomic stasis and recombination in strains from Burkina-Faso. *Infection, Genetics and Evolution*, *9*(6), 1320–1328. doi:10.1016/j.meegid.2009.08.003 PMID:19712754

Martinez, D., Maillard, J. C., Coisne, S., Sheikboudou, C., & Bensaid, A. (1994). Protection of goats against heartwater acquired by immunisation with inactivated elementary bodies of Cowdria ruminantium. *Veterinary Immunology and Immunopathology*, *41*(1/2), 153–163. doi:10.1016/0165-2427(94)90064-7 PMID:8066991

Martinez, D., Vachiéry, N., Stachurski, F., Kandassamy, Y., Raliniaina, M., Aprelon, R., & Gueye, A. (2004). Nested PCR for detection and genotyping of Ehrlichia ruminantium: Use in genetic diversity analysis. *Annals of the New York Academy of Sciences*, *1026*(1), 106–113. doi:10.1196/annals.1307.014 PMID:15604477

Matheron, G., Barre, N., Camus, E., & Gogue, J. (1987). Genetic resistance of Guadeloupe native goats to heartwater. *The Onderstepoort Journal of Veterinary Research*, *54*(3), 337–340. PMID:3448569

Molia, S., Frebling, M., Vachiéry, N., Pinarello, V., Petitclerc, M., Rousteau, A., Martinez, D., & Lefrançois, T. (2008). Amblyomma variegatum in cattle in Marie Galante, French Antilles: Prevalence, control measures, and infection by Ehrlichia ruminantium. *Veterinary Parasitology*, *153*(3/4), 338–346. doi:10.1016/j.vetpar.2008.01.046 PMID:18406061

Morand, S., & Frutos, R. (2010). Clonal origin of emerging populations of Ehrlichia ruminantium in Burkina Faso. *Infection, Genetics and Evolution*, *10*(7), 903–912. doi:10.1016/j.meegid.2010.05.011 PMID:20553966

Neitz, W. O. (1968). Heartwater. *Bulletin - Office International des Epizooties*, *70*(1), 329–336. PMID:5761407

Neitz, W. O., & Alexander, R. A. (1941). The immunisation of calves against heartwater. *Journal of the South African Veterinary Medical Association*, *12*(4), 103–111.

Norval, R. A. I., Andrew, H. R., & Yunker, C. E. (1990). Infection rates with Cowdria ruminantium of nymphs and adults of the bont tick Amblyomma hebraeum collected in the field in Zimbabwe. *Veterinary Parasitology*, *36*(3-4), 277–283. doi:10.1016/0304-4017(90)90039-E PMID:2399648

Obereb, P. T., & Benzuidenhout, J. D. (1987). Heartwater in hosts other than domestic ruminants. *The Onderstepoort Journal of Veterinary Research*, *54*, 271–275. PMID:3329317

Peixoto, C. C., Marcelino, I., Vachiéry, N., Bensaid, A., Martinez, D., Carrondo, M. J. T., & Alves, P. M. (2005). Quantification of Ehrlichia ruminantium by real time PCR. *Veterinary Microbiology*, *107*(3/4), 273–278. doi:10.1016/j.vetmic.2005.02.001 PMID:15863287

Petney, Horak, & Rechav. (1987). The ecology of the African vectors of heartwater, with particular reference to Amblyomma hebraeum and Amblyomma variegatum. *Onderstepoort Journal of Veterinary Research, 54*(3), 381-395.

Petney, T. N., & Horak, I. G. (1988). Comparative host usage by Amblyomma hebraeum and Amblyomma marmoreum, (Acari, Ixodidae) the South African vectors of the disease heartwater. *Journal of Applied Entomology*, *105*(5), 490–495. doi:10.1111/j.1439-0418.1988.tb00215.x

Pienaar, J. G. (1970). Electron microscopy of Cowdria (Rickettsia) ruminantium (Cowdry, 1926) in the endothelial cells of the vertebrate host. *The Onderstepoort Journal of Veterinary Research*, *37*(1), 67–78. PMID:4115395

Plessis, Camus, Oberem, & Malan. (1987). Heartwater serology: Some problems with the interpretation of results. *The Onderstepoort Journal of Veterinary Research*, *54*(3), 327–329. PMID:3329323

Plessis & Malan. (1987). Problems with the interpretation of epidemiological data in heartwater: A study on 23 farms. *The Onderstepoort Journal of Veterinary Research*, *54*(3), 427–433. PMID:3448573

Plessis, Gas, Olivier, & Bezuidenhout. (1989). The heterogenicity of Cowdria ruminantium stocks: Cross-immunity and serology in sheep and pathogenicity to mice. *The Onderstepoort Journal of Veterinary Research*, *56*(3), 195–201. PMID:2812704

Postigo, M., Bell-Sakyi, L., Paxton, E., & Sumption, K. (2002). Kinetics of experimental infection of sheep with Ehrlichia ruminantium cultivated in tick and mammalian cell lines. Experimental and Applied Acarology. *Ticks and Tick-Borne Pathogens. Proceedings of the 4th International Conference on Ticks and Tick-Borne Pathogens*, *28*(1/4), 187-193. 10.1023/A:1025390215007

Purchase, H. S. (1945). A simple and rapid method for demonstrating Rickettsia ruminantium (Cowdry, 1925) in heartwater brains. *The Veterinary Record*, *57*(36), 413–414.

Pypekamp & Prozesky. (1987). Heartwater. An overview of the clinical signs, susceptibility and differential diagnosis of the disease in domestic ruminants. *The Onderstepoort Journal of Veterinary Research*, *54*(3), 263–266. PMID:3329315

Srvan & Oberem. (1987). The treatment of heartwater. *The Onderstepoort Journal of Veterinary Research*, *54*(3), 475–479. PMID:3329331

Steck, W. (1928). *Pathological studies on heartwater*. 13th and 14th Report of the Director of Veterinary Education and Research.

Stewart, C. G. (1987). Specific immunity in farm animals to heartwater. *The Onderstepoort Journal of Veterinary Research*, *54*, 341–342. PMID:3448570

Stewart, C. G. (1989). Cowdria ruminantium stocks: How significant are they? *Zimbabwe Veterinary Journal*, *20*, 149–154.

Steyn, H. C., Pretorius, A., McCrindle, C. M. E., Steinmann, C. M. L., & Van Kleef, M. (2008). A quantitative real-time PCR assay for Ehrlichia ruminantium using pCS20. *Veterinary Microbiology*, *131*(3/4), 258–265. doi:10.1016/j.vetmic.2008.04.002 PMID:18511221

Synge, B. A. (1978). Brain biopsy for the diagnosis of heartwater. *Tropical Animal Health and Production*, *10*(1), 45–48. doi:10.1007/BF02235303 PMID:625799

Uilenberg, G. (1981). Heartwater disease. In M. Ristic & I. McIntyre (Eds.), *Diseases of Cattle in the Tropics* (pp. 345–360). Martinus Nijhoff Publishers.

Uilenberg, G. 1982. Disease problems associated with the introduction of European cattle in the tropics. *Proceedings of the 12th World Congress on Diseases of Cattle*, 1025-1032.

Uilenberg, G. (1983). Heartwater (Cowdria ruminantium infection): Current status. *Advances in Veterinary Science and Comparative Medicine*, *27*, 427–480. PMID:6359836

Uilenberg, G., Živković, D., Dwinger, R. H., Ter Huurne, A. A. H. M., & Perié, N. M. (1983). Cross immunity between strains of Cowdria ruminantium. *Research in Veterinary Science*, *35*(2), 200–205. doi:10.1016/S0034-5288(18)32179-9 PMID:6195703

Weisss, K. E., Haig, D. A., & Alexander, R. A. (1952). Aureomycin in the treatment of heartwater. *The Onderstepoort Journal of Veterinary Research*, *25*, 41–50.

Zweygarth, E., & Josemans, A. I. (2005). An attenuated Ehrlichia ruminantium (Welgevonden stock) vaccine protects small ruminants against virulent heartwater challenge. *Vaccine, 23*(14), 1695-1702. https://www.sciencedirect.com/science/journal/0264410X

Chapter 12
Bovine Babesioses and Control

Leonard Muchenditsi Khaluhi
Kenya Agricultural and Livestock Research Organisation, Kenya

Erick Ouma Mungube
Kenya Agricultural and Livestock Research Organisation, Kenya

James M. Wanjohi
Kenya Agricultural and Livestock Research Organisation, Kenya

Sebastian K. Waruri
Kenya Agricultural and Livestock Research Organisation, Kenya

ABSTRACT

Bovine babesiosis is considered one of the most economically important tick-borne diseases in tropical and subtropical areas. Thus, it is important to understand the disease in terms causative agent, vectors, hosts, and distribution. Babesia is the causative agent of the disease and has a life cycle both in the invertebrate and vertebrate hosts. The major bovine species are B. bovis, B. bigemina, found in the tropics and sub tropics while B. divergens and B. major are found in the temperate regions. Bovine babesiosis is transmitted by the one-host cattle ticks, formerly in the genus Boophilus and now Rhipicephalus. In endemic areas are tick management, immunization, and anti-babesia drugs or by a combination. Chemotherapy is important for controlling the disease either to treat field cases or to control artificially induced infections. Conventional drugs for treating bovine babesiosis are Imidocarb dipropionate and Diminazene aceturate. Ticks are controlled by application of acaricides, but tick resistance and environmental pollution are drawbacks.

INTRODUCTION

In 1893, Kilborne and Smith described Texas cattle fever, initially called *pyrosoma* which was later described as *Babesia bigemina* (Kjemtrup & Conrad, 2000). Parasites of the genus *Babesia* infect a wide variety of domestic and wild mammals as well as man. According to the classification of protozoa by (Levine et al., 1980), the genus *Babesia* belongs to the phylum of the *Apicomplexa*, order *Eucoccidiorida*,

DOI: 10.4018/978-1-7998-6433-2.ch012

suborder *Piroplasmorina* and family *Babesidae.* The genus of *Babesia* as mentioned above comprises several species, parasitizing domestic and wild animals. All the species are transmitted by ticks after completing part of their cycle in these hosts (Mehlhorn & Schein, 1985a).

Babesioses with synonyms Red water, Texas fever, Tick fever, Piroplasmosis, Tristeza are infectious, non-contagious diseases, affecting most domestic and wild mammals, mainly but not exclusively in (sub)-tropical regions (Fakhar et al., 2012). Fever, haemolytic anaemia, haemoglobinuria and in severe cases, mortality, are the general characteristics of the disease. Some *Babesia* species can lead to shock, often accompanied by intravascular thromboses (Wodaje et al., 2019; Wright et al., 1989)

Bovine babesiosis is considered one of the most economically important tick-borne diseases in tropical and subtropical areas (Bock et al., 2004; Tavassoli et al., 2013). The economic losses in cattle babesiosis are attributed to mortality, decreased milk or meat production, abortions, reduction of draft power, cost due to control measures, including increased cost of management to maintain ill animals in addition it has impact on international trade. In controlling the disease, a lot of acaricides are used. This has effect on both the animal and the environment. A brief analysis shows that the ability of acaricidal substances is retained in the soil, translocated into plants, deposited in the organs and tissues of animals, excreted in milk and all these leads to contamination with residues of these substances in animal feed and plant and animal products (Osmonov, 2019). In addition the disease causes annual loss of varying degree in various countries. For example in Kenya the annual loss is estimated at 5.1 US dollars (Bock et al., 2004). In India, clinical babesiosis in crossbred cattle decreased milk production for 30 days with total loss for the period estimated as 51.6 litres of milk (Laha et al., 2012). The aim of the chapter is to reassess the above issues and suggest a way forward

BACKGROUND

Bovine babesiosis or Tick fever or cattle fever as mentioned above is economically the most important arthropod-borne disease of cattle worldwide. This is attributed to the fact that it has a vast distribution in areas of Australia, Africa, South and Central America and the United States. Tick fever was the first disease for which transmission by an arthropod to a mammal was implicated at the turn of the twentieth century (Bock et al., 2004). The importance of this disease cannot be further emphasized and to get a more understanding of the condition this section will be discussed under Aetiology as described in different species and Epidemiology of the disease

Aetiology

The causative agent, *Babesia* is divided into two groups namely small and large *Babesia* based on morphology. Small Babesia averagely measure 1.0–2.5 µm long and comprises *Babesia bovis*, *Babesia gibsoni*, *Babesia. Microti*, *Babesia rodhaini Babesia divergens* among others.

The large *Babesia* measure 2.5–5.0 µm long and include *Babesia bigemina*, *Babesia caballi, Babesia canis among others* (Laha et al., 2015). The size to a great extend determines how the parasites appear in red blood cells (RBC) on a giemsa stained blood smear. Large pyriform parasites meet at their pointed ends at an acute angle to each other and small forms make an obtuse angle to each other (Ruprah, 1985). Some animals affected with *Babesia* will show overt disease while others show no clinical disease.

Figure 1. Babesia bovis and Babesia caballi

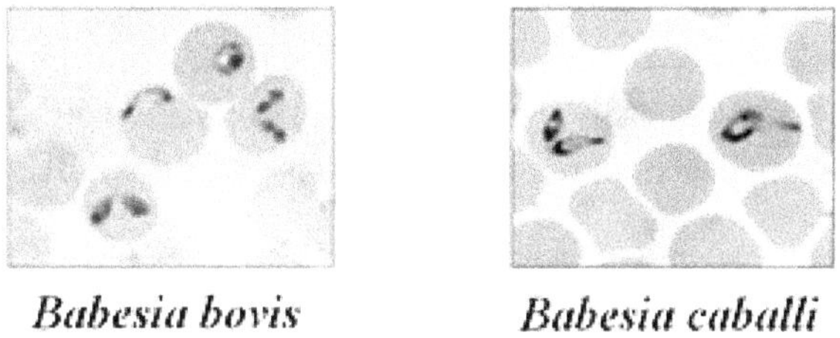

Figure 2. Babesia canis

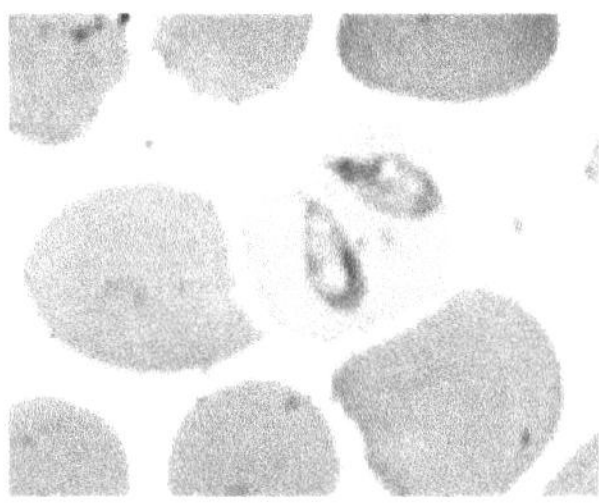

There are more than 100 species of *Babesia* species with potential to infect diverse mammalian species (Hunfeld et al., 2008).

Based on the morphology of different *Babesia* species the following are some species of animals affected by the different types of Babesia.

Cattle and Buffaloes

Generally, both large and small Babesia are responsible for causing clinical disease in cattle and buffaloes. Among these are: *B. bigemina* which measures which averagely measure 4.5 µm. The parasites are characteristically pear shaped. Other forms may be found in the stained smear referred to as anaplasmoid forms. For example round forms 2–3 µm in diameter, oval or irregularly shaped forms may also be found (Laha et al., 2015). Other species are, *B. bovis* which is a small form and measures 2.0 µm and is slightly larger than *B. divergens*. Vacuolated signet ring forms *B. bovis* are particularly common (Soulsby, 1986). *B. divergens* fall in the category of small forms and measure 1.5 µm and usually remain as paired forms and superficially lie on the RBCs.

Canine

Canine babesiosis is caused by two species of *Babesia*; *B. canis* and *B. gibsoni*, which are morphologically differentiated on the basis of their size (Laha et al., 2015).

Ovine and Caprine

The causative agents are *Babesia motasi* and *Babesia ovis species. Babesia motasi* are the large forms measuring (2.5–4.5 µm). The organisms are pyriform in shape. *Babesia ovis* are small forms measuring (1.0–2.5 µm in length) round in shape and remain at the margin of the RBCs (Friedhoff, 1997)

Equine

Equine babesiosis is caused by the following: *Babesia caballi* which are the large forms of *Babesia*. Commonly occur as a pair. They have a pyriform shape and measure 2.5–4 µm long. Round or oval forms with 1.5–3.0 µm in diameter may be found (Laha et al., 2015). *Babesia equi* are small forms measuring 2.0 µm. The piroplasms characteristically form a Maltese cross of four organisms (Onyiche et al., 2019). *Babesia equi* is also named as *Theileria equi* due to its preerythrocytic development as recorded in other *Theileria* species (Kumar et al., 2009).

Swine

Swine babesiosis is caused by two main species. *Babesia trautmanni* is the large form measuring 2.5–4.5 µm. Frequently occurs in pairs, but oval, amoeboid, and ring forms may also be found. Multiple infection in single RBC (up to 4 parasites) may be found. *Babesia perroncitoi* is smaller, the majority of parasites being pleomorphic, annular and measuring 0.7 to 2 µm in diameter (De Waal, 2019).

Feline

Feline babesiosis is caused mainly by *Babesia cati* which are large forms and *Babesia felis* the small form. *Babesia* species in a cat from Kerala with a mean size of 2.4 µm × 1.3 µm has been reported (Sabu et al., 2013).

Epidemiology

Epidemiology of babesiosis generally depends on several parameters such as availability of host, presence of vector ticks for transmission of infections, presence of parasites within vectors, and environmental conditions. These parameters are responsible for the spread of infections (Laha et al., 2015). Absence of any one parameter will discontinue the spread of infections. Babesiosis has a worldwide distribution both in the temperate and the tropical regions (Fakhar et al., 2012). However, the distribution of the causative protozoa is governed by the geographical and seasonal distribution of the tick vectors (Radostits et al., 2006; Yusuf, 2017).

Vectors of Babesia

The one-host cattle ticks, formerly in the genus *Boophilus* and now *Rhipicephalus*, are all vectors that transmit the pathogenic *Babesia* species. *Rhipicephalus microplus* is evidently a species of Asian origin (Madder et al., 2011) but now also colonizes Central and South America and parts of Africa (tropics and sub tropics) (Bock et al., 2004; Muhanguzi et al., 2020; Penzhorn & Verwoerd, 2005). *Rhipicepha-*

lus annulatus colonizes semiarid areas in most countries of the Mediterranean basin, with a broken distribution in Sahel region of Africa (Penzhorn & Verwoerd, 2005). This species was introduced with livestock of the early Spanish colonialists into Central America, where it colonizes dry zones of Mexico and localized regions in the southern United States, replacing *R. microplus* in drier areas (Estrada et al., 2006). *Rhipicephalus australis*, previously confused with *R. microplus* but recently reinstated as a species (Estrada et al., 2012), occurs in Australia, New Caledonia, and countries in Southeast Asia. In view of the long-lasting confusion between *R. microplus* and *R. australis*, reexamination of the ecological requirements of these two species is essential (Gray et al., 2019). The related *R. decoloratus* is exclusively African and occurs south of the Sahara (Morzaria et al., 1977)

Several *Babesia* vectors occur in the temperate regions and are characterized by life cycles with marked seasonality imposed by climate. The most common tick in Europe is *Ixodes ricinus*, a vector of both bovine and human babesiosis, and is distributed in Ireland, from northern Sweden to northern Spain and the Atlantic coast of Portugal (Medlock et al., 2013). It has also been recorded in Northern Africa, but this may be a sister species. *Ixodes ricinus* requires a relative humidity of at least 80% to survive during its off-host periods and is therefore restricted to areas with vegetation that retain high humidity levels. Typical habitats vary across Europe and include deciduous and coniferous woodland, heathland, moorland, and rough pasture (Gray, 1999; Gray et al., 2009).

Endemic Stability

Generally endemic stability refers to a state of a host-tick-pathogen interaction with a high level of challenge to calves by infected ticks without clinical disease and a high level of immunity in adult cattle resulting in low incidence of clinical disease (Jonsson et al., 2012; Mahoney & Ross, 1972; Perry et al., 1998). For endemic stability to occur, the proportion of infected ticks must be greater than 0.005 and the number of infected ticks required to achieve this can be estimated (Bock et al., 2004; Mahoney & Ross, 1972.). This state can be achieved with Babesia infection because of its characteristics where the concept of inverse age immunity exist (most of the calves in a herd are resistant to clinical disease until about 12 months). Thus it has been observed that if 75% of calves are exposed to infection with *Babesia bovis* (indicated by 75% seroprevalence) before nine months of age, then clinical disease should be rare (Mahoney, 1974).

Good examples of this state is *Rhipicephalus microplus–Babesia* system which was used in Australia in the 1970s and 1980s where predominantly taurine cattle genotypes were used (Jonsson et al., 2012). Also the large-scale disruption to compulsory tick control dipping schemes which occurred in Zimbabwe during the country's civil war in the 1960s and 1970s resulted in the death of an estimated one million cattle (Gray et al., 2019). However, by the end of the war, 15 years later in 1980, enzootic stability had emerged in some areas, with the majority of surviving cattle immune to babesiosis (Norval et al., 1983). This observation led to a reassessment of the value of radically reducing tick populations (unless the aim is eradication) and to the integration of strategic dipping with vaccination and the use of tick-resistant cattle (Petney, 1997).

Disease Propagation

For disease propagation the vector, host and environment/climate system must be in a certain balance. The parasite *Babesia* itself is the weakest point of this system of spread of infection as it needs both vectors and host for its survival and thus dependent on them (Alonso et al., 1992). The weak position of *Babesia* in the biological system is evident when one considers that 74-90% of tick larvae fixed on susceptible cattle do not complete their cycle (Sutherst & Utech, 1981). Mortality of the larvae occurs in the first 24 hours after tick fixation on the host (Koudstaal et al., 1978; Roberts, 1968) and this might be contributed by cattle resistance to the ticks (Tabor et al., 2017; Yessinou et al., 2018). Moreover, climatic factors may result in larval mortality and prevent oviposition or normal development of the eggs (Harley, 1966; Owen, 1975; Panda et al., 1992) while cattle can also avoid places with high larval infestation (Okello et al., 1999; Tatchell, 1992). Therefore, in order to be maintained in the surrounding environment, *Babesia* must generate tick infection rates high enough to exceed the tick mortality rate and infect cattle. In cattle, *Babesia* must overcome the immunological reaction of animals and maintain a sufficient parasitaemia to infect ticks before they drop off from the host (Brown et al., 2006).

The second weakest point is the vector which depends on the host and the climatic conditions as mentioned above in relation to distribution. It should be noted that different species and breeds of animals have different resistance to ticks (Tabor et al., 2017). Accordingly, seasonal variations and biology of the ticks influence the time of infections and reinfections and even the species involved and thus the current climate change has great impact on the epidemiology of the disease (Dantas, 2015; Gray et al., 2009)

Finally the host which supports these two parameters to spread the infection but is not dependent on these two parameters. Although, in some situations, babesiosis is considered more pathogenic in calves (Karunakaran et al., 2011; Venu et al., 2015), *Babesia* infection is generally more severe in cattle over nine months old (Parker et al., 1985; Smith et al., 1978; Taylor et al., 2013). This was confirmed in Australia, where 76% of the mortality induced by *B. bovis* occurred in cattle over one year of age. The recovered cattle are protected against another challenge from *Babesia* for about four years (Joyner & Donnelly, 1979). The greater resistance of young cattle leads to the concept of enzootic stability and instability. In the case of enzootic stability, the interaction of the tick, *Babesia* and local cattle is significant enough to produce an early infection of cattle and prevent cases of babesiosis. In an unstable situation, however, many calves reach critical age without infection and express babesiosis later. In the case of *B. bovis* and most other *Babesia* the tick usually becomes infective within 2-3 days after larval ticks attach (Mosqueda et al., 2012; Riek, 1964) and it does not persist in *R. microplus* after the larval stage (Howell et al., 2007). But for *B. bigemina,* the tick becomes infective approximately 9 days after larval attachment (Potgieter & Els, 1977b; Riek, 1964), the parasite is only transmitted by nymphs and adult (Buscher, 1988). Bos indicus cattle tend to have a less severe disease than Bos taurus (Parker et al., 1985)

BOVINE BABESIOSIS

Bovine babesiosis (BB) is a tick-borne disease of cattle caused by the protozoan parasites of the genus *Babesia* (Levine et al., 1980). The main species of *Babesia* that cause bovine babesiosis include *B. bovis, B. bigemina* and *B. divergens*. Other *Babesia* that can infect cattle include *B. major, B. ovata, B. occultans* and *B. jakimovi* (Gray et al., 2019; OIE, 2014). The disease is one of the most important tick borne diseases of cattle and thus contributes heavily to food insecurity (Yusuf, 2017)

Transmission

The major vectors for *B. bigemina* and *B. bovis* are *Rhipicephalus microplus* (formerly *Boophilus microplus*) and *R. decoloratus (*formerly *Boophilus decoloratus).* In some areas, *R. annulatus* (formerly *Boophilus annulatus*) (OIE, 2014) acts as the vector. The vectors are one-host ticks that complete their life cycle on a single host, and preferentially feed on cattle. Additional members of *Rhipicephalus* and some ticks in other genera have also been suggested as vectors in some regions (Gray et al., 2019).

Life Cycle

Babesia are transmitted transovarially (Buscher, 1988; Howell et al., 2007; Morzaria et al., 1977; Uilenberg, 2006) although in limited cases transstadial transmission can occur (Chauvin et al., 2009). Babesia are stimulated to undergo their final maturation when an infected tick attaches to the host and starts feeding (Bock et al., 2004). Temperature increase is involved in *B. bovis* and *B. bigemina* differentiation, but various factors may also be associated with the start of the blood meal, for example hypertrophy and increased metabolic and synthetic activities within the salivary glands (Dalgliesh & Stewart, 1978)

Figure 3. Life cycle of Babesia spp. Adapted from 'Mehlhorn, H., & Schein, E. (1985': The Piroplasms: Life Cycle and Sexual Stages. In Advances in Parasitology (Vol. 23, pp. 37–103)

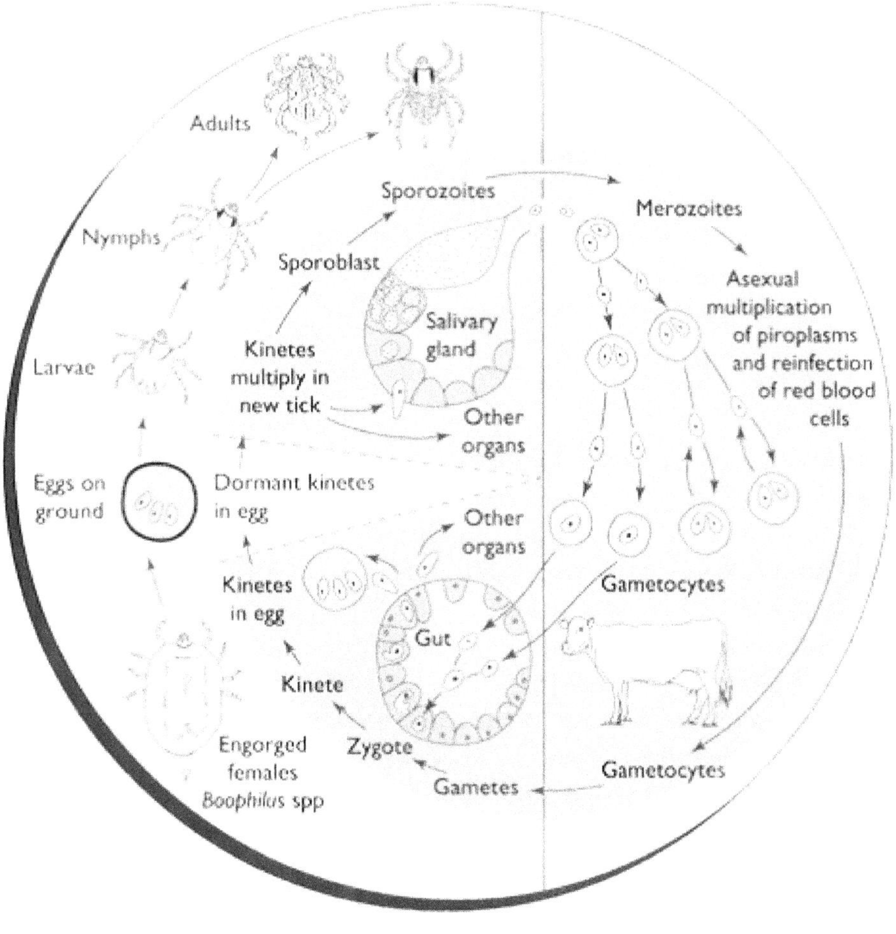

Life Cycle in the Tick

There has been a lot of discussion on variation in the general life cycle of *Babesia* across the genus, but a basic scheme (**Figure 1**) above has generally been agreed upon (Gray et al., 2019) and this was based on many studies (Gray et al., 2019; Mehlhorn & Schein, 1985a; Mehlhorn et al., 1980; Rudzinska et al., 1983; Shortt, 1973).

Vector ticks become infected while ingesting blood of the parasitized vertebrate host. Circular bodies, (gametocytes) are liberated in the mid-gut and develop into elongated cells with bristle cytoplasmic extensions dubbed, 'rayed bodies' (Gough et al., 1998). These rayed bodies are considered as being the gametes. Electronic microscopy allows differentiating between cells with dense and with less dense cytoplasm leading to the hypothesis of the existence of a sexual cycle (gametogony) (Rudzinska et al., 1983). Cells with dense cytoplasm unite with those with less dense cytoplasm and their nuclei fuse together (Gough et al., 1998), resulting in a zygote (Friedhoff, 1988). This zygote has a smooth wall, initially it is spherical but it becomes elongated, measuring 7 to 8 μm. These zygotes penetrate the intestinal cells of the tick and multiply by a series of binary fissions to form sporokinetes (Mackenstedt et al., 1995, 1990). These vermiform sporokinetes (also named by some authors, kinetes or merozoites) pass into the haemolymph of the body cavity (Agbede et al., 1986) and from there they invade the tick ovaries, where they penetrate the eggs leading to transovarial transmission. Dormant kinetes remain in the eggs and continue their development in the subsequent larval stage of the tick. During the first meal of the tick, the kinetes invade the salivary glands and develop into sporoblasts in which sporozoites, infective to the new cattle-host, are formed. In all species, sporozoite development usually only begins when the infected tick attaches to the vertebrate host (Bock et al., 2004). In principle, transovarial transmission takes place through successive generations without ticks necessarily feeding on infected hosts, but there are many variations - including on the stage of the tick responsible for the transmission - depending on the vector and the *Babesia* species involved. Ticks feeding on heavily infected animals may suffer disruptions such as reduction of egg production and loss of permeability of the intestinal wall. In addition, ambient temperature has an influence on the vectorial capacity of the ticks (Dalgliesh & Stewart, 1982), which is variable according to the species involved e.g. the optimal temperature for the complex *Boophilus microplus* - *Babesia bovis* is 28°C, whereas for *Ixodes ricinus* - *Babesia divergens* is 20°C.

Life Cycle in Cattle

Most *Babesia* species do not parasitise any vertebrate host cell other than red blood cells (Friedhoff, 1988) with the exception of *B. equi* and *B Microti* (Mehlhorn & Schein, 1985b). The vertebrate host is infected through injection of sporozoites with the saliva of a parasitized tick but only after minimum of 3 days of attachment for *B bovis* (Mosqueda et al., 2012; Riek, 1964). In the case of *B. bigemina*, the inoculation occurs after 8 days of attachment (Potgieter & Els, 1977a; Riek, 1964). After inoculation, sporozoites invade RBCs through penetration in a complex process, aided by a complex of organelles, typical for all Apicomplexa (Potgieter & Els, 1977b, 1979). This apical complex is located at the anterior end of the parasites and comprises rhoptries, dense granules and micronemes. Secretion of these rhoptries facilitates host cell attachment, invasion and the formation of a parasitophorous vacuole membrane. For *Babesia* species, this membrane soon gets dissolved after entry. The parasite is then in direct contact with the erythrocyte cytoplasm. Once inside the RBCs, it transforms into a trophozoite from which two merozoites develop by a process of merogony (binary fission) (Friedhoff, 1988; Potgieter & Els, 1977b,

1979). In some species, a subsequent division leads to the formation of tetrads with 4 merozoites. These merozoites will break out of the RBCs and enter new cells. This asexual multiplication can continue indefinitely, and the host remains a carrier for life.

Pathogenesis

Generally the course of pathogenesis produced by two principle mechanism; haemolysis and circulatory disturbance (Carlton et al., 1995). During tick bite, sporozoites invade RBCs and develop into piroplasms inside the infected erythrocytes resulting in two or sometimes four daughter cells that leave the host cell to infect other erythrocytes (Hunfeld et al., 2008). The invasion of the RBCs causes intravascular and extravascular haemolysis (Carlton et al., 1995) due to rapid destruction of the erythrocytes with accompanying haemoglobinaemia, haemoglobinuria and fever. This may be so acute as to cause death within a few days, during which the packed cell volume (PCV) falls below 20% (Yusuf, 2017) which leads to anaemia. Parasitaemia is detectable once the clinical signs appear and may involve between 0.2% up to 45% of the red cells, depending on the species of *Babesia* involved (Taylor et al., 2016). Overproduction of cytokines and other pharmacologically active agents contributes to disease progress causing vasodilation, hypotension, increased capillary permeability, oedema, vascular collapse, coagulation disorders, endothelial damage and circulatory stasis (Ahmed, 2002). For *B. bovis*, coagulation disorders, cytoadherence and hypotension effects are experienced with the involvement of the brain and lungs (Bock et al., 2004). This sometimes can result in cerebral babesiosis and a respiratory distress syndrome associated with infiltration of neutrophils, vascular permeability and oedema (Wright et al., 1989). *Babesia bovis* infected erythrocytes that remain sequestered do not circulate in the blood and avoid phagocytosis, leading to establishment of persistent infections (Al-Khedery & Allred, 2006; Hutchings et al., 2007)

DIAGNOSIS

Thin and Thick Blood Smears

Thin blood smear samples are usually collected from the capillary blood (with anticoagulant if to be examined later) from ear margin or tip of the tail and smeared on a clean glass slide, air-dried, fixed with methanol, and stained with Giemsa or a similar stain for more than 10 minutes. The slide is then washed thoroughly and dried. Intraerythrocytic parasites are observed under a microscope using a 100X objective and a drop of immersion oil (Mosqueda et al., 2012). *Babesia* can be detected for up to one infected erythrocyte per ten thousand cells, requiring the analysis of 100-200 fields, the equivalent to 0.5 μl of blood (Morzaria et al., 1992). The observation of paired Intraerythrocytic merozoites is indicative of infection.

Thick smears are majorly used to detect low levels of parasitaemia, especially in cases where *B. bovis* is involved or in cases of suspected subclinical disease (Bose et al., 1995). A small drop of blood is placed in the center of a clean slide and it is fixed by heat, without smearing it, then it is stained routinely, washing it carefully to avoid losing the tissue while the excess of stain is removed. Once dry, it is examined similarly as the thin smear (Bishop & Adams, 1973). This technique allows a large number of erythrocytes to be analyzed in a reduced amount of space, therefore the probability of finding infected cells is ten times higher than in the thin smear (Bose et al., 1995).

Brain Smear

This is carried at post mortem particularly when an animal has shown nervous clinical signs and in cases where *B. bovis* is suspected as the cause of disease. Small sample of gray matter of the cerebral cortex is placed on a slide and the tissue is smeared using another slide (Mosqueda et al., 2012). The brain tissue is fixed and stained as mentioned above. Diagnosis is based upon observation of brain capillaries filled with infected erythrocytes (Ristic, 1981). Almost one hundred percent of erythrocytes present in the brain capillaries are infected (Uilenberg, 1972). Smears of other organs as the kidney or liver can also be carried out (Mosqueda et al., 2012)

Indirect Fluorescent Antibody Test (IFAT)

When parasitaemia is too low this too can be used mainly for detection of carriers. The tool has a good level of specificity and sensitivity (Burridge et al., 1973; Vercammen et al., 1995) and is based on the recognition of parasite antigens by serum antibodies in the blood of the tested animal.

Enzyme-Linked Immunosorbent Assay (ELISA)

ELISA has the advantage of non-subjectivity, capacity to read a large number of samples easily and presents higher specificity than IFAT. There are several ELISA versions for the detection of several *Babesia* species including *B. bovis, B. bigemina, and B. divergens* (Waltisbuhl et al., 1987). Modern recombinant protein ELISA and competitive ELISA are now available for diagnosis of babesiosis (Dominguez et al., 2012).

Immunochromatography Test (ICT)

The immunochromatographic test is a rapid diagnostic device that detects antibodies against a specific antigen in a small amount of serum by means of specific antibody and a recombinant antigen both impregnated on a nitrocellulose membrane-based test strip. ICT is very easy to perform and read (Valecha et al., 1998). It does not use any special equipment, so it can be implemented in the field and has a low cost comparable with other techniques. The technique is fast, it takes only ten to fifteen minutes to complete and it is very stable under different temperatures.

Molecular Methods

Molecular methods are very useful tools in detecting parasites in infections involving *Babesia* in cattle. They are considered indirect methods since they detect nucleic acids of the parasites. The sensitivity and specificity of these methods are very high and over the past years many different approaches have been developed to detect *Babesia.* Some of these approaches include but are not limited to the following: DNA Probes, Polymerase Chain Reaction (PCR), Reverse Line Blot Hybridization (RLB), Real Time PCR (RT-PCR), Loop Mediated Isothermal Amplification (LAMP) (Mosqueda et al., 2012).

DNA Probes

Deoxyribonucleic acid probes was the first method developed to detect babesial DNA from parasitized blood (Mosqueda et al., 2012). For this technique, DNA is isolated, purified and cloned while adding a marker, which initially was radioactive. The marked single stranded DNA is then used to hybridize to DNA present in a tissue, or a membrane. The high specificity of DNA alignment allows only the hybridization to complementary DNA, thus recognition of the desired sequence and a very specific diagnosis.

Polymerase Chain Reaction (PCR)

The first descriptions of the polymerase chain reaction to detect *Babesia* were reported in 1992 for *B. bovis* (Fahrimal et al., 1992), *B. bigemina* (Figueroa et al., 1992) and *B. microti* (Persing et al., 1992). An adaptation of the technique is the nested polymerase chain reaction (nested PCR), a format in which two pairs of primers are used in two successive PCR amplifications where the second pair is intended to amplify a secondary target within the first amplified product. A nested PCR has been effective for the detection of carrier animals infected with *B. bigemina* and *B. microti* and the sensitivity has been reported to be as low as one and three infected erythrocytes, respectively (Figueroa et al., 1992; Persing et al., 1992). When nested PCR has been combined with labelled probes, the sensitivity increases several logs of magnitude. For example, nested PCR for *B. bovis* was reported to detect 1 parasite in 10^6 erythrocytes, but when the resulted amplified product was detected with a DNA probe, the sensitivity increased to 1 in 10^9 (Figueroa et al., 1994). Nested PCR is a very sensitive tool but it is more expensive, takes more time and the risk of self-contamination is higher than the one-step PCR (Mosqueda et al., 2012)

Reverse Line Blot Hybridization (RLB)

Reverse Line Blot is a tool for detection of multiple genera, species or strains in a sample of blood or tissue

Real Time PCR (RT-PCR)

Real Time PCR is a technique that amplifies and quantifies a specific DNA fragment. The DNA amplified is quantified as it is being generated (in “real time”), therefore it determines whether a specific sequence is present in the sample, and it also determines the number of copies of that sequence. Real Time -PCR does this by detecting a fluorescent signal emitted during the PCR reaction as an indicator of the production of the sequence being generated in each cycle; this is opposite to what happens in during the end-point PCR, where the detection of the product is at the end of the reaction.

Loop Mediated Isothermal Amplification (LAMP)

LAMP is a detection method that amplifies DNA under isothermal conditions with high efficiency, specificity and speed. Loop Mediated Isothermal Amplification is based on the use of four primers specifically designed to amplify six different sequences on the same target DNA, with the aid of an isothermal DNA polymerase. By using four primers to amplify the same target sequence, the specificity of the amplification in increased, solving in part the background amplification observed in most nucleic acid amplification methods. Because LAMP produces a large amount of DNA, it can be analysed by direct observation (Notomi et al., 2015).

CONTROL

In Endemic Conditions

Endemic stability is reached when the contact with infected ticks is sufficient to infect 80 to 100% of the calves when they are less than 9 months. Hence when their age at first contact is very low, no prophylactic measure should be necessary. Obviously in this case the impact of an excessive number of ticks or the possibility of transmission of other infections cannot be disregarded and therefore will need strategic control.

If cattle are to be introduced from non-endemic regions, they should be vaccinated (see below) beforehand or be protected by an injection of imidocarb dihydrocloride/dipropionate if they are allowed on the pastures without tick control.

In *Babesia*-free regions: Upon accidental introduction of ticks with the import of animals from endemic regions it is necessary:

- to apply rigorous tick control aiming at eradication;
- to apply babesicidal treatment (sterilization dose) of all sick and seropositive animals;
- that all cattle movements out the affected region are prohibited;
- That all the imported animals are placed in quarantine and preferably with sterilizing treatment.

Non Endemic Conditions

Control of bovine babesiosis can either be by tick management, immunization, and anti-Babesia drugs or by a combination of these approaches (Pipano & Hadani, 1984; Suarez & Noh, 2011). Chemotherapy is important for controlling the disease either to treat field cases or to control artificially induced infections. Conventional drugs for treating bovine babesiosis are Imidocarb dipropionate and Diminazene aceturate

Vector Control

There are several ways to control ticks but chemical control remains the main method (Bram, 1983; Drummond, 1970; Rajput et al., 2006). It is done by repeated treatment of cattle with acaricides. in areas of high tick challenge; such treatment may require to be carried out in order to kill the tick before the infective sporozoite develop in the salivary gland (Taylor et al., 2016). Acaricides used to control ticks on livestock should be applied in such a manner that the ticks will be killed, the treatments will not harm livestock or applicators, the tissues of treated animals will not contain illegal residues, and the environment will not be adversely affected (Rogoff, 1961). The two main methods of acaricide application are dipping and spraying (Barnett, 1961). The use of acaricides has disadvantages, such as the selection of resistant tick populations, particularly one host tick (Rodriguez-Vivas et al., 2018) and harmful effects on the animals, human beings and the environment (García-García et al., 2000). In addition the cost of acaricides is very high. The development of new acaricides is a long and expensive process, which reinforces the need for alternative approaches to control tick infestations (Graf et al., 2004).

IMMUNIZATION

Immunization Against *Babesia*

Bovines can be immunized in different ways. When immunization is based on the presence of living parasites remaining in the host (carrier), i.e. concomitant immunity, there is no sterile immunity. Immunized bovines may remain a source of infection for ticks and naive and susceptible cattle in the herd and area. Thus Inherent disadvantages of vaccine containing parasites derived from blood of animals are the risk of reactions or contamination, sensitization against blood groups, tick transmissibility of vaccine strains and the need for a cold chain (Callow et al., 1997). In general, vaccine containing one attenuated strain of *B. bovis* or *B. bigemina* affords durable protection against local field challenges (Vos & Bock, 2006) and the immunity lasts for four years (Bock & Vos, 2001).

Infection and Treatment

This method is used in the Mediterranean basin and in America. Local non-attenuated strains of *Babesia* species are used. The infection is achieved by injecting citrated parasitized blood and the treatment is not done before the appearance of clinical symptoms. The treatment dose should be carefully adjusted to avoid complete clearance of parasites to help in continued boosting of the immunity (Benavides et al., 2006; IDRC, 2018; Montenegro-James, 1992; Todorovic et al., 1975).

Natural Infection Controlled by Imidocarb/Treatment of Clinical Cases

This method can be used to protect newly introduced animals in an endemic environment, assuming that animals will come into contact with the parasite while still under protection of imidocarb. For treatment Imidocarb is used. It is the principal babesiacide used in animals, because it consistently clears the host of parasites (Kuttler, 1980; Suarez & Noh, 2011) and for over 20 years, it has been used in the treatment and prophylaxis of babesiosis and anaplasmosis (Kuttler, 1980, 2018).

Live Attenuated Vaccines

Since 1965, strains of attenuated *B. bigemina* and *B. bovis* have been used in Australia for vaccinations (Alonso et al., 1994; Alvarez- et al., 2017). For both species, attenuation was achieved by continuous passages in splenectomized calves, resulting in less virulent, annular-shaped parasites. This principle continues to be used in other countries. The vaccine is injected intramuscularly or subcutaneously as diluted blood, containing 10^7 parasites. Currently, in Australia, a single vaccination is preferred instead of one every year as used to be done before (Florin-Christensen et al., 2014; Shkap et al., 2007). In general when the vaccine is used, the reaction is weak and the only observation could be a transient fever which disappears even without treatment

Disadvantages are:

- preservation time of the vaccines is rather short (one week at 4°C)
- reversal of attenuated forms towards virulent forms cannot be excluded
- given the wide antigenic variation within *Babesia*-species, a vaccine does not necessarily protect against all strains from different geographical regions
- occasionally, post - vaccinal reactions might require treatment with low doses of imidocarb avoiding interference with the development of immunity
- The risk of transmission of undesirable pathogenic agents, present in blood from the donor calf.

The remedy to these disadvantages lies in the development of novel vaccines against babesiosis (IDRC, 2018)

Vaccine Based on Exoantigens

Exoantigens of *B. bovis* or *B.bigemina* can be recovered from in vitro culture media. Administration of these antigens with an adjuvant (primary vaccine + booster) allows for at least 12 months protection against the homologous strains and in case of *B. bovis*, often against heterologous strains (James, 1989; Montenegro et al., 1992)

Immunization Against Ticks

It has been known that bovines can develop some immunity against ticks and to some extent it can explain the differences in susceptibility of certain breeds. Immunization against *Boophilus microplus* was first achieved in Australia with "Tickguard" and "Tickguard Plus" and later in Cuba with "Gavac" (Andreotti, 2006; Fuente et al., 2007; Merino et al., 2013). They are based on a recombinant antigen (Bm86) composed of glycoproteins, located on the surface of the intestinal cells of ticks (Fuente et al., 2007). These vaccines, in combination with an oil-adjuvant, yield good results, reducing tick loads by at least 76% (Fuente et al., 2007; Merino et al., 2013). When ticks ingest the anti-Bm86 antibodies present in the blood of vaccinated hosts, these antibodies bind to their intestinal cell surface. This causes cell lysis and a subsequent flow of the intestinal contents into the haemolymph leading to death between feeding and oviposition. Thus the effects of these vaccines lies in the reduction of the population of ticks through death or through reduction of the reproductive capacity (Merino et al., 2013).

FUTURE RESEARCH DIRECTIONS

- Live *Babesia* vaccine production with wider cross protection
- Recombinant antigens against *Babesia* parasites with consideration of incorporation more antigens
- Combination of vaccines for wider protection
- Further research in tick vaccines with an aim of identification of more recombinant antigens in the tick

CONCLUSION

Bovine Babesiosis still remains an economically important disease across the world and its management remains a challenge. Currently In the developing world management of the disease can be achieved through development of endemic stability as explained above or treatment of any clinical disease and tick management, while further research is carried out for novel control strategies. For the developed world in addition to the above control strategies, immunization against the parasite or the vector tick can be done. It should be noted that eradication of the disease can be achieved but it requires a lot of resources. In general further research to produce quality vaccines against the parasite or the vector tick or both will be the appropriate approach.

REFERENCES

Agbede, R., Kemp, D., & Hoyte, H. (1986). *Secretory and digest cells of female Boophilus microplus: Invasion and development of Babesia bovis; light and electron microscope studies*. Academic Press.

Ahmed, J. (2002). The role of cytokines in immunity and immunopathogenesis of pirolasmoses. *Parasitology Research*, *88*(S1), S48–S50. doi:10.100700436-001-0573-4 PMID:12051611

Al-Khedery, B., & Allred, D. R. (2006). Antigenic variation in Babesia bovis occurs through segmental gene conversion of the ves multigene family, within a bidirectional locus of active transcription. *Molecular Microbiology*, *59*(2), 402–414. doi:10.1111/j.1365-2958.2005.04993.x PMID:16390438

Alonso, M., Arellano-Sota, C., Cereser, V. H., Cordoves, C. O., Guglielmone, A. A., Kessler, R., ... Solari, M. A. (1992). Epidemiology of bovine anaplasmosis and babesiosis in Latin America and the Caribbean. *Revue Scientifique et Technique (International Office of Epizootics)*, *11*(3), 713–733. doi:10.20506/rst.11.3.623 PMID:1472722

Alonso, M., Blandino, T., Mendoza, E., Savon, L., & Camacho, M. (1994). Development of a Babesia bovis live attenuated vaccine. *Archives of Medical Research*, *25*(2), 273–277. PMID:7919825

Alvarez, J., Rojas, C., & Jose, J. (2017). *Development of a Live Attenuated Vaccine for the Control of Bovine Babesiosis in Mexico*. Academic Press.

Andreotti, R. (2006). Performance of two Bm86 antigen vaccine formulation against tick using crossbreed bovines in stall test. Revista Brasileira De Parasitologia Veterinaria = Brazilian Journal of Veterinary Parasitology. *Orgao Oficial Do Colegio Brasileiro De Parasitologia Veterinaria*, *15*(3), 97–100.

Barnett, S. (1961). The control of ticks on livestock. *FAO Agricultural Studies*, 54.

Benavides, E., Vizcaino, O., Britto, C. M., Romero, A., & Rubio, A. (2006). Attenuated Trivalent Vaccine against Babesiosis and Anaplasmosis in Colombia. *Annals of the New York Academy of Sciences*, *916*(1), 613–616. https://doi.org/10.1111/j.1749-6632.2000.tb05346.x

Bishop, J. P., & Adams, L. G. (1973). Combination thin and thick blood films for the detection of Babesia parasitemia. *American Journal of Veterinary Research*, *34*(9), 1213–1214.

Bock, R., Jackson, L., de Vos, A., & Jorgensen, W. (2004). Babesiosis of cattle. *Parasitology, 129*, S247-269. doi:10.10170031182004005190

Bock, R., & Vos, A. (2001). Immunity following use of Australian tick fever vaccine: A review of the evidence. *Australian Veterinary Journal*, *79*(12), 832–839. https://doi.org/10.1111/j.1751-0813.2001.tb10931.x

Bose, R., Jorgensen, W. K., Dalgliesh, R. J., Friedhoff, K. T., & de Vos, A. J. (1995). Current state and future trends in the diagnosis of babesiosis. *Veterinary Parasitology, 57*(1–3), 61–74. (94)03111-9 doi:10.1016/0304-4017

Bram, R. (1983). Bram, R. A. (1983). Tick-borne livestock diseases and their vectors. *FAO Animal Production and Health Paper*, 36.

Brown, W. C., Norimine, J., Knowles, D. P., & Goff, W. L. (2006). Immune control of Babesia bovis infection. *Veterinary Parasitology*, *138*(1–2), 75–87. https://doi.org/10.1016/j.vetpar.2006.01.041

Burridge, M. J., Kimber, C. D., & McHardy, N. (1973). Detection of antibodies to Babesia bigemina in dried blood samples using the indirect fluorescent antibody test. *Annals of Tropical Medicine and Parasitology*, *67*(2), 191–195. https://doi.org/10.1080/00034983.1973.11686876

Buscher, G. (1988). The infection of various tick species with Babesia bigemina, its transmission and identification. *Parasitology Research*, *74*(4), 324–330. https://doi.org/10.1007/bf00539453

Callow, L. L., Dalgliesh, R. J., & De Vos, A. J. (1997). Development of effective living vaccines against bovine babesiosis—The longest field trial? *International Journal for Parasitology, 27*(7), 747–767. (97)00034-9 doi:10.1016/S0020-7519

Carlton, W., McGavin, M. D., & Thomson, R. G. (Eds.). (1995). *Thomson's special veterinary pathology* (2nd ed.). Mosby.

Chauvin, A., Moreau, E., Bonnet, S., Plantard, O., & Malandrin, L. (2009). Babesia and its hosts: Adaptation to long-lasting interactions as a way to achieve efficient transmission. *Veterinary Research*, *40*(2), 37. https://doi.org/10.1051/vetres/2009020

Dalgliesh, R. J., & Stewart, N. P. (1978). The extraction of infective Babesia bovis and Babesia bigemina from tick eggs and B. bigemina from unfed larval ticks. *Australian Veterinary Journal*, *54*(9), 453–454. https://doi.org/10.1111/j.1751-0813.1978.tb05582.x

Dalgliesh, R. J., & Stewart, N. P. (1982). Some effects of time, temperature and feeding on infection rates with Babesia bovis and Babesia bigemina in Boophilus microplus larvae. *International Journal for Parasitology, 12*(4), 323–326. (82)90035-2 doi:10.1016/0020-7519

Dantas, T. (2015). Climate change, biodiversity, ticks and tick-borne diseases: The butterfly effect. *International Journal for Parasitology. Parasites and Wildlife*, *4*(3), 452–461. https://doi.org/10.1016/j.ijppaw.2015.07.001

De Waal, D. (2019). *Porcine Babesiosis*. Anipedia.

Dominguez, M., Echaide, I., de Echaide, S. T., Wilkowsky, S., Zabal, O., Mosqueda, J. J., … Florin-Christensen, M. (2012). Validation and Field Evaluation of a Competitive Enzyme-Linked Immunosorbent Assay for Diagnosis of Babesia bovis Infections in Argentina. *Clinical and Vaccine Immunology, 19*(6), 924–928. doi:10.1128/CVI.00015-12

Drummond, R. (1970). *Current worldwide research on control of ticks involved in animal diseases*. Entomol Soc Amer Misc Publ.

Estrada, A., Bouattour, A., Camicas, J.-L., Guglielmone, A., Horak, I., Jongejan, F., … Walker, A. R. (2006). The known distribution and ecological preferences of the tick subgenus Boophilus (Acari: Ixodidae) in Africa and Latin America. *Experimental & Applied Acarology, 38*(2–3), 219–235. doi:10.100710493-006-0003-5

Estrada-Peña, A., Venzal, J. M., Nava, S., Mangold, A., Guglielmone, A. A., Labruna, M. B., & de la Fuente, J. (2012). Reinstatement of Rhipicephalus (Boophilus) australis (Acari: Ixodidae) with redescription of the adult and larval stages. *Journal of Medical Entomology*, *49*(4), 794–802. https://doi.org/10.1603/me11223

Fahrimal, Y., Goff, W., & Jasmer, D. (1992). Detection of Babesia bovis carrier cattle by using polymerase chain reaction amplification of parasite DNA. *Journal of Clinical Microbiology*, *30*(6), 1374–1379. https://europepmc.org/articles/PMC265295

Fakhar, M., Hajihasani, A., Maroufi, S., Alizadeh, H., Shirzad, H., Piri, F., & Pagheh, A. S. (2012). An epidemiological survey on bovine and ovine babesiosis in Kurdistan Province, western Iran. *Tropical Animal Health and Production*, *44*(2), 319–322. https://doi.org/10.1007/s11250-011-0023-y

Figueroa, J., Chieves, L., Johnson, G., & Buening, G. (1992). Detection of Babesia bigemina-infected carriers by polymerase chain reaction amplification. *Journal of Clinical Microbiology*, *30*(10), 2576–2582. https://europepmc.org/articles/PMC270481

Figueroa, J. V., Chieves, L. P., Johnson, G. S., Goff, W. L., & Buening, G. M. (1994). Polymerase chain reaction-based diagnostic assay to detect cattle chronically infected with Babesia bovis. *Revista Latinoamericana de Microbiologia*, *36*(1), 47–55.

Florin-Christensen, M., Suarez, C. E., Rodriguez, A. E., Flores, D. A., & Schnittger, L. (2014). Vaccines against bovine babesiosis: Where we are now and possible roads ahead. *Parasitology*, *141*(12), 1563–1592. https://doi.org/10.1017/S0031182014000961

Friedhoff, K. T. (1988). Transmission of Babesia. In *Babesiosis of Domestic Animals and Man*. CRC Press, Inc.

Friedhoff, K. T. (1997). Tick-borne diseases of sheep and goats caused by Babesia, Theileria or Anaplasma spp. *Parassitologia*, *39*(2), 99–109.

de la Fuente, J., Almazán, C., Canales, M., Pérez de la Lastra, J. M., Kocan, K. M., & Willadsen, P. (2007). A ten-year review of commercial vaccine performance for control of tick infestations on cattle. *Animal Health Research Reviews*, *8*(1), 23–28. https://doi.org/10.1017/S1466252307001193

García-García, J. C., Montero, C., Redondo, M., Vargas, M., Canales, M., & Boue, O. … de la Fuente, J. (2000). Control of ticks resistant to immunization with Bm86 in cattle vaccinated with the recombinant antigen Bm95 isolated from the cattle tick, Boophilus microplus. *Vaccine, 18*(21), 2275–2287. (99)00548-4 doi:10.1016/S0264-410X

Gough, J. M., Jorgensen, W. K., & Kemp, D. H. (1998). Development of Tick Gut Forms of Babesia bigemina In Vitro. *The Journal of Eukaryotic Microbiology*, *45*(3), 298–306. https://doi.org/10.1111/j.1550-7408.1998.tb04540.x

Graf, J.-F., Gogolewski, R., Leach-Bing, N., Sabatini, G. A., Molento, M. B., Bordin, E. L., & Arantes, G. J. (2004). Tick control: An industry point of view. *Parasitology*, *129*(S1), S427–S442. https://doi.org/10.1017/S0031182004006079

Gray, J. S. (1999). The ecology of Lyme borreliosis vector ticks. In J. Bruin, L. P. S. van der Geest, & M. W. Sabelis (Eds.), *Ecology and Evolution of the Acari. In Entomologica* (Vol. 55, pp. 269–276). Springer.

Gray, J. S., Dautel, H., Estrada-Peña, A., Kahl, O., & Lindgren, E. (2009). Effects of Climate Change on Ticks and Tick-Borne Diseases in Europe. *Interdisciplinary Perspectives on Infectious Diseases*, *2009*, 1–12. https://doi.org/10.1155/2009/593232

Gray, J. S., Estrada-Peña, A., & Zintl, A. (2019). Vectors of Babesiosis. *Annual Review of Entomology*, *64*(1), 149–165. https://doi.org/10.1146/annurev-ento-011118-111932

Harley, K. (1966). Studies on the survival of the non-parasitic stages of the cattle tick Boophilus microplus in three climatically dissimilar districts of North Queensland. *Australian Journal of Agricultural Research*, *17*(3), 387. https://doi.org/10.1071/AR9660387

Howell, J. M., Ueti, M. W., Palmer, G. H., Scoles, G. A., & Knowles, D. P. (2007). Transovarial Transmission Efficiency of Babesia bovis Tick Stages Acquired by Rhipicephalus (Boophilus) microplus during Acute Infection. *Journal of Clinical Microbiology*, *45*(2), 426–431. https://doi.org/10.1128/JCM.01757-06

Hunfeld, K.-P., Hildebrandt, A., & Gray, J. S. (2008). Babesiosis: Recent insights into an ancient disease. *International Journal for Parasitology*, *38*(11), 1219–1237. https://doi.org/10.1016/j.ijpara.2008.03.001

Hutchings, C. L., Li, A., Fernandez, K. M., Fletcher, T., Jackson, L. A., Molloy, J. B., … Cooke, B. M. (2007). New insights into the altered adhesive and mechanical properties of red blood cells parasitized by Babesia bovis. *Molecular Microbiology, 65*(4), 1092–1105. doi:10.1111/j.1365-2958.2007.05850.x

IDRC. (2018). Development and deployment of a novel vaccine against babesiosis. United States Department of Agriculture, Monash University, and Biosciences Eastern and Central Africa-International Livestock Research Institute (BecA-ILRI) Hub.

James, M. A. (1989). Application of exoantigens of Babesia and Plasmodium in vaccine development. *Transactions of the Royal Society of Tropical Medicine and Hygiene, 83*, 67–72. (89)90607-X doi:10.1016/0035-9203

Jonsson, N. N., Bock, R. E., Jorgensen, W. K., Morton, J. M., & Stear, M. J. (2012). Is endemic stability of tick-borne disease in cattle a useful concept? *Trends in Parasitology*, *28*(3), 85–89. https://doi.org/10.1016/j.pt.2011.12.002

Joyner, L. P., & Donnelly, J. (1979). The Epidemiology of Babesial Infections. In *Advances in Parasitology* (Vol. 17, pp. 115–140). (08)60550-3 doi:10.1016/S0065-308X

Karunakaran, S., Pillai, U., & Kuriakose, A. (2011). *Babesia bigemina infection in a 20 day old calf.* Academic Press.

Kjemtrup, A. M., & Conrad, P. A. (2000). Human babesiosis: An emerging tick-borne disease. *International Journal for Parasitology, 30*(12–13), 1323–1337. (00)00137-5 doi:10.1016/S0020-7519

Koudstaal, D., Kemp, D. H., & Kerr, J. D. (1978). Boophilus microplus: Rejection of larvae from British breed cattle. *Parasitology*, *76*(3), 379–386. https://doi.org/10.1017/s0031182000048241

Kumar, S., Kumar, R., & Sugimoto, C. (2009). A perspective on Theileria equi infections in donkeys. *The Japanese Journal of Veterinary Research*, *56*(4), 171–180.

Kuttler, K. L. (1980). Pharmacotherapeutics of drugs used in treatment of anaplasmosis and babesiosis. *Journal of the American Veterinary Medical Association, 176*(10), 1103–1108.

Kuttler, K. L. (2018). World-wide impact of babesiosis. In *Babesiosis of domestic animals and man* (pp. 1–22). CRC Press.

Laha, R., Das, M., Goswami, A., & Singh, P. (2012). *Losses of milk production due to Babesia bigemina infection in a cross bred cow: A case study*. Academic Press.

Laha, R., Das, M., & Sen, A. (2015). Morphology, epidemiology, and phylogeny of Babesia: An overview. *Tropical Parasitology*, *5*(2), 94. https://doi.org/10.4103/2229-5070.162490

Levine, N. D., Corliss, J. O., Cox, F. E. G., Deroux, G., Grain, J., Honigberg, B. M., … Wallace, F. G. (1980). A Newly Revised Classification of the Protozoa: The Committee On Systematics Evolution Of The Society Of Protozoologists. *The Journal of Protozoology, 27*(1), 37–58. doi:10.1111/j.1550-7408.1980.tb04228.x

Mackenstedt, U., Gauer, M., Fuchs, P., Zapf, F., Schein, E., & Mehlhorn, H. (1995). DNA measurements reveal differences in the life cycles of Babesia bigemina and B. canis, two typical members of the genus Babesia. *Parasitology Research*, *81*(7), 595–604. https://doi.org/10.1007/bf00932027

Mackenstedt, U., Gauer, M., Mehlhorn, H., Schein, E., & Hauschild, S. (1990). Sexual cycle of Babesia divergens confirmed by DNA measurements. *Parasitology Research*, *76*(3), 199–206. https://doi.org/10.1007/bf00930815

Madder, M., Thys, E., Achi, L., Touré, A., & De Deken, R. (2011). Rhipicephalus (Boophilus) microplus: A most successful invasive tick species in West-Africa. *Experimental & Applied Acarology*, *53*(2), 139–145. https://doi.org/10.1007/s10493-010-9390-8

Mahoney, D. F. (1974). *The application of epizootiological principals in the control of babesiosis in cattle*. Academic Press.

Mahoney, D. F., & Ross, D. (n.d.). *Epizootiological factors in the control of bovine babesiosis*. Academic Press.

Mahoney, D. F., & Ross, D. R. (1972). Epizootiological factors in the control of bovine babesiosis. *Australian Veterinary Journal*, *48*(5), 292–298. https://doi.org/10.1111/j.1751-0813.1972.tb05160.x

Medlock, J. M., Hansford, K. M., Bormane, A., Derdakova, M., Estrada-Peña, A., George, J.-C., … Van Bortel, W. (2013). Driving forces for changes in geographical distribution of Ixodes ricinus ticks in Europe. *Parasites & Vectors, 6*(1), 1. doi:10.1186/1756-3305-6-1

Mehlhorn, H., & Schein, E. (1985a). The Piroplasms: Life Cycle and Sexual Stages. In *Advances in Parasitology* (Vol. 23, pp. 37–103). (08)60285-7 doi:10.1016/S0065-308X

Mehlhorn, H., & Schein, E. (1985b). The Piroplasms: Life Cycle and Sexual Stages Supported by DFG. In J. R. Baker & R. Muller (Eds.), *Advances in Parasitology* (Vol. 23, pp. 37–103). (08)60285-7 doi:10.1016/S0065-308X

Mehlhorn, H., Schein, E., & Voigt, W. P. (1980). Light and electron microscopic study on developmental stages of Babesia canis within the gut of the tick Dermacentor reticulatus. *The Journal of Parasitology*, *66*(2), 220–228.

Merino, O., Alberdi, P., Pérez de la Lastra, J. M., & de la Fuente, J. (2013). Tick vaccines and the control of tick-borne pathogens. *Frontiers in Cellular and Infection Microbiology, 3*. doi:10.3389/fcimb.2013.00030

Montenegro-James, S. (1992). Prevalence and control of Babesiosis in the Americas. *Memorias do Instituto Oswaldo Cruz*, *87*(suppl 3), 27–36. https://doi.org/10.1590/S0074-02761992000700003

Montenegro-James, S., Toro, M., Leon, E., & Guillen, A. T. (1992). Field evaluation of an exoantigen-containing Babesia vaccine in Venezuela. *Memorias do Instituto Oswaldo Cruz*, *87*(suppl 3), 283–288. https://doi.org/10.1590/S0074-02761992000700048

Morzaria, S., Katende, J., Kairo, A., Nene, V., & Musoke, A. (1992). New methods for the diagnosis of Babesia bigemina infection. *Memorias do Instituto Oswaldo Cruz*, *87*(suppl 3), 201–205. https://doi.org/10.1590/S0074-02761992000700033

Morzaria, S. P., Young, A. S., & Hudson, E. B. (1977). Babesia bigemina in Kenya: Experimental transmission by Boophilus decoloratus and the production of tick-derived stabilates. *Parasitology*, *74*(3), 291–298. https://doi.org/10.1017/s0031182000047910

Mosqueda, J., Olvera-Ramirez, A., Aguilar-Tipacamu, G., & Canto, G. J. (2012). Current Advances in Detection and Treatment of Babesiosis. *Current Medicinal Chemistry*, *19*(10), 1504–1518. https://doi.org/10.2174/092986712799828355

Mudaliar, S. V., Achary, G. R., & Alwar, V. S. (1950). On a species of Babesia in an Indian wild cat (Felis catus). *The Indian Veterinary Journal*, *26*(5), 392–395.

Muhanguzi, D., Byaruhanga, J., Amanyire, W., Ndekezi, C., Ochwo, S., Nkamwesiga, J., … Jongejan, F. (2020). Invasive cattle ticks in East Africa: Morphological and molecular confirmation of the presence of Rhipicephalus microplus in south-eastern Uganda. *Parasites & Vectors, 13*(1), 165. doi:10.118613071-020-04043-z

Norval, R. A., Fivaz, B. H., Lawrence, J. A., & Daillecourt, T. (1983). Epidemiology of tick-borne diseases of cattle in Zimbabwe. I. Babesiosis. *Tropical Animal Health and Production*, *15*(2), 87–94. https://doi.org/10.1007/bf02239802

Notomi, T., Mori, Y., Tomita, N., & Kanda, H. (2015). Loop-mediated isothermal amplification (LAMP): Principle, features, and future prospects. *Journal of Microbiology (Seoul, Korea)*, *53*(1), 1–5. https://doi.org/10.1007/s12275-015-4656-9

OIE. (2014). *Manual of Diagnostic Tests and Vaccines for Terrestrial Animals*. OIE.

Okello, J., Tukahirwa, E. M., Perry, B. D., Rowlands, G. J., Nagda, S. M., Musisi, G., … Opuda-Asibo, J. (1999). Population dynamics of ticks on indigenous cattle in a pastoral dry to semi-arid rangeland zone of Uganda. *Experimental & Applied Acarology, 23*(1), 79–88. doi:10.1023/A:1006058317111

Onyiche, T. E., Suganuma, K., Igarashi, I., Yokoyama, N., Xuan, X., & Thekisoe, O. (2019). A Review on Equine Piroplasmosis: Epidemiology, Vector Ecology, Risk Factors, Host Immunity, Diagnosis and Control. *International Journal of Environmental Research and Public Health*, *16*(10). https://doi.org/10.3390/ijerph16101736

Osmonov, Y. D. (2019). *Methods for Acaricidal Solute Disinfection and Technical Means for their Implementation.* Academic Press.

Owen, I. (1975). Survival of the Non-Parasitic Stages of the Cattle Tick, Boophilus Microplus, Under Simulated Field Conditions in Papua New Guinea. *Australian Journal of Zoology*, *23*(2), 219. https://doi.org/10.1071/ZO9750219

Panda, D. N., Ansari, M. Z., & Sahai, B. N. (1992). Studies on the development and survival periods of the non-parasitic stages of Boophilus microplus (Canestrini), in the climatic conditions of Ranchi (India). *Veterinary Parasitology, 44*(3–4), 275–283. (92)90122-P doi:10.1016/0304-4017

Parker, R. J., Shepherd, R. K., Trueman, K. F., Jones, G. W., Kent, A. S., & Polkinghorne, I. G. (1985). Susceptibility of Bos indicus and Bos taurus to Anaplasma marginale and Babesia bigemina infections. *Veterinary Parasitology, 17*(3), 205–213. (85)90032-9 doi:10.1016/0304-4017

Penzhorn, B. L., & Verwoerd, D. W. (2005). Infectious diseases of livestock, 2nd edition: Book review. *Journal of the South African Veterinary Association, 76*(4), 232. doi:10.4102/jsava.v76i4.435

Perry, B., Chamboko, T., Mahan, S., & Medley, G. (1998). *The economics of integrated tick and tick-borne disease control on commercial farms in Zimbabwe*. Academic Press.

Persing, D. H., Mathiesen, D., Marshall, W. F., Telford, S. R., Spielman, A., Thomford, J. W., & Conrad, P. A. (1992). Detection of Babesia microti by polymerase chain reaction. *Journal of Clinical Microbiology*, *30*(8), 2097–2103.

Petney, T. N. (1997). Ecological implications of control strategies: Arthropods of domestic and production animals. *International Journal for Parasitology, 27*(2), 155–165. (96)00145-2 doi:10.1016/S0020-7519

Pipano, E., & Hadani, A. (1984). Control of bovine babesiosis. In M. Ristic, P. Ambroise-Thomas, & J. Kreier (Eds.), Malaria and Babesiosis (pp. 263–303). https://doi.org/10.1007/978-94-009-6042-8_15.

Potgieter, F. T., & Els, H. J. (1977a). Light and electron microscopic observations on the development of Babesia bigemina in larvae, nymphae and non-replete females of Boophilus decoloratus. *The Onderstepoort Journal of Veterinary Research*, *44*(4), 213–231.

Potgieter, F. T., & Els, H. J. (1977b). The fine structure of intra-erythrocytic stages of Babesia bigemina. *The Onderstepoort Journal of Veterinary Research*, *44*(3), 157–168.

Potgieter, F. T., & Els, H. J. (1979). An electron microscopic study of intra-erythrocytic stages of Babesia bovis in the brain capillaries of infected splenectomized calves. *The Onderstepoort Journal of Veterinary Research*, *46*(1), 41–49.

Radostits, O., Gay, C., Hinchcliff, K., & Constable, P. (2006). *Veterinary Medicine:A textbook of the diseases of cattle, horses, sheep, pigs and goats* (10th ed.). Saunders Elsevier.

Rajput, Z. I., Hu, S., Chen, W., Arijo, A. G., & Xiao, C. (2006). Importance of ticks and their chemical and immunological control in livestock. *Journal of Zhejiang University. Science. B.*, *7*(11), 912–921. https://doi.org/10.1631/jzus.2006.B0912

Riek, R. (1964). The life cycle of Babesia bigemina (Smith and Kilborne, 1893) in the tick vector Boophilus microplus (Canestrini). *Australian Journal of Agricultural Research*, *15*(5), 802. https://doi.org/10.1071/AR9640802

Ristic, M. (1981). Babesiosis. In M. Ristic & W. I. M. McIntyre (Eds.), Diseases of Cattle in the Tropics (pp. 443–468). https://doi.org/10.1007/978-94-011-9034-3_35.

Roberts, J. A. (1968). Resistance of cattle to the tick boophilus microplus (canestrini). II. Stages of the life cycle of the parasite against which resistance is manifest. *The Journal of Parasitology*, *54*(4), 667–673.

Rodriguez-Vivas, R. I., Jonsson, N. N., & Bhushan, C. (2018). Strategies for the control of Rhipicephalus microplus ticks in a world of conventional acaricide and macrocyclic lactone resistance. *Parasitology Research*, *117*(1), 3–29. https://doi.org/10.1007/s00436-017-5677-6

Rogoff, W. M. (1961). Chemical control of insect pests of domestic animals. *Advances in Pest Control Research*, *4*, 153.

Rudzinska, M. A., Spielman, A., Lewengrub, S., Trager, W., & Piesman, J. (1983). Sexuality in piroplasms as revealed by electron microscopy in Babesia microti. *Proceedings of the National Academy of Sciences of the United States of America*, *80*(10), 2966–2970. https://doi.org/10.1073/pnas.80.10.2966

Ruprah, N. (1985). *Textbook of Clinical Protozoology*. Oxonian Press Pvt. Ltd.

Sabu, L., Sreekrishnan, R., Devada, K., Rejitha, T., & Lakshmanan, B. (2013). *Babesia species in a cat-A case report*. Academic Press.

Shkap, V., de Vos, A. J., Zweygarth, E., & Jongejan, F. (2007). Attenuated vaccines for tropical theileriosis, babesiosis and heartwater: The continuing necessity. *Trends in Parasitology*, *23*(9), 420–426. https://doi.org/10.1016/j.pt.2007.07.003

Shortt, H. E. (1973). Babesia canis: The life cycle and laboratory maintenance in its arthropod and mammalian hosts. *International Journal for Parasitology*, *3*(2), 119–148. (73)90019-2 doi:10.1016/0020-7519

Smith, R. D., Osorno, B. M., Brener, J., De La Rosa, R., & Ristic, M. (1978). Bovine babesiosis: Severity and reproducibility of Babesia bovis infections induced by Boophilus microplus under laboratory conditions. *Research in Veterinary Science*, *24*(3), 287–292.

Soulsby, E. (1986). *Helminths, Arthropods and Protozoa of Domesticated Animals* (7th ed.). Bailliere Tindall.

Suarez, C. E., & Noh, S. (2011). Emerging perspectives in the research of bovine babesiosis and anaplasmosis. *Veterinary Parasitology*, *180*(1–2), 109–125. https://doi.org/10.1016/j.vetpar.2011.05.032

Sutherst, R., & Utech, K. (1981). *Controlling livestock parasites with host resistance. CRC Handbook of pest management in agriculture*. CRC Press Inc.

Tabor, A. E., Ali, A., Rehman, G., Rocha Garcia, G., Zangirolamo, A. F., Malardo, T., & Jonsson, N. N. (2017). Cattle Tick Rhipicephalus microplus-Host Interface: A Review of Resistant and Susceptible Host Responses. *Frontiers in Cellular and Infection Microbiology*, *7*, 506. https://doi.org/10.3389/fcimb.2017.00506

Tatchell, R. J. (1992). Ecology in relation to integrated tick management. *International Journal of Tropical Insect Science*, *13*(04), 551–561. https://doi.org/10.1017/S1742758400016131

Tavassoli, M., Tabatabaei, M., Mohammadi, M., Esmaeilnejad, B., & Mohamadpour, H. (2013). PCR-based Detection of Babesia spp. Infection in Collected Ticks from Cattle in West and North-West of Iran. *Journal of Arthropod-Borne Diseases*, *7*(2), 132–138.

Taylor, M. A., Coop, R. L., & Wall, R. (2016). *Veterinary parasitology* (4th ed.). John Wiley and Sons, Inc.

Taylor, M. A., Coop, R. L., & Wall, R. L. (2013). *Veterinary Parasitology*. Retrieved from http://qut.eblib.com.au/patron/FullRecord.aspx?p=1158636

Todorovic, R. A., Lopez, L. A., Lopez, A. G., & Gonzalez, E. F. (1975). Bovine babesiosis and anaplasmosis: Control by premunition and chemoprophylaxis. *Experimental Parasitology, 37*(1), 92–104. (75)90056-9 doi:10.1016/0014-4894

Uilenberg, G. (1972). Diagnosis of Babesia Argentina Infection in Cattle Using Brain Smears. *Australian Veterinary Journal*, *48*(9), 534–534. https://doi.org/10.1111/j.1751-0813.1972.tb02327.x

Uilenberg, G. (2006). Babesia—A historical overview. *Veterinary Parasitology*, *138*(1–2), 3–10. https://doi.org/10.1016/j.vetpar.2006.01.035

Valecha, N., Sharma, V. P., & Devi, C. U. (1998). A Rapid Immunochromatographic Test (ICT) for Diagnosis of Plasmodium falciparum. *Diagnostic Microbiology and Infectious Disease, 30*(4), 257–260. (98)00003-0 doi:10.1016/S0732-8893

Venu, R., Sailaja, N., Srinivasa Rao, K., Jayasree, N., & Vara Prasad, W. L. N. V. (2015). Babesia bigemina infection in a 14-day old Jersey crossbred calf: A case report. *Journal of Parasitic Diseases: Official Organ of the Indian Society for Parasitology*, *39*(2), 264–265. https://doi.org/10.1007/s12639-013-0338-x

Vercammen, F., De Deken, R., & Maes, L. (1995). Clinical and serological observations on experimental infections with Babesia canis and its diagnosis using the IFAT. *Parasite (Paris, France)*, *2*(4), 407–410.

Vos, A. J., & Bock, R. E. (2006). Vaccination against Bovine Babesiosis. *Annals of the New York Academy of Sciences*, *916*(1), 540–545. https://doi.org/10.1111/j.1749-6632.2000.tb05333.x

Waltisbuhl, D. J., Goodger, B. V., Wright, I. G., Commins, M. A., & Mahoney, D. F. (1987). An enzyme linked immunosorbent assay to diagnose Babesia bovis infection in cattle. *Parasitology Research*, *73*(2), 126–131. https://doi.org/10.1007/BF00536468

Wodaje, A., Adudna, B., & Hamid, M. (2019). *A review on Bovine Basiosis. International Journal of Advanced Research in Biological Sciences*. IJARBS.

Wright, I. G., Goodger, B. V., Buffington, G. D., Clark, I. A., Parrodi, F., & Waltisbuhl, D. J. (1989). Immunopathophysiology of babesial infections. *Transactions of the Royal Society of Tropical Medicine and Hygiene, 83*, 11–13. (89)90596-8 doi:10.1016/0035-9203

Yessinou, R. E., Adoligbe, C., Akpo, Y., Adinci, J., Youssao Abdou Karim, I., & Farougou, S. (2018). Sensitivity of Different Cattle Breeds to the Infestation of Cattle Ticks Amblyomma variegatum, Rhipicephalus microplus, and Hyalomma spp. On the Natural Pastures of Opkara Farm, Benin. *Journal of Parasitology Research, 2018*, 2570940. doi:10.1155/2018/2570940

Yusuf, J. (2017). *Review on Bovine Babesiosis and its Economic Importance*. Academic Press.

KEY TERMS AND DEFINITIONS

Acaricides: Chemicals substances that kill ticks and mites.

Aetiology: Is the causative agent of a particular disease or condition.

Babesioses: A complex of a number of diseases caused by *Babesia spp* that affect various animals and humans.

(Bm86): Recombinant antigen composed of glycoproteins, located on the surface of the intestinal cells of ticks.

Endemic Stability: Is a host-vector-pathogen interaction with a high level of challenge to young hosts without clinical disease and a high level of immunity in adult hosts resulting in low incidence of clinical disease.

Haemoglobinuria: Abnormally high concentration of haemoglobin in urine.

Transovarially: Is a process where there is passage of parasites or disease causing agents from maternal body to eggs in the ovary or ovaries.

APPENDIX: FURTHER READING

Diagnosis and Treatment of Babesia-By James Schaller
Babesiosis of Domestic Animals and Man-By Miodrag Ristic
Vaccines against bovine babesiosis: where we are now and possible roads ahead (Florin-Christensen et al., 2014)

Chapter 13
Canine Ehrlichiosis in Africa: Epidemiology, Diagnosis, and Control

Jahashi Saidi Nzalawahe
Sokoine University of Agriculture, Tanzania

Erick . V. G. Komba
Sokoine University of Agriculture, Tanzania

Athumani Msalale Lupindu
Sokoine University of Agriculture, Tanzania

Adrian Evance Materu
Sokoine University of Agriculture, Tanzania

Abdul S. Katakweba
Sokoine University of Agriculture, Tanzania

Ladslaus L. Mnyone
Sokoine University of Agriculture, Tanzania

ABSTRACT

Canine ehrlichiosis is an infection of canids causing a clinical disease in domestic dogs and asymptomatic infections in wild canids. Currently, the disease assumes a cosmopolitan distribution. This chapter summarises published information on the disease from across Africa. Some studies were able to demonstrate experimental infections in these canids. Different diagnostic methods, cell culture, direct microscopy, serology, hematology, and molecular methods were employed in different studies for detection of ehrlichiosis. Treatment of the disease mainly involved use of oxytetracycline, doxycycline, imidocarb disproportionate, and levamisole. In severe cases, management has involved administration of supportive therapy such as blood transfusion. Generally, though available, the information on different aspects of the disease in the Africa is scant and fragmented. There is still a need to generate more information on the epidemiology, diagnosis, clinical aspects, and treatment of the disease.

DOI: 10.4018/978-1-7998-6433-2.ch013

INTRODUCTION

African countries are rich in infectious diseases of animals and humans. This situation is largely attributable to an environment, which favours perpetuation of reservoir animals and arthropod vectors, as well as the spread of disease agents. As in other animals, the health of domestic and wild canids is affected by vector-borne and other diseases, some of which are transmissible to humans (Maggi & Krämer, 2019). Despite the growing population of companion animals and disease transmission risks, information on the occurrence of such diseases worldwide, particularly in Africa, which suffers the highest burden, is either limited or non-existent. Ehrlichiosis is a typical example of companion animal diseases, transmissible to humans, whose data are limited and non-existent. This chapter collates and discussed essential information on canine ehrlichiosis that is currently available across Africa, including disease prevalence rates, diagnostic methods, and control/treatment options. Details on the documented diagnostic and control methods are provided. Information gaps and areas requiring further research have been highlighted. The bottom-line is that the collated information informs the development of appropriate disease surveillance and control strategies in Africa and beyond.

BACKGROUND

Description of Canine Ehrlichiosis: Definition, Aetiology, and Epidemiology

Canine ehrlichiosis, also known as canine rickettsiosis, canine haemorrhagic fever, tracker dog disease, canine tick typhus, Nairobi bleeding disorder, and tropical canine pancytopenia, is an important tick-borne disease that is a member of the family Canidae, which is particularly found in dogs (Dubie et al., 2014). Canine ehrlichiosis was reported for the first time in Algeria in the year 1935 (Donatien & Lestoquard, 1937). Since then, the disease has distributed worldwide (Beall et al., 2012), and Africa suffers the highest burden. The geographical distribution of causative agents corresponds to distribution of their vector tick, *Rhipicephalus sanguineus*. Several species have been reported to cause ehrlichiosis in domestic dogs and wild canids: *Ehrlichia canis*, *E. chaffeensis*, *E. ewingii,* and *E. ruminantium* (Kelly, 2000; Ndip et al., 2010; Parola et al., 2003; Pretorius & Kelly, 1998). *E. canis* is responsible for most of the cases worldwide (Dhliwayo et al., 2019; Penzhorn et al., 2018). *E. canis* is transmitted transstadially and intrastadially by *R. sanguineus* (Bremer et al., 2005). The infective stages, elementary bodies, are transmitted to new hosts via tick bites. These infective stage bodies enter the host's monocytes, or other types of leucocytes, by phagocytosis. They then replicate inside phagosomes by binary fission, forming clusters of tightly packaged elementary bodies termed "initial bodies" (Forbes et al., 2007), which develop further to form morulae. The rupture of affected host cells release elementary bodies, which then infect new cells (Kruss et al., 2003).

Description of the Disease Vector: Ecology and Life Cycle

Rhipicephalus sanguineus is the most widespread tick species in the world, and it is found more commonly in warmer climates. The tick displays both questing- (ambush) and host seeking-behaviours (hunting). Domestic dogs are the most preferential host of *R. sanguineus* (Dantas-Torres et al., 2009). Adult ticks are mostly found in the dog's ears, along the neck and shoulders, and between the toes. Nymphs are mostly found in the ears and longhaired areas of the neck. Larvae are mostly found on the stomach and flanks (Estrada Pena & Jongejan, 1999). Being a typical three-host tick, *R. sanguineus* spends most of its lifetime in the environment rather than on the host.

The tick is more prevalent in free-ranging dogs compared to the domiciled dogs (Dantas-Torres et al., 2004). The tick occasionally infests other domestic/wild hosts, including cats, rodents, birds, and humans (Dantas-Torres et al., 2010; Estrada-Peña & Jongejan, 1999; Iori et al., 1996). Parasitism of humans by *R. sanguineus* mainly occurs during high temperatures (Parola et al., 2008). In tropical and subtropical areas, the ticks are prevalent throughout the year (Jacobs et al., 2001; Louly et al., 2006). In temperate regions, ticks are most prevalent during the summer (Papadopoulos et al., 1996). They can overwinter on the environment and even infest dogs during the winter in some regions (Koch, 1982), however successful oviposition, egg hatching, as well as larval and nymphal moulting, are unlikely at low temperatures (Inokuma et al., 1996; Koch & Tuck, 1986). Under optimal conditions, the tick often completes 3-4 generations (Louly et al., 2007; Silveira et al., 2009).

All developmental stages of *R. sanguineus*, larvae, nymphs, and adults, feed on a host, fall off, and moult on the environment. The adult female remains on the host for five to 21 days after which it drops to digest blood and lay eggs in a sheltered place (Pegram et al., 1987). The incubation of eggs ranges from 1-3 weeks (Jittapalapong et al., 2000). Larvae feed on the host for 3-10 days before they drop off and moult into nymphs, and nymphs feed for 3-11 days before they drop off and develop into adults (Pegram et al., 1987). Unfed larvae can survive for approximately 8 months, whereas unfed nymphs and adults can survive for 6 and 19 months, respectively (Goddard, 1987). Under favourable conditions, the entire lifecycle takes 63–91 days (Louly et al., 2007).

MAIN FOCUS OF THE CHAPTER

This chapter focuses on the epidemiology, diagnosis, and control of canine ehrlichiosis in the African region. On epidemiology, the chapter highlights the disease status with a focus on prevalence rates. The section on diagnosis includes a brief description of existing diagnostic techniques, a detailed discussion of each technique, and an indication of the studies that deployed such techniques. The section on disease control highlights the different types of control strategies available to date. The most widely deployed control strategies in different parts of Africa have also been narrated. Finally, the chapter is culminated with a general discussion, conclusions, recommendations, and research priorities.

The Status of Canine Ehrlichiosis in Africa

Reports of canine ehrlichiosis from 1991 to 2020, from 22 African countries have shown a prevalence that ranges from 0.4% to 88.8%, which is summarized in Table 1.

Diagnosis of Canine Ehrlichiosis

Canine ehrlichiosis can be diagnosed using different techniques: clinical picture, microscopy, haematology, serology, cell culture, and polymerase chain reaction, which make use of different samples from the infected animal. Blood and serum are the most used diagnostic samples. Other diagnostic samples include lung, spleen, liver, and kidney, collected at post-mortem (Harrus & Waner, 2011).

Table 1. Prevalence of canine ehrlichiosis in African countries over a period of 30 years

Country	Prevalence (%)	N	Reference
Tanzania	25	100	Muhairwa et al., 2012
Zimbabwe	52.2, 34–75.2	105–117, respectively	Matthewman et al., 1993; Kelly et al., 1994; Dhliwayo et al., 2019
Uganda	29.5	105	Proboste et al., 2015
Ethiopia	10.66	150	Kirade et al., 2018
Nigeria	10.3–46.1	116–100, respectively	Happi et al., 2018; Kamani et al., 2013
Sudan	88.8	78	Inokuma et al., 2006
Malawi	22	100	Alvåsen et al., 2016
Senegal	18.8–50	142–68	Brouqui et al., 1991; Dahmani et al., 2019
Tunisia	2.14–54.2	106, 68, 10, 216, 286 & 180	Tagorti, 2019; Brouqui et al., 1991; Ghorbel et al., 1994; M`ghirbi et al., 2009
Namibia	23.5–53.8	106	Stuben, 2004; Manyarara et al., 2015
Zambia	0.4	247	Qiu et al., 2018
Angola	5.8	103	Cardoso et al., 2016
Botswana	21	223	Thompson et al., 2020
Cameroon	32–16.3	104	Ndip et al., 2005
South Africa	3–42	1138–161	Matjila et al., 2008; Pretorius & Kelly, 1998
Kenya	15.5	148	Alexandra et al., 1993
Egypt	33	374	Botros et al., 1995
Chad	50	68	Brouqui et al., 1991
Gabon	3.1–84	256 & 111	Roqueplo et al., 2009; Davoust et al., 2006
Ivory Coast	67.8–74	137 & 76	Roqueplo et al., 2009; Davoust et al., 2006
Cape Verde	3.3 - 26.2	130 &153	Götsch et al., 2009; Lauzi et al., 2016
Algeria	6.4	110	Dahmani et al., 2015

Clinical Picture

Canine ehrlichiosis can manifest symptomatically and asymptomatically. Asymptomatic infections occur in both experimentally- and naturally-infected dogs. The manifestation of symptomatic infections ranges from mild (non-myelosuppressive) to life threatening disease (myelosuppressive) (Waner et al., 2013). Clinical presentations are based on retrospective studies of disease records in veterinary practices in Kenya, South Africa, and Namibia (Kitaa et al., 2014; Manyarara et al. 2015; Van Heerden, 1982, 1979). The clinical signs documented across Africa include fever, listlessness, depression, tenesmus, lethargy, harsh lung sounds, vomiting, weakness, inappetence, pale mucous membranes, congestion of mucous membrane, ulcers in the external nares, ocular discharge, panting, vomiting, splenomegaly, lymphadenomegaly, tender abdomen, diarrhoea, haemorrhage, hind limb weakness, pounding heart assumed a squatting-position, and regurgitation in response to begging gestures. Such signs have also been reported elsewhere outside Africa (Mylonakis et al., 2011; Shipov et al., 2008). However, studies outside Africa documented other signs including hypothermia, ulcerative stomatitis, necrotic glossitis, hind limb and/or scrotal edema, bacterial pyoderma, icterus, and central nervous system signs (Kaewmongkol et al., 2016; Mylonakis et al., 2011).

Certain studies associated ehrlichiosis with bleeding diathesis occurrence in the chronic phase and in dogs with concurrent bleeding-predisposing conditions; manifested as cutaneous and mucosal petechiae and ecchymoses, hyphema, epistaxis, hematuria, melena, prolonged bleeding from venipuncture sites, or intraoperative bleeding (Harrus et al., 1997; Lanza-Perea et al., 2014). When diagnosis is based on clinical manifestation, the following differential diagnoses should be considered for acute stage: Rocky Mountain spotted fever, brucellosis, blastomycosis, endocarditis, systemic lupus erythematosus, and lymphosarcoma. Differential diagnoses for a chronic stage include: estrogenic toxicity, myelopathies, immune-mediated pancytopenia, and multisystemic diseases associated with specific organ dysfunction. Other causes of anaemia, haemorrhage, fever, and skin conditions, should also be considered (Waner et al., 2001). Diagnosis of ehrlichiosis becomes complicated in co-infection with babesiosis, Lyme disease, or Rocky Mountain spotted fever (De Tommasi et al., 2013) and bartonellosis.

Direct Microscopy

Diagnosis of ehrlichiosis by direct microscopy is considered simple and cheap, and it involves demonstration of Ehrlichia spp. in Romanowsky (Giemsa or Diff-Quik) stained peripheral-blood buffy coat smears. The organisms appear as clusters of round, dark, purple-stained, small dots or clusters of dots "mulberry blotches" in the leucocytes (Rikihisa, 1991). Intra-cytoplasmic inclusions can be seen in impression smears prepared from the lungs, spleen, liver, and kidneys of infected dogs (Greene et al., 1984; Huxoll, 1976). Occasionally, direct microscopy may involve use of cerebrospinal fluid smears. Direct microscopy works well in acute ehrlichiosis caused by species that primarily infect granulocytes or lymphocytes (Rikihisa, 1991). It is therefore considered as a means for establishing a definitive diagnosis of acute CME (Harrus et al., 2012). E. canis is more frequently found in blood monocytes than other species, thus a positive blood smear provides a definitive diagnosis. The diagnostic sensitivity of cytology has been observed to vary for smears made from different samples/organs (Faria et al., 2010). Cytology can detect co-infections and is capable of supporting diagnosis in acute cases, even before infected dogs seroconvert (Mylonakis et al., 2010). Limitations of direct microscopy include low sensitivity, especially in subclinical and chronic cases, labour-intensiveness, particularly when fewer mononuclear cells are infected, and unreliable specificity attributable to its inability to identify the responsible *Ehrlichia* spp, and some incidences where some artefacts may resemble morulae (Raskin & Messick, 2012; Mylonakis et al., 2010). Many studies in Africa employed direct microscopy for diagnosis of canine ehrlichiosis (Happi et al. 2018; Huxoll, 1976; Kirade et al. 2018; Van Heerden & Immelman, 1979). In Ethiopia, Kirade et al. (2018) adopted a Modified Wright's stain of blood smear and observed it under a microscope at 100×. In Nigeria, Happi et al. (2018) used giemsa-stained whole blood and buffy coat smears. Microscopy establishes definitive diagnosis of acute canine ehrlichiosis and can detect co-infections with other common hemoprotozoans of dogs in Africa (Harrus & Waner, 2011; Harrus et al., 2012; Mylonakis et al., 2010; Mylonakis et al., 2011).

Haematology

The *Ehrlichia canis* infection can also be diagnosed based on haematological abnormalities. Infected dogs show reduced platelets, red blood cells (anaemia), and/or white blood cells (Nakagh et al., 2008). Reduced platelets appear in more than 80% of the cases (Harrus et al., 1998a). However, canine ehrlichiosis should not be ruled out solely based on normal platelet count (Mylonakis et al., 2011; Sainz et al., 2015). Other haematological derangements include non-regenerative anaemia, leukopenia, neutropenia, and lymphopenia or mild lymphocytosis (De Castro et al., 2004; Gianopoulos et al., 2016). Subclinical and chronic phases may be associated with granular lymphocytosis with T-cell expansion imitating lymphocytic leukaemia (Qurollo et al., 2013). Aplastic pancytopenia is a typical feature of myelosuppressive canine ehrlichiosis (Frezoulis et al., 2017). Pancytopenia with normocellular BM may occur in acute canine ehrlichiosis (Mylonakis et al., 2011). Two studies conducted in South Africa diagnosed ehrlichiosis based on haematology (Van Heerden, 1979; Rautenbach et al., 1991).

Serology

Serology is considered the mainstay for the confirmation of exposure to *E. canis* (Harrus & Waner, 2011). In-direct fluorescent antibody (IFA) is considered the "gold standard" for the detection and titration of anti-*E. canis* antibodies (Harrus et al., 2002). Generally, serology is considered to be simple, reliable, and cost-effective (Troy & Forrester, 1990). However, it is limited by its inability to make a species-specific diagnosis due to antigenic cross-reactivity with other *Ehrlichia* species. Antibodies develop 7-35 days post-infection and do not correlate with current carrier status, duration of infection, as well as presence and severity of clinical disease (Harrus & Waner, 2011; Hegarty et al., 2009). In acute infections, clinical illness and derangements in haematological parameters precede sero-conversion (Gaunt et al., 2010; Mylonakis et al., 2010). Therefore, diagnosis of CME in acute cases should not be ruled out based on serology alone. Several ELISA tests are commercially available and they have been calibrated to become positive at an antibody level equivalent to an IFA titer of approximately 1:320 or higher. This could imply a relatively low sensitivity particularly in acute disease cases (Harrus & Waner, 2011). Western immune blotting may distinguish infections that display cross-reactivity, while the chronicity of *E. canis* infection may be inferred based on immunoblot patterns (Suksawat et al., 2000). Serological techniques have been employed to detect *E. canis* in several studies across Africa including ELISA in Tanzania (Muhairwa et al., 2012), Namibia (Manyarara, et al., 2015), Zimbabwe, Malawi (Alvåsen et al., 2016; Dhliwayo, et al., 2019), Botswana (Thompson et al., 2020), IFA in Senegal (Parzy et al., 1991), Zimbabwe (Kelly et al., 2000; Matthewman et al., 1993), Uganda (Proboste et al., 2015), Cameroon (Ndip et al., 2005), Gabon, the Ivory Coast (Davoust et al., 2006), Egypt (Botros et al., 1995), South Africa (Rautenbach et al., 1991; Van Heerden et al., 1995), Tunisia (Ghorbel et al., 2001; M'ghirbi et al., 2009), and Kenya (Alexander et al., 1993).

Cell Culture Technique

This technique involves isolation of the *Ehrlichia* species through inoculation, into the cell culture of the buffy coat or mononuclear cell fractions of the peripheral blood of affected animals (Rikihisa et al., 1991). Misao and Kobayashi, (1954) were able to isolate *E. sennetsu* by intra-peritoneal injection of blood from infected human into mice. The cell culture technique has a higher sensitivity than direct observation of blood smears. Cell culture can detect infections even in samples containing a single organism. Notably, cell culture was used to detect *E. canis* in free-living jackals *(Canis mesomelas)* and domestic dogs in Kenya (Price & Karstad, 1980).

Polymerase Chain Reaction

Polymerase chain reaction (PCR) is most robust, sensitive, and rapid diagnostics for detection of ehrlichiosis (Happi et al., 2018; McBride et al., 1996). The method addresses several limitations of other techniques such as serology, which confirms exposure rather than current infection, and cytology, whose diagnostic sensitivity is low. PCR is useful for the early detection, characterization, and quantification of ehrlichial organisms (Sainz et al., 2015). Also, PCR is better than serology in detection of concurrent infection with multiple species (Kordick et al., 1999). Several studies employed PCR alone (Dahmani et al., 2019; Mtshali et al., 2017; Qiu et al., 2018; Rautenbach et al., 2018) or in combination with other diagnostic methods (Happi et al., 2018; Ndip et al., 2005) for detection of ehrlichiosis. These studies were conducted in Algeria (Dahmani et al., 2015), Cape Verde (Götsch, et al., 2009; Lauzi et al., 2016), Nigeria (Happi et al., 2018), Angola (Cardoso et al., 2016), South Africa (Matjila et al., 2008; Mtshali et al., 2017; Rautenbach et al., 2018), Sudan (Inokuma, et al., 2006), Senegal (Dahmani et al., 2019), Cameroon (Ndip et al., 2005), and Zambia (Qiu et al., 2018).

CONTROL OF CANINE EHRLICHIOSIS

There is no commercial vaccine available to date for the control of canine ehrlichiosis. A strict tick control program for dogs and premises, and treatment of infected animals remain the most widely used control approaches (Ettinger et al., 1995; Rudoler et al., 2012).

Vector Control

Control of *R. sanguineus* in dogs and the environment can be achieved by using acaricides and non-chemical methods (Beugnet & Franc, 2012; Dantes-Torres, 2008; Dixit et al., 2012; Roopali et al., 2018). The non-chemical methods include removal of ticks by hand (Blagburn & Dryden, 2009) and minimizing refugia, such as floor cracks, garden waste, and long grass. Acaricides are mostly deployed during heavy infestations of dogs and the environment (Dantes-Torres, 2008). The application of acaricides on dogs can be achieved by dipping or spraying using hand-sprays.

Alternatively, "pour-on" formulations, acaricide-impregnated collars, and systemic acaricides, can be used. In endemic areas with high tick populations, the use of low doses of tetracycline (0.45 kg of body weight) or doxycycline (0.45 kg of body weight) every 24 hours during high tick season is recommended (Dantas-Torres, 2008; Straten van der, 2008). Coincidentally, control measures directed against *R. sanguineus* can prevent several tick-borne haemo-parasites of dogs (Weerathunga et al., 2019).

Treatment of Infected Dogs

The drugs of choice for treatment of canine ehrlichiosis are tetracycline and doxycycline (Eddlestone et al., 2006; Lanza-Perea et al., 2014; McClure et al., 2010; Schaefer et al., 2008). The recommended dosage for doxycycline in dogs is 5 mg/kg to 10 mg/kg, per os for 28 days; it can be administered either once or twice daily (Neer & Harrus, 2006). Despite the existence of several studies that have addressed the therapeutic efficacy of tetracycline hydrochloride (Adawa et al., 1992; Davidson et al., 1978) or doxycycline hyclate (Bartsch & Greene, 1996; Ristic et al., 1972; Van Heerden & Immelman, 1979), the effectiveness of tetracycline derivatives for clearing E. canis infection remains controversial. In an experimental infection, treatment with doxycycline hyclate for 7 days resulted in a carrier status in some dogs (Iqbal & Rikihisa, 1994). The drug has been very effective in achieving clinical and/or clinico-pathologic recovery in the vast majority of experimentally or naturally infected dogs, but it has not been invariably effective in eliminating E. canis infection (Harrus et al., 1998b; Harrus et al., 2004; Eddlestone et al., 2007; McClure et al., 2010; Schaefer et al., 2007; Schaefer et al., 2008). Despite the clinical and hematologic recovery of the dogs, and negative blood PCR in the majority of treated dogs, *R. sanguineus* ticks fed on such dogs after doxycycline treatment (xenodiagnosis) and became PCR-positive for *E. canis* (McClure et al., 2010). These results imply that *E. canis* infection may persist even after prolonged doxycycline treatment. Doxycycline is also ineffective in dogs with profound aplastic pancytopenia (Mylonakis et al., 2004).

There exist reports of using imidocarb dipropionate as an alternative treatment of *E. canis* (Eddlestone et al., 2006; Matthewman et al., 1993; Ogunkoya et al., 1981; Sainz et al., 2000; Van Heerden & Van Heerden, 1981) in which authors reveal variable outcomes. Some authors assert that imidocarb dipropionate is not effective in achieving clinical and hematologic remission or in clearing the infection (Neer et al., 1999; Eddlestone et al., 2006; Rudoler et al., 2012). This observation indicates limitation in the use of imidocarb dipropionate except in cases of co-infection with *Babesia canis* (Eddlestone et al., 2006; Sainz et al., 2015).

Rifampicin has recently received attention as a potential alternative drug to doxycycline. In an in vitro study, rifampicin was as effective as doxycycline against E. canis (Branger et al., 2004). Rifampicin given (15 mg/kg/12 h orally for 7 days) to two sub-clinically infected and moderately pancytopenic dogs, resolved pancytopenia and cleared the infection (Schaefer et al., 2008). The same rifampicin regimen, given to two asymptomatic *E. canis*-infected dogs 700 days post-inoculation, after an ineffective course with doxycycline, appeared to clear the infection in one of the two dogs, based on xeno diagnosis with ticks (McClure et al., 2010). Provided that the safety profile of rifampicin in the dog will be sufficiently refined (Bajwa et al., 2013), it might be a promising alternative to doxycycline in CME, as the clinical experience in human ehrlichiosis currently suggests (Dumler et al., 2007; Ismail & McBride 2017).

In dogs with CME-induced aplastic pancytopenia, supportive treatment is critically important if the limited chances for survival are to be pursued. It includes the administration of balanced crystalloid solutions and/or the periodic blood-typed and cross-matched, packed red blood cells or whole blood transfusions, and prophylactic or therapeutic use of antibiotics (Mylonakis et al., 2010).

A few studies have reported protocols that were adopted for treatment of canine ehrlichiosis in Africa. In Kenya, documentation of canine ehrlichiosis over a period of 13 years (Kitaa et al., 2014) revealed the following treatment protocols: intramuscular injection of imidocarb dipropionate (5 mg/kg) with a repeat after two weeks; oral doxycycline (10 mg/kg daily) for two weeks; oral tetracycline (22 mg/kg three times a day) for two weeks; or a combination of imidocarb dipropionate with either tetracycline or doxycycline. Van Heerden (1982) conducted a retrospective review of 120 cases in South Africa and revealed the following treatment protocols: (i) Oxytetracycline (Terramycin capsules, Pfizer) administered orally at a dosage rate of 100 mg/kg once daily or 50 mg/kg twice daily for not less than 10 days; (ii) Oxytetracycline (Liquamycin, Pfizer) administered intravenously at a dosage rate of 5-10 mg/kg once daily for not less than six days. In this protocol, if the drug was given for less than six days, it was followed by oral administration of oxytetracycline for over 10 days; (iii) Doxycycline (Doxyvet, Milvet) administered either orally or intravenously at a dose of 5-10 mg/kg for not less than 10 days. Depending on the severity of the disease, one or more of the following supportive therapy was administered in addition to the drugs: a blood transfusion, intravenous fluids, such as Sodium Chloride and Dextrose, and essential phospholipids.

DISCUSSION

Tick-borne disease agents, particularly hemoparasites, are among the most important causes of morbidity and mortality in dogs worldwide. Among them is canine ehrlichiosis, a life-threatening disease of dogs, mostly occurring in tropical and subtropical countries, including those in Africa. The reported frequency of occurrence of canine ehrlichiosis across Africa is more or less similar to other parts of the world (Dahmani et al., 2019; Qiu et al., 2018; Thompson et al., 2020). This underscores the assertion that the spread of canine ehrlichiosis worldwide correlates with distribution of the principal vector tick, *R. sanguineus*, which is the most widespread and well-recognized vector of many pathogens affecting dogs, other animals, and occasionally humans (Ayoob et al., 2010; Dantas-Torres et al., 2010; Harrus & Waner, 2011).

Although *E. canis* is the most predominant causative agent, natural infections with other species, *E. ewingii, E. chaeensis, and E. ruminantium*, have also been reported in different parts of the world (Beall et al., 2012). In Africa, however, only *E. canis* and *E. ruminantium* are known to occur (Hegarty et al., 1997; Kelly et al., 1994; Matthewman et al., 1994; Pretorius & Kelly, 1998). This information emanates from a few studies that were conducted in the southern part of the continent. Presumably, more studies across the continent would have potentially disclosed other species. All this information is necessary for efforts to devise surveillance and control strategies against the disease.

The diagnosis of canine ehrlichiosis can be achieved by several methods (Rotondano et al., 2012), which vary greatly in terms of their sensitivity and specificity. Cell culture re-isolation is known to be the most sensitive and definitive for early diagnosis of ehrlichiosis, however its long turn-around time makes it inconvenient. PCR, on the other hand, is convenient, direct, and quick, in detecting *E. canis* DNA, and is as sensitive as the other well-established methods, including serology. Most of the authors in Africa adopted serological techniques for detection of ehrlichiosis in domestic and wild canids. Partly, this could be explained by the reasonable cost associated with the technique.

Clinical features of canine ehrlichiosis have been reported by several authors globally (Kaewmongkol et al., 2016; Lanza-Perea et al., 2014; Mylonakis et al., 2011; Waner et al., 2013). Similar signs have been detected in *E. canis* infections in dogs in Africa. The most commonly encountered signs of the disease include: fever, listlessness, depression, tenesmus, lethargy, harsh lung sounds, vomiting, weakness, inappetance, pale mucous membranes, congestion of mucous membrane, ulcers in the external nares, ocular discharge, panting, vomiting, splenomegaly, lymphadenomegaly, tender abdomen, diarrhoea, haemorrhage, hind limb weakness, pounding heart when in a squatting position, and regurgitation in response to begging gestures. Documentation of these signs would greatly help in syndromic surveillance of ehrlichiosis in areas without access to laboratory facilities.

Canine ehrlichiosis has significant health and socio-economic consequences, mainly attributable to the loss of companions, as well as disease treatment and control costs. The zoonotic potential of the disease has been reported in certain parts of the world (Barrantes-González et al., 2016; Gutierrez et al., 2008). This makes infections caused by these organisms to be of both veterinary and public health concern, thus emphasizing collaborative control efforts between veterinary and public health professionals.

The effective control of vector-borne diseases requires a thorough understanding of the aetiological agents and their epidemiology, including their vectors and hosts. This book chapter has revealed limited information on the vector ecology, distribution, and dynamics of *E. canis* infections, under different eco-climatic conditions of the African continent. Such epidemiological information would provide very useful inputs in planning appropriate surveillance and control strategies.

The control of canine ehrlichiosis can be achieved by either targeting the vector tick or targeting the aetiological agent present in an infected dog through treatment. Though, not formerly reported, dipping of dogs to control ticks is very common in several African communities. The use of acaricides to control *R. sanguineus* in dogs is constrained by the development resistance, as reported in other parts of the world (Coles & Dryden, 2014). In Africa, research reports on the effectiveness of commonly used acaricides against dog ticks are very limited. Therefore, future research on this area is necessary to support informed decision on strategies to control tick-borne diseases in Africa. The use of acaricide-impregnated dog collars for control of ticks has been reported to be more effective compared to dipping. This method, however, is mostly deployed outside Africa. Its deployment in Africa is as of yet extremely low. The collars are much safer and easier to use (Koch & Popham, 1987). Nevertheless, dips may be more cost-effective if several dogs have to be treated, especially in the African continent where dog owners prefer to use commercial dips.

The use of medicinal plants across Africa is skewed to in-vitro acaricidal and repellent activities against immature and adult stages of ticks on cattle (Adenubi et al., 2016; Wanzala et al., 2012). This development provides a good alternative for tick control in cattle and other animal species, as Africa is rich in natural products, and development of resistance against natural products occurs at a much slower rate.

Most African countries are rich in infectious diseases, yet they are still lacking well-established surveillance systems. This has been clearly revealed based on many studies the authors reviewed in this chapter. Very few surveillance studies were found. This fact makes a thorough characterization of the magnitude of the disease and associated impact difficult, thus compromising efforts aimed at developing disease control strategies.

FUTURE RESEARCH DIRECTIONS

Although the available information on canine ehrlichiosis in Africa covers several aspects of the disease, there are still several research gaps that need to be addressed. Epidemiological aspects that require further research include the actual distribution of the disease, seasonal trends, breed and sex predisposition, co-occurrence with other diseases, predictors for its occurrence and persistence, reservoir populations, the role of wildlife on disease epidemiology, responsible vectors, and their distribution. Clinical aspects that require further research include the most common manifestations, responses to different treatment regimes, and prognosis in different dog populations. Infection outcomes in wild canids should also be well investigated, especially among endangered species populations.

CONCLUSION

Information on canine ehrlichiosis in Africa is scant and fragmented. However, the available information provides evidence of the presence of the responsible organism both in domestic canids and in reservoir wildlife canids. In domestic canids, the organisms cause a diverse range of clinical, haematological, and pathological features observed by authors elsewhere outside of Africa.

SOLUTIONS AND RECOMMENDATIONS

Despite the existence of reports on canine ehrlichiosis from across Africa, the information is scant and fragmented. This limits the efforts aimed at development of control strategies for the disease. Therefore, there is a dire need to conduct more studies to complement the available information. Consideration should be given to the use of more robust and sensitive techniques and extend to a wider population. The local population needs to be educated on tick control and the disease as a whole. Clinicians should be made aware of the potential severity of ehrlichiosis to ensure timely and proper treatment of the disease. Furthermore, because ehrlichia infections may result in non-specific clinical manifestations, clinicians must be regularly updated on the various clinical observations that may be exhibited by infected dogs in order to make a rational decision on the diagnosis and management.

ACKNOWLEDGMENT

This research received no specific grant from any funding agency in the public, commercial, or not-for-profit sectors.

REFERENCES

Adawa, D. A. Y., Hassan, A. Z., Abdullah, S. U., Ogunkoya, A. B., Adeyanju, J. B., & Okoro, J. E. (1992). Clinical trial of long-acting oxytetracycline and piroxicam in the treatment of canine ehrlichiosis. *The Veterinary Quarterly*, *15*(3), 118–120. doi:10.1080/01652176.1992.9694345 PMID:1413442

Adenubi, O. T., Fasina, F. O., McGaw, L. J., Eloff, J. N., & Naidoo, V. (2016). Plant extracts to control ticks of veterinary and medical importance: A review. *South African Journal of Botany*, *105*, 178–193. doi:10.1016/j.sajb.2016.03.010

Alexander, K. A., Conrad, P. A., Gardner, I. A., Parish, C., Appel, M., Levy, M. G., Lerche, N., & Kat, P. (1993). serologic survey for selected microbial pathogens in African wild dogs (Lycaon pictus) and sympatric domestic dogs (*Canis familiaris*) in Maasai Mara, Kenya. *Journal of Zoo and Wildlife Medicine*, *24*(2), 140–144.

Alvåsen, K., Johansson, S. M., Höglund, J., Ssuna, R., & Emanuelson, U. (2016). A field survey on parasites and antibodies against selected pathogens in owned dogs in Lilongwe, Malawi. *Journal of the South African Veterinary Association*, *87*(1), a1358. doi:10.4102/jsava.v87i1.1358 PMID:27543039

Annaline Van Heerden, J., & Heerden, V. (1981). Attempted treatment of canine ehrlichiosis with imidocarbdipropionate. *Journal of the South African Veterinary Association*, *52*(3), 173–175. PMID:7310786

Ayoob, A. L., Hackner, S. G., & Prittie, J. (2010). Clinical management of canine babesiosis. *Journal of Veterinary Emergency and Critical Care (San Antonio, Tex.)*, *20*(1), 77–89. doi:10.1111/j.1476-4431.2009.00489.x PMID:20230437

Bajwa, J., Charach, M., & Duclos, D. (2013). Adverse effects of rifampicin in dogs and serum alanine aminotransferase monitoring recommendations based on a retrospective study of 344 dogs. *Veterinary Dermatology*, *24*(6), 570–e135. doi:10.1111/vde.12083 PMID:24164464

Barrantes-González, A. V., Jiménez-Rocha, A. E., Romero-Zuñiga, J. J., & Dolz, G. (2016). Serology, molecular detection and risk factors of Ehrlichia canis infection in dogs in Costa Rica. *Ticks and Tick-Borne Diseases*, *7*(6), 1245–1251. doi:10.1016/j.ttbdis.2016.07.006 PMID:27440400

Beall, M. J., Alleman, A. R., Breitschwerdt, E. B., Cohn, L. A., Couto, C. G., Dryden, M. W., Guptill, L. C., Iazbik, C., Kania, S. A., Lathan, P., Little, S. E., Roy, A., Sayler, K. A., Stillman, B. A., Welles, E. G., Wolfson, W., & Yabsley, M. J. (2012). Seroprevalence of Ehrlichia canis, Ehrlichia chaffeensis and Ehrlichia ewingii in dogs in North America. *Parasites & Vectors*, *5*(1), 29. doi:10.1186/1756-3305-5-29 PMID:22316160

Beugnet, F., & Franc, M. (2012). Insecticide and acaricide molecules and/or combinations to prevent pet infestation by ectoparasites. *Trends in Parasitology*, *28*(7), 267–279. doi:10.1016/j.pt.2012.04.004 PMID:22627116

Blagburn, B. L., & Dryden, M. W. (2009). Biology, treatment and control of flea and tick infestations. *Veterinary Clinics: Small Animal Practice*, *39*(6), 1173–1200. doi:10.1016/j.cvsm.2009.07.001 PMID:19932369

Botros, B. A. M., Elmolla, M. S., Salib, A. W., Calamaio, C. A., Dasch, G. A., & Arthur, R. R. (1995). Canine ehrlichiosis in Egypt: Sere-epidemiological survey. *The Onderstepoort Journal of Veterinary Research*, *62*, 41–43. PMID:8539033

Branger, S., Rolain, J. M., & Raoult, D. (2004). Evaluation of antibiotic susceptibilities of Ehrlichia canis, Ehrlichia chaffeensis, and Anaplasma phagocytophilum by real-time PCR. *Antimicrobial Agents and Chemotherapy*, *48*(12), 4822–4828. doi:10.1128/AAC.48.12.4822-4828.2004 PMID:15561862

Bremer, W. G., Schaefer, J. J., Wagner, E. R., Ewing, S. A., Rikihisa, Y., Needham, G. R., Jittapalapong, S., Moore, D. L., & Stich, R. W. (2005). Transstadial and intrastadial experimental transmission of Ehrlichia canis by male Rhipicephalus sanguineus. *Veterinary Parasitology*, *131*(1-2), 95–105. doi:10.1016/j.vetpar.2005.04.030 PMID:15941624

Brouqui, P., Davoust, B., Haddad, S., Vidor, E., & Raoult, D. (1991). Serological evaluation of Ehrlichia canis infections in military dogs in Africa and Reunion Island. *Veterinary Microbiology*, *26*(1-2), 103–105. doi:10.1016/0378-1135(91)90046-I PMID:2024432

Cardoso, L., Oliveira, A. C., Granada, S., Nachum-Biala, Y., Gilad, M., Lopes, A. P., Sousa, S. R., Vilhena, H., & Baneth, G. (2016). Molecular investigation of tick-borne pathogens in dogs from Luanda, Angola. *Parasites & Vectors*, *9*(1), 1–6. doi:10.118613071-016-1536-z PMID:27160839

Coles, T. B., & Dryden, M. W. (2014). Insecticide/acaricide resistance in fleas and ticks infesting dogs and cats. *Parasites & Vectors*, *7*(1), 1–10. doi:10.1186/1756-3305-7-8 PMID:24393426

Dahmani, M., Davoust, B., Sambou, M., Bassene, H., Scandola, P., Ameur, T., Raoult, D., Fenollar, F., & Mediannikov, O. (2019). Molecular investigation and phylogeny of species of the Anaplasmataceae infecting animals and ticks in Senegal. *Parasites & Vectors*, *12*(1), 1–5. doi:10.118613071-019-3742-y PMID:31640746

Dahmani, M., Loudahi, A., Mediannikov, O., Fenollar, F., Raoult, D., & Davoust, B. (2015). Molecular detection of Anaplasma platys and Ehrlichia canis in dogs from Kabylie, Algeria. *Ticks and Tick-Borne Diseases*, *6*(2), 198–203. doi:10.1016/j.ttbdis.2014.12.007 PMID:25583345

Dantas-Torres, F. (2008). The brown dog tick Rhipicephalus sanguineus (Latreille, 1806) (Acari:Ixodidae) from taxonomy to control. *Veterinary Parasitology*, *152*(3-4), 173–185. doi:10.1016/j.vetpar.2007.12.030 PMID:18280045

Dantas-Torres, F., Ferreira, D. R., de Melo, L. M., Lima, P. A. C., Siqueira, D. B., Rameh-de-Albuquerque, L. C., de Melo, A. V., & Ramos, J. A. C. (2010). Ticks on captive and free-living wild animals in northeastern Brazil. *Experimental & Applied Acarology*, *50*(2), 181–189. doi:10.100710493-009-9296-5 PMID:19693679

Dantas-Torres, F., Figueredo, L. A., & Faustino, M. A. G. (2004). Ectoparasitos de cães provenientes de alguns municípios da região metropolitana do Recife, Pernambuco, Brasil. *Revista Brasileira de Parasitologia Veterinária*, *13*(4), 151–154.

Dantas-Torres, F., Melo, M. F., Figueredo, L. A., & Brandão-Filho, S. P. (2009). Ectoparasite infestation on rural dogs in the municipality of São Vicente Férrer, Pernambuco, Northeastern Brazil. *Revista Brasileira de Parasitologia Veterinária*, *18*(3), 75–77. doi:10.4322/rbpv.01803014 PMID:19772781

Davidson, D. E., Dill, G. S., Tingpalapong, M., Premabutra, S., Nguen, P. L., Stephenson, E. H., & Ristic, M. (1978). Prophylactic and therapeutic use of tetracycline during an epizootic of ehrlichiosis among military dogs. *Journal of the American Veterinary Medical Association*, *172*(6), 697. PMID:640933

Davoust, B., Bourry, O., Gomez, J., Lafay, L., Casali, F., Leroy, E., & Parzy, D. (2006). Surveys on Seroprevalence of Canine Monocytic Ehrlichiosis among Dogs Living in the Ivory Coast and Gabon and Evaluation of a Quick Commercial Test Kit Dot-ELISA. *Annals of the New York Academy of Sciences*, *1078*(1), 464–469. doi:10.1196/annals.1374.086 PMID:17114755

De Castro, M. B., Machado, R. Z., de Aquino, L. P., Alessi, A. C., & Costa, M. T. (2004). Experimental acute canine monocytic ehrlichiosis: Clinicopathological and immunopathological findings. *Veterinary Parasitology*, *119*(1), 73–86. doi:10.1016/j.vetpar.2003.10.012 PMID:15036578

De Tommasi, A. S., Otranto, D., Dantas-Torres, F., Capelli, G., Breitschwerdt, E. B., & Caprariis, D. (2013). Are vector-borne pathogen co-infections complicating the clinical presentation in dogs? *Parasites & Vectors*, *6*(1), 97. doi:10.1186/1756-3305-6-97 PMID:23587324

Dhliwayo, S., Chihambakwe, B., Taonezvi, K., Chikerema, S. M., Tivapasi, M. T., & Pfukenyi, D. M. (2019). Seroprevalence of Canine Ehrlichiosis and Microscopic Screening for Canine Babesiosis in Dogs in Harare, Zimbabwe, 2016-2017. *Veterinary Medicine International*, *2019*, 4130210. doi:10.1155/2019/4130210 PMID:31885847

Dixit, A. K., Dixit, P., & Shukla, P. C. (2012). Canine Monocytic Ehrlichiosis and its therapeutic management in a dog. *Intas Polivet*, *13*(1), 140–141.

Donatien, A., & Lestoquard, F. (1937). State of the present knowledge concerning rickettsiosis of animals. *Archives. Institut Pasteur d'Algerie*, *15*, 142–187.

Dubie, T., Mohammed, Y., Terefe, G., Muktar Y., & Tesfaye, J. (2014). An insight review on canine ehrlichiosis with emphasis on its epidemiology and pathogenesity importance. *Global Journal of Veterinary Medicine and Research, 2*(4), 59–67.

Dumler, J. S., Madigan, J. E., Pusterla, N., & Bakken, J. S. (2007). Ehrlichioses in humans: Epidemiology, clinical presentation, diagnosis, and treatment. *Clinical Infectious Diseases*, *45*(1, Supplement_1), 45–51. doi:10.1086/518146 PMID:17582569

Eddlestone, S. M., Diniz, P. P., Neer, T. M., Gaunt, S. D., Corstvet, R., Cho, D., Hosgood, G., Hegarty, B., & Breitschwerdt, E. B. (2007). Doxycycline clearance of experimentally induced chronic Ehrlichia canis infection in dogs. *Journal of Veterinary Internal Medicine*, *21*(6), 1237–1242. doi:10.1111/j.1939-1676.2007.tb01944.x PMID:18196732

Eddlestone, S. M., Neer, T. M., Gaunt, S. D., Corstvet, R., Gill, A., Hosgood, G., Hegarty, B., & Breitschwerdt, E. B. (2006). Failure of imidocarb dipropionate to clear experimentally induced Ehrlichia canis infection in dogs. *Journal of Veterinary Internal Medicine*, *20*(4), 840–844. doi:10.1111/j.1939-1676.2006.tb01795.x PMID:16955806

Estrada-Peña, A., & Jongejan, F. (1999). Ticks feeding on humans: A review of records on human-biting Ixodoidea with special reference to pathogen transmission. *Experimental & Applied Acarology*, *23*(9), 685–715. doi:10.1023/A:1006241108739 PMID:10581710

Ettinger, S., Feldman, F., & Edward, C. (1995). *Textbook of Veterinary Internal Medicine* (4th ed.). W.B. Saunders Company.

Faria, J. L. M., Dagnone, A. S., Munhoz, T. D., João, C. F., Pereira, W. A. B., Machado, R. Z., & Tinucci-Costa, M. (2010). Ehrlichia canis morulae and DNA detection in whole blood and spleen aspiration samples. *Revista Brasileira de Parasitologia Veterinária*, *19*(2), 98–102. doi:10.1590/S1984-29612010000200006 PMID:20624346

Forbes, B. A., Sahm, D. F., & Weissfeld, A. S. (2007). Obligate intracellular and nonculturable bacterial agents. In Bailey and Scott's diagnostic microbiology (12th ed.). Miss. Mosby.

Frezoulis, P. S., Angelidou, E., Karnezi, D., Oikonomidis, I. L., Kritsepi-Konstantinou, M., Kasabalis, D., & Mylonakis, M. E. (2017). Canine pancytopoenia in a Mediterranean region: A retrospective study of 119 cases (2005 to 2013). *The Journal of Small Animal Practice*, *58*(7), 395–402. doi:10.1111/jsap.12647 PMID:28247969

Gaunt, S. D., Beall, M. J., Stillman, B. A., Lorentzen, L., Diniz, P. P. V. P., Chandrashekar, R., & Breitschwerdt, E. B. (2010). Experimental infection and co-infection of dogs with Anaplasma platys and Ehrlichia canis: Hematologic, serologic and molecular findings. *Parasites & Vectors*, *3*(1), 33. doi:10.1186/1756-3305-3-33 PMID:20377870

Ghorbel, A., Ben Ayed, M., Diwan, F., Ghram, A., Landolsi, F., Messaadp, L., Zrellp, S., & Chabchoub, A. (2001). Incidence and seroprevalence of canine ehrlichiosis in the Medjez El Bab region (north-western Tunisia) during 1994, 1995 and 1996. *Archives de l'Institut Pasteur de Tunis*, *78*(1-4), 41–47. PMID:14658238

Ghorbel, A., Clerc, B., & Djaiem, A. (1994). Canine ehrlichiosis in Tunisia. Seroepidemiological survey. *Revue d'elevage et de medecine veterinaire despays tropicaux. Revue d'Élevage et de Médecine Vétérinaire des Pays Tropicaux*, *47*(3), 271–275. PMID:7709025

Gianopoulos, A., Mylonakis, M. E., Theodorou, K., & Christopher, M. M. (2016). Quantitative and qualitative leukocyte abnormalities in dogs with experimental and naturally occurring acute canine monocytic ehrlichiosis. *Veterinary Clinical Pathology*, *45*(2), 281–290. doi:10.1111/vcp.12359 PMID:27142117

Götsch, S., Leschnik, M., Duscher, G., Burgstaller, J. P., Wille-Piazzai, W., & Joachim, A. (2009). Ticks and haemoparasites of dogs from Praia, Cape Verde. *Veterinary Parasitology*, *66*(1-2), 171–174. doi:10.1016/j.vetpar.2009.08.009 PMID:19729247

Greene, C. E., & Harvey, J. W. (1984). Canine ehrlichiosis. In C. E. Greene (Ed.), *Clinical microbiology and infectious diseases of the dog and cat*. The W. B. Saunders Company.

Gutierrez, C. N., Martinez, M., Sanchez, E., De Vera, M., Rojas, M., Ruiz, J., & Triana-Alonso, F. J. (2008). Cultivation and molecular identification of Ehrlichia canis and Ehrlichia chaffeensis from a naturally co-infected dog in Venezuela. *Veterinary Clinical Pathology*, *37*(3), 258–265. doi:10.1111/j.1939-165X.2008.00046.x PMID:18761516

Happi, A. N., Toepp, A. J., Ugwu, C. A., Petersen, C. A., & Sykes, J. E. (2018). Detection and identification of blood-borne infections in dogs in Nigeria using light microscopy and the polymerase chain reaction. *Veterinary Parasitology. Regional Studies and Reports*, *11*, 55–60. doi:10.1016/j.vprsr.2017.12.002 PMID:31014619

Harrus, S., Alleman, A. R., Bark, H., Mahan, S. M., & Waner, T. (2002). Comparison of three enzyme-linked immunosorbant assays with the indirect immunofluorescent antibody test for the diagnosis of canine infection with Ehrlichia canis. *Veterinary Microbiology*, *86*(4), 361–368. doi:10.1016/S0378-1135(02)00022-6 PMID:11955786

Harrus, S., Kass, P. H., Klement, E., & Waner, T. (1997). Canine monocytic ehrlichiosis: A retrospective study of 100 cases, and an epidemiological investigation of prognostic indicators for the disease. *The Veterinary Record*, *141*(14), 360–363. doi:10.1136/vr.141.14.360 PMID:9351183

Harrus, S., Kenny, M., Miara, L., Aizenberg, I., Waner, T., & Shaw, S. (2004). Comparison of simultaneous splenic sample PCR with blood sample PCR for diagnosis and treatment of experimental Ehrlichia canis infection. *Antimicrobial Agents and Chemotherapy*, *48*(11), 4488–4490. doi:10.1128/AAC.48.11.4488-4490.2004 PMID:15504892

Harrus, S., & Waner, T. (2011). Diagnosis of canine monocytotropic ehrlichiosis (Ehrlichia canis): An overview. *Veterinary Journal (London, England)*, *187*(3), 292–296. doi:10.1016/j.tvjl.2010.02.001 PMID:20226700

Harrus, S., Waner, T., Aizenberg, I., & Bark, H. (1998b). Therapeutic effect of doxycycline in experimental subclinical canine monocytic ehrlichiosis: Evaluation of a 6-week course. *Journal of Clinical Microbiology*, *36*(7), 2140–2142. doi:10.1128/JCM.36.7.2140-2142.1998 PMID:9650986

Harrus, S., Waner, T., Aizenberg, I., Foley, J. E., Poland, A. M., & Bark, H. (1998a). Amplification of ehrlichial DNA from dogs 34 months after infection with Ehrlichia canis. *Journal of Clinical Microbiology*, *36*(1), 73–76. doi:10.1128/JCM.36.1.73-76.1998 PMID:9431923

Harrus, S., Waner, T., & Neer, T. M. (2012). *Ehrlichia canis infection. Infectious diseases of the dog and cat* (4th ed.). Elsevier Saunders.

Hegarty, B. C., de Paiva Diniz, P. P., Bradley, J. M., Lorentzen, L., & Breitschwerdt, E. (2009). Clinical relevance of annual screening using a commercial enzyme-linked immunosorbent assay (SNAP 3Dx) for canine ehrlichiosis. *Journal of the American Animal Hospital Association*, *45*(3), 118–124. doi:10.5326/0450118 PMID:19411647

Hegarty, B. C., Levy, M. G., Gager, R. F., & Breitschwerdt, E. B. (1997). Immunoblot analysis of the immunoglobulin G response to *Ehrlichia canis* in dogs: An international survey. *Journal of Veterinary Diagnostic Investigation*, *9*(1), 32–38. doi:10.1177/104063879700900106 PMID:9087922

Huxoll, D. L. (1976). Canine ehrlichiosis (tropical canine pancytopaenia): A review. *Veterinary Parasitology*, *2*(1), 49–60. doi:10.1016/0304-4017(76)90052-2

Inokuma, H., Oyamada, M., Davoust, B., Boni, M., Dereure, J., Bucheton, B., Hammad, A., Watanabe, M., Itamoto, K., Okuda, M., & Brouqui, P. (2006). Epidemiological survey of Ehrlichia canis and related species infection in dogs in eastern Sudan. *Annals of the New York Academy of Sciences*, *1078*(1), 461–463. doi:10.1196/annals.1374.085 PMID:17114754

Inokuma, H., Tamura, K., & Onishi, T. (1996). Seasonal occurrence of Rhipicephalus sanguineus in Okayama Prefecture, Japan and effect of temperature on development of the tick. *The Journal of Veterinary Medical Science*, *58*(3), 225–228. doi:10.1292/jvms.58.225 PMID:8777229

Iori, A., Lanfranchi, P., & Manilla, G. (1996). Contribution to the knowledge of Ixodidae ticks of wild mammals of Somalia. *Parassitologia*, *38*(3), 571–573. PMID:9257347

Ismail, N., & McBride, J. W. (2017). Tick-borne emerging infections: Ehrlichiosis and anaplasmosis. *Clinics in Laboratory Medicine*, *37*(2), 317–340. doi:10.1016/j.cll.2017.01.006 PMID:28457353

Jacobs, P. A. H., Fourie, L. J., Kok, D. J., & Horak, I. G. (2001). Diversity, seasonality and sites of attachment of adult ixodid ticks on dogs in the central region of the Free State Province. Academic Press.

Jittapalapong, S., Stich, R. W., Gordon, J. C., Wittum, T. E., & Barriga, O. O. (2000). Performance of female Rhipicephalus sanguineus (Acari: Ixodidae) fed on dogs exposed to multiple infestations or immunization with tick salivary gland or midgut tissues. *Journal of Medical Entomology*, *37*(4), 601–611. doi:10.1603/0022-2585-37.4.601 PMID:10916303

Kaewmongkol, G., Maneesaay, P., Suwanna, N., Tiraphut, B., Krajarngjang, T., Chouybumrung, A., Kaewmongkol, S., Sirinarumitr, T., Jittapalapong, S., & Fenwick, S. G. (2016). First detection of *Ehrlichia canis* in cerebrospinal fluid from a nonthrombocytopenic dog with meningoencephalitis by broad-range PCR. *Journal of Veterinary Internal Medicine*, *30*(1), 255–259. doi:10.1111/jvim.13788 PMID:26610407

Kamani, J., Baneth, G., Mumcuoglu, K. Y., Waziri, N. E., Eyal, O., Guthmann, Y., & Harrus, S. (2013). Molecular detection and characterization of tick-borne pathogens in dogs and ticks from Nigeria. *PLoS Neglected Tropical Diseases*, *7*(3), e2108. doi:10.1371/journal.pntd.0002108 PMID:23505591

Kelly, P. J. (2000). Canine ehrlichiosis: An update. *Journal of the South African Veterinary Association*, *71*(2), 77–86.

Kelly, P. J., Matthewman, L. A., Mahan, S. M., Semu, S., Peter, T., Mason, P. R., Brouqui, P., & Raoult, D. (1994). Serological evidence for antigenic relationships between Ehrlichia canis and Cowdria ruminantium. *Research in Veterinary Science*, *56*(2), 170–174. doi:10.1016/0034-5288(94)90100-7 PubMed

Kirade, P., Shukla, S., Shrivastava, N., Audarya, S. D., & Jatav, G. P. (2018). Rickettsial and haemoparasitic infections in dogs of malwa region, Madhya Pradesh-An etiopathological study. *Intas Polivet*, *19*(1), 3–6.

Kitaa, J. M. A., Mulei, C. M., Mande, J. D., & Wabacha, J. (2014). A retrospective study of canine ehrlichiosis in Kenya. *International Journal of Veterinary Science*, *3*(3), 122–124.

Koch, H. G. (1982). Seasonal incidence and attachment sites of ticks (Acari: Ixodidae) on domestic dogs in southeastern Oklahoma and northwestern Arkansas, USA. *Journal of Medical Entomology*, *19*(3), 293–298. doi:10.1093/jmedent/19.3.293 PMID:7120309

Koch, H. G., & Popham, T. W. (1987). Evaluation of acaricide-impregnated plastic collars for control of lone star ticks, Amblyomma americanum (L.)(Acari: Ixodidae) on domestic dogs in Oklahoma, U.S.A. *Preventive Veterinary Medicine*, *5*(1), 63–67. doi:10.1016/0167-5877(87)90007-9

Koch, H. G., & Tuck, M. D. (1986). Molting and survival of the brown dog tick (Acari: Ixodidae) under different temperatures and humidities. *Annals of the Entomological Society of America*, *79*(1), 11–14. doi:10.1093/aesa/79.1.11

Kordick, S. K., Breitschwerdt, E. B., Hegarty, B. C., Southwick, K. L., Colitz, C. M., Hancock, S. I., Bradley, J. M., Rumbough, R., Mcpherson, J. T., & MacCormack, J. N. (1999). Co-infection with multiple tick-borne pathogens in a walker hound kennel in North Carolina. *Journal of Clinical Microbiology*, *37*(8), 2631–2638. doi:10.1128/JCM.37.8.2631-2638.1999 PMID:10405413

Kruss, H. A., Weber, A., Appel, M., Enders, B., Isenberg, H. D., Schiefer, H. G., Slenczka, W., von Graevenitz, A., & Zahner, H. (2003). *Zoonoses: infectious diseases transmissible from animals to humans*. ASM Press. doi:10.1128/9781555817787

Lanza-Perea, M., Zieger, U., Qurollo, B. A., Hegarty, B. C., Pultorak, E. L., Kumthekar, S., Bruhl-Day, R., & Breitschwerdt, E. B. (2014). Intraoperative bleeding in dogs from Grenada seroreactive to Anaplasma platys and Ehrlichia canis. *Journal of Veterinary Internal Medicine*, *28*(6), 1702–1707. doi:10.1111/jvim.12442 PMID:25274547

Lauzi, S., Maia, J. P., Epis, S., Marcos, R., Pereira, C., Luzzago, C., Santos, M., Puente-Payo, P., Giordano, A., Pajoro, M., Sironi, G., & Faustino, A. (2016). Molecular detection of Anaplasma platys, Ehrlichia canis, Hepatozoon canis and Rickettsia monacensis in dogs from Maio Island of Cape Verde archipelago. *Ticks and Tick-Borne Diseases*, *7*(5), 964–969. doi:10.1016/j.ttbdis.2016.05.001 PMID:27177475

Louly, C. C. B., Fonseca, I. N., Oliveira, V. F., & Borges, L. M. F. (2006). Occurrence of Rhipicephalus sanguineus in workers of veterinary clinics and kennels, in the city of Goiânia, GO. *Brazilian Animal Science*, *7*(1), 103–106.

Louly, C. C. B., Fonseca, I. N., Oliveira, V. F. D., Linhares, G. F. C., Menezes, L. B. D., & Borges, L. M. F. (2007). Seasonal dynamics of Rhipicephalus sanguineus (Acari: Ixodidae) in dogs from a police unit in Goiânia, Goias, Brazil. *Ciência Rural*, *37*(2), 464–469. doi:10.1590/S0103-84782007000200026

M'ghirbi, Y., Ghorbel, A., Amouri, M., Nebaoui, A., Haddad, S., & Bouattour, A. (2009). Clinical, serological, and molecular evidence of ehrlichiosis and anaplasmosis in dogs in Tunisia. *Parasitology Research Journal*, *104*(4), 767–774. doi:10.100700436-008-1253-4 PMID:19018569

Maggi, R. G., & Krämer, F. (2019). A review on the occurrence of companion vector-borne diseases in pet animals in Latin America. *Parasites & Vectors*, *12*(1), 145. doi:10.118613071-019-3407-x PMID:30917860

Manyarara, R., Tubbesing, U., Soni, M., & Noden, B. H. (2015). Serodetection of Ehrlichia canis amongst dogs in central Namibia. *Journal of the South African Veterinary Association*, *86*(1), 272. doi:10.4102/jsava.v86i1.1272 PMID:26244587

Matjila, P. T., Leisewitz, A. L., Jongejan, F., & Penzhorn, B. L. (2008). Molecular detection of tick-borne protozoal and ehrlichial infections in domestic dogs in South Africa. *Veterinary Parasitology*, *155*(1-2), 152–157. doi:10.1016/j.vetpar.2008.04.012 PMID:18502588

Matthewman, L. A., Kelly, P. J., Mahan, S. M., Semu, D., Tagwira, M., Bobade, P. A., Brouqui, P., Mason, P. R., & Raoult, D. (1993). Western blot and indirect fluorescent antibody testing for antibodies reactive with Ehrlichia canis in sera from apparently healthy dogs in Zimbabwe. *Journal of the South African Veterinary Association*, *64*(3), 111. PMID:8176682

McBride, J. W., Corstvet, R. E., Gaunt, S. D., Chinsangaram, J., Akita, G. Y., & Osburn, B. I. (1996). PCR detection of acute Ehrlichia canis infection in dogs. *Journal of Veterinary Diagnostic Investigation*, *8*(4), 441–447. doi:10.1177/104063879600800406 PMID:8953528

McClure, J. C., Crothers, M. L., Schaefer, J. J., Stanley, P. D., Needham, G. R., Ewing, S. A., & Stich, R. W. (2010). Efficacy of a doxycycline treatment regimen initiated during three different phases of experimental ehrlichiosis. *Antimicrobial Agents and Chemotherapy*, *54*(12), 5012–5020. doi:10.1128/AAC.01622-09 PMID:20921310

Misao, T., & Kobayashi, Y. (1954). Studies on infectious mononucleosis. I. Isolation of etiologic agent from blood, bone marrow, and lymph node of a patient with infectious mononucleosis by using mice. *Tokyo Iji Shinshi*, *71*, 683–686.

Mtshali, K., Nakao, R., Sugimoto, C., & Thekisoe, O. (2017). Occurrence of *Coxiella burnetii, Ehrlichia canis, Rickettsia* species and *Anaplasma phagocytophilum*-like bacterium in ticks collected from dogs and cats in South Africa. *Journal of the South African Veterinary Association, 88*(0), a1390. doi:10.4102/jsava.v88i0.1390 PMID:28582983

Muhairwa, A. P., Mwakijungu, E. O., Msoffe, P. L. M., & Mtambo, M. M. A. (2012). Seroprevalence and factors affecting canine monocytic ehrlichiosis and canine brucellosis in Tanzania. *Research Opinions in Animal & Veterinary Sciences, 2*(3), 181–188.

Mylonakis, M. E., Ceron, J. J., Leontides, L., Siarkou, V. I., Martinez, S., Tvarijonaviciute, A., Koutinas, A. F., & Harrus, S. (2011). Serum acute phase proteins as clinical phase indicators and outcome predictors in naturally occurring canine monocytic ehrlichiosis. *Journal of Veterinary Internal Medicine, 25*(4), 811–817.

Mylonakis, M. E., Koutinas, A. F., Breitscwerdt, E. B., Hegarty, B. C., Billinis, C. D., Leontides, L. S., & Kontos, V. S. (2004). Chronic canine ehrlichiosis (Ehrlichia canis): A retrospective study of 19 natural cases. *Journal of the American Animal Hospital Association, 40*(3), 174–184. doi:10.5326/0400174 PMID:15131097

Mylonakis, M. E., Kritsepi-Konstantinou, M., Dumler, J. S., Diniz, P. P., Day, M. J., Siarkou, V. I., Breitschwerdt, E. B., Psychas, V., Petanides, T., & Koutinas, A. F. (2010). Severe hepatitis associated with acute Ehrlichia canis infection in a dog. *Journal of Veterinary Internal Medicine, 24*(3), 633–638. doi:10.1111/j.1939-1676.2010.0501.x PubMed

Mylonakis, M. E., Siarkou, V. I., & Koutinas, A. F. (2010). Myelosuppressive canine monocytic ehrlichiosis (Ehrlichia canis): An update on the pathogenesis, diagnosis and management. *Israel Journal of Veterinary Medicine, 65*(4), 129–135.

Nakaghi, A. C. H., Machado, R. Z., Costa, M. T., André, M. R., & Baldani, C. D. (2008). Clinical, hematological, serological and molecular survey of canine ehrlichiosis. *Ciência Rural, Santa Maria, 38*(3), 776–770. doi:10.1590/S0103-84782008000300027

Ndip, L. M., Ndip, R. N., Esemu, S. N., Dickmu, V. L., Fokam, E. B., Walker, D. H., & McBride, J. W. (2005). Ehrlichial infection in Cameroonian canines by Ehrlichia canis and Ehrlichia ewingii. *Veterinary Microbiology, 111*(1-2), 59–66. doi:10.1016/j.vetmic.2005.08.010 PMID:16181750

Ndip, L. M., Ndip, R. N., Esemu, S. N., Walker, D. H., & McBride, J. W. (2010). Predominance of Ehrlichia chaffeensis in Rhipicephalus sanguineus ticks from kennel-confined dogs in Limbe. *Experimental & Applied Acarology, 50*(2), 163–168. doi:10.100710493-009-9293-8 PMID:19693681

Neer, T. M., Eddlestone, S. M., Gaunt, S. D., & Corstvet, R. E. (1999). Efficacy of enrofloxacin for the treatment of experimentally induced *Ehrlichia canis* infection. *Journal of Veterinary Internal Medicine, 13*(5), 501–504. doi:10.1111/j.1939-1676.1999.tb01470.x PMID:10499737

Neer, T. M., & Harrus, S. (2006). *Canine monocytotropic ehrlichiosis and neorickettsiosis. E. canis, E. chaffeensis, E. ruminantium, N. sennetsu.* Academic Press.

Ogunkoya, A. B., Adeyanju, J. B., & Aliu, Y. O. (1981). Experiences with the use of Imizol in treating canine blood parasites in Nigeria. *The Journal of Small Animal Practice*, *22*(12), 775–777. doi:10.1111/j.1748-5827.1981.tb00585.x PMID:7339205

Papadopoulos, B., Morel, P. C., & Aeschlimann, A. (1996). Ticks of domestic animals in the Macedonia region of Greece. *Veterinary Parasitology*, *63*(1-2), 25–40. doi:10.1016/0304-4017(95)00877-2 PMID:8792578

Parola, P., Cornet, J. P., Sanogo, Y. O., Miller, R. S., Van Thien, H., Gonzalez, J. P., Raoult, D., Telford, S. R. III, & Wongsrichanalai, C. (2003). Detection of Ehrlichia spp, Anaplasma spp, Rickettsia spp, and other eubacteria in ticks from the Thai-Myanmar border and Vietnam. *Journal of Clinical Microbiology*, *41*(4), 1600–1608. doi:10.1128/JCM.41.4.1600-1608.2003 PMID:12682151

Parola, P., Socolovschi, C., Jeanjean, L., Bitam, I., Fournier, P. E., Sotto, A., Labauge, P., & Raoult, D. (2008). Warmer weather linked to tick attack and emergence of severe rickettsioses. *PLoS Neglected Tropical Diseases*, *2*(11), e338. doi:10.1371/journal.pntd.0000338 PMID:19015724

Parzy, D., Davoust, B., Raphenon, G., & Vidor, E. (1991). Canine ehrlichiosis in Senegal: human and canine seroepidemiological survey in Dakar. *Medicina tropical: revue du Corps de sante colonial, 51*(1), 59–63.

Pegram, R. G., Keirans, J. E., Clifford, C. M., & Walker, J. B. (1987). Clarification of the Rhipicephalus sanguineus group (Acari, Ixodoidea, Ixodidae). II. R. sanguineus (Latreille, 1806) and related species. *Systematic Parasitology*, *10*(1), 27–44. doi:10.1007/BF00009100

Penzhorn, B. L., Netherlands, E. C., Cook, C. A., Smit, N. J., Vorster, I., Harrison-White, R. F., & Oosthuizen, M. C. (2018). Occurrence of Hepatozoon canis (Adeleorina: Hepatozoidae) and Anaplasma spp. (Rickettsiales: Anaplasmataceae) in black-backed jackals (Canis mesomelas) in South Africa. *Parasites & Vectors*, *11*(1), 1–7. doi:10.118613071-018-2714-y PMID:29554933

Pretorius, A. M., & Kelly, P. J. (1998). Serological survey for antibodies reactive with Ehrlichia canis and E. chaffeensis in dogs from the Bloemfontein area of South Africa. *Journal of the South African Veterinary Association*, *64*(4), 126–128. doi:10.4102/jsava.v69i4.840 PMID:10192085

Price, J. E., & Karstad, L. H. (1980). Free-living jackals (*Canis mesomelas*) - potential reservoir hosts for *Ehrlichia canis* in Kenya. *Journal of Wildlife Diseases*, *16*(4), 469–473. doi:10.7589/0090-3558-16.4.469 PMID:7463598

Proboste, T., Kalema-Zikusoka, G., Altet, L., Solano-Gallego, L., Fernández de Mera, I. G., Chirife, A. D., Muro, J., Bach, E., Piazza, A., Cevidanes, A., Blanda, V., Mugisha, L., de la Fuente, J., Caracappa, S., & Millán, J. (2015). Infection and exposure to vector-borne pathogens in rural dogs and their ticks, Uganda. *Parasites & Vectors*, *8*(1), 306. doi:10.118613071-015-0919-x PMID:26043771

Qiu, Y., Kaneko, C., Kajihara, M., Ngonda, S., Simulundu, E., Muleya, W., Thu, M. J., Hang'ombe, M. B., Katakura, K., Takada, A., Sawa, H., Simuunza, M., & Nakao, R. (2018). Tick-borne haemoparasites and Anaplasmataceae in domestic dogs in Zambia. *Ticks and Tick-Borne Diseases*, *9*(1), 988–995. doi:10.1016/j.ttbdis.2018.03.025 PMID:29622515

Qurollo, B. A., Davenport, A. C., Sherbert, B. M., Grindem, C. B., Birkenheuer, A. J., & Breitschwerdt, E. B. (2013). Infection with Panola Mountain E. hrlichia sp. in a Dog with Atypical Lymphocytes and Clonal T-Cell Expansion. *Journal of Veterinary Internal Medicine*, *27*(5), 1251–1255. doi:10.1111/jvim.12148 PMID:23875820

Raskin, R. E., & Messick, J. B. (2012). Bone marrow cytologic and histologic biopsies: Indications, techniques, and evaluation. *Veterinary Clinics: Small Animal Practice*, *42*(1), 23–42. doi:10.1016/j.cvsm.2011.10.001 PMID:22285155

Rautenbach, G. H., Boomker, J., & De Villiers, I. L. (1991). A descriptive study of the canine population in a rural town in southern Africa. *Journal of the South African Veterinary Association*, *62*(4), 158–162. doi:10.4102/jsava.v62i4.1778 PMID:1770490

Rautenbach, R., Schoeman, J., & Goddard, A. (2018). Prevalence of canine Babesia and Ehrlichia co-infection and the predictive value of haematology. *The Onderstepoort Journal of Veterinary Research*, *85*(1), 1–5. doi:10.4102/ojvr.v85i1.1626 PMID:30326715

Rikihisa, Y. (1991). The tribe Ehrlichieae and ehrlichial diseases. *Clinical Microbiology Reviews*, *4*(3), 286–308. doi:10.1128/CMR.4.3.286 PMID:1889044

Rikihisa, Y., Stills, H., & Zimmerman, G. (1991). Isolation and continuous culture of Neorickettsia helminthoeca in a macrophage cell line. *Journal of Clinical Microbiology*, *29*(9), 1928–1933. doi:10.1128/JCM.29.9.1928-1933.1991 PMID:1774317

Ristic, M., Huxsoll, D. L., Weisiger, R. M., Hildebrandt, P. K., & Nyindo, M. B. A. (1972). Serological diagnosis of tropical canine pancytopenia by indirect immunofluorescence. *Infection and Immunity*, *6*(3), 226–231. doi:10.1128/IAI.6.3.226-231.1972 PMID:4118045

Roopali, B., Vivek, R., Kasaralikar, N. A., Patil, R. B. G., Sandeep, H., & Dilipkumar, D. (2018). Clinico, haemato-biochemical changes and therapeutic management of canine ehrlichiosis. *Pharma Innovation Journal*, *7*(9), 1–6.

Roqueplo, C., Cheminel, V., Bourry, O., Gomez, J., Pre'vosto, J. M., Parzy, D., & Davoust, B. (2009). Canine ehrlichiosis in the Ivory Coast and Gabon: Alteration of biochemical blood parameters based on Ehrlichia canis serology. *Clinical Microbiology and Infection*, *15*(2), 41–42. doi:10.1111/j.1469-0691.2008.02168.x PMID:19793128

Rotondano, T. E. D. F., de Almeida, A. M. P., Lustosa, E. M. C., Cordeiro, A. A., Camboim, E. K. A., de Azevedo, S. S., de Andrade, P. P., & de Melo, M. A. (2012). An assessment of whole blood and fractions by nested PCR as a DNA source for diagnosing canine ehrlichiosis and anaplasmosis. *TheScientificWorldJournal*, *2012*, 2012. doi:10.1100/2012/605743 PMID:22973174

Rudoler, N., Baneth, G., Eyal, O., van Straten, M., & Harrus, S. (2012). Evaluation of an attenuated strain of Ehrlichia canis as a vaccine for canine monocytic ehrlichiosis. *Vaccine*, *31*(1), 226–233. doi:10.1016/j.vaccine.2012.10.003 PMID:23072894

Sainz, A., Roura, X., Miró, G., Estrada-Peña, A., Kohn, B., Harrus, S., & Solano-Gallego, L. (2015). Guideline for veterinary practitioners on canine ehrlichiosis and anaplasmosis in Europe. *Parasites & Vectors*, *8*(1), 75. doi:10.118613071-015-0649-0 PMID:25649069

Sainz, A., Tesouro, M. A., Amusategui, I., Rodríguez, F., Mazzucchelli, F., & Rodríguez, M. (2000). Prospective comparative study of 3 treatment protocols using doxycycline or imidocarb dipropionate in dogs with naturally occurring ehrlichiosis. *Journal of Veterinary Internal Medicine*, *14*(2), 134–139. doi:10.1111/j.1939-1676.2000.tb02226.x PMID:10772483

Sarih, M., M'Ghirbi, Y., Bouattour, A., Gern, L., Baranton, G., & Postic, D. (2005). Detection and Identification of *Ehrlichia* spp. in Ticks Collected in Tunisia and Morocco. *Journal of Clinical Microbiology*, *43*(3), 1127–1132. doi:10.1128/JCM.43.3.1127-1132.2005 PMID:15750072

Schaefer, J. J., Kahn, J., Needham, G. R., Rikihisa, Y., Ewing, S. A., & Stich, R. W. (2008). Antibiotic clearance of Ehrlichia canis from dogs infected by intravenous inoculation of carrier blood. *Annals of the New York Academy of Sciences*, *1149*(1), 263–269. doi:10.1196/annals.1428.087 PMID:19120226

Schaefer, J. J., Needham, G. R., Bremer, W. G., Rikihisa, Y., Ewing, S. A., & Stich, R. W. (2007). Tick acquisition of Ehrlichia canis from dogs treated with doxycycline hyclate. *Antimicrobial Agents and Chemotherapy*, *51*(9), 3394–3396. doi:10.1128/AAC.00358-07 PMID:17606682

Shipov, A., Klement, E., Reuveni-Tager, L., Waner, T., & Harrus, S. (2008). Prognostic indicators for canine monocytic ehrlichiosis. *Veterinary Parasitology*, *153*(1-2), 131–138. doi:10.1016/j.vetpar.2008.01.009 PMID:18289794

Silveira, J. A., Passos, L. M., & Ribeiro, M. F. (2009). Population dynamics of Rhipicephalus sanguineus (Latrielle, 1806) in Belo Horizonte, Minas Gerais state, Brazil. *Veterinary Parasitology*, *161*(3-4), 270–275. doi:10.1016/j.vetpar.2009.01.028 PMID:19339119

Straten van der, G. (2008). *A survey of ticks, Ehrlichia canis and current control methods on dogs on the island of Aruba*. Research report, Utrecht Centre of Tick-borne diseases.

Stuben, J. (2004). *Investigations on the prevalence of ehrlichiosis and babesiosis in dogs in Namibia with special consideration of the housing conditions* [Unpublished Master's Dissertation]. Free University.

Suksawat, J., Hegarty, B. C., & Breitschwerdt, E. B. (2000). Seroprevalence of Ehrlichia canis, Ehrlichia equi, and Ehrlichia risticii in sick dogs from North Carolina and Virginia. *Journal of Veterinary Internal Medicine*, *14*(1), 50–55. doi:10.1111/j.1939-1676.2000.tb01499.x PMID:10668817

Tagorti, G. (2019). Disease prevalence among young dogs in Grand Tunis, Tunisia: A retrospective study. *Veterinary World*, *12*(4), 489–495. doi:10.14202/vetworld.2019.489-495 PMID:31190703

Thompson, R., Adams, H., Odoi, A., & Kennedy, M. (2020). Seroprevalence of viral and vector-borne bacterial pathogens in domestic dogs (Canis familiaris) in northern Botswana. *PLoS One*, *15*(1), e0220593. doi:10.1371/journal.pone.0220593 PMID:31914123

Troy, G. C., & Forrester S. D. (1990). Canine ehrlichiosis. *Infectious diseases of the dog and cat*, 404–418.

Van Heerden, J. (1979). The transmission of canine ehrlichiosis to the wild dog lycaon pictus (temminck) and blackbacked jackal Canis mesomelas Schreber. *Journal of the South African Veterinary Association*, *50*(4), 245–248. PMID:553960

Van Heerden, J. (1982). A retrospective study on 120 natural cases of canine ehrlichiosis. *Journal of the South African Veterinary Association*, *53*(1), 17–22. PMID:7097704

Van Heerden, J., MilIs, M. G. L., Van Vuuren, M. J., Kelly, P. J., & Dreyer, M. J. (1995). An investigation into the health status and diseases of wild dogs (Lycaon *pictus)* in the Kruger National Park. *Journal of the South African Veterinary Association*, *66*(1), 18–27. PMID:7629782

Van Heerden, J., & Immelman, A. (1979). The use of doxycycline in the treatment of canine ehrlichiosis. *Journal of the South African Veterinary Association*, *50*(4), 241–244. PMID:553959

Waner, T., & Harrus, S. (2013). Canine monocytic ehrlichiosis: From pathology to clinical manifestations. *Israel Journal of Veterinary Medicine*, *68*(1), 12–18.

Wanzala, W., Takken, W., Mukabana, W. R., Pala, A. O., & Hassanali, A. (2012). Ethnoknowledge of Bukusu community on livestock tick prevention and control in Bungoma district, western Kenya. *Journal of Ethnopharmacology*, *140*(2), 298–324. doi:10.1016/j.jep.2012.01.021 PMID:22306288

Weerathunga, D., Amarasinghe, A., Iddawel, D., & Wickramasinghe, S. (2019). Prevalence of canine tick-borne haemoparasites in three Divisional Secretariat Divisions (Rambewa, Tirappane, and Galenbidunuwewa) in the Anuradhapura district, Sri Lanka. *Sri Lankan Journal of Infectious Diseases*, *9*(2), 111–119. doi:10.4038ljid.v9i2.8254

KEY TERMS AND DEFINITIONS

Binary Fission: Asexual reproduction by a separation of the body into two new bodies.

Lethargy: Is tiredness, weariness, fatigue, or lack of energy. It can be accompanied by depression, decreased motivation, or apathy.

Myelosuppressive or Bone Marrow Suppression: A decrease in the ability of the bone marrow to produce blood cells, which may result in a lack of red blood cells, white blood cells, and platelets.

Pancytopenia: A reduction in all blood lines (reduction in the number of red and white blood cells as well as platelets).

Phagocytosis: A fundamental process of cells to capture and ingest foreign particles.

Phagosomes: Vesicles formed around a particle engulfed by a phagocyte via phagocytosis. Professional phagocytes include macrophages, neutrophils, and dendritic cells.

Tenesmus: A continual or recurrent inclination to evacuate the bowels, caused by disorder of the rectum or other illness.

Chapter 14
Tick Borne Viruses:
Nairobi Sheep Disease/Ganjam Disease

Moses Okoth Olum
Kenya Agricultural and Livestock Research Organisation, Kenya

Michael Kiraithe Muthamia
Kenya Veterinary Vaccines Production Institute, Kenya

ABSTRACT

Nairobi sheep disease virus (NSDV) is a one of the viruses transmitted by tick vectors and causes Nairobi sheep disease (NSD) – a disease of small ruminants. The disease was first identified as a problem in sheep and goats in Kenya in 1910. The virus belongs to the genus Nairovirus and family Bunyaviridae. An equally pathogenic variant of NSDV is Ganjam virus, which is found in India and Sri Lanka. Both viruses are closely related to the human pathogen Crimean-Congo haemorrhagic fever virus (CCHFV), a fever transmitted through infected tick bites or through contact with infected animal blood or tissues during and immediately after slaughter. Other members of the genus include Dugbe virus, Hazara virus, and Kupe virus, isolated from cattle ticks in East Africa. The virus is spread by hard (Ixodid) ticks and appears to be dependent on the tick vector for dissemination, with no direct transmission between animals. The transmission of tick-borne viruses (TBV) occurs primarily during tick feeding and is a complex process, known to be promoted by tick saliva constituents.

INTRODUCTION

Nairobi sheep disease virus (NSDV) is a one of the viruses transmitted by tick vectors and causes Nairobi sheep disease (NSD) – a disease of small ruminants. The disease was first identified as a problem in sheep and goats in Kenya in 1910. The virus belongs to the genus *Nairovirus* and family *Bunyaviridae* (Fenner, 2017). An equally pathogenic variant of NSDV is Ganjam virus (GANV), which is found in India and Sri Lanka (bin Tarif et al ., 2012). Both viruses are closely related to the human pathogen Crimean-Congo haemorrhagic fever virus (CCHFV) (Burt et al.,1996) a fever transmitted through infected tick bites or through contact with infected animal blood or tissues during and immediately after

DOI: 10.4018/978-1-7998-6433-2.ch014

slaughter. Other members of the genus include *Dugbe virus*, *Hazara virus* and *Kupe virus*, both isolated from cattle ticks in East Africa. The virus is spread by hard (Ixodid) ticks, and appears to be dependent on the tick vector for dissemination, with no direct transmission between animals (bin Tarif et al., 2012).

The transmission of Tick Borne Viruses TBV occurs primarily during tick feeding and is a complex process, known to be promoted by tick saliva constituents. However, the underlying molecular mechanisms of TBV transmission are poorly understood. Immuno-modulatory properties of tick saliva helping overcome the first line of defense to injury and early interactions at the tick-host skin interface appear to be essential in successful TBV transmission and infection of susceptible vertebrate hosts. The local host skin site of tick attachment, modulated by tick saliva, is an important focus of virus replication (Kazimírová et al., 2017).

Tick borne viruses affect animals, humans and other vertebrate hosts causing various diseases in different parts of the world. The viruses are found in six different virus families (Asfarviridae, Reoviridae, Rhabdoviridae, Orthomyxoviridae, Bunyaviridae, and Flaviviridae) Bunyavirales (International Committee on Taxonomy of Viruses ICTV, 2018). and in at least 9 genera (Fenner F, Peter A. Bachmann, 1987) . From recent studies, some yet to be assigned tick-borne viruses are thought to a seventh family, the Arenaviridae. With only one exception (African swine fever virus, family Asfarviridae) all tick-borne viruses (as well as all other arboviruses) are RNA viruses (Nuttall & Labuda, 2004). Until recently, the NSDV virus was only known to exist in parts of Africa and on the Indian subcontinent. However, viral RNA was recently identified in ticks in China identified by viral metagenomic analysis of ticks collected from the northeast region of the People's Republic of China (Gong et al., 2015).

NSDV is the most pathogenic tick borne viral diseases of small ruminants. In susceptible animals, this tick-borne viral infection results in a hemorrhagic gastroenteritis with very high morbidity and mortality rates of up to 90%. Other accompanying clinical manifestations include Fever, reduction in white blood cells, rapid respiration, anorexia, profound depression, diarrhea, drop in body temperature, pregnant animals frequently abort, death - in some cases within 12 hours of the onset of the fever (Montgomery, 1917).

In humans, NSDV infection can cause febrile illness, headache, nausea, and vomiting (Yadav et al., 2011). Vaccines have been developed which showed considerable promise when tested in the laboratory, but there is no demand for these in the field. The obligate vector nature of ticks causing NSD may explain why the virus is not seen as a major economic threat, since young animals in endemic areas tend to be protected by maternal antibodies through the period where they are first exposed to the virus via a bite from an infected tick, after which they have their own immune protection. The disease tends to be only noticed on introduction of naive livestock into an endemic area, e.g. for the purposes of improving local breeds by crossing (bin Tarif et al., 2012). This is due to the presence of a steady enzootic state which ensures that high levels of immunity are present in the small-ruminant populations at risk. No interventions such as vaccination, or tick control are indicated except during outbreaks when tick control is advised. There is no effective treatment, and eradication is generally not feasible once the virus has become established in ticks (Calisher & Fenner, 2000; Cobo, 2014; Craighead, 2000; Payne, 2012; Strauss & Strauss, 2008)

NAIROBI SHEEP DISEASE VECTORS AND VIRUS TRANSMISSION

The NSD virus is transmitted most commonly by the brown ear tick, *Rhipicephalus appendiculatus* in Africa (Daubney & Hudson, 1931; Lewis, 1946). The tick has a wide distribution and abundance directly affected by seasonal dynamics thereby determining the epidemiology of NSD. *R. appendiculatus* has been sampled from several countries of Central, Eastern and Southern Africa in which its presence is associated with theileriosis occurrence. They are three-host ticks and feed on three separate animal hosts as larva, nymph and finally adult. The hosts can be of the same species or different species.

Larvae feed indiscriminately and therefore feed on a variety of hosts while nymphs show a preference for ungulates. The adult tick feeds primarily on cattle and buffaloes. Experimental infections have shown that infection of any instar results in transmission by the next succeeding stage, which then loses its infection during the next moult. A female tick infected in this manner as an adult would pass infection through the eggs to the larvae of the next generation. Further it was demonstrated that the transmitting stage could be re-infected if the reaction of the host animal commenced before the infecting meal was completed, thus ensuring the carrying on of infection for a further stage.

Other ticks suspected to transmit this infection in East Africa include *R. pulchellus, R. evertsi,* and *R. simus*. Also, the bont tick *A. variegatum* has been implicated to transmit the infection (Terpstra, 1990).

Figure 1. Rhipicephalus appendiculatus female and male, dorsal view
Source: https://commons.wikimedia.org/wiki/File:Rhipicephalus-appendiculatus-female-male-dorsal.jpg

Other tick species transmit variants of NSDV. In India, the Ganjam variant is mainly transmitted by *Haemaphysalis intermedia* (Flat-inner-spurred haemaphysalid) (University of Michigan, 2020) It is found in India and Sri Lanka. It is an obligate ectoparasite of mammals. It is a potential vector of Ganjam virus as well as Kyasanur forest disease (KFD) which is a tick-borne viral haemorrhagic fever endemic to South India.

The NSDV variant in Asia is thought to be majorly transmitted by *Haemaphysalis longicornis* - Asian longhorned tick (ALT). The ALT is native to eastern Asia and has spread to Australia, New Zealand, and several Pacific islands, where it is a threat to the health and welfare of wild and domestic animals. This tick species is also established in the United States. The hosts preferred of all active life stages can be found on sheep, humans, and vegetation (Hutcheson, Dergousoff, & Lindsay, 2019).

Figure 2. Top and under side of the Haemaphysalis longicornis - Asian longhorned tick" (ALT)
Source: https://www.cdc.gov/ticks/longhorned-tick/index.html

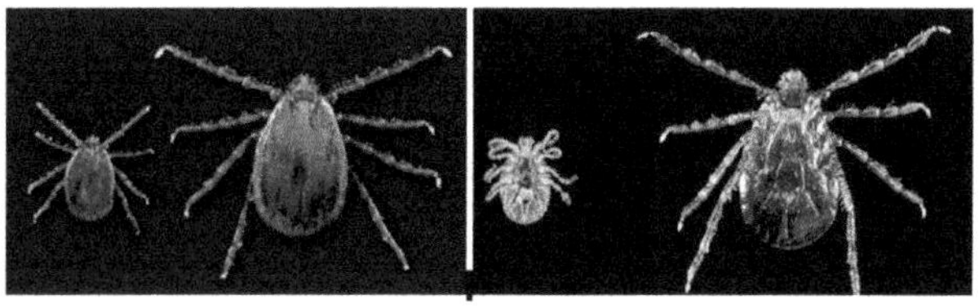

Nairobi sheep disease transmission has been largely studied in *R. appendiculatus*. The virus is transmitted trans-stadially by this Ixodid tick. During feeding of different instars (stages) upon immune, susceptible or non-susceptible hosts, they can transmit the infection. Ticks in a vertically infected series can still transmit the virus at the adult stage after feeding as larvae and nymphs upon immune sheep. Transovarian transmission of the virus is thought to be dependent upon an infective feed at the adult stage. Patent infection in eggs has only been demonstrated from the fourth day of oviposition (Davies and Mwakima, 1982). Transovarian transmission has however only been evidenced in *R. appendiculatus*. A blood meal is not key for infectivity at any stage since an unfed larva can infect a host 6 months post hatching while unfed nymphs and adults are infective for 12 and 28 months respectively (Terpstra, 1990). Different instars of non infected ticks can transmit the infection on feeding on a viremic host in the succeeding stage of development and all stages thereafter remain infective irrespective of the tick having fed upon an immune, susceptible or non-susceptible host.

In Africa, NSDV is primarily transmitted by *R. appendiculatus* ticks. In South Asia (India and Sri Lanka), GANV has been isolated from other ticks (*H. intermedia*, *H. wellingtoni*, and *R. haemaphysaloides*) (Joshi et al., 2005; Rajagopalan et al, 1970), mosquitoes (Dandawate et al, 1969), sheep and humans; *H. intermedia* ticks are considered the main vector for the virus (Gong et al., 2015)

Table 1. Isolation of GANV in India (Sudeep et al, 2009)

Source of isolation	Number of Isolates	Locality	Year of collection
Haemaphysalis intermedia	1	Ganjam District, Orissa	1954 – 55
H. intermedia	18	Hubli, Barur and Sagar, Karnataka	1961 – 62
H. wellingtoni	2	Mysore, Karnataka	1970
H. intermedia	2	Chittoor district, Andhra Pradesh	1981
H. intermedia	5	Pune city	2004 – 05
Rhipicephalus haemaphysaloides	1	Pune city	2004 – 05
Culex vishnui complex	1	North Arcot district, Andhra Pradesh	1955 – 57
Sheep	2	Chittoor district, Andhra Pradesh	1981
Humans	1	Vellore, Tamil Nadu	1969

TICK PATHOGEN INTERACTION

Ticks have to consume a blood meal so as to moult into the succeeding instar. To achieve this, they have evolved several unique parasitic features that ensure survival on the host. Such features include, prolonged life-span and complex development, ability to stay inactive for many months without a blood meal, haematophagy in all post-embryonic life stages, long feeding periods, and blood digestion within midgut cells – that contribute to their success as vectors of viruses (Kazimírová et al., 2017). In order to competently propagate and transmit pathogens, various parameters have to be achieved by both the vector and the pathogen in the transmission cycle. Such include: (i) the virus is acquired by a tick during blood-feeding on an infected host; (ii) the virus is transmitted to a host by a tick that takes its blood-meal after it has molted to the next developmental stage. The period between virus acquisition and virus transmission has been termed the "extrinsic incubation period" during which the tick is not able to transmit the virus (Nuttall, 2009)

On ingestion during feeding in any instar, the virus is picked in infected blood then enters the midgut. In order to be transmitted to the next host, the pathogen exits the digestive system through the gut wall, disseminates in the tick body tissues finally establishing itself in the salivary glands (SG). An infective dose has to be achieved for every host to which the infection will be successfully transmitted to, in order to achieve this; the virus undergoes amplification creating the prerequisite infective dose for the subsequent host. To transmit the infection, the virus then undergoes reverse movement to the host. The barriers overcome initially to achieve amplification have to be maneuvered outwards this is intrinsically initiated within the tick during attachment, saliva deposition and feeding. The barriers to be crossed include the midgut infection barrier, midgut release barrier, midgut escape barrier, SG infection barrier, and SG release barrier (Nuttall, 2014). With the various infection barriers overcome, the virus is then transmitted into a susceptible or non susceptible host and the infection can progress to clinical or no infection on the host. Susceptible hosts then become the source of infection for uninfected ticks.

Molecular Biology of NSDV/GANV

Nairobi sheep disease virus/ Ganjam virus (NSDV/GANV is a member of the genus Nairovirus, in the family Bunyaviridae; one of the largest and most diverse of all viral families. Bunyaviruses are enveloped,

Figure 3. Schematic illustration of a nairovirus particle showing the transmebrane proteins (Gn and Gc), the large, medium and small ribonucleprotein complexs (L RNP, M RNP and SRNP).
Adapted from (Garrison et al., 2020)

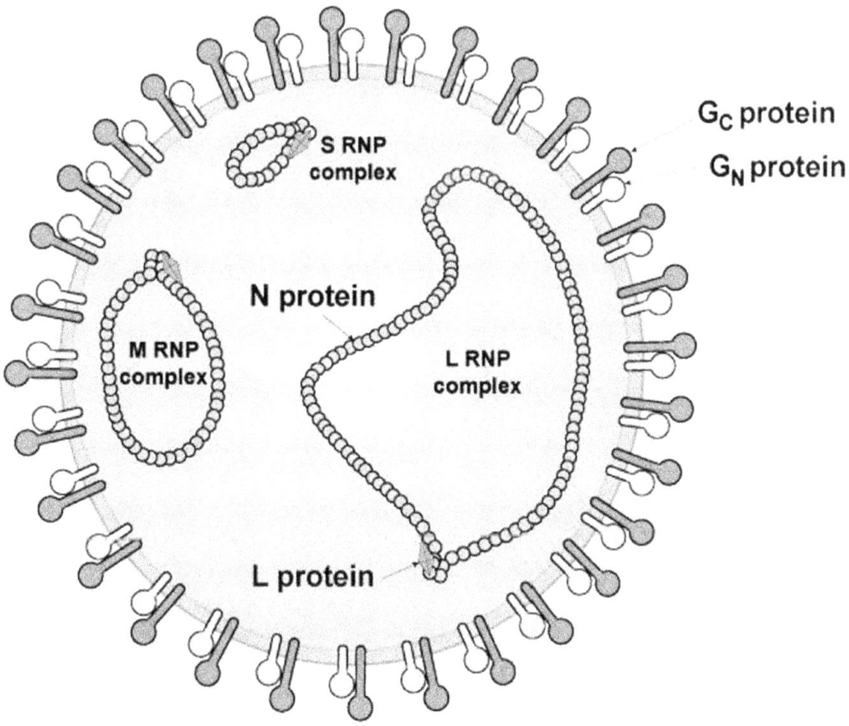

segmented, negative-strand RNA viruses. Bunyaviruses are spherical, enveloped particles measuring 80–120 nm in diameter (Whitehouse, Kuhn, Wada, & Ergunay, 2015). The family has over 400 member viruses that are classified predominantly into five genera: *Hantavirus*, *Nairovirus*, *Orthobunyavirus*, *Phlebovirus*, and *Tospovirus* — which include 35 sero-groups with at least 304 types and subtypes (Samuel, 1996; Whitehouse et al., 2015). Virion structure is relatively simple; structural proteins include a nucleocapsid (N) protein and two trans membrane glycoproteins (Gn and Gc). The L polymerase

Figure 4. Phylogenetic tree of genus nairovirus based on the alignment of viral homologue of the ovarian tunor protease. The species groupings are indicated in colored ovals. Nairoviruses are designated by the following abbreviations: CCHFV, Crimean-Congo hemorrhagic fever virus; GANV, Ganjam virus; NSDV, Nairobi sheep disease virus; DUGV, Dugbe virus; KUPEV, Kupe virus; HAZV, Hazara virus; TFLV, Tofla virus; TAGV, Taggert virus; TILLV, Tillamook virus; SAKV, Sakhalin virus; PRMV, Paramushir virus; AVAV, Avalon virus; ARTSV, Artashat virus; TFAV, Thiafora virus; ERVEV, Erve virus; HUGV, Hughes virus; FARV, Farallon virus; RAZAV, Raza virus; PSV, Punta Salinas virus; ZIRV, Zirqa virus; SOLV, Soldado virus; GRSV, Great Saltee virus; CASV, Caspiy virus; AHV, Abu Hammad virus; DGKV, Dera Ghazi Khan virus; SAPV, Sapphire II virus; WzTV, Wēnzhōu tick virus; BURV, Burana virus; HpTV-1, Huángpí tick virus 1; TcTV-1, Tachéng tick virus 1; TDYV, Tamdy virus; YOGV, Yogue virus; LPHV, Leopards Hill virus; QYBV, Qalyub virus; GERV, Geran virus; CHIMV, Chim virus; GOSV, Gossas virus; ISKV, Issyk-kul virus; UZAV, Uzun-Agach virus; KTRV, Keterah virus.
Adapted from(Dzimianski et al., 2019

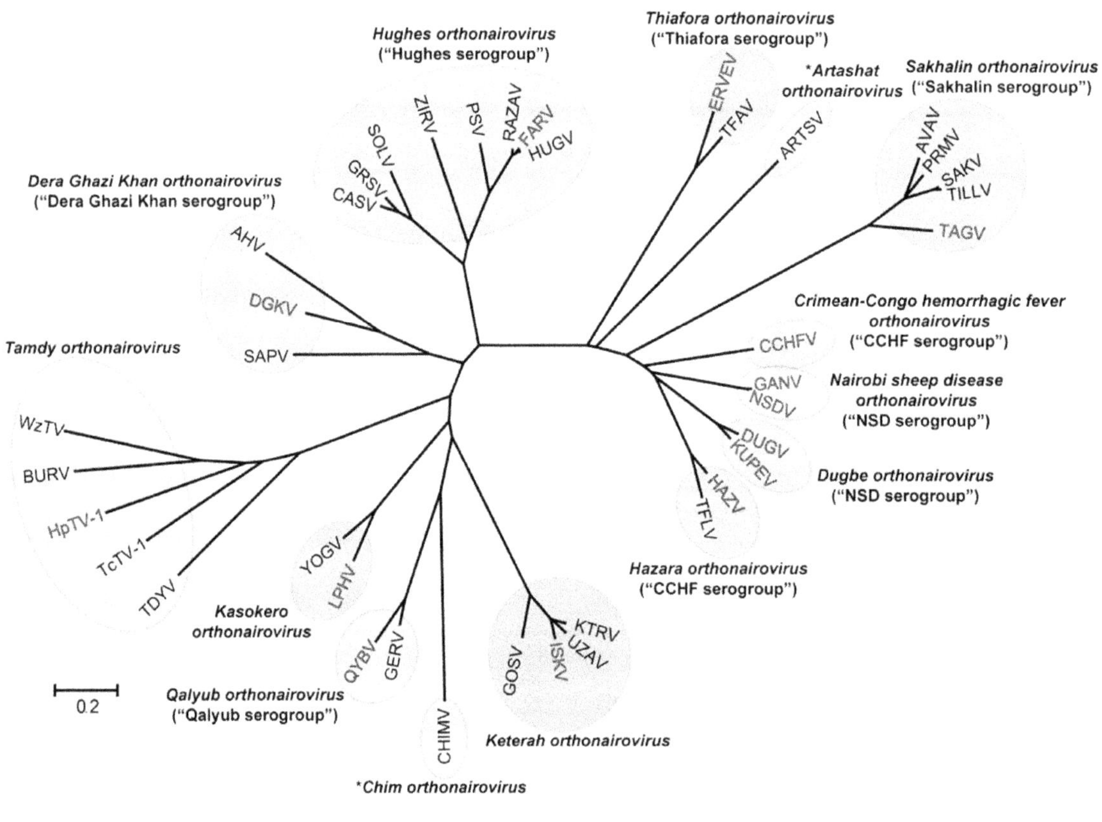

(RNA-dependent RNA polymerase) is also present in the virion. Bunyaviruses have a replication cycle similar to that of other negative-strand RNA viruses; however, some bunyaviruses have ambisense genome segments (Payne, 2012). Bunyaviruses are divided into arthropod-borne viruses (arboviruses) and rodent-borne viruses (roboviruses).

TNSDV/GANV is a prototype member of the genus Nairovirus. A unique characteristic of the viruses of the genus Nairovirus is that they are all mainly transmitted by ticks with only a few isolations made from culicoides lies and mosquitoes (Kuhn et al., 2016; Lasecka & Baron, 2014a). Nairoviruses are classified into seven serogroups, including the Crimean-Congo haemorrhagic fever (CCHF) group, which includes the human pathogen Crimean-Congo haemorrhagic fever virus (CCHFV), and hazara virus, the Nairobi sheep disease serogroup which includes the viruses NSDV/GANV, dugbe virus and kupe virus (Lasecka & Baron, 2014a). The other sero groups are Hughes group, Dera Ghazi Khan group, Qalyub group, Sakhalin group, and Thiafora group (Kuhn et al., 2016; Lasecka & Baron, 2014a). Most of the viruses in the genus Nairovirus have not been well characterized. The only one that has been characterized to a fairly high level is CCHFV which is an important human pathogen that causes hemorrhagic fever with mortality rates of up to 30% (Lasecka & Baron, 2014a).

As stated above, NSDV was first recognized in 1910 in Nairobi, Kenya while GANV was recognized 1969 in Ganjam, India (Dandawate & Shah, 1969; Montgomery, 1917). NSDV/GANV is an enveloped virus with a diameter of approximately 100nm (Munz, Göbel, Krolopp, Reimann, & Davies, 1981). NSDV and GANV were initially categorized as two different viruses based on their aetiology and the association with different tick host but they were later characterized to be the same virus. Serological assays using antibody panels indicated that NSDV and GANV were antigenically closely related and they even postulated that they were the same virus (Davies et al, 1978). More recently, analysis of their sequences determined that indeed they were the same virus. In particular, analysis of the S, M, and L genomic segments revealed that NSDV and GANV differ by 9.40–10.40% and 3.2–4.10% for nucleotide and amino acid respectively at the complete S segment, while for the complete L segment the difference at the nucleotide and amino acid was 9.90% and 2.70%. In addition, M segment NSDV and GANV differed by 22.90% and 14.70% for nucleotide and amino acid respectively (Yadav et al., 2011). These findings clearly demonstrated that NSDV and GANV were different isolates of the same virus.

In both the NSDV and GANV isolates the S segment is 1590nt in length and codes for nucleocapsid (N) protein that is 482 amino acids in length and it is approximately 53kDa (Kuhn et al., 2016; Yadav et al., 2011). In the virion, N protein is the most abundant protein, and it encapsidates the viral genomic RNA (vRNA), and complementary RNA (cRNA.. The M segment is 5063 and 5033 nucleotide (nt) long and codes for a glycoprotein precursor that is 1627 and 1624 amino acids in NSDV and GANV respectively (Kuhn et al., 2016; Yadav et al., 2011). The glycoprotein precursor is further processed into two envelope glycoproteins (Gn and Gc) and yet to be identified non-structural protein(s) (Lasecka & Baron, 2014a). Detailed characterization of the glycoprotein precursor and the structural glycoproteins has not been done, but through comparative studies with other viruses of the Nairovirus genus and Bunyavirus family, researchers have been able to identify and map important functional parts of the proteins. In particular, similar to Crimean-Congo hemorrhagic fever virus (CCHV) precursor glycoprotein, five trans-membrane regions have been identified in the NSDV/GANV glycoprotein precursor(Kuhn et al., 2016; Yadav et al., 2011). Furthermore, a putative signal peptide has been identified in the N-terminal 24 and 23 amino acids of the in the NSDV and GANV glycoprotein precursor respectively (Kuhn et al., 2016; Yadav et al., 2011). In addition, in the N-terminal region of the NSDV/GANV glycoprotein precursor several serine and threonine sites have been identified as possible O-glycosylation sites (Yadav et al., 2011). Notably, possible SKI-1 or SKI-1 like cleavage sites have been identified in NSDV/GANV glycoprotein precursor. These have been mapped at amino acid 453 and 449 (RRLM) and amino acid 976 and 9726 (RRLL) for NSDV and GANV respectively, indicating the likelihood of SKI-1-like proteases in the post-translational processing of the glycoprotein precursor (Kuhn et al., 2016; Yadav et al., 2011). The NSDV and GANV L segment is 12064 and 12080 nt long, and codes for a RdRP which is 3991 amino acids long and 453kDa for both isolates (Yadav et al., 2011). Of note, ovarian tumour like (OTU)-like domains have been identified at aa positions 29–158 and 35–157 of the RdRp protein (Yadav et al., 2011). Furthermore, there is a Zinc finger motif present at aa positions 624-627 for both NSDV and GANV isolates. The region of RdRp involved in RNA dependent RNA polymerase spans aa 2082-2823 for both NSDV and GANV isolates (Yadav et al., 2011).

The two membrane glycoproteins (Gn and Gc) initiate infection of susceptible cells by attaching a yet to be identified one or more cellular receptors (Lasecka & Baron, 2014a). Interestingly, while Gc-specific antibodies have been shown to interfere with the infectivity of CCHV and Dugbe viruses, antibodies against the Gn glycoprotein do not inhibit the infection of susceptible cells (Lasecka & Baron, 2014a). This indicates that the Gc glycoprotein is essential for attachment and entry to susceptible cells by CCHV and Dugbe virus and it is likely that NSDV/GANV Gc glycoprotein plays a similar role. The entry of NSDV/GANV into cells is probably through a clathrin-mediated extocytosis, after which the nucleocapsid enclosing the vRNA and cRNA is released into the cytoplasm where viral replication takes place (Lasecka & Baron, 2014a). The viral RNAs serve as a template for the transcription of mRNA by the RdRp and also for the replication of cRNA. The cRNA acts as template for the synthesis of viral RNA segments and the newly synthesized viral RNA and viral structural proteins are assembled into viral particles (Lasecka & Baron, 2014a).

Pathogenesis of the NSDV/GANV Disease

The natural mammalian host for NSDV/GANV is goats and sheep; other domestic animals do not develop disease or viremia even when experimentally infected (Dandawate, 1977; Davies, 1978a; Mugera & Chema, 1967). Importantly, cases of human infections through laboratory accidents have been reported previously, with the infected individuals presenting with a mild febrile illness (Dandawate, Work, Webb, & Shah, 1969). In areas where the NSDV/GANV is endemic such as India, East and Central Africa, the virus does not cause an identifiable disease probably because young animals are initially protected by maternal antibodies and they also develop their own protection upon exposure to the virus by an infected tick (M. D. Baron & Holzer, 2015). Notably, when naïve animals are moved into an endemic areas, NSDV/GANV cause a severe disease that his highly pathogenic with high mortality rates of up to 90%, with the animals that survive the infection animals developing immunity against later infections.

The exact pathogenic mechanisms for NSDV/GANV have not been well characterized. Infected animals have been found to have inflammation and hemorrhage in the gastrointestinal tract and enlarged spleen and lymph nodes (Mugera & Chema, 1967; Weinbren, Gourlay, Lumsden, & Weinbren, 1958). Furthermore, staining tissue sections from infected animals with antibodies against the NSDV/GANV N and P proteins established that the virus infects and replicates in the mucosal lining of the lungs, liver, spleen, mesenteric lymph nodes(Lasecka et al., 2015). Notably, staining of the tissue sections from infected animals established that NSDV/GANV does not infect endothelial and epithelial cells, correlating with previous findings of poor growth of a pathogenic NSDV/GANV strain in endothelial and epithelial cell cultures(Lasecka et al., 2015). Taken together these results indicate that endothelial and epithelial cells are not targeted by NSDV/GANV and the pathology observed during an infection, namely hemorrhage in the gums, gastrointestinal tract, heart and spleen might be due to immunopathology and not a direct action of the virus on the cells. Interestingly, NSDV/GANV through the precursor glycoprotein has been observed to be involved in the redistribution of protein disulphide isomerase (PDI) from the lumen of the endoplasmic reticulum to the cell membrane and also to upregulate PDI secretion by infected cells (Lasecka & Baron, 2014b). It is thus possible that the one of the mechanisms for NSDV/GANV pathogenesis might be through the increased extracellular PDI which in turn leads to the upregulation in the activation of integrins and tissue factor which are both involved in pro inflammatory immune responses and intravascular coagulation (Lasecka & Baron, 2014b) both of which have been observed in NSDV/GANV infected animals.

. Importantly, NSDV/GANV has been observed to be inhibiting the induction of interferon-β by infected cell (Holzer, Bakshi, Bridgen, & Baron, 2011). In particular, the NSDV/GANV inhibits the induction of interferon stimulated genes through interference with the phosphorylation of STAT I and STAT II which is a key step in the signaling pathway activated by interferon-α and interferon-β (Holzer et al., 2011). Other mechanisms through which NASDV/GANV is able to evade detection by the innate immune system include encapsidation of vRNA and cRNA by the N protein and thus preventing the formation of dsRNA which can be detected by PRRs activated by double stranded RNA such as MDA-5, TLR-3, protein kinase R and 2'5'-oligoadenylate synthetase (Akira, Uematsu, & Takeuchi, 2006; Kawai & Akira, 2011; Lasecka & Baron, 2014a). Interestingly, CCHFV has been found to avoid activation of RIG-I through having 5' cap structure similar to host cell mRNA though it remains to be confirmed if NSDV/GANV utilizes such a mechanism (Lasecka & Baron, 2014a). Interestingly, NSDV/GANV has been established to be down-regulating total cellular ubiquitination and the conjugation of ISG15 to cellular proteins (ISG15ylation) (Holzer et al., 2011). Notably, ubiquitin is involved in modulation of innate immune signaling through activation of NF-κβ by targeting its inhibitor (I-κβ) for degradation, while ISG15 enhances NF-κβ activity by conjugating and inhiting protein phosphatase 2Cb and thus suppressing the dephosphorylation of I-κβ (Holzer et al., 2011; Kawai & Akira, 2006; Loo & Gale, 2011). Additionally, ISG15ylation has been reported to be involved in modulation of several proteins involve in innate immune response including PKR, MxA, STAT1, RIG-I, Janus kinase 1 (JAK1), and IRF-3 (Lasecka & Baron, 2014a). Additionally, the downregulation of total ubiquitination might inhibit the activation of several molecules involved in IFN-β pathway including g RIG-I, mitochondrial antiviral signalling protein (MAVS), TANK-binding kinase-1 (TBK1), IjB kinase-e (IKK-e), tumour necrosis factor receptor-associated factor 3 (TRAF3) and TRAF6 all of which become activated when their inhibitors are degraded (Lasecka & Baron, 2014a)

Through the down regulation of ubiquitination and ISG15ylation NSDV/GANV virus has impact on the immune response by inhibiting the signaling pathway activated by interferon-α and interferon-β (Holzer et al., 2011). Additionally, through the down-regulation of ubiquitination, NSDV/GANV will inhibit the processing and presentation of antigens by antigen presentation by major histocompatibility complex (MHC) class I and II proteins (Cresswell et al., 2005; Hansen & Bouvier, 2009). $CD8^+$ and $CD4^+$ T cells recognize specific peptides processed from pathogen-derived proteins and presented in the grooves of MHC Class I and II molecules respectively. The viral proteins targeted for proteolysis are first ubiquitinated before being cleaved into smaller peptides by proteosome and by down-regulating ubiquitination NSDV/GANV is thus able to inhibit the processing and presentation of viral antigens (Cresswell et al., 2005). The ubiquitination and ISG15ylation activities of NSDV/GANV have been mapped to the OUT-like domain of the RdRp at the N-terminal (Holzer et al., 2011).

EPIDEMIOLOGY AND RISK FACTORS

The typical Nairobi Sheep disease infection which manifest with fever, acute hemorrhagic gastroenteritis abortion, and high mortality in sheep and goats is endemic in East and Central Africa (Edelsten, 1975). In susceptible flocks, it has a high mortality rate of (70- 90%) (Marczinke et al, 2002). Other livestock (e.g. cattle, horses) are refractory to the disease. The countries in which the disease is endemic in includes Kenya, Uganda, Tanzania, Somalia, Ethiopia, Botswana, Mozambique, and Republic of Congo (Aiello et al, 2016). Nairobi sheep disease virus is however not transmitted between animals by direct contact but strictly by the vectors in natural infection (Davies, 1978b; Lasecka & Baron, 2014b).

The phylogenetically related variant – Ganjam virus is endemic in India and Sri Lanka among other south Asian countries but is widely distributed in India (Yadav et al., 2011). The virus was first isolated from *Haemaphysalis intermedia* ticks collected from goats, suffering from lumbar paralysis from Orissa, India, during 1954-55 and named after the place of isolation – Ganjam District (Sudeep et al, 2009).

The virus is highly pathogenic to exotic and crossbred sheep and goats causing high morbidity and mortality. Just like NSDV, GANV has a zoonotic potential. However, an illness associated with GANV in humans has never been reported at an epidemic level. There has been a case of natural infection in a 12 yr old European boy in Vellore, Tamil Nadu, and a few laboratory acquired cases have been reported (Sudeep et al., 2009)

NSDV and Ganjam viruses have been shown to affect domestic, laboratory and wild animals including rodents. Among domesticated and laboratory animals, only sheep and goats can be infected readily by NSDV/ Ganjam virus, and only these species appear to be important as reservoir hosts (Davies, 1978b). A few fatal cases of Nairobi sheep disease have been reported among blue duikers (*Cephalophus monticola*) in zoos or in the wild. However, African wildlife are not thought to be important in maintaining this virus. Experimental infections with NSDV and Ganjam virus, respectively, have been established in the African field rat (*Arvicathus abysinicus nubilans*) and langur monkeys (*Presbytis entellus*). Bonnet monkeys (*Macaca radiata*) inoculated with Ganjam virus did not become viremic (Neitz, 1966)

Serological surveys of NSDV antibodies have been found to correspond to the ecological distribution of *R. Appendiculatus* across East Africa and most sheep and goats are positive for NSDV corroborating the fact that this tick is the main transmitter of this disease. This has rendered most parts of the region enzootic to NSD and therefore limited losses occur due to this infection. There is evidence of maternal antibodies being present and protective in kids and lambs reflecting on exposure to the infection due to the presence of the disease vector. Due to the ensuing enzootic stability, outbreaks of NSD are rare and with only incidental losses observed. Outbreaks however do occur with movement of naive animals into and NSD enzootic area or spread of infective ticks to an NSD free area or flocks by movement of infective animals. Spread of vectors into new areas may occur due to change in climatic or ecological conditions and therefore favouring the incursion and establishment the vectors. There have been rare occasions when NSD outbreaks have been recorded in enzootic areas due to a breakdown in tick control measures.

Outbreaks of NSD can be classified in two forms; enzootic and epizootic. Enzootic out breaks occur in areas where the tick vector is present in extremely high numbers, which is typically in the high altitude hot and wet regions. The small-ruminant populations show high levels of immunity. No control strategy is necessary for NSD here. The levels of flock immunity remain high, wherever high levels of larval survival ensure persistent populations of the tick vectors, these areas are characterized by low mortality due to the disease (Davies et al., 1978) Disease is signaled here whenever susceptible breeds of sheep and goats are introduced into the enzootic areas to improve the production potential of the indigenous breeds. Disease also follows the movement of sheep and goats from non-enzootic semi-arid zones to the enzootic areas for trade purposes. Any attempts at tick control in the small-ruminant populations may allow the development of susceptible population groups, which are at risk when the tick control activities cease or fail for any reason Epizootic areas: This is in areas where the tick vector may be present, but in only extremely small numbers, which is typically in the Acacia savannahs of agroclimatic zone III in East Africa. These populations have, however, been seen to increase dramatically following the periodic changes in rainfall pattern, when for 6-18 months there may be prolonged and heavy rainfall with heavy cloud cover. This results in the development of more ground vegetation, which allows huge increases in the tick populations to occur. This results in the development of more ground vegetation, which allows huge increases in the tick populations to occur. The history of epizootic NSD in Somalia suggests a similar epidemiological basis. The outbreaks have followed periods of severe drought, which resulted in much movement of livestock (Davies, 1978b)

Breed-related differences in susceptibility have been reported. In Africa, Nairobi sheep disease is less likely to be fatal in non-native breeds, such as Romney and Corriedale sheep, with reported mortality rates of 30-40% in such exotic breeds and crossbred animals. Indigenous African breeds are more susceptible; the mortality rate can be 75% or higher in East African hair sheep and Persian fat tailed sheep. The opposite situation occurs in India, where Ganjam virus infections are mild or subclinical in indigenous breeds, but at least two severe outbreaks have been seen in imported sheep. Dorset Horn and Suffolk sheep were affected in one of these outbreaks, and Madras red sheep in the other (Davies, 1997)

CLINICAL SIGNS AND DIAGNOSIS

It is hard to differentiate animals infected with NSDV/GANV from other infections such as Rift Valley Fever, or Peste des Petits Ruminants since they tend to have a similar clinical picture. Animals infected with NSDV/GANV have been observed to develop a fever for about three to seven days. In addition the animals have a bloody diarrhea and hemorrhagic lesions can be observed in the gums. The disease progresses rapidly and the animals die about a week after the onset of fever. Postmortem examination find signs hemorrhage in the respiratory mucosa, caecum and duodenum, and the mesenteric lymph nodes and spleen are swollen.

Laboratory diagnosis can be done using reverse transcriptase reverse- transcription polymerase chain reaction (RT-PCR) which has been reported to be highly sensitive and can even detect the virus after the febrile period is over and the temperatures are back to normal (bin Tarif et al., 2012). RT-PCR also has the advantages of taking a shorter time although it's of limited use in resource poor countries because it requires machines which are not readily available. There are currently no commercial kits to detect viral antigens in clinical samples but researchers have previously developed and described the use of complement fixation tests, Agar Gel Immunidiffusion and ELISA to either detect antibodies against the virus or viral antigens in clinical specimens (Davies, Mungai, & Taylor, 1977; Lasecka et al., 2015; Munz, Reimann, & Meier, 1983). NSDV/GANV virus can also be isolated from the plasma of infected animals during febrile stage of the disease. NSDV/GANV readily grows in baby hamster-derived cell line BHK21 clone 13 cells (BHK21/13) and it has clear cytopathic effects. After the isolation of virus in cell lines, proper identification can then be done using RT-PCR, ELISA or complement fixation using virus-specific antiserum.

TREATMENT AND CONTROL

Being a viral infection, No specific antiviral agent is available for treatment (Susan et al, 2016). However, supportive treatment, good shelter and quality feed may improve survival.

To disinfect premises and farms, the NSDV/GANV is reported to be sensitive to lipid solvents and detergents or disinfectants, and high or low Ph solutions. The pathogen is killed by most disinfectants listed for other Bunyaviridae viruses which include glutaraldehyde, hypochlorite, 70% alcohol, peracetic acid, hydrogen peroxide and iodophors. They are also sensitive to heat and UV light (Davies, 1978b).

PREVENTION

In areas where Nairobi sheep disease is not endemic, the disease might be eradicated by movement controls, quarantines and slaughter of infected animals, together with tick control measures. All available acaricide classes can be used to control ticks that transmit NSDV/GANV. It is however advisable to carry out tests to check for acaricide resistance developed by ticks which leads to disease transmission despite acaricide treatment of flocks. Poor acaricide application techniques have also been implicated in the spread of tick borne pathogens. NSDV can persist in an infected tick for more than two years, and eradication is generally not feasible once the virus becomes established in vector populations. Sheep and goats can be protected from tick vectors by dipping and/or spraying with an acaricide. These treatments are generally used short-term in areas bordering enzootic regions when extension of the ticks' range is expected or seen, or to help protect unaffected animals in a flock. In enzootic areas, the use of tick controls may weaken immunity, and animals are usually allowed to develop resistance via tick bites (although a few deaths may occur). Experimental vaccines have been developed for use in naive animals entering enzootic areas, or to protect animal populations when the tick vector expands its geographic range. However, these vaccines are not produced commercially and their availability is unclear. Strict quarantine is not necessary in enzootic areas, as the infection is not transmitted by casual contact. Dead animals should be buried or incinerated.

FUTURE RESEARCH DIRECTIONS

To better understand and approach the control of NSD, more research is needed on the spread of the NSD virus and related viruses in the world order to put in place proper measures to control tick transmitted viruses. This will be strengthened by further study of the tick species shown to carry and transmit NSDV which will be imperative as the distribution of these species changes with environmental factors and increasing international trade. As the ticks' distribution changes, so too should surveillance strategies. Additionally, many countries that could potentially be at risk for NSDV have never had cross-sectional studies carried out on their livestock or tick species of interest, which would hinder their ability to quickly and effectively respond to a disease outbreak. As an OIE reportable disease, NSDV can cause significant trade implications for countries and could be underreported as a result of limited knowledge on the disease in some countries as well as lack of preparedness to control NSD outbreaks.

CONCLUSION

NSD/GANV is an important disease of small ruminants with significant economic effects on naive herds and in non endemic zones but in endemic zones, it is of limited economic significance due to endemic stability. The disease must be considered in areas where new exotic breeds of small ruminants are to be introduced as well as ensuring that areas free from the disease remain free through routine disease surveillance and control of livestock movement. It is also a disease to consider whenever genetic improvement of native small ruminants through breeding is an option. Since this disease is widely studied and has been around for a long time, the pathogen forms a basis for research on therapeutics and vaccines on related pathogens of zoonotic importance. This therefore creates impetus in one health approach to research on related pathogens as well as leading to development of novel interventions on infections such as CCHFV.

Ganjam virus (GANV) - a member of genus Nairovirus of family Bunyavirdae is of considerable veterinary importance in India.

REFERENCES

Aiello, Moses, & Allen. (2016). The Merck veterinary manual (11th ed.). Kenilworth, NJ: Merck & Co., Inc.

Akira, S., Uematsu, S., & Takeuchi, O. (2006). Pathogen Recognition and Innate Immunity. *Cell*, *3*(4), 783–801. doi:10.1016/j.cell.2006.02.015 PMID:16497588

Baron, M. D., & Holzer, B. (2015). Nairobi sheep disease virus/Ganjam virus. *Revue Scientifique et Technique (International Office of Epizootics)*, *34*(2), 411–417. doi:10.20506/rst.34.2.2367 PMID:26647464

bin Tarif, A., Lasecka, L., Holzer, B., & Baron, M. D. (2012). Ganjam virus/Nairobi sheep disease virus induces a pro-inflammatory response in infected sheep. *Veterinary Research*, *43*(1), 71. doi:10.1186/1297-9716-43-71

Burt, F. J., Spencer, D. C., Leman, P. A., Patterson, B., & Swanepoel, R. (1996). Investigation of tick-borne viruses as pathogens of humans in South Africa and evidence of Dugbe virus infection in a patient with prolonged thrombocytopenia. *Epidemiology and Infection*, *116*(3), 353–361. https://doi.org/10.1017/S0950268800052687

Calisher, C. H., & Fenner, F. J. (2000). Macroecology and Microecology of Viruses of Terrestrial Mammals. In Viral Ecology (pp. 493–517). Elsevier. https://doi.org/10.1016/B978-012362675-2/50014-8.

Cobo, F. (2014). Viral haemorrhagic fevers. In Imported Infectious Diseases (pp. 243–255). Elsevier. https://doi.org/10.1533/9781908818737.243.

Craighead, J. E. (2000). Hemorrhagic Fever Viruses. In Pathology and Pathogenesis of Human Viral Disease (pp. 277–296). Elsevier. https://doi.org/10.1016/B978-012195160-3/50020-0.

Cresswell, P., Ackerman, A. L., Peaper, D. R., Wearsch, P. A., Peaper, D. R., & Wearsch, P. A. (2005). Mechanisms of MHC class I-restricted antigen processing and cross- presentation. *Immunological Reviews*, *207*, 145–157.

Dandawate, C. N. (1977). Studies on Ganjam virus with reference to viremia in vertebrate hosts & development of antibodies. *Indian Journal of Experimental Biology*, *15*(11), 1058–1059.

Dandawate, C. N., Rajagopalan, P. K., Pavri, K. M., & Work, T. H. (1969). Virus isolations from mosquitoes collected in North Arcot district, Madras state, and Chittoor district, Andhra Pradesh between November 1955 and October 1957. *The Indian Journal of Medical Research*, *57*(8), 1420–1426. https://www.ncbi.nlm.nih.gov/pubmed/4395257

Dandawate, C. N., & Shah, K. V. (1969). Ganjam virus: A new arbovirus isolated from ticks Haemaphysalis intermedia Warburton and Nuttall, 1909 in Orissa, India. *The Indian Journal of Medical Research*, *57*(5), 799–804.

Dandawate, C. N., Work, T. H., Webb, J. K., & Shah, K. V. (1969). Isolation of Ganjam virus from a human case of febrile illness: A report of a laboratory infection and serological survey of human sera from three different states of India. *The Indian Journal of Medical Research*, *57*(6), 975–982.

Daubney, R., & Hudson, J. R. (1931). Nairobi Sheep Disease. *Parasitology*, *23*(4), 507–524. https://doi.org/10.1017/S0031182000013895

Davies, F. G. (1978a). Nairobi sheep disease in Kenya. The isolation of virus from sheep and goats, ticks and possible maintenance hosts. *The Journal of Hygiene*, *81*(2), 259–265. https://doi.org/10.1017/s0022172400025092

Davies, F. G. (1978b). Nairobi sheep disease in Kenya. The isolation of virus from sheep and goats, ticks and possible maintenance hosts. *The Journal of Hygiene*, *81*(2), 259–265. https://doi.org/10.1017/S0022172400025092

Davies, F. G. (1997). Nairobi sheep disease. *Parassitologia*, *39*(2), 95–98. https://doi.org/10.1201/9780429280276-10

Davies, F. G., Casals, J., Jesset, D. M., & Ochieng, P. (1978). The serological relationships of Nairobi sheep disease virus. *Journal of Comparative Pathology*, *88*(4), 519–523. https://doi.org/10.1016/0021-9975(78)90005-1

Davies, F. G., Mungai, J. N., & Taylor, M. (1977). The laboratory diagnosis of Nairobi sheep disease. *Tropical Animal Health and Production*, *9*(2), 75–80. https://doi.org/10.1007/bf02236383

Davies, F. G., & Mwakima, F. (1982). Qualitative studies of the transmission of Nairobi sheep disease virus by Rhipicephalus appendiculatus (ixodoidea, ixodidae). *Journal of Comparative Pathology*, *92*(1), 15–20. https://doi.org/10.1016/0021-9975(82)90039-1

Dzimianski, J. V., Beldon, B. S., Daczkowski, C. M., Goodwin, O. Y., Scholte, F. E. M., Bergeron, É., & Pegan, S. D. (2019). Probing the impact of nairovirus genomic diversity on viral ovarian tumor domain protease (vOTU) structure and deubiquitinase activity. *PLoS Pathogens* (Vol. 15). doi:10.1371/journal.ppat.1007515

Edelsten, R. M. (1975). The distribution and prevalence of Nairobi Sheep disease and other tick-borne infections of sheep and goats in northern Somalia. *Tropical Animal Health and Production*, *7*(1), 29–34. https://doi.org/10.1007/BF02383239

Fenner, F., Peter, A., & Bachmann, D. O. W. (1987). *Veterinary Virology*. Elsevier. doi:10.1016/C2013-0-10647-2

Fenner, F. J. (2017). Fenner's Veterinary Virology (N. J. M. E. J. Dubov, Ed.; 4th ed.). Elsevier., https://doi.org/10.1016/C2013-0-06921-6.

Garrison, A. R., Alkhovsky Альховский Сергей Владимирович, S. V., Avšič-Županc, T., Bente, D. A., Bergeron, É., Burt, F., … Consortium, I. R. (2020). ICTV Virus Taxonomy Profile: Nairoviridae. *The Journal of General Virology, 101*(8), 798–799. doi:10.1099/jgv.0.001485

Gong, S., He, B., Wang, Z., Shang, L., Wei, F., Liu, Q., & Tu, C. (2015). Nairobi Sheep Disease Virus RNA in Ixodid Ticks, China, 2013. *Emerging Infectious Diseases*, *21*(4), 718–720. https://doi.org/10.3201/eid2104.141602

Hansen, T. H., & Bouvier, M. (2009). MHC class I antigen presentation : Learning from viral evasion strategies. *Nature Reviews. Immunology*, *9*(7), 503–513. https://doi.org/10.1038/nri2575

Holzer, B., Bakshi, S., Bridgen, A., & Baron, M. D. (2011). Inhibition of interferon induction and action by the nairovirus Nairobi sheep disease virus/Ganjam virus. *PLoS One*, *6*(12), e28594. https://doi.org/10.1371/journal.pone.0028594

Hutcheson, H. J., Dergousoff, S. J., & Lindsay, L. R. (2019). Haemaphysalis longicornis: A tick of considerable veterinary importance, now established in North America. *The Canadian Veterinary Journal =. The Canadian Veterinary Journal. La Revue Veterinaire Canadienne*, *60*(1), 27–28. https://www.ncbi.nlm.nih.gov/pubmed/30651646

Joshi, M. V., Geevarghese, G., Joshi, G. D., Ghodke, Y. S., Mourya, D. T., & Mishra, A. C. (2005). Isolation of Ganjam Virus from Ticks Collected off Domestic Animals Around Pune, Maharashtra, India. *Journal of Medical Entomology*, *42*(2), 204–206. https://doi.org/10.1093/jmedent/42.2.204

Kawai, T., & Akira, S. (2006). *Innate immune recognition of viral infection*. doi:10.1038/ni1303

Kawai, T., & Akira, S. (2011). Review Toll-like Receptors and Their Crosstalk with Other Innate Receptors in Infection and Immunity. *Immunity*, *34*(5), 637–650. https://doi.org/10.1016/j.immuni.2011.05.006

Kazimírová, M., Thangamani, S., Bartíková, P., Hermance, M., Holíková, V., Štibrániová, I., & Nuttall, P. A. (2017). Tick-Borne Viruses and Biological Processes at the Tick-Host-Virus Interface. *Frontiers in Cellular and Infection Microbiology, 7*. doi:10.3389/fcimb.2017.00339

Kuhn, J. H., Wiley, M. R., Rodriguez, S. E., Bào, Y., Prieto, K., & Travassos da Rosa, A. P. A. … Palacios, G. (2016). Genomic Characterization of the Genus Nairovirus (Family Bunyaviridae). *Viruses, 8*(6). doi:10.3390/v8060164

Lasecka, L., & Baron, M. D. (2014a). The molecular biology of nairoviruses, an emerging group of tick-borne arboviruses. *Archives of Virology*, *159*(6), 1249–1265. https://doi.org/10.1007/s00705-013-1940-z

Lasecka, L., & Baron, M. D. (2014b). The Nairovirus Nairobi Sheep Disease Virus/Ganjam Virus Induces the Translocation of Protein Disulphide Isomerase-Like Oxidoreductases from the Endoplasmic Reticulum to the Cell Surface and the Extracellular Space. *PLoS One*, *9*(4), e94656. https://doi.org/10.1371/journal.pone.0094656

Lasecka, L., Bin-Tarif, A., Bridgen, A., Juleff, N., Waters, R. A., & Baron, M. D. (2015). Antibodies to the core proteins of Nairobi sheep disease virus/Ganjam virus reveal details of the distribution of the proteins in infected cells and tissues. *PLoS One*, *10*(4), e0124966. https://doi.org/10.1371/journal.pone.0124966

Lewis, E. A. (1946). Nairobi sheep disease: The survival of the virus in the tick Rhipicephalus appendiculatus. *Parasitology*, *37*(1–2), 55–59. https://doi.org/10.1017/S0031182000013159

Loo, Y., & Gale, M. (2011). Review Immune Signaling by RIG-I-like Receptors. *Immunity*, *34*(5), 680–692. https://doi.org/10.1016/j.immuni.2011.05.003

Marczinke, B. I., & Nichol, S. T. (2002). Nairobi Sheep Disease Virus, an Important Tick-Borne Pathogen of Sheep and Goats in Africa, Is Also Present in Asia. *Virology*, *303*(1), 146–151. https://doi.org/10.1006/viro.2002.1514

Montgomery, E. (1917). On a Tick-borne Gastroenteritis of Sheep and Goats occurring in British East Africa. *The Journal of Comparative Pathology and Therapeutics*, *30*(1), 28–57.

Montgomery, R. M. (1917). On a tick-borne gastroenteritis of sheep and goats occurring in Kenya. *The Journal of Comparative Pathology and Therapeutics*, *30*(1), 28–57.

Mugera, G. M., & Chema, S. (1967). Nairobi sheep disease: A study of its pathogenesis in sheep, goats and suckling mice. *Bulletin of Epizootic Diseases of Africa. Bulletin Des Epizooties En Afrique*, *15*(4), 337–354.

Munz, E., Göbel, E., Krolopp, C., Reimann, M., & Davies, F. G. (1981). Electron microscopic studies on the morphology of Nairobi sheep disease virus. *Zentralblatt Fur Veterinarmedizin. Reihe B. Journal of Veterinary Medicine. Series B, 28*(7), 553–563. doi:10.1111/j.1439-0450.1981.tb01773.x

Munz, E., Reimann, M., & Meier, K. (1983). An enzyme-linked immunosorbent assay (ELISA) for the detection of antibodies to Nairobi sheep disease virus in comparison with an indirect immunofluorescent and hemagglutination test. I. Development and testing of the ELISA with sera of experimentally inf. *Zentralblatt Fur Veterinarmedizin. Reihe B. Journal of Veterinary Medicine. Series B, 30*(7), 473–479. doi:10.1111/j.1439-0450.1983.tb01870.x

Neitz, W. 0. (1966). Nairobi Sheep Disease. In *Encyclopedic Reference of Parasitology* (pp. 362–362). Berlin: Springer Berlin Heidelberg. doi:10.1007/3-540-29835-5_2337

Nuttall, P. A. (2009). Molecular characterization of tick-virus interactions. *Frontiers in Bioscience,* (14), 2466. doi:10.2741/3390

Nuttall, P. A. (2014). Tick-borne viruses. In Biology of Ticks (vol. 2, pp. 180–210). New York, NY: Oxford University Press.

Nuttall, P. A., & Labuda, M. (2004). *Tick-host interactions: saliva-activated transmission*. doi:10.1017/S0031182004005633

Payne, S. (2012). Bunyaviridae. In Virus Taxonomy (pp. 725–741). Elsevier. https://doi.org/10.1016/B978-0-12-384684-6.00059-8.

Rajagopalan, P. K., Sreenivasan, M. A., & Paul, S. D. (1970). Isolation of Ganjam virus from the bird tick Haemaphysalis wellingtoni Nuttall and Warburton 1907. *The Indian Journal of Medical Research*, *58*(9), 1195–1196. https://www.ncbi.nlm.nih.gov/pubmed/5534070

Samuel, B. (1996). Bunyaviruses. In S. Baron (Ed.), *Medical Microbiology* (4th ed.). University of Texas Medical Branch at Galveston. Retrieved from https://www.ncbi.nlm.nih.gov/books/NBK8004/

Strauss, J. H., & Strauss, E. G. (2008). Minus-Strand RNA Viruses. In *Viruses and Human Disease* (pp. 137–191). Elsevier. doi:10.1016/B978-0-12-373741-0.50007-6

Sudeep, A. B., Jadi, R. S., & Mishra, A. C. (2009). Ganjam virus. *The Indian Journal of Medical Research*, *130*(5), 514–519. https://www.ncbi.nlm.nih.gov/pubmed/20090098

Terpstra, C. (1990). Nairobi Sheep Disease Virus. In Virus Infections of Ruminants (pp. 495–500). Elsevier. https://doi.org/10.1016/B978-0-444-87312-5.50065-5.

University of Michigan. (2020). *Animal Diversity Web*. Retrieved from https://animaldiversity.org/accounts/Haemaphysalis_intermedia/classification/#Haemaphysalis_intermedia

Weinbren, M. P., Gourlay, R. N., Lumsden, W. H., & Weinbren, B. M. (1958). An epizootic of Nairobi sheep disease in Uganda. *Journal of Comparative Pathology*, *68*(2), 174–187. https://doi.org/10.1016/s0368-1742(58)80018-1

Whitehouse, C. A., Kuhn, J. H., Wada, J., & Ergunay, K. (2015). Family Bunyaviridae. In Global Virology I - Identifying and Investigating Viral Diseases (pp. 199–246). Springer New York. https://doi.org/10.1007/978-1-4939-2410-3_10.

Yadav, P. D., Vincent, M. J., Khristova, M., Kale, C., Nichol, S. T., Mishra, A. C., & Mourya, D. T. (2011). Genomic analysis reveals Nairobi sheep disease virus to be highly diverse and present in both Africa, and in India in the form of the Ganjam virus variant. *Infection, Genetics and Evolution*, *11*(5), 1111–1120. https://doi.org/10.1016/j.meegid.2011.04.001

KEY TERMS AND DEFINITIONS

Bunyaviridae: Is a family of arthropod-borne or rodent-borne, spherical, enveloped RNA viruses.

Gastroenteritis: Inflammation of the gastrointestinal tract—the stomach and small intestine.

Immuno-Modulatory: A chemical agent that modifies the immune response or the functioning of the immune system.

Immunopathology: A branch of medicine and the manifestation of conditions concerned with immune responses associated with the production of disease.

Metagenomics: Is the study of genetic material recovered directly from environmental samples.

Nairoviruses: Are arthropod-borne bunyaviruses (genus Nairovirus, family Bunyaviridae)

Nucleocapsid: A viral protein coat that surrounds the genome (either DNA or RNA).

Oviposition: Means expulsion of the egg from the oviduct to the external environment.

Phylogenetic: Is the scientific study that pertains to the evolutionary history of a taxonomic group of organisms.

Transovarian: Transmission (transmission from parent to offspring via the ovaries) occurs in certain arthropod vectors as they transmit pathogens from parent arthropod to offspring arthropod.

Viremia: Is a medical term for viruses present in the bloodstream.

Virion: Is an entire virus particle consisting of an outer protein shell called a capsid and an inner core of nucleic acid.

Chapter 15
Control of African Swine Fever and Avian Spirochaetosis

Charles Muleke Inyagwa
Department of Veterinary Pathology, Microbiology and Parasitology, Faculty of Veterinary Medicine and Surgery, Egerton University, Kenya

Erick O. Mungube
Kenya Agricultural and Livestock Research Organization, Veterinary Science Research Institute, Kenya

ABSTRACT

Ticks are distributed worldwide and have an enormous medical and veterinary importance owing to the direct damage they cause and indirectly as vectors of a large variety of human and animal pathogens. The family Argasidae (soft ticks) comprises five genera and with about 193 species. Among all the argasid ticks, only four Argas and two Ornithodoros species are competent to transmit diseases. This chapter describes the various ticks of the argasidae family, diseases they transmit, and strategies for their control. A description of the two important genera, Ornithodorus and Argas, that belong to the family argasidae are provided. Emphasis is on the mammalian hosts affected, tick species involved, morphological features (with relevant pictorials), geographic distribution, life cycle, and economic importance. A detailed description of the two most important diseases transmitted by argasidae ticks namely African swine fever (ASF) and avian spirochaetosis is given. Emphasis is laid on the historical background, epidemiology, clinical signs, and strategies for their control.

INTRODUCTION

The family Argasidae comprises about 193 known tick species. The phylogeny and taxonomy of the ticks in the argasidae family is still controversial. Classification at the genus-level is far less settled than that of the Ixodidae. Most species of Argasidae can be assigned to more than one genus (Estrada-Peña *et al.*, 2010). Argasids tend to be endophilic/nidicolousparasites that colonize the nests and burrows of their hosts and feed when the host arrives. Argasids are very resistant to starvation and can survive for several years without feeding (Manzano-Román *et al.*, 2012). Argasid distribution is cosmopolitan

DOI: 10.4018/978-1-7998-6433-2.ch015

since they are found throughout the world with the exception of Sub-Antarctic bio-geographical regions (Estrada-Peña *et al.*, 2010). Regarding the pathogens transmitted by argasid ticks, they are mainly viruses together with a number of bacterial species, and they cause severe diseases in humans and animals (Manzano-Román *et al.*, 2012).

Aserious swine disease transmitted by argasids is African swine fever. This disease jumped between continents in the 60´s and 70´s and was recently reported in Germany (Sanchez-Cordon *et al.*,2018) and in China (Normile, 2018; Lin Zhou *et al.*, 2019), exemplifying the growing possibility that tick-borne infectious diseases can emerge and colonize previously uninfected areas because of the potential distribution of the vector transcontinental. Other viruses transmitted by soft ticks are Alkhurma hemorrhagic Fever virus; Crimean Congo hemorrhagic fever and Qalyub virus. The argasid-borne bacteria are almost exclusively borreliae, which causes relapsing fever in humans (Raúl Manzano *et al.*, 2012). The fact that argasids serve as vectors for many pathogens require greater effort for broad control measures including epidemiological and parasite-surveillance programs (Raúl Manzano *et al.*, 2012). Since there is no single ideal solution for the control of argasid ticks, an integrated control approach is the most effective. This chapter describes the strategies for the control of African Swine Fever and Avaianspirochaetos is transmitted by tick vectors in the family argasidae.

BACKGROUND

African swine fever (ASF) is a devastating viral hemorrhagic fever transmitted by argarsids that can kill up to 100% of domestic pigs, for which there is still no vaccine or treatment. The ability to survive for long periods in uncooked pork confersthe virus with a very high capacity for trans boundary spread over great distances, demonstrated by its introduction into Europe. It is also currently prevalent in most countries in sub-Saharan Africa (SSA) wherever pigs are kept. The last decades have seen exponential growth in the pig sector in SSA but also a striking increase in the number of ASF outbreaks. The control for ASF described herein provides a roadmap for progressive control of ASF in SSA. It takes into account the complexities of both the epidemiology of the disease and the diverse and largely disorganized pig sector in the region. Eradication of the virus from the region is not an option owing to the involvement of African wild suids and ticks of the *Ornithodoros moubata* complex. The main focus of the control strategy is by preventing outbreaks in domestic pigs by improved organization of the pig sector and mitigating the risks throughout the pig value chains. In most SSA countries up to 90% of pigs are kept in traditional scavenging systems. An account of the control methods for Avian spirochetosis caused by the bacteria *Borrelia anserine* and transmitted by the *Argaspersicus* is provided.

FAMILY: ARGASIDAE

The family argasidae generally refers to what are called soft ticks. Unlike hard ticks, soft ticks lack a scutum; and the males and females are very close to the same size. Their mouthparts are not visible when viewed from above. The significant morphological features of the nymph or adult stage, distinguishing Argasidae ticks from the Ixodidae are the ventrally located head, invisible from the dorsal side, and the lack of chitin dorsal shield (scutum). Unlike hard ticks, pulvilli are not present at the tip of tarsi in adult and nymphal stages.

Life Cycle of the Argasidae Ticks

Members of the family Argasidae have two or more nymphal stages, each of which requires a blood meal. This pattern is referred to as the multi-host life cycle, Fig. 1.

Figure 1. The Multi-host life cycle of Argasidae ticks (Adopted and Modified from CDC-2017)

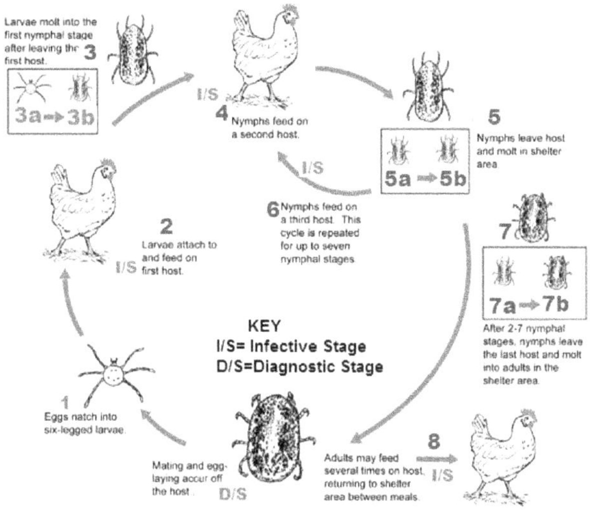

After mating egg-laying usually always occurs off the host in a sheltered area preferably in an animal nest. The laid eggs hatch into six-legged larvae in the parents sheltered area (**1**). The larvae will search for a host in the vicinity of the sheltered area. Once a suitable host is found, they feed for about one hour to several days depending on the tick species (**2**). After feeding, the larvae leave the host and molt into the first nymphal instars in a sheltered area **(3a-3b).** Thenymphs' quest for and feed on the second host (**3**) rapidly (usually about an hour). The second host is usually the same species and often the same individual as the first host. The first nymphal instars leave the host and molt into the next nymphal instars in the sheltered area **(5a-5b).**This cycle can continue to accommodate up to seven nymphal instars (**6**) depending on the species. After the last nymphal instar has fed, it leaves the host and molts into an adult **(7a-7b)** under shelter. Adults may continue to feed on the host (**8**), feeding rapidly and detaching after each blood meal. Females of some species lay egg batches after each meal. It is worth noting that humans are usually only incidental hosts for argasid ticks and may be fed upon by any of the stages.

Ornithodorus Species

The genus *Ornithodorus* (synonym *Ornithodoros*) is constituted by soft ticks called hut/sand tampans or the eyeless tampan which are found in sandy soils, primitive housing or shaded areas around trees. There are five known species of the genus *Ornthidorous* which include:

a. *Ornithodorus moubata*
b. *Ornithodorus savignyi*

c. *Ornithodorus turicate*
d. *Ornithodorus hermsi*
e. *Ornithodorus turicata*

Morphological Features of Ornithodorus Ticks

The surface of *Ornithodorus* is leather-like, wrinkled, and covered by small nodules. *Ornithodorus moubata* has no eyes and lack a "suture line" at the border of dorsal and ventral surfaces. It has long mouthparts (hypostomum), equipped with long and backward-pointing chitin barbs, whichare usedforpiercingthe skin of the host animal. The female is about 10 mm and the male 8mm long, Fig. 2.

Figure 2. Adult Ornithodorus moubata (Courtesy: Shivairo et al., 2018)

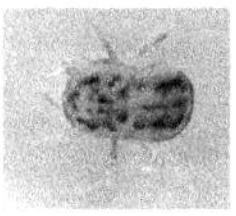

Geographic Distribution and Transmitted Diseases

Ticks of the genus *Ornithodorus* have wide distribution and are found in Africa, the Middle East and US among other places (Table1). Two species of public health concern; *Ornithodoros hermsi* and *O. turicata* which are vectors of tick-borne relapsing fever (TBRF) and spirochetesare found in the United States. In Africa and Asia, *O. moubata* is the main tick vector for ASF (Pigs).

Table 1. Geographic Distribution and Transmitted Diseases

Species	Geographic Distribution	Transmitted Diseases
Ornithodoros moubata	Africa and Middle East	African Swine Fever (pigs)
Ornithodoros Savignyi	Africa and Middle East	Alkhurma Hemorrhagic Fever Virus (camels, man)
Ornithodoros turicate	Africa and Middle East	Q - fever and tick paralysis (man)
Ornithodoros hermsi	USA	Tick-Borne Relapsing Fever (TBRF) (man)

DISEASES TRANSMITTED BY SOFT TICKS AND THEIR CONTROL

The description of important diseases (African swine fever (ASF) and avian spirochaetosis) transmitted by argasidae ticks and their control strategies are as follows:

African Swine Fever (ASF)

African swine fever (ASF) is a highly contagious hemorrhagic viral disease of domestic and wild pigs, which is responsible for serious economic and production losses. It is caused by a large DNA virus of the *Asfarviridae* family, which also infects ticks of the genus *Ornithodoros (*Dixon *et al.*, 2020). It is the only virus with a double-stranded DNA genome known to be transmitted by arthropods (Dixon*et al.*, 2020). ASF is a notable diseaselisted in the World Organization for Animal Health (OIE) Terrestrial Animal Health Code and must be reported to the OIE (OIE WAHID., 2014). TheASF-virus causes a hemorrhagic fever with high mortality rates in domestic pigs; some isolates can cause death of animals as quickly as a week after infection. It persistently infects its natural hosts, warthogs, bush pigs, and soft ticks of the genus Ornithodoros, which likely act as a vector, with no disease signs, and it does not cause disease in humans (Dixon *et al.*, 2013).ASF is present in regions of Asia, Europe and Africa and is endemic to sub-Saharan Africa where it exists in the wild through a cycle of infection between ticks and wild pigs, bush pigs, and warthogs (Dixon *et al.*, 2013).

Historical Background of ASF

African swine fever (ASF) was first diagnosed in Kenya in 1910 and reported in 1921 (Montgomery, 1921) were it manifested as an acute hemorrhagic fever, causing mortality approaching 100% in domestic pigs (Anderson *et al.*, 1998). Reports of ASF emanated from South Africa in 1928, (Steyn, 1928), Malawi in 1931, (Turnbull, (1932) and Angola in 1932 (Penrith *et al.*, 2013). The corporate database of the Food and Agriculture Organization of the United Nations (FAOSTAT) indicates that four countries in the Eastern African region do not have pigs (Comoros including Mayotte, Djibouti, Eritrea and South Sudan(FAO, 2017). Seychelles, Somalia and the French island of Réunion have small pig populations and have never experienced ASF(Léopold *et al.*, 2019). All the remaining Eastern African countries have reported ASF, including two Indian Ocean island nations, Madagascar and Mauritius (Jori *et al.*, 2013; Penrith *et al.*, 2013). Zimbabwe has a relatively small pig population, with traditional and smallholder pig production largely confined to particular districts (Chiduwa *et al.*, 2008; Mashatise *et al.*, 2005). Following these early descriptions, ASF was reported in East and Southern African countries, where the virus is recognized to have been present in wildlife hosts for a very long time (Léopold *et al*, 2019). The endemic occurrence of ASF in domestic pigs has been documented in several Eastern African countries: Kenya (Lichoti *et al.*, 2016; Okoth *et al.*, 2013), Madagascar (Roger*et al.*, 2001), Malawi (Haresnape *et al.*,1987); Mozambique (Quembo *et al.*, 2016), Tanzania (Misinzo *et al.*, 2014), Uganda (Barongo *et al.*, 2015) and the Eastern Province of Zambia (Simulundu *et al.*, 2018).The first report of ASF from Central Africa emanated from Angola in 1932 (Gago da Camara, 1933), although the virus from the Angolan outbreaks was only confirmed as ASF many years later (Mendes, 1994). In Central Africa, introductions into São Tomé Principe in 1979 and Cameroon in 1982 were officially reported (Léopold *et al.*, 2019). In West Africa, ASF was diagnosed in Senegal for the first time in 1959 (Etter *et al.*, 2011), and regular reports of outbreaks were submitted to the OIE from at least 1980 (OIE WAHID., 2014; Brown *et al.*, 2018). Senegal remained the only West African country reporting ASF to OIE until 1996 when Côte d'Ivoire reported its first outbreak (Brown *et al.*, 2018). Subsequent decades witnessed an upsurge in the number of countries newly infected with ASF (Benin and Togo in 1997, Nigeria in 1998, Ghana in 1999 and Burkina Faso in 2003) (Brown *et al.*, 2018). ASF was introduced into Madagascar (1998) and Mauritius (2007) (AU-IBAR, 2013). Ethiopia reported an outbreak to AU-IBAR in 2011

(AU-IBAR, 2011), which was confirmed as ASF and further outbreaks were experienced in 2013 and 2014 (Achenbach *et al.*, 2017). ASF 1957 was introduced into Portugal, probably from Angola (Sandra *et al.*, 2020). This was rapidly eradicated, but a second introduction occurred in 1960. It later spread in Europe and across to the Caribbean and Brazil from where it was eradicated with the exception of the Italian island of Sardinia, where it remains endemic (Sandra *et al.*, 2020). Since 2007, the disease has spread through Russia and some countries in Eastern Europe, where it is still endemic (Sandra *et al.*, 2020). The official monthly disease report data submitted to AU-IBAR in the past several years reveals that ASF is an endemic disease in four of the five regions of Africa, namely the West, Central, East and Southern Africa (Brown *et al.*, 2018; Léopold *et al.*, 2019). The number of countries reporting ASF has increased yearly expanding its geographical focus to almost all of SSA. In 2008, 18 countries reported the presence of ASF (AU-IBAR, 2013). The number increased during subsequent years to 19 countries in 2009, 21 in 2010, 23 in 2011 and 25 in 2012 (Léopold *et al.*, 2019; Brown *et al.*, 2018). By 2013, the OIE WAHIS database indicated that 22 countries in SSA regularly reported ASF, four more countries have reported incursions since 2005, and three more have experienced one or more outbreaks prior to 2005, two of which have the sylvatic cycle in warthogs (OIE WAHID, 2014; Brown *et al.*, 2018; Léopold *et al.*, 2019). This means that 33 SSA countries have at some time experienced ASF (OIE WAHID, 2014). This evidence underscores the importance of the disease and the critical need for concerted action to stop further spread in the continent and beyond its borders. Only three African countries have successfully eradicated ASF after a single introduction (São Tomé e Principe, Mauritius and Côte d'Ivoire) (FAO, 2017).

Epidemiology of ASF

The epidemiology of ASF is complex and varies depending on the environment, types of pig production systems, presence/absence of competent tick vectors, human behavior and the presence/absence of wild pigs (Jori *et al.*, 2013). Only pigs which belong to the family Suidae are susceptible to the infection. Pigs are infected mainly through the oro-nasal route after contact with infected pigs or after feeding on virus containing pork or contaminated products (Bellini *et al.*, 2016). All excretions and secretions of infected pigs such as blood, feaces, urine or saliva can contain virus, and ASFV may remain viable in blood and tissues for long periods (Sandra Blome, *et al.*, 2020). The host range of ASFV is very narrow with suids as sole vertebrate hosts, and soft ticks of the genus Ornithodoros as competent arthropod vector (Jori and Bastos, 2009). ASF has no zoonotic potential and there are no indications that this might change (Dixon *et al.*, 2020). Once introduced into the domestic pig population, the virus does not need its arthropod vector for transmission (Jori and Bastos, 2009). ASFV can be transmitted by direct contact between infected and susceptible animals and by indirect contact to contaminated objects or feed (Chenais *et al.*, 2019). Contaminated pork (swill feeding) and also blood products used as protein source can play an important role (Wen *et al.*, 2019). Fomites such as clothing, trucks and veterinary equipment (especially vaccine pistols and similar objects) may act as source of infection (Wen *et al.*, 2019). In the wild boar habitat, carcasses are crucial in maintaining infection cycles (Depner *et al.*, 2016). Persistently infected carrier animals have been incriminated to play a role in the viral maintenance, especially in an endemic situation (Depner *et al.*, 2016). The role of such persistently infected animals in long-term transmission is still controversially discussed (Stahl *et al.*, 2019). While it is very clear that soft ticks play a major role in the disease transmission, the role of mechanical vectors including carrion eating predators, birds and other blood-sucking arthropods is not yet known (Sandra Blome *et al.*, 2020)

Three cycles of transmission have been confirmed in sub-Saharan Africa: two cycles that involve soft ticks of the *Ornithodoros moubata* complex and a cycle involving circulation among domestic pigs (Pedro Jose Sanchez-Cordon *et al.*, 2018). The virus is believed to have evolved in Eastern and Southern Africa in an ancient sylvatic cycle between common warthogs and argasid ticks of the *Ornithodoros moubata* complex that live in their burrows (warthog-tick cycle) (Plowright*et al.*, 1994). Recent studies suggest that the sylvatic cycle originated in the early 1700s in Africa (Alkhamis *et al.*, 2018). The ability of ticks of the genus Ornithodoros to transmit African swine fever virus (ASFV) to pigs was first demonstrated in Spain after the introduction of the virus in 1960 (Sánchez Botija, 1963). The involvement of ticks from warthog burrows in the sylvatic cycle was first confirmed in Kenya, Uganda and Tanzania (Plowright *et al.*, 1969a, 1969b). The *Ornithodoros moubata* ticks are competent biological vectors for ASFV, transmitting the virus transovarial, trans-staidly and sexually and to pigs when feeding on their blood via the saliva and the coxal fluid that is excreted while feeding (Plowright*et al.*, 1974). Transmission of the virus from warthogs to ticks is explained by circulation of the virus maintained between neonatal warthogs that spend the first 4–6 weeks of life in the burrow, where the ticks inhabiting the burrow feed on them (Plowright*et al.*, 1974). Although no clinical signs of disease have been observed in young warthogs, they were reported to develop sufficiently high viraemia to be able to infect ticks (Thomson, 1985; Thomson *et al.*, 1983).

Warthogs are resistant to the pathogenic effects of the virus, so the warthog–tick cycle is not characterized by overt disease or mortality (Thomson *et al.*, 1983). On the contrary, the disease has devastating socio-economic consequences in domestic pigs due to its high lethality. A domestic pig–tick cycle between domestic pigs and Ornithodoros ticks that live in their shelters (pig–tick cycle) has been demonstrated in Malawi and Mozambique and may be more widespread (Simulundu *et al.*, 2017). A domestic cycle dependent on direct or indirect transmission amongst domestic pig populations without involvement of wild hosts or vectors occurs throughout the region.

Countries in Eastern and Southern Africa are home to all three cycles and the area remains important in the maintenance and spread of the disease. The introductions of ASFV into countries outside the African region as well as West Africa, where the warthog–tick cycle is not present, have been traced through molecular phylogenetic studies to viruses that were circulating in the Eastern, Central and Southern Africa (ECSA) region (Sánchez-Vizcaíno, 2013). The presence of the disease in Africa is seen as a significant continuous threat to countries with industrialized pig farming that derive high revenues from pork exports (Sánchez-Vizcaíno, 2013; Vergne *et al.*, 2017). By the 1960s ASF outbreaks had been reported from the majority of Eastern and Southern African countries (Sanchez-Cordon *et al.*, 2018). The epidemiology varies across the African continent depending on the availability of wild suids, arthropod vectors and pig production system. The role of wild suids is well described for warthogs in East and Southern Africa (Anderson *et al.*, 1998). In West Africa incidental infections in warthog and a Red River hog in Nigeria are widely regarded as the result of spill over from domestic pigs (Luther *et al.*, 2007). In the endemic area in Senegal suggested neither warthogs nor Ornithodorus ticks have been implicated in ASFV maintenance or transmission (Léopold *et al.*, 2019). The dynamics of ASF in Africa and the role of wildlife in its spread are not yet fully understood. However, it is possible that the virus couldjump from one cycle to another for instance from wild suids and ticks to domestic pigs and vice versa (Léopold *et al.*, 2019). It has been demonstrated that biosecurity lapses in the pig value chain (production, transportation, marketing and processing) rank as the priority risk factors for the outbreaks of ASF and its maintenance in Africa (Bellini *et al.*, 2016; FAO, 2017). Studies on the pig sector in three SSA countries revealed that lack of biosecurity on farms, production systems, slaughtering under hygienic conditions, insecure marketing systems increased the risk to the infection (Penrith *et al.*, 2013). Proximity to the pig slaughter facility and lack of proper slaughter facilities were also reported to be as high risk for exposure to ASF (Penrith *et al.*, 2007). The human role in the epidemiology of the disease (movement of people across borders, moving infected pork products and live pigs) increase the risk of trans-boundary spread (Penrith *et al.*, 2013).

Clinical Signs of ASF

Clinical signs and mortality rates vary according to virulence and the type/species of pig kept. Acute forms of ASF are characterized by high fever, depression, anorexia and loss of appetite, hemorrhages in the skin (redness of skin on ears, abdomen and legs), abortion in pregnant sows, cyanosis, vomiting, diarrhoea and death within 6-13 days (or up to 20 days). Mortality rates may be as high as 100%. Sub-acute and chronic forms are due to low virulent viruses, which produce less intense clinical signs that can be expressed for much longer periods. Mortality rates are lower, but can still range from 30-70%. Chronic disease symptoms include loss of weight, intermittent fever, respiratory signs, chronic skin ulcers and arthritis.

Prevention and Control

To date, safe and efficacious vaccines against ASF are still lacking and production of most vaccine candidates is massively hampered bythe simultaneous lack of a permanent cell line that is sufficiently susceptible to ASFV and does not force further genetic adaptations within the ASFV genome (Blome *et al.*, 2020). However, vaccine design is still feasible due to the fact that animals recovering from acute ASF are protected against homologous challenge infection with closely related strains (King *et al.*, 2011). Problems in developing ASFV vaccine are partly due to the fact that antibodies produced by the affected animal cannot really render the virus harmless (no complete neutralization) (Blome *et al.*, 2020). However, with the pandemic spread of ASF, research towards vaccine development has been intensified with very promising results (Bosch-Camós *et al.*, 2020; Dixon*et al.*, 2020; Sang*et al.*, 2020). However, vaccines will not replace the need for biosecurity, behavior changes, improved management, diagnostic approaches and targeted culling measures (Dixon *et al.*, 2020).

Prevention in countries free of the disease depend on implementation of appropriate import policies and biosecurity measures, ensuring that neither infected live pigs nor pork products are introduced into areas free of ASF (Penrith *et al.*, 2013; Bellini *et al.*, 2016). This include ensuring proper disposal of waste food from aircraft, ships or vehicles coming from affected countries and policing illegal imports of live pigs and pork products from affected countries. Classic sanitary measures may be employed including early detection and humane killing of animals (with proper disposal of carcasses and waste); thorough cleansing and disinfection; zoning and movement controls; surveillance and detailed epidemiological investigation; strict biosecurity measures on farms (Penrith *et al.*, 2004; Bellini *et al.* 2016). In Europe and Asia, the spread of ASF depend on the wild boar population and its interaction with low-biosecurity pig production systems (Blome *et al.*, 2020). Management of the wild boar population and coordination among the veterinary services, wildlife and forestry authorities and control of the soft tick vector are necessary for successfully ASF controlprogram.

Strategy for the Control of ASF in Africa

Components of the Strategy for the Control of ASF in Africa

The different components of the strategy are illustrated, Fig. 3, and are described within the three founding principles/pillars as follows:

Developing Knowledge Base on ASF in Africa

Update Knowledge Base on the Epidemiology and Control of ASF

This involves collecting, collating, updating and analysis of the existing knowledge relevant on epidemiology and control ASF in Africa. The use of biosciences capacity for diagnosis and epidemiological surveillance within the African continent and formation of consortium linking epidemiological research, state of the art diagnostics, mathematical modeling to better understand the epidemiology of the disease, studies on the economics of production and control and social networks studies to generate the data and make the information available for practical application. This process will promote capacity development of staff working in national systems to ensure sustainability of surveillance and control in individual states

Figure 3. The strategy for the control of ASF in Africa (Adopted and modified from FAO, 2017)

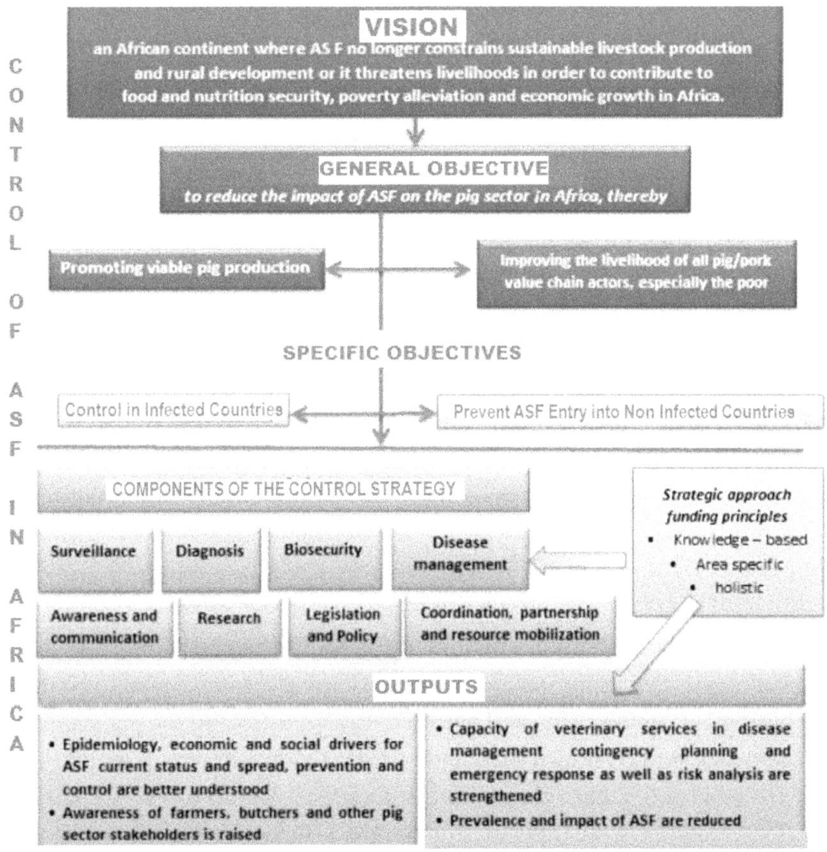

Surveillance of the ASF

Disease monitoring and surveillance are key elements in the management of transboundary animal diseases (TADs), including ASF. For both the active and passive disease monitoring and surveillance system and real time confirmation of disease to be effective the system must include all possible stakeholders. These comprise the producers, extension services, traders, national veterinary services, and research, academic and training institutions. Organizing such all-inclusive networks would enable capturing and sharing of all relevant data and information among actors and for informed decision making on the control. In most parts of Africa, the epidemio-surveillance systems are weak due to a lack of resources (funds, personnel and infrastructure). Information on diseases is generated through community-based surveillance as described in the OIE Terrestrial Animal Health Code (OIE WAHID, 2014). This type of surveillance is seriously constrained by under- and delayed reporting, due to the lack of incentive, i.e., absence of an adequate compensation policy for income lost through quarantine and sales bans and sometimes culling of pigs. Moreover, there is often no feedback to disease reports. Lack of resources has led to failure in sustaining surveillance systems and institutions established. This will need to be addressed through resource mobilization at country level and partnerships with the private sector.

Diagnosis for ASF

Laboratory diagnosis is an essential component of animal disease surveillance systems. However, in some African countries, facilities for diagnostic testing for ASF are not yet well established. For practical epidemiological surveillance system, all national laboratories in Africa should have access to all key diagnostic tests including serology, molecular diagnostics, virus isolation, and histopathology. To improve the capacity of diagnostic laboratories, training of laboratory staff is of great importance in addition to introduction of new diagnostic technologies, techniques, standards and quality control. Twinning with regional and global reference laboratories should be encouraged with the view of enhancing capacities of sub-regional reference laboratories in all regions of Africa. Strengthening laboratory diagnostic capacities would enhance early detection, accurate case classification as well as support structured disease surveys and risk analyses. This is a fundamental prerequisite to developing rational control and prevention strategies against ASF and other animal diseases. Field symptomatic diagnosis is a prerequisite to laboratory diagnosis because it directs the decision to conduct a laboratory investigation. For this purpose, it is necessary to train field officers, including extension officers and farmers, to recognize the signs of ASF and take appropriate samples for laboratory testing. Field diagnosis should be supported by a pen side test to detect ASFV antigen or DNA, since there are a number of differential diagnoses for ASF.

Research

Increasing capacity for field and laboratory diagnosis for ASF through training and information sharing is important. Anthropologic, economic, social and behavioral researches need to be integrated with epidemiological and biological studies. This will help scientists understand the pig sector, value chains and factors that determine how ASF is transmitted and spread in domestic pig populations. This will allow the design of more feasible and better accepted prevention and control tools. A comprehensive analysis of the role of social interactions in the transmission of ASFV is required using social network studies. In addition, disease control measures, especially on poor smallholder farmers, need to be assessed. More research is also needed on alternative feeds (what grows or can be grown, how to better to utilize by-products or to render kitchen-waste/food scraps microbiologically safe). It is important to identify locally available materials for constructing low cost but pig proof housing and knowledge of pig genetics and epigenetics to support formulation of a balanced production systems, productivity, health and feeding. The complex epidemiology of ASF requires the identification of risk factors and critical control points in the pig value chains for targeting prevention and control measures in specific ecosystems. Molecular characterization of ASFV isolates has contributed to strengthening surveillance and provided insight into the geographic origin of infections and their spread, but requires a well-structured and systematic approach for it to add value to control strategies and the future development of effective vaccines. As demonstrated for other diseases, having a vaccine against ASFV will be very useful in the control of the disease and reducing global spread. Hence, there is need for continuous research to develop effective vaccines and design efficient delivery mechanisms tailored to address the needs and realities of Africa.

Developing Regional and System-Specific ASF Control and Prevention Plan

Prevention of ASF is the most crucial element of this strategy since it eliminates the losses due to morbidity, mortality and costs of outbreak control. Its success depends on risk mitigation throughout the pig value chain, which will involve various actors from the producer to the retailer.

PIG SECTOR ORGANIZATION AND DEVELOPMENT

Constraints such as animal diseases, scarce feed resources and poor performance of markets and services affect pork market access and thus result in unsustainable pig value chains. The ASF control strategy involves:

- Promote mapping of market opportunities for pork and investigation of the multiple factors preventing pig producers from exploiting pig production.
- Promote organization of producer associations, trader networks and infrastructure including slaughter facilities at various levels.
- Establishment of a regional pig sector platform that will facilitate participation in the implementation of the strategy by all pig value chain actors.

Biosecurity

Biosecurity remains the most crucial aspect in the control of ASF (Bellini *et al.*, 2016). For it to be effective, good practices of improved husbandry and disease prevention need to be advocated through a participatory process. To ensure compliance with the required level of biosecurity, it will be critical to develop realistic, cost-effective and sustainable measures through dialogue and consensus with the stakeholders who will implement them on-farm. Implementing biosecurity should be affordable and comprise cheap locally available materials for housing. The design of recommended husbandry and biosecurity practices has to be based on a deep understanding of the pig sector and its value chains, particularly those of backyard pigs. Best practices and farmer's awareness on biosecurity issues should be applied along the entire value chain including the application of Hazard Analysis and Critical Control

Points (HACCP) in the slaughterhouses and on waste management procedures (FAO, 2011). At farm level, an all-inclusive approach should be adopted for the improved hygienic practices in the piggeries, housing, feeding and breeding. Pig farmers field schools (PFS), study groups using household and village level ASF risk assessment and mitigation methods could be a useful approach to promote biosecurity measures at production level by combining commercial and village producers. This will help in sharing experiences among farmers and progressively change scavenging systems into confined systems. The concept of compartmentalization of highly bio-secure commercial farms should be encouraged so that safe trade from compartments can continue even in the face of outbreaks in the area and the deaths or destruction of larger pig herds can be avoided. The principle has been applied in the South African control zone where strict separation between domestic and wild pig populations was essential (FAO/OIE/WB, 2010). As a result, accredited compartments have been able to continue providing pork for export in the region during outbreaks of foot and mouth disease and ASF.

Awareness and Communication

In Africa, awareness and communication for the control and eradication of TADs are challenging in many ways, with ASF being a classical example. The weak points include inadequate risk communication among value-chain actors; lack of communication strategy on ASF, cultural and religious sensitivities and biases, lack of readily available information packages for dissemination, low recognition of the disease, as well as inadequate knowledge about the disease among animal health workers. There is a real danger of creating ability to recognize the disease that can result in panic clandestine selling of potentially infected pigs. An organized public awareness campaign is probably the most important aid in preventing and controlling ASF and must be an intrinsic part of a strategy in every country at all times, be it at risk or already infected (Léopold *et al.*, 2019). Countries' specific characteristics determine the type of campaign that will succeed best, but certain basic rules apply to all countries. The involvement of public/community opinion leaders is essential for any awareness campaign to succeed and will be helpful throughout the sector transformation process. Core areas should be identified and supported by communication, advocacy, education, extension and social mobilization interventions. This should be done by reporting to veterinary authorities, establishing and promoting biosecurity as professional and social rules along the value chain, promoting community-based surveillance and reporting and advocacy to ensure greater interaction and coordination amongst all stakeholders both at private and public levels. Enhancing communication and information sharing efforts include the sustainable adoption of up-to-date technology on reporting and database management at the country, regional and global levels. Where possible, the principle of farmer-to-farmer extension should be applied, as nobody enjoys greater credibility than a successful fellow farmer. Experience has shown that when providing information on ASF, pig farmers accept it better if it is combined with general information about pig diseases and their management (FAO, 2017). ASF messages provide an opportunity to inform the pig owners about the importance of other diseases, in particular zoonotic diseases, e.g., *Taenia solium*, and vice versa. Politicians who make the policies and decisions are very important targets for awareness creation about ASF. The strategy needs to emphasize communication with policy- and decision-makers as part of the awareness and communication.

Outbreak Management

Management of ASF outbreaks in endemic and newly infected areas is the biggest challenge due to the lack of cost-effective control measures in the African context. As there is no vaccine against ASF, the traditional control tools include increased awareness, cleaning and disinfection procedures, movement control, quarantine, destruction of infected and exposed animals (stamping out) and trade bans on import of pigs and pork from infected countries/areas. As funds for compensation are not available in African countries, the strategy is to promote a collection of actions. These include stamping out where possible. It is important to promote measures to prevent incursion of ASF into new areas and ASF free countries. The application of stamping out depends on the availability of funds for compensation or insurance schemes in both endemic and newly infected countries. Applying stamping out in the absence of compensation policy and sufficient funds promotes underreporting and further spread of the disease through uncontrolled movements and should be avoided. Prompt notification of disease occurrences is an important ingredient for early warning and rapid response mechanisms. Early reporting of disease events by producers and other value-chain actors should trigger a swift response from veterinary services to contain and eventually eliminate disease agents and outbreaks. Timely reporting to the authorities should be encouraged through awareness campaigns via appropriate communication tools e.g., manuals, trainings, posters, radio and TV advertisements, leaflets and other promotional materials. In the event of disease detection, movement control, quarantine, stamping-out, proper carcass disposal as well as cleaning and disinfection will be applied in all situations where this is feasible. Whenever a case of ASF is confirmed in an epidemiological unit, the veterinary authorities should swiftly apply movement restrictions on the infected herd to stop further spread of the disease. The progressive transformation of breeding systems to more controllable systems (from free-ranging to more and more confined systems) should form an additional measure where disease facts may have made the community more receptive to awareness creation activities. The responsibility of managing disease through the implementation of appropriate disease control measures lies with the Chief Veterinary Officers (CVO) based on the global, regional and country strategies. However, the involvement and cooperation of all the actors in the pig sector is essential.

HOLISTIC TRANSFORMATION OF THE PIG SECTOR TO CONTROL ASF

Legislation and Policy

Various pig sub-sectors, the complex value chains, the socio-economic impacts, the different epidemiological cycles, translate into numerous policy challenges related to ASF prevention and control, which need to be reviewed and linked to existing strategies for animal disease prevention and control. In many SSA countries the major challenge is that livestock is not even included in agricultural policies, so that very limited funds are available for livestock development. At a higher level, the Comprehensive African Agriculture Development Program (CAADP) has omitted the livestock component. This has been corrected at continental level but has not been reflected in the national compacts. In many countries the pig sector is small and receives little attention from the public livestock sector, yet it has high potential for increasing food security and reducing poverty. The main regulatory constraints to ASF control in Africa include poor enforcement of regulations, obsolete legal framework and lack of a compensation scheme.

Poor enforcement is because the legal system is overloaded, so priority is not given to prosecution of breaches of animal disease control regulations and the costs involved are too high to be worth the effort. An overarching problem in most SSA countries is the wrong policies of privatization and decentralization of veterinary services which resulted in the loss of the clear chain of command from the Director of Veterinary Services to local county level, as recommended by OIE. Addressing this problem is beyond the scope of this control strategy and should be raised in appropriate forums. As there is no vaccine or treatment so far available for the control of ASF, stamping out with compensation and strict sanitary measures (biosecurity) remain as the main outbreak control options available. This control strategy advocates for inclusion of an effective compensation scheme in country disease policies and legislation. Other innovations such as insurance schemes and interest-free loans to affected farmers should also be explored. To this effect, this strategy proposes review and enforcement of the existing legislations and policies on disease control including across national borders and call upon the commitment and political will of national and regional authorities.

Creation of an Enabling Environment to Support Safe, Sustainable Pig Production

The success of disease control programs depends on the environment in which they are implemented. Various studies have revealed that up to 90% of the pigs produced in Africa are raised in the informal sector, characterized by a high degree of disorganization throughout the pig value chains(FAO, 2017). The sector has immense potential to significantly contribute to food security, improved nutrition and poverty alleviation, except its development is constrained by lack of financial support, weak marketing strategies and consequently lack of incentives to improve production and health of the pigs. Lack of appropriate infrastructure for the hygienic slaughter of pigs poses threats to public health and is prejudicial to animal welfare, Diseases like ASF flourish in this environment and unless the situation is addressed any efforts to control them will be doomed to failure. In this control strategy, reforms in the pig sector should be envisaged that will result in better organization, sanitary standards and profitability in the sector. Organization of the pig sector requires accurate information on the numbers and distribution of pigs. This is facilitated by identification and traceability system applied at least at the herd level. This control strategy therefore recommends implementation of identification and traceability in the commercial sector and to be phased in throughout the sector.

Coordination, Partnership and Resource Mobilization

The lessons learned from Rinderpest eradication and the fights against highly pathogenic avian influenza (HPAI) have underscored the importance of regional approaches and partnerships amongst national, regional and continental stakeholders for effective management of TADs. Due to their cross-border implication, TADs require the coordination of prevention and control actions among African member states, including the sharing of information. Collaboration and partnerships with initiatives to control other diseases, in particular zoonosis, in the region would have beneficial synergic effects and will be helpful in obtaining the required political support. To respond to the current ASF threat, the national, regional and international community'sneedto step up their efforts through an agreed upon coordinated approach and financing frameworks. Coordination is one of the most important and effective response tools for ASF control and should be sustained as a critical mechanism to ensure synergy of contributions by the multiple actors. Involvement of the private sector in implementation of the strategy including provision of resources and active participation in the evolution of the strategy are essential if success is to be achieved.

FAMILY: ARGASIDAE

Genus: Argas

Argaspersicus, also known as fowl tick or poultry tick, is a small soft-bodied tick that is found primarily on domestic fowl such as chickens, ducks, geese, turkeys and wild birds. It was first recorded by Lorenz Oken in 1818 in Mianeh, Persia, and named *Rhynochoprionpersicum* (CDC, 2017). Three *Argas* tick species exist namely:

i. Argaspersicus
ii. Argaspersicus complex- (Argasreflexus; A. miniatus; A. radiates)

Morphological Features

Argaspersicus is a dark red to yellowish-brown when starved tick, turning slatey-blue (often called the blue tick) when fully fed/engorged. The outer surface of the tick is irregular with a creased appearance. The female is about 8–10 mm in length and larger than the male (5 mm) with leathery, mamillated and wrinkled dorsal and ventral surfaces. *Argaspersicus* can be found on the host or in their nests or in cracks and crevices of commercial poultry buildings and sheds.

Figure 4. Argaspersicus–adult (Courtesy: Center for Disease Control and Prevention, 03 May 2016. Web. 15 Apr. 2020).

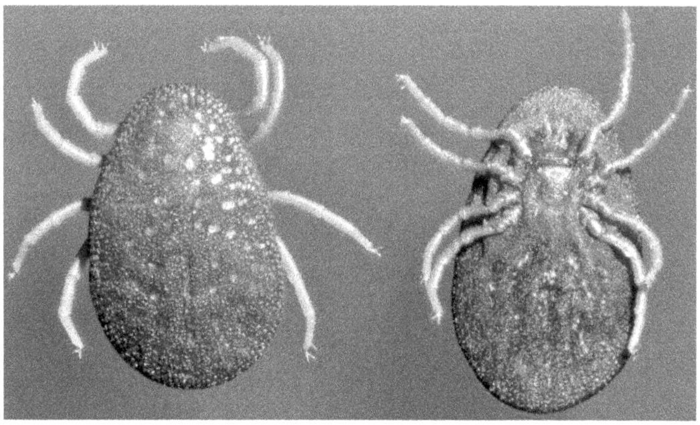

Table 2. Geographic Distribution and Public Health Importance.

Species	Geographic Distribution	Vector for/disease
Argas persicus	Worldwide in tropical and subtropical countries	a) Avians pirochetosis Bacteria *Borrelia anserina* b) Aegyptianellosis in Poultry Rickettsia infection c) *Aegyptianella pullorum* Anemia in pigeons.
Argas reflexus	Worldwide in tropical and subtropical countries	
Argas miniatus	USA	*Argas persicus complex*
Argas sanchezi	USA	*Argas persicus complex*
Argas radiatus	USA	*Argas persicus complex*

Life Cycle of *A. persicus*

Argaspersicus breeds in cracks in poultry sheds. It is active at night, feeding on the stock. The female is capable of producing large numbers of eggs. Female ticks often lay eggs in clutches of 25–100 off the host after each blood meal (CDC, 2017). Mating and the laying of eggs occur off the host, particularly in bird nests. Six-legged larvae hatch from the eggs in 1–4 weeks. The larvae draw several large blood mealsfora few hours up to 5 daysprior to dropping off their hosts and returning to cracks in the poultry sheds(CDC, 2017). The larvae undergo three molts intothe first-instar nymphs in about a week. These willseek a second host (same individual as the first host and likely the same species) to feed usually at night. There is one larval and at least two nymphal stages (although from three up to seven stages may occur) with the last stage molting to the adult. The life cycle from egg to adult can be completed within 30 daysunder optimal warm and humid conditions. In case there is a suitable hostfrom which to takeblood meals from, life cycle may be shortened otherwise it getsprolonged. Larvae and nymphs can survive for months without a blood meal while the adult may survive for more than 2 years (CDC, 2017). They transmit the spirochete *Borreliaanserina*, the causative agent of avian spirochetosis. Other *Argas* tick species that parasitize avian hosts include *A. reflexus*, the pigeon tick, and *A. walkerae*, the chicken tick.

Geographic Distribution and Transmitted Diseases by Argas ticks

The fowl tick (*Argaspersicus)* has worldwide distribution although it prefers the warmer tropical climates (Table2). In Kenya, *A periscus* is quite prevalent in the ASAL regions (Agro-ecological zones 4 and 5) which have dry and hot conditions favorable for the survival of this parasite.

Avian Spirochetosis

This is a disease of poultry caused by the bacteria *Borreliaanserine*. The disease has a worldwide distribution and tends to occur in places where the tick vector is present. The fowl tick has preference for a dry and warm climate (Lisbôa *et al.*, 2017). The lower United States especially has a large distribution of this soft bodied tick due to the warm climate (Rosenstein, 2017). *Argaspersicus* is a natural reservoir for Bacteria *Borreliaanserina* as it helps these spirochetes survive longer (CDC, 2017). If the pathogen infects humans due to *A. persicus*, it would likely be as an incidental host but humans may be fed upon by any of the tick infective stages (CDC, 2017). *Borreliaanserina is* transmitted in *Argaspersicus* saliva and can be seen in blood samples taken from infected birds (Lisbôa *et al.*, 2017).

Young birds arequitesusceptible to the larvae of *A. persicus* since they are naïve. Although domestic birds do not fully develop immunity against the paralysis inducing toxins produced, a weak protective mechanism may result from having been infected (Lisbôa *et al.*, 2017). Avian spirochetosis manifests with various symptoms including drop in egg production, depression, polydipsia (increased thirst), drowsiness, anorexia, loss of appetite, green diarrhea, ruffled feathers, pale combs, weight loss, paralysis of the legs and wings (flaccid paralysis) and abrupt death. Before death, birds tend to become recumbent or will lie down stretched out. This is especially true when heavily infested by the larval stage of *A. persicus* (Lisbôa *et al.*, 2017).

Tick paralysis is included in avian spirochetosis and exhibit similar symptoms. The flaccid paralysis ascends throughout the body. In coordination first occurs followed by paralysis of the hind limbs, forelimbs, and then leads to respiratory distress and failure followed by death. Blood loss during infection could occur as well and result in clinical anemia (Lisbôa *et al.*, 2017). Targeting the larval stages of *A. persicus* for control is important since they act as vectors for *B. anserina*. Repeated treatments with malathion spray help to reduce the ticks which assists the birds to recover. Penicillin injections appear to have no effect in treatment of the bacteria (Lisbôa *et al.*, 2017). The elimination of tick larvae is key to reversing the effects of avian spirochetosis. Hydration is also extremely important as death most often results due to dehydration from prolonged paralysis (Rosenstein, 2017)

Avian Aegyptianellosis

This is a rickettsia infection of birds caused by *Aegyptianella pullorum* (Synonyms: *Balfouriagallinarum, Balfouria anserine*). The disease is also calledavian piroplasmosis and has a wide range of hosts including chicken, goose, duck, and turkey with chicken being the most important host. *Aegyptinella pullorum* infects the erythrocytes of birds. The disease occurs in North Africa, South Africa, Indonesia, China, India, USSR (Transcaucasia), southeast Europe.The natural vector of *Aegyptianella pullorum*is the fowl tick, *Argaspersicus*. In the tick vector, the trophozoites are small, ranging in size from 0.5 to 3 sometimes up to 4 μ, depending upon the stage of development. They are round, oval or piriform, sometimes with a vacuole. They multiply by schizogony, producing a variable number - up to 20 - of very small merozoites. Transmission does not take place either through the egg or stage-to-stage. After the adult tick becomes infected by feeding on an infected bird, it takes 26 days or more before it is able to transmit it to another bird (CDC, 2017). Ticks can remain infective for as long as 162 days. The stages of development in the tick have not been described. *Aegyptianellosis (Egyptianella pullorum)* can be transmitted experimentally by intravenous, intraperitoneal, subcutaneous or intramuscular injection or by scarification. The incubation period in chickens is 10 to 15 or more days (CDC, 2017).

Pathogenesis

Aegyptianella pullorum may cause either a latent, chronic, subacute or acute disease in chickens. The acute form occurs primarily in young or imported birds in endemic regions, while the chronic and latent forms occur primarily in adult birds in endemic regions. Severe outbreaks have been reported in chickens in Algeria, Egypt, South Africa and Greece. Ducks and geese are apparently less seriously affected.

Symptoms: The principal signs are anemia, fever, icterus, diarrhea and anorexia. Necropsy findings include splenomegaly, liver degeneration, characteristic greyish yellow kidneys, intestinal congestion, petechial hemorrhages on the serosa, and sometimes pericarditis. Adult birds usually recover.

***Immunity*:** Birds which have recovered from infection are pre-immunized, but their latent infections can be reactivated by splenectomy or by intercurrent disease.

Diagnosis: *Aegyptianella pullorum* infections can be diagnosed by identifying the parasites in stained blood smears. They are difficult to stain, however, so the staining time must be prolonged. Affected birds are often simultaneously infected with *Borreliaanserina*, the cause of fowl spirochetosis, which is also transmitted by *Argaspersicus*.

TREATMENT AND CONTROL OF ARGASIDAE TICKS

Chemotherapy

Controlling the larvae is important as it acts as a vector for *B. anserina*. Repeated treatments with malathion spray on poultry houses and chicken wings helps to reduce the ticks and helps the birds recover. Penicillin injections appear to have no effect in treatment of the bacteria. The elimination of tick larvae is key in reversing the effects of avian spirochetosis. Hydration is also extremely important as death most often results due to dehydration from prolonged paralysis.

A variety of chemicals are used for the control of tick infestation in poultry either by application to the birds or poultry houses. These include topical application of N, N- diethyl- 2-phenylacetamide (DEPA), N, N- diethyl –3- methyl benzamide (DEET) and Dimethyl phthalate (DMP). DEET and DEPA are more effective ((Abbas *et al.*, 2004). Feeding of chicks with 100gIvermectin per kg body weight (Mousa *et al.*, 1988), dipping of tick infested chicken legs in 0.05% deltamethrin (Abbas *et al.*, 2004), bromocyclen spray to the birds and treatment of their houses with 0.6% bromocyclen ((Abbas *et al.*, 2004), topical application of 1:20 dilution of pestoban (a liquid concentrate of various non –poisonous plants) (Abbas *et al.*, 2004), application of 0.5% trichlorfon, 0.25% crotoxyphos or 0.25% carbaryl @ 25-50ml per bird and spraying the poultry houses with 1-2% trichlorfon, 0.25% diazinon, 0.02% dichlorvos, 0.5% crotoxyphos, 0.1% carbaryl or 0.25% dicresyl ((Abbas *et al.*, 2004), phosalone, tetrachlorvinphos, porpoxur and fenethacarb (Frolov *et al.*, 1978), application of 1-1.5% ftalofos (phosmet) to poultry houses at the rate of 150–200 mL/sq.m (Frolov *et al.*, 1972), ectocide, a preparation containing 10g of pyrethrin, 15g piperonylbutoxide, 3g benzalkonium chloride and water to make 10g Aerosol of 2%carbamate preparation, and pyrethroids (Abbas *et al.*, 2004).

Biological Control

Biological control of tick infestation in poultry includes, the cessation of egg laying by females of *A. persicus* after exposure to irradiation (Abbas *et al.*, 2004). Entobakterin and Dendrobacillin (preparations of spores and endotoxin of *Bacillus thuringiensis*), and Boverin (conidiophores of the fungus *Beauvariabassiana*). These can be used alone or in combination with chemical pesticides in the control of arthropods (Frolov, 1974) and the activity of *Bacillus thuringiensis* against *A. persicus* (Hassanain *et al.*, 1997).

Appropriate Housing

The design and construction of chicken housing units must be done carefully to reduce cracks, crevices and other suitable sites where the *A. persicus* hide. Where chicken houses are constructed using locally sources trees and other vegetation, remove the barks since after drying the bark detaches and creating spaces where the soft ticks hide.

Enhancing Biosecurity Measures

Whenever chickens for breeding purchased from the markets or farms prone to infestation with *A. persicus*, they should first be treated with effective acaricides before they are mixed with the rest of the flock. Such birds should be isolated and closely observed before they are eventually admitted to the flocks.

Anti-Tick Vaccines

Research in the post-genomic era is leading to the development of new control measures such as the recombinant vaccines. Tick infestation is rarely a one-species issue and therefore, anti-tick vaccines should aim at a more global protection against the main pathogen species of economical and epidemiological interest. The great rise of acaricide resistance, still asks for the implementation of an effective vaccine. The discovery of potential antigens against tick's bit and tick-pathogens proteins should result in improved production of vaccines, more advantageous in an overall approach to control tick-borne diseases (Domingos *et al.*, 2013).

FUTURE RESEARCH DIRECTION

The future fight against African swine fever must even more focus on enhanced classical control measures like early detection, strict hygiene and biosecurity measures, culling of infected farms, compensation of losses, stand still measures, epidemiological tracing and wild boar control programs. This is particularly necessary since an applicable, safe and effective vaccine that could be produced in large quantities and also be used in the wild animal reservoir will probably not be available in the short term. Therefore, future research should concentrate ASFV virology and functional genomics studies to support vaccine discovery research with important additional data sets (e.g., protein structure and function, and challenge models to identify candidate vaccines in immunization/challenge trials. Control measures for affected wild boar populations also need further research and taking different regional conditions into account. The potential role of mechanical vectors including arthropods, birds, and predatorsfor ASFV transmission also need further research. The controversial potential role of survivors and carriers in disease epidemiology should also be studied in more detail.

CONCLUSION

African swine fever is still a threat to pig production worldwide. Spread of the disease to other parts of the world have ultimately originated in Africa and therefore managing ASF better at its source is imperative, not only to enable the full potential of pig production to improve nutrition and lift African families out of poverty, but also to protect the rest of the world. The ability to manage a disease depends on the knowledge of its epidemiology, capacity for rapid outbreak detection and diagnosis, bio-security and an effective and affordable vaccine. Research on ASF in Eastern and Southern Africa historically focused strongly on the sylvatic cycle and on the biological characteristics of the virus. Recent research has continued to focus on both these aspects, using molecular genetics technology to further clarify the epidemiology and also to contribute to improved diagnostic tests and the continuing efforts to develop a vaccine. Whether or not a sylvatic cycle exists in the country, human behavior appears to play the major role in maintenance and spread of ASF virus. The possibility that argasids might serve as vectors for many more pathogens than expected requires a greater effort in implementing control measures, such as the search for new protective antigens to be included in a broad-spectrum anti-tick vaccine as well as specific coordinated and urgent epidemiological and parasite-surveillance programs. Since there is no single ideal solution for the control of ticks, an integrated control approach is probably the most effective. Vector and reservoir surveillance are an important component of such a strategy.

REFERENCES

Abbas, H. S., Muhammed, N. K., Zafar, I., & Muhammad, S. S. (2004). Review of Tick Infestation in Poultry. *International Journal of Agriculture and Biology*. www.ijab.org

Achenbach, J. E., Gallardo, C., Nieto-Pelegrín, E., Rivera-Arroyo, B., Degefa-Negi, T., Arias, M., & Sánchez-Vizcaíno, J. M. (2017). Identification of a new genotype of African swine fever virus in domestic pigs from Ethiopia. *Transboundary and Emerging Diseases*, *64*(5), 1393–1404. doi:10.1111/tbed.12511 PMID:27211823

Alkhamis, M. A., Gallardo, C., Jurado, C., Soler, A., Arias, M., & Sánchez-Vizcaíno, J. M. (2018). Phylodynamic and evolutionary epidemiology of African swine fever p72-CVR genes in Eurasia and Africa. *PLoS One*, *13*(2), e0192565. doi:10.1371/journal.pone.0192565 PMID:29489860

Anderson, E. C., Hutchings, G. H., Mukarati, N., & Wilkinson, P. J. (1998). African swine fever virus infection of the bush pig (*Potamochoerusporcus*) and its significance in the epidemiology of the disease. *Veterinary Microbiology*, *62*(1), 1–15. doi:10.1016/S0378-1135(98)00187-4 PMID:9659687

AU-IBAR. (2011). *Pan African Animal Health Yearbook 2011*. AU-IBAR.

AU-IBAR. (2013). *Pan African Animal Resources Yearbook 2013*. AU-IBAR.

AU-IBAR. (2014). *Pan African Animal Resources Yearbook*. AU-IBAR.

AU-IBAR. (2017). *Pan African Animal Resources Yearbook*. AU-IBAR.

Barongo, M. B., Ståhl, K., Bett, B., Bishop, R. P., Fèvre, E. M., Aliro, T., & Ssematimba, A. (2015). Estimating the basic reproductive number (R0) for African swine fever virus (ASFV) transmission between pig herds in Uganda. *PLoS ONE*, *10*. doi:10.1371/journal.pone.0125842

Bosch-Camós, L., Lopez, E., & Rodriguez, F. (2020). African swine fever vaccines: A promising work still in progress. *Porcine Health Management*, *6*(1), 14. doi:10.118640813-020-00154-2 PMID:32626597

Brown, A. A., Penrith, M. L., Fasina, F. O., & Beltran-Alcrudo, D. (2018). The African swine fever epidemic in West Africa, 1996–2002. *Transboundary and Emerging Diseases*, *65*(1), 64–76. doi:10.1111/tbed.12673 PMID:28710823

CDC - Laboratory Identification of Parasitic Diseases of Public Health Concern. (2017). Centers for Disease Control and Prevention.

Chenais, E., Depner, K., Guberti, V., Dietze, K., Viltrop, A., & Stahl, K. (2019). Epidemiological considerations on African swine fever in Europe 2014-2018. *Porcine Health Management*, *5*(1), 6. doi:10.118640813-018-0109-2 PMID:30637117

Chenais, E., Sternberg-Lewerin, S., Boqvist, S., Emanuelson, U., Aliro, T., Tejler, E., & Ståhl, K. (2015). African swine fever in Uganda: Qualitative evaluation of three surveillance methods with implications for other resource-poor settings. *Frontiers in Veterinary Science*, *2*, 51. doi:10.3389/fvets.2015.00051 PMID:26664978

Chiduwa, G., Chimonyo, M., Halimani, T. E., Chisambara, S. R., & Dzama, K. (2008). Herd dynamics and contribution of indigenous pigs to the livelihoods of rural farmers in a semi-arid area of Zimbabwe. *Tropical Animal Health and Production*, *40*(2), 125–136. doi:10.100711250-007-9071-8 PMID:18422255

Depner, D., Staubach, C., Probst, C., Globig, A., Blome, S., & Dietze, K., C., S.-L, Conraths, F. (2016). African Swine fever – epidemiological considerations and consequences for disease control. *Tierärztliche Umschau Heft*, *3/2016*, 72–78.

Dixon, L. K., Chapman, D. A. G., Netherton, C. L., & Upton, C. (2013). African swine fever virus replication and genomics. *Virus Research*, *173*(1), 3–14. doi:10.1016/j.virusres.2012.10.020 PMID:23142553

Dixon, L. K., Stahl, K., Jori, F., Vial, L., & Pfeiffer, D. U. (2020). African swine fever epidemiology and control. *Annual Review of Animal Biosciences*, *8*, 221–246.

Domingos, A., Antunes, S., Borges, L., & Do Rosario, V. E. (2013). Approaches towards tick and tick-borne diseases control. *Revista da Sociedade Brasileira de Medicina Tropical*, *46*(3), 265–269.

Estrada-Peña, A., Mangold, A. J., Nava, S., Venzal, J. M., Labruna, M. B., & Guglielmone, A. A. (2010). A systematic review of the tick family argasidae (Ixodida). Acarologia, 50(3), 317–333.

Estrada-Peña, A., Venzal, J. M., González-Acuña, D., & Guglielmone, A. A. (2003). *Argas* (*Persicargas) keiransin*. sp. (Acari: Argasidae), a parasite of the Chimango, *Milvago c. chimango* (Aves: Falconiformes). *Journal of Medical Entomol*ogy, *40*(6), 766-769.

Etter, E. M. C., Seck, I., Grosbois, V., Jori, F., Blanco, E., Vial, L., & Akakpo, A. J., Alhambedji, R., Kone, P., & Roger, F. L. (2011). Sero-prevalence of African swine fever in Senegal. *Emerging Infectious Diseases*, *17*(1), 49–54.

FAO. (2011). *A value chain approach to animal diseases management. Technical foundations and practical framework for field application. FAO Animal Production and Health Guidelines No. 4*. Food and Agriculture Organization of the United Nations.

FAO/OIE/WB. (2010). *Good practices for biosecurity in the pig sector – Issues and options in developing and transition countries. FAO Animal Production and Health Paper No. 169*. Food and Agriculture Organization of the United Nations.

Food and Agriculture Organization of the United Nations and African Union-Inter-African Bureau for Animal Resources and International Livestock Research Institute. (2017). *Regional strategy for the control of African Swine Fever in Africa.* www.fao.org/publications

Frolov, B. A. (1974). Chemicaland biological methods of controlling poultry ecto-parasites. *Aria, 12*, 66–8.

Gago da Câmara, N. J. (1933). História da pestesuínaem Angola. *Pecuária: Anais Dos Serviços Pecuários Da Colónia De Angola*, *1932*(1), 25–40.

Gallardo, M. C., & de la Torre Reoyo, A. (2015). Fernández-Pinero. J., Iglesias I., Munoz M. J., Arias M. L. (2015). African swine fever: A global view of the current challenge. *PHM*, *1*, 21. doi:10.118640813-015-0013-y

Haresnape, J. M., Lungu, S. A. M., & Mamu, F. D. (1987). An updated survey of African swine fever in Malawi. *Epidemiology and Infection*, *99*, 723–732.

Hassanain, M. A., El-Garhy, M. F., Abdul-Ghaffar, F. A., El-Sharaby, A., & Abdul Megeed, K. N. (1997). Biological control studies of soft and hard ticks. The effect of *Bacillus thuringiensis*varieties on soft and hard ticks (Ixodidae). *Parasitology Research*, *83*, 208–1.

Jori, F., & Bastos, A. D. S. (2009). Role of wild suids in the epidemiology of African swine fever. *EcoHealth*, *6*, 296–310. https://doi.org/10.1007/s10393-009-0248-7

Jori, F., Vial, L., Penrith, M. L., Pérez-Sánchez, R., Etter, E., Albina, E., & Roger, F. (2013). Review of the sylvatic cycle of African swine fever in sub-Saharan Africa and the Indian ocean. *Virus Research*, *173*, 212–227. https://doi.org/10.1016/j.virusres.2012.10.005

King, K., Chapman, D., Argilaguet, J. M., Fishbourne, E., Hutet, E., Cariolet, R., Hutchings, G., Oura, C. A., Netherton, C. L., Moffat, K., Taylor, G., Le Potier, M. F., Dixon, L. K., & Takamatsu, H. H. (2011). Protection of European domestic pigs from virulent African isolates of African swine fever virus by experimental immunization. *Vaccine*, *29*, 4593–4600.

Lichoti, J. K., Davies, J., Kitala, P. M., Githigia, S. M., Okoth, E., Maru, Y., & Bishop, R. P. (2016). Social network analysis provides insights into African swine fever epidemiology. *Preventive Veterinary Medicine*, *126*, 146210.

Lisbôa, R. S., Teixeira, R. C., Rangel, C. P., Santos, H. A., Massard, C. L., & Fonseca, A. H. (2017). *Avian spirochetosis in chickens following experimental transmission of Borrelia anserina by Argas (Persicaargas) Miniatus.* National Library.

Luther, N. J., Majiyagbe, K. A., Shamaki, D., Lombin, L. H., Antiagbong, J. A., Bitrus, Y., & Owolodun, O. (2007). Detection of African swine fever virus genomic DNA in a Nigerian Red River hog (*Potamochoerusporcus*). *The Veterinary Record*, *160*, 58–59.

Manzano-Román, Díaz-Martín, de la Fuente, & Pérez-Sánchez. (2012). *Soft ticks as pathogen vectors: distribution, surveillance and control, parasitology.* In Tech. Available from: https://www.intechopen.com/books/parasitology/softticks-as-pathogen-vectors-distribution-surveillance-and-control

Mashatise, E., Hamudikuwanda, H., Dzama, K., Chimonyo, M., & Kanengoni, A. (2005). Socio-economic roles, traditional management systems and reproductive patterns of Mukota pigs semi-arid north-eastern Zimbabwe. *Bunda Journal of Agriculture. Environmental Science & Technology*, *3*, 97–105.

Mendes, A. M. (1994). A história da pestesuínaafricanaem Angola. *Revista Portuguesa de Ciências Veterinárias*, *89*, 110–120.

Misinzo, G., Jumapili, F., Ludosha, M., Mafie, E., Silialis, J., Mushi, R., & Van Doorsselaere, J. (2012). Genotyping of African swine fever virus from an outbreak in Tanzania in 2009. *Research Opinions in Animal & Veterinary Sciences*, *2*, 334–338.

Montgomery, R. E. (1921). On a form of swine fever occurring in British East Africa (Kenya Colony). *Journal of Comparative Pathology*, *34*(159–191), 243– 262.

Mulumba-Mfumu, L. K., Saegerman, C., Dixon, L. K., Madimba Eric Kazadi, K. C., Mukalakata, N. T., Oura, C. A. L., Chenais, E., & Masembe Karl Ståhl, C. (2019, July). African swine fever: Update on Eastern, Central and Southern Africa. *Transboundary and Emerging Diseases*, *66*(4), 1462–1480. https://doi.org/10.1111/tbed.13187

Normile, D. (2018). Arrival of deadly pig disease could spell disaster for China. *Science*, *361*, 741. https://science.sciencemag.org/

OIE WAHID. (2014). *Office International Des Epizooties–World Animal Health Information Database (WAHID) Interface*. http://www.oie.int/wahis/public. php? page=home

Okoth, E., Gallardo, C., Macharia, J. M., Omore, A., Pelayo, V., Bulimo, D. W., & Bishop, R. P. (2013). Comparison of African swine fever virus prevalence and risk in two contrasting pig farming systems in South west and Central Kenya. *Preventive Veterinary Medicine*, *110*, 198205. https://doi.org/10.1016/j.prevetmed.2012.11.012

Penrith, M. L., Lopes Pereira, C., Lopes Da Silva, M. M. R., Quembo, C., Nhamusso, A., & Banze, J. (2007). African swine fever in Mozambique: Review, risk factors and considerations for control. *The Onderstepoort Journal of Veterinary Research*, *74*, 149–160.

Penrith, M. L., & Thomson, G. R. (2004). Special factors affecting the control of livestock diseases in sub-Saharan Africa. In J. A. W. Coetzer & R. C. Tustin (Eds.), *Infectious diseases of livestock* (2nd ed.). Oxford University Press.

Penrith, M. L., Vosloo, W., Jori, F., & Bastos, A. D. S. (2013). African swine fever virus eradication in Africa. *Virus Research*, *173*, 228–246.

doi:10.1016/j.virusres.2012.10.011

Plowright, W., Parker, J., & Peirce, M. A. (1969b). The epizootiology of African swine fever in Africa. *The Veterinary Record*, *85*, 668–674.

Plowright, W., Parker, J. A., & Peirce, M. A. (1969a). African swine fever virus in ticks (*Ornithodorosmoubata* Murray) collected from animal burrows in Tanzania. *Nature*, *221*, 1071–1073. https://doi.org/10.1038/2211071a0

Plowright, W., Perry, C. T., & Greig, A. (1974). Sexual transmission of African swine fever virus in the tick, *Ornithodorosmoubataporcinus* Walton. *Research in Veterinary Science*, *17*, 106–113.

Plowright, W., Thomson, G. R., & Neser, J. A. (1994). African swine fever. In J. A. W. Coetzer, G. R. Thomson, & R. C. Tustin (Eds.), *Infectious diseases of Livestock, with special reference to Southern Africa, 1* (pp. 568–599). Oxford University Press South Africa.

Quembo, C., Jori, F., Heath, L., Pérez Sánchez, R., & Vosloo, W. (2016). Investigation into the epidemiology of African swine fever virus at the wildlife domestic interface of the Gorongosa National Park, Central Mozambique. *Transboundary and Emerging Diseases*, *63*, 443–451. https://doi.org/10.1111/tbed.12289

Roger, F., Ratovonjato, J., Vola, P., & Uilenberg, G. (2001). *Ornithodorosporcinus* ticks, bush pigs, and African swine fever in Madagascar. *Experimental & Applied Acarology*, *25*, 263–269.

Rosenstein, M. (1976). Paralysis in chickens caused by larvae of the poultry tick, ArgasPersicus. *Avian Diseases*, *20*(2), 407–409.

Sánchez Botija, C. (1963). Reservoirs of African swine fever virus/ a study of the African swine fevere virus in arthropods by means of haemadsorption. *Bulletin - Office International des Epizooties*, *60*, 895–899.

Sanchez-Cordon, P. J., Reis, A. L., & Dixon, L. K. (2018). African swine fever: A re-emerging viral disease threatening the global pig industry. *Veterinary Journal (London, England)*, *233*. Advance online publication. doi:10.1016/j.tvjl.2017.12.025

Sanchez-Vizcaino, J. M., Mur, L., & Martinez-Lopez, B. (2013). African swine fever (ASF): Five years around Europe. *Veterinary Microbiology*, *165*, 45–50.

Sandra, B., Kati, F., & Martin, B. (2020). Review of African swine fever – A review of current knowledge. *Virus Research*, *287*, 198099. www.elsevier.com/locate/virusre

Sang, H., Miller, G., Lokhandwala, S., Sangewar, N., Waghela, S. D., Bishop, R. P., & Mwangi, W. (2020). Progress Toward Development of Effective and Safe African Swine Fever Virus Vaccines. *Frontiers in Veterinary Science*, *7*, 84.

Shivairo, R. S., Muleke, C. I., & Lubembe, D. M. (2018). Text Book of Veterinary Parasitology and Parasitic Diseases. Academic Press.

Silvia, B., Rutili, D., & Guberti, V. (2016). Preventive measures aimedat minimizing the risk of African swine fever virus spread in pig farming systems. *Acta Veterinaria Scandinavica*, *58*(1), 82. doi:10.118613028-016-0264-x PMID:27899125

Simulundu, E., Chambaro, H. M., Sinkala, Y., Kajihara, M., Ogawa, H., Mori, A., & Mweene, A. S. (2018). Co-circulation of multiple genotypes of African swine fever viruses among domestic pigs in Zambia (2013–2015). *Transboundary and Emerging Diseases*, *65*, 114–122. https://doi.org/10.1111/tbed.12635

Simulundu, E., Lubaba, C., van Heerden, J., Kajihara, M., Mataa, L., Chambaro, H., & Mweene, A. (2017). The epidemiology of African swine fever in "non-endemic" regions of Zambia (1989–2015): Implications for disease prevention and control. *Viruses*, *9*, 9.

Stahl, K., Sternberg-Lewerin, S., Blome, S., Viltrop, A., Penrith, M. L., & Chenais, E. (2019). Lack of evidence for long term carriers of African swine fever virus - a systematic review. *Virus Research*, *272*, 197725.

Steyn, D. G. (1928). *Preliminary report on a South African virus disease amongst pigs. 13th and 14th Reports of the Director of Veterinary Education and Research, Union of South Africa*. Google Scholar.

Thomson, G. R. (1985). The epidemiology of African swine fever: The role of free-living hosts in Africa. *The Onderstepoort Journal of Veterinary Research*, *52*, 201–209.

Turnbull, D. O. (1932). *Nyasaland. Annual Report of the Veterinary Department, 1931. Swine fever*. Google Scholar.

Vergne, T. C., Fu, C., Li, S., Cappelle, J., Edwards, J., Martin, V., & Roger, F. L. (2017). Pig empire under infectious threat: Risk of African swine fever introduction into the People's Republic of China. *Veterinary Record*, *181*, 117. doi:10.1136/vr.103950

Wen, X., He, X., Zhang, X., Zhang, X., Liu, L., Guan, Y., Zhang, Y., & Bu, Z. (2019). Genome sequences derived from pig and dried blood pig feed samples provide important insights into the transmission of African swine fever virus in China in 2018. *Emerging Microbes & Infections*, *8*, 303–306.

Zhou, L., Evan, Y. W. Yu., Wang, S., & Changpo, S. (2019). African swine fever epidemic in China, Veterinary Record. *Journal of the British Veterinary Association*. doi:10.1136/vr.l4026

ADDITIONAL READING

FAO. (2011). *A value chain approach to animal diseases management. Technical foundations and practical framework for field application. FAO Animal Production and Health Guidelines No. 4*. Food and Agriculture Organization of the United Nations.

FAO/OIE/WB. (2010). *Good practices for biosecurity in the pig sector – Issues and options in developing and transition countries. FAO Animal Production and Health Paper No. 169*. Food and Agriculture Organization of the United Nations.

OIE. (2019). African swine fever (infection with African swine fever virus). In W. O. F. A. Health (Ed.), *Manual of Diagnostic Tests and Vaccines for Terrestrial Animals*. Off. Int. Epiz.

Compilation of References

doi:10.1016/j.virusres.2012.10.011

Abbas, H. S., Muhammed, N. K., Zafar, I., & Muhammad, S. S. (2004). Review of Tick Infestation in Poultry. *International Journal of Agriculture and Biology*. www.ijab.org

Abbas, R. Z., Zaman, M. A., Colwell, D. D., Gilleard, J., & Iqbal, Z. (2014). Acaricide resistance in cattle ticks and approaches to its management: The state of play. *Veterinary Parasitology*, *203*(1-2), 6–20. doi:10.1016/j.vetpar.2014.03.006 PMID:24709006

Abba, Y., Jesse, F. F. A., Sadiq, M. A., Ibrahim, H. H., Chung, E. L. T., Bitrus, A. A., Lila, M. A. M., & Hambali, I. U. (2016). Clinical management and gross pathology findings of a severe anaplamosis in a dairy cow. *Journal of Advanced Veterinary and Animal Research*, *3*(2), 195–199. doi:10.5455/javar.2016.c150

Abd-Alla, A., Meki, I., Bourtzis, K., & Parker, A. (2018). *Standard Operating Procedures for Identification of Tsetse Species from Wild Populations and Laboratory Colonies*. FAO/IAEA.

Abdisa, T. (2019). Epidemiology of Bovine Anaplasmosis. *SOJ Vet Sci*, *5*(1), 1–6. doi:10.15226/2381-2907/5/1/00165

Abebe, G., & Jobre, Y. (1996). Trypanosomosis: A threat to cattle production in Ethiopia. *Rev. Méd. Vét.*, *147*, 897–902.

Abenga, J. N., Enwezor, F. N. C., Lawani, F. A. G., Osue, H. O., & Ikemereh, E. C. D. (2004). Trypanosome prevalence in cattle in Lere Area in Kaduna State, North central Nigeria. *Revue d'Élevage et de Médecine Vétérinaire des Pays Tropicaux*, *57*(1-2), 45–48. doi:10.19182/remvt.9904

Abenga, J. N., & Vuza, D. (2005). About factors that determine trypanotolerance and prospects for increasing resistance against trypanosomosis. *African Journal of Biotechnology*, *4*(13), 13.

Abenga, J. N., & Vuza, D. (2005). About factors that determine trypanotolerance and prospects for increasing resistance against trypanosomosis. *African Journal of Biotechnology*, *4*(13), 1563–1567.

Abera, Z., Usmane, A., & Zelalem, A. Z. (2015). Review on Camel Trypanosomosis: Its Epidemiology and Economic Importance. *Acta Parasitologica Globalis*, *6*(2), 117–128.

Abila, P., Kiendrebeogo, M., Mutika, G., Parker, A., & Robinson, A. (2003). The effect of age on the mating competitiveness of male Glossina fuscipes fuscipes and G. palpalis palpalis. *Journal of Insect Science*, *3*(13), 1–8. doi:10.1673/031.003.1301 PMID:15841229

Achenbach, J. E., Gallardo, C., Nieto-Pelegrín, E., Rivera-Arroyo, B., Degefa-Negi, T., Arias, M., & Sánchez-Vizcaíno, J. M. (2017). Identification of a new genotype of African swine fever virus in domestic pigs from Ethiopia. *Transboundary and Emerging Diseases*, *64*(5), 1393–1404. doi:10.1111/tbed.12511 PMID:27211823

Adam, Y., Cecchi, G., Kgori, P., Marcotty, T., Mahama, C., Abavana, M., Anderson, B., Paone, M., Mattioli, R., & Bouyer, J. (2013). The Sequential Aerosol Technique: A Major Component in an Integrated Strategy of Intervention against Riverine Tsetse in Ghana. *PLoS Neglected Tropical Diseases*, *7*(3), e2135. doi:10.1371/journal.pntd.0002135 PMID:23516662

Adam, Y., Marcotty, T., Cecchi, G., Mahama, C. I., Solano, P., Bengaly, Z., & Van den Bossche, P. (2012). Bovine trypanosomosis in the Upper West Region of Ghana: Entomological, parasitological and serological cross-sectional surveys. *Research in Veterinary Science*, *92*(3), 462–468. doi:10.1016/j.rvsc.2011.04.004 PMID:21550616

Adawa, D. A. Y., Hassan, A. Z., Abdullah, S. U., Ogunkoya, A. B., Adeyanju, J. B., & Okoro, J. E. (1992). Clinical trial of long-acting oxytetracycline and piroxicam in the treatment of canine ehrlichiosis. *The Veterinary Quarterly*, *15*(3), 118–120. doi:10.1080/01652176.1992.9694345 PMID:1413442

Adehan, S. B., Adakal, H., Gbinwoua, D., Yokossi, D., Zoungrana, S., Toé, P., Ouedraogo, M., Gbaguidi, A. M., Adoligbé, C., Fandohan, A. B., Hounmanou, G., Kakaï, R. G., Farougou, S., & De Clercq, E. M. (2018). West African cattle farmers' perception of tick-borne diseases. *EcoHealth*, *15*(2), 437–449. doi:10.100710393-018-1323-8 PMID:29536206

Adehan, S. B., Biguezoton, A., Adakal, H., Assogba, M. N., Zoungrana, S., Gbaguidi, A. M., Tonouhewa, A., Kandé, S., Achi, L., Kagone, H., Adehan, R., Mensah, G. A., De Deken, R., Madder, M., & Farougou, S. (2016). Acaricide resistance of *Rhipicephalus microplus* ticks in Benin. *African Journal of Agricultural Research*, *11*(14), 1199–1208. doi:10.5897/AJAR2015.10619

Adenubi, O. T., Fasina, F. O., McGaw, L. J., Eloff, J. N., & Naidoo, V. (2016). Plant extracts to control ticks of veterinary and medical importance: A review. *South African Journal of Botany*, *105*, 178–193. doi:10.1016/j.sajb.2016.03.010

Adungo, F., Mokaya, T., Makwaga, O., & Mwau, M. (2020). Tsetse distribution, trypanosome infection rates, and smallholder livestock producers' capacity enhancement for sustainable tsetse and trypanosomiasis control in Busia, Kenya. *Tropical Medicine and Health*, *48*(1), 62. doi:10.118641182-020-00249-0 PMID:32760194

Agbede, R., Kemp, D., & Hoyte, H. (1986). *Secretory and digest cells of female Boophilus microplus: Invasion and development of Babesia bovis; light and electron microscope studies*. Academic Press.

Agnew, A. (1996). *The graze and browse resource at Olkiramatian Group Ranch. Fourth report to KETRI/NRI on landscape/vegetation monitoring at Olkiramatian group ranch. Report to the ODA on a Visit to the Kenya Trypanosomosis Research Institute (Nguruman Station)*. Academic Press.

Agyemang, D. R. H., Little, D. A., & Rowlands. (1997). *Village N'Dama cattle production in West Africa. Six years of research in the Gambia*. ILRI. https://agris.fao.org/agris-search/search.do?recordID=QM1998000001

Ahmed, J. (2002). The role of cytokines in immunity and immunopathogenesis of pirolasmoses. *Parasitology Research*, *88*(S1), S48–S50. doi:10.100700436-001-0573-4 PMID:12051611

Ahmed, S., Rahman, A., Hassan, M., Salih, S., Paone, M., & Cecchi, G. (2016). An atlas of tsetse and bovine trypanosomosis in Sudan. *Parasites & Vectors*, *9*(1), 194. doi:10.118613071-016-1485-6 PMID:27056678

Aiello, Moses, & Allen. (2016). The Merck veterinary manual (11th ed.). Kenilworth, NJ: Merck & Co., Inc.

Akira, S., Uematsu, S., & Takeuchi, O. (2006). Pathogen Recognition and Innate Immunity. *Cell*, *3*(4), 783–801. doi:10.1016/j.cell.2006.02.015 PMID:16497588

Aksoy, E., Telleria, E., Echodu, R., Wu, Y., Okedi, L., Weiss, B., Aksoy, S., & Caccone, A. (2014). Analysis of multiple tsetse fly populations in Uganda reveals limited diversity and species-specific gut microbiota. *Applied and Environmental Microbiology*, *80*(14), 4301–4312. doi:10.1128/AEM.00079-14 PMID:24814785

Aksoy, S. (1995). *Wigglesworthia* gen. nov. and *Wigglesworthia glossinidae* sp. nov., taxa consisting of the mycetocyte-associated primary endosymbionts of tsetse flies. *International Journal of Systematic Bacteriology*, *45*(4), 848–851. doi:10.1099/00207713-45-4-848 PMID:7547309

Aksoy, S., Maudlin, I., Dale, C., Robinson, A. S., & O'Neill, S. L. (2001). Prospects for control of African trypanosomiasis by tsetse vector manipulation. *Trends in Parasitology*, *17*(1), 29–35. doi:10.1016/S1471-4922(00)01850-X PMID:11137738

Alam, U., Medlock, J., Brelsfoard, C., Pais, R., Lohs, C., Balmand, S., Carnogursky, J., Heddi, A., Takac, P., Galvani, A., & Aksoy, S. (2011). Wolbachia symbiont infections induce strong cytoplasmic incompatibility in the tsetse ñy Glossina morsitans. *PLoS Pathogens*, *7*(12), e1002415. Advance online publication. doi:10.1371/journal.ppat.1002415 PMID:22174680

Alberta Agriculture and Rural Development. (2008). *A guide for the control of flies in Alberta confined feeding operations*. Author.

Albert, M., Wardrop, N., Atkinson, P., Torr, S., & Welburn, S. (2015). Tsetse Fly (G.f. fuscipes) Distribution in the Lake Victoria Basin of Uganda. *PLoS Neglected Tropical Diseases*, *9*(4), e0003705. doi:10.1371/journal.pntd.0003705 PMID:25875201

Alemu, T., Luckins, A. G., Philips, L. P., Reid, S. W. J., & Holmes, P. H. (1997). The use of ELISA to investigate the prevalence of T. equiperdum in Ethiopian horses. *Veterinary Parasitology*, *71*, 239–250. doi:10.1016/S0304-4017(96)01139-9 PMID:9299693

Alexander, K. A., Conrad, P. A., Gardner, I. A., Parish, C., Appel, M., Levy, M. G., Lerche, N., & Kat, P. (1993). serologic survey for selected microbial pathogens in African wild dogs (Lycaon pictus) and sympatric domestic dogs (*Canis familiaris*) in Maasai Mara, Kenya. *Journal of Zoo and Wildlife Medicine*, *24*(2), 140–144.

Alexander, R. A. (1931). Heartwater - the present state of our knowledge of the disease. 17th Report of the Director of Veterinary Services and Animal Industry. *Union of South Africa*, *17*(1), 89–150.

Alexander, R., Neitz, W. O., & Adelaar, T. F. (1946). Heartwater. *Farming in South Africa*, *21*, 548–552.

Ali, M., & De Castro, J. J. (1993). Host resistance to ticks (Acari: Ixodidae) in different breeds of cattle at Bako, Ethiopia. *Tropical Animal Health and Production*, *25*(4), 215–222. doi:10.1007/BF02250871 PMID:8109054

Alingu, R., Muhanguzi, D., MacLeod, E., Waiswa, C., & Fyfe, J. (2014). Bovine trypanosome species prevalence and farmers' trypanosomiasis control methods in south-western Uganda. *Journal of the South African Veterinary Association*, *28*(85(1)). Advance online publication. doi:10.4102/jsava.v85i1.1094 PMID:25686358

Alkhamis, M. A., Gallardo, C., Jurado, C., Soler, A., Arias, M., & Sánchez-Vizcaíno, J. M. (2018). Phylodynamic and evolutionary epidemiology of African swine fever p72-CVR genes in Eurasia and Africa. *PLoS One*, *13*(2), e0192565. doi:10.1371/journal.pone.0192565 PMID:29489860

Al-Khedery, B., & Allred, D. R. (2006). Antigenic variation in Babesia bovis occurs through segmental gene conversion of the ves multigene family, within a bidirectional locus of active transcription. *Molecular Microbiology*, *59*(2), 402–414. doi:10.1111/j.1365-2958.2005.04993.x PMID:16390438

Allison, A. C. (1981). Cellular immune responses in theileriosis. In A. D. Irvin, M. P. Cunnigham, & A. S. Young (Eds.), *Advances in the control of theileriosis* (pp. 273–288). Martinus Nijhoff Publishers. doi:10.1007/978-94-009-8346-5_49

Allred, D. R., McGuire, T. C., Palmer, G. H., Leib, S. R., Harkins, T. M., McElwain, T. F., & Barbet, A. F. (1990). Molecular basis for surface antigen size polymorphisms and conservation of a neutralization-sensitive epitope in *Anaplasma marginale. Proceedings of the National Academy of Sciences of the United States of America*, *87*(8), 3220–3224. doi:10.1073/pnas.87.8.3220 PMID:1691504

Allsopp, B. A. (2010). Natural history of *Ehrlichia ruminantium. Veterinary Parasitology*, *167*(2-4), 123–135. doi:10.1016/j.vetpar.2009.09.014 PMID:19836892

Allsopp, B. A., Baylis, H. A., Allsopp, M. T. E. P., Cavalier-Smith, T., Bishop, R. P., Carrington, D. M., Sohanpal, B., & Spooner, P. (1993). Discrimination between six species of Theileria using oligonucleotide probes which detect small ribosomal RNA sequencies. *Parasitology*, *107*(2), 157–165. doi:10.1017/S0031182000067263 PMID:8414670

Allsopp, B. A., Bezuidenhout, J. D., & Prozesky, L. (2004). Heartwater. In J. A. W. Coetzer & R. C. Tustin (Eds.), *Infectious Diseases of Livestock* (2nd ed., pp. 507–535). Oxford University Press Southern Africa.

Allsopp, B. A., Carrington, M., Baylis, H., Sohal, S., Dolan, T. T., & Iams, K. (1989). Improved characterization of Theileria parva isolates using the polymerase chain reaction and oligonucleotide probes. *Molecular and Biochemical Parasitology*, *35*(2), 137–148. doi:10.1016/0166-6851(89)90116-3 PMID:2570351

Allsopp, M. T. E. P., Theron, J., Coetzee, M. L., Dunsterville, M. T., & Allsopp, B. A. (1999). The occurrence of *Theileria* and *Cowdria* parasites in African buffalo (*Syncerus caffer*) and their associated *Amblyomma hebraeum* ticks. *The Onderstepoort Journal of Veterinary Research*, *66*, 245–249. PMID:10631711

Allsop, R. (1984). Control of tsetse flies (Diptera: Glossinidae) using insecticides: a review and future prospects. *Bulletin of Entomological Research*, *74*(1), 1–23. doi:10.1017/S0007485300009895

Alonso, M., Arellano-Sota, C., Cereser, V. H., Cordoves, C. O., Guglielmone, A. A., Kessler, R., ... Solari, M. A. (1992). Epidemiology of bovine anaplasmosis and babesiosis in Latin America and the Caribbean. *Revue Scientifique et Technique (International Office of Epizootics)*, *11*(3), 713–733. doi:10.20506/rst.11.3.623 PMID:1472722

Alonso, M., Blandino, T., Mendoza, E., Savon, L., & Camacho, M. (1994). Development of a Babesia bovis live attenuated vaccine. *Archives of Medical Research*, *25*(2), 273–277. PMID:7919825

Alvarez, J., Rojas, C., & Jose, J. (2017). *Development of a Live Attenuated Vaccine for the Control of Bovine Babesiosis in Mexico*. Academic Press.

Alvåsen, K., Johansson, S. M., Höglund, J., Ssuna, R., & Emanuelson, U. (2016). A field survey on parasites and antibodies against selected pathogens in owned dogs in Lilongwe, Malawi. *Journal of the South African Veterinary Association*, *87*(1), a1358. doi:10.4102/jsava.v87i1.1358 PMID:27543039

Amrit, P. S., Rajesh, K. G., & Tajpreet, K. (2011). Mechanisms pertaining to arsenic toxicity. *Toxicology International Journal*, *18*(2), 87–93. doi:10.4103/0971-6580.84258 PMID:21976811

Anderson, H., Brodsky, I.E., & Mangalmurti, N.S. (2018). The Evolving Erythrocyte: Red Blood Cells as Modulators of Innate Immunity. *Journal of Immunology, 201*(5), 1343–1351. doi:10.4049/jimmunol.1800565

Anderson, E. C., Hutchings, G. H., Mukarati, N., & Wilkinson, P. J. (1998). African swine fever virus infection of the bush pig (*Potamochoerusporcus*) and its significance in the epidemiology of the disease. *Veterinary Microbiology*, *62*(1), 1–15. doi:10.1016/S0378-1135(98)00187-4 PMID:9659687

Andreotti, R. (2006). Performance of two Bm86 antigen vaccine formulation against tick using crossbreed bovines in stall test. Revista Brasileira De Parasitologia Veterinaria = Brazilian Journal of Veterinary Parasitology. *Orgao Oficial Do Colegio Brasileiro De Parasitologia Veterinaria*, *15*(3), 97–100.

Andrew, H. R., & Norval, R. A. I. (1989). The carrier status of sheep, cattle and African buffalo recovered from heartwater. *Veterinary Parasitology*, *34*(3), 261–266. doi:10.1016/0304-4017(89)90056-3 PMID:2617830

Anene, B. M., Chime, A. B., Jibike, G. I., & Anika, S. M. (1991). Prevalence of trypanosomiasis in Zebu cattle at Obudu ranch— A tsetse-free zone in Nigeria. *Preventive Veterinary Medicine*, *10*(4), 257–260. doi:10.1016/0167-5877(91)90011-P

Angus, B. M. (1996). The history of the cattle tick *Boophilus microplus* in Australia and achievement in its control. *International Journal for Parasitology*, *26*(12), 1341–1355. doi:10.1016/S0020-7519(96)00112-9 PMID:9024884

Annaline Van Heerden, J., & Heerden, V. (1981). Attempted treatment of canine ehrlichiosis with imidocarbdipropionate. *Journal of the South African Veterinary Association*, *52*(3), 173–175. PMID:7310786

Anosa, V. O. a. O., T. U. (1980). Haematological studies on domestic animals in Nigeria. The effect of age, breed and haemoglobin type on bovine haematology and anaemia. *Bulletin of Veterinary Medicine, 27,* 773 – 788.

Anosa, V. O., & Isoun, T. T. (1980). Haematological studies on Trypanosoma vivax infection of goats and intact and splenectomized sheep. *Journal of Comparative Pathology*, *90*(1), 155–168. doi:10.1016/0021-9975(80)90036-5 PMID:7391277

Aregawi, W. G., Agga, G. E., Abdi, R. D., & Büscher, P. (2019). Systematic review and meta-analysis on the global distribution, host range, and prevalence of Trypanosoma evansi. *Parasites & Vectors*, *12*(1), 67. Advance online publication. doi:10.118613071-019-3311-4 PMID:30704516

Arulkanthan, A., Brown, W. C., McGuire, T. C., & Knowles, D. P. (1999). Biased immunoglobulin G1 isotype responses induced in cattle with DNA expressing *msp*1a of *Anaplasma marginale. Infection and Immunity*, *67*(7), 3481–3487. doi:10.1128/IAI.67.7.3481-3487.1999 PMID:10377129

Ashcroft, M. T. (1959). The importance of African wild animals as a reservoir of trypanosomiasis. *East African Medical Journal*, *36*, 289–297. PMID:13672101

Aubry, P. & Geale, D. W. (2010). *Transboundary and Emerging Diseases. A Review of Bovine Anaplasmosis*. Academic Press.

Aubry, P., & Geale, D. W. (2011). A review of bovine anaplasmosis. *Transboundary and Emerging Diseases*, *58*(1), 1–30. doi:10.1111/j.1865-1682.2010.01173.x PMID:21040509

Augustinos, A. A., Meki, I. K., Demirbas-Uzel, G., Ouédraogo, G. M. S., Saridaki, A., Tsiamis, G., Parker, A. G., Abd-Alla, A. M. M., & Bourtzis, K. (2018). Nuclear and Wolbachia-based multimarker approach for the rapid and accurate identification of tsetse species. *BMC Microbiology*, *18*(1), 147. doi:10.118612866-018-1295-4 PMID:30470190

AU-IBAR. (2011). *Pan African Animal Health Yearbook 2011*. AU-IBAR.

AU-IBAR. (2013). *Pan African Animal Resources Yearbook 2013*. AU-IBAR.

AU-IBAR. (2014). *Pan African Animal Resources Yearbook*. AU-IBAR.

AU-IBAR. (2017). *Pan African Animal Resources Yearbook*. AU-IBAR.

Auty, H., Cleaveland, S., Malele, I., Masoy, J., Lembo, T., & Bessell, P. (2016). Quantifying Heterogeneity in Host-Vector Contact: Tsetse (Glossina swynnertoni and G. pallidipes) Host Choice in Serengeti National Park, Tanzania. *PLoS ONE, 11*, e0161291. doi: .pone.0161291 doi:10.1371/journal

Auty, H. K., Picozzi, K., Malele, I., Torr, S. J., Cleaveland, S., & Welburn, S. (2012). Using molecular data for epidemiological inference: Assessing the prevalence of Trypanosoma brucei rhodesiense in tsetse in Serengeti, Tanzania. *PLoS Neglected Tropical Diseases*, *6*(1), e1501. doi:10.1371/journal.pntd.0001501 PMID:22303496

Ayoob, A. L., Hackner, S. G., & Prittie, J. (2010). Clinical management of canine babesiosis. *Journal of Veterinary Emergency and Critical Care (San Antonio, Tex.)*, *20*(1), 77–89. doi:10.1111/j.1476-4431.2009.00489.x PMID:20230437

Baars, R. M. T. (1999). The effect of rangeland fires on cattle tick infestation in Western Zambia. *Tropical Animal Health and Production*, *31*(5), 275–278. doi:10.1023/A:1005299605959 PMID:10509420

Babo Martins, S., Di Giulio, G., Lynen, G., Peters, A., & Rushton, J. (2010). Assessing the impact of East Coast fever immunization in Tanzanian pastoralists systems. *Preventive Veterinary Medicine*, *97*, 175–182. doi:10.1016/j.prevetmed.2010.09.018 PMID:20974501

Bailey, K. P. (1960). Notes on the rearing of *Rhipicephalus appendiculatus* and their infection with *Theileria parva* for experimental transmission. *Bulletin of Epizootiological Diseases of Africa*, *8*, 33–43.

Bajwa, J., Charach, M., & Duclos, D. (2013). Adverse effects of rifampicin in dogs and serum alanine aminotransferase monitoring recommendations based on a retrospective study of 344 dogs. *Veterinary Dermatology*, *24*(6), 570–e135. doi:10.1111/vde.12083 PMID:24164464

Baldacchino, F., Desquesnes, M., Mihok, S., Foil, L., Duvallet, G., & Jittapalapong, S. (2014). Tabanids: Neglected subjects of research, but important vectors of disease agents. *Infection, Genetics and Evolution*, *28*, 596–615. doi:10.1016/j.meegid.2014.03.029 PMID:24727644

Baldacchino, F., Muenworn, V., Desquesnes, M., Desoli, F., Charoenviriyaphap, T., & Duvallet, G. (2013). Transmission of pathogens by Stomoxys flies (Diptera, Muscidae): A review. *Parasite (Paris, France)*, *20*, 20–26. doi:10.1051/parasite/2013026 PMID:23985165

Baldwin, C. L., Black, S. J., Brown, W. C., Conrad, P. A., Goddeeris, B. M., Kinuthia, S. W., Lalor, P. A., MacHugh, N. D., Morrison, W. I., Morzaria, S. P., Naessens, J., & Newson, J. (1988). Bovine T cells, B cells and null cells are transformed by the protozoan parasite *Theileria parva. Infection and Immunity*, *56*(2), 462–467. doi:10.1128/IAI.56.2.462-467.1988 PMID:3123392

Baneth G. (2014). Tick-borne infections of animals and humans: a common ground. *International Journal. Parasitol*ogy, *44*(9), 591-6. doi: .011 doi:10.1016/j.ijpara.2014.03

Barbet, A. F., Byrom, B., & Mahan, S. M. (2009). Diversity of *Ehrlichia ruminantium* major antigenic protein 1-2 in field isolates and infected sheep. *Infection and Immunity*, *77*(6), 2304–2310. doi:10.1128/IAI.01409-08 PMID:19307215

Barnett, S. (1961). The control of ticks on livestock. *FAO Agricultural Studies*, 54.

Barnett, S. F. (1960). Connective tissue reactions in acute fatal East Coast fever (*Theileria parva* of cattle). *The Journal of Infectious Diseases*, *107*(3), 253–282. doi:10.1093/infdis/107.3.253 PMID:13687137

Barnett, S. F., & Brocklesby, D. W. (1966). The passage of *Theileria lawrencei* (Kenya) through cattle. *The British Veterinary Journal*, *122*(9), 396–409. doi:10.1016/S0007-1935(17)40406-4

Barongo, M. B., Ståhl, K., Bett, B., Bishop, R. P., Fèvre, E. M., Aliro, T., & Ssematimba, A. (2015). Estimating the basic reproductive number (R0) for African swine fever virus (ASFV) transmission between pig herds in Uganda. *PLoS ONE*, *10*. doi:10.1371/journal.pone.0125842

Baron, M. D., & Holzer, B. (2015). Nairobi sheep disease virus/Ganjam virus. *Revue Scientifique et Technique (International Office of Epizootics)*, *34*(2), 411–417. doi:10.20506/rst.34.2.2367 PMID:26647464

Barrantes-González, A. V., Jiménez-Rocha, A. E., Romero-Zuñiga, J. J., & Dolz, G. (2016). Serology, molecular detection and risk factors of Ehrlichia canis infection in dogs in Costa Rica. *Ticks and Tick-Borne Diseases*, *7*(6), 1245–1251. doi:10.1016/j.ttbdis.2016.07.006 PMID:27440400

Barré, N. (1989). *Biology and ecology of the tick Amblyomma variegatum (Acarina: Ixodina) and Guadeloupe (French West Indies)* (Doctoral thesis in science). University of Paris-Sud-Orsay.

Barré, N., & Uilenberg, G. (2010). Spread of parasites transported with their hosts: Case study of two species of cattle tick. *Revue Scientifique et Technique (International Office of Epizootics)*, *29*(1), 149–160. PMID:20617654

Batista, J., Riet-Correa, F., Teixeira, M. M. G., Madruga, C. R., Simoes, S. D. V., & Maia, T. F. (2006). Trypanosomiasis by Trypanosoma vivax in cattle in the Brazilian semiarid: Description of an outbreak and lesions in the nervous system. *Veterinary Parasitology*, *143*(2), 174–181. doi:10.1016/j.vetpar.2006.08.017 PMID:16965857

Bauer, B., Holzgrefe, B., Mahama, C. I., Baumann, M. P., Mehlitz, D., & Clausen, P. H. (2011). Managing tsetse transmitted trypanosomosis by insecticide-treated nets-an an affordable and sustainable method for resource-poor pig farmers in Ghana. *PLoS Neglected Tropical Diseases*, *5*(10), e1343. doi:10.1371/journal.pntd.0001343 PMID:22022625

Beall, M. J., Alleman, A. R., Breitschwerdt, E. B., Cohn, L. A., Couto, C. G., Dryden, M. W., Guptill, L. C., Iazbik, C., Kania, S. A., Lathan, P., Little, S. E., Roy, A., Sayler, K. A., Stillman, B. A., Welles, E. G., Wolfson, W., & Yabsley, M. J. (2012). Seroprevalence of Ehrlichia canis, Ehrlichia chaffeensis and Ehrlichia ewingii in dogs in North America. *Parasites & Vectors*, *5*(1), 29. doi:10.1186/1756-3305-5-29 PMID:22316160

Bekure, S. (1991). *Maasai herding: An analysis of the livestock production system of Maasai pastoralists in eastern Kajiado District, Kenya* (Vol. 4). ILRI (aka ILCA and ILRAD).

Bell, L. J. (1983). *Development of Theileria in tick tissue culture* (MPhil Thesis). University of Edinburgh.

Benavides, E., Vizcaino, O., Britto, C. M., Romero, A., & Rubio, A. (2006). Attenuated Trivalent Vaccine against Babesiosis and Anaplasmosis in Colombia. *Annals of the New York Academy of Sciences*, *916*(1), 613–616. https://doi.org/10.1111/j.1749-6632.2000.tb05346.x

Bengaly, Z., Sidibe, I. R. G., Desquesnes, M., Boly, H., & Sawadogo, L. (2002). Comparative pathogenicity of three genetically distinct types of Trypanosoma congolense in cattle: Clinical observations and haematological changes. *Veterinary Parasitology*, *108*(1), 1–19. doi:10.1016/S0304-4017(02)00164-4 PMID:12191895

Benoit, J. B., Attardo, G. M., Baumann, A. A., Michalkova, V., & Aksoy, S. (2015). Adenotrophic viviparity in tsetse flies: Potential for population control and as an insect model for lactation. *Annual Review of Entomology*, *60*(1), 351–371. doi:10.1146/annurev-ento-010814-020834 PMID:25341093

Benzuidenhout, J. D., Peterson, C. L., & Bernard, B. J. H. (1985). Article. *The Onderstepoort Journal of Veterinary Research*, *52*, 113–120. PMID:3900854

Berthier, D., Peylhard, M., Dayo, G.-K., Flori, L., Sylla, S., Bolly, S., Sakande, H., Chantal, I., & Thevenon, S. (2015). A Comparison of Phenotypic Traits Related to Trypanotolerance in Five West African Cattle Breeds Highlights the Value of Shorthorn Taurine Breeds. *PLoS One*, *10*(5), e0126498. doi:10.1371/journal.pone.0126498 PMID:25954819

Bett, B., Orenge, C., Irungu, P., & Munga, L. K. (2004). Epidemiological factors that influence time-to-treatment of trypanosomosis in Orma Boran cattle raised at Galana Ranch, Kenya. *Veterinary Parasitology*, *120*(1–2), 43–53. doi:10.1016/j.vetpar.2003.12.011 PMID:15019142

Bett, M. K., Saini, R. K., & Hassanali, A. (2015). Repellency of tsetse-refractory waterbuck (Kobus defassa) body odour to Glossinapallidipes (Diptera: Glossinidae): assessment of the relative contribution of different classes and individual constituents. *Acta Tropica*, *146*, 17–24. doi:10.1016/j.actatropica.2015.02.017 PMID:25746973

Beugnet, F., & Franc, M. (2012). Insecticide and acaricide molecules and/or combinations to prevent pet infestation by ectoparasites. *Trends in Parasitology*, *28*(7), 267–279. doi:10.1016/j.pt.2012.04.004 PMID:22627116

Bezuidenhout, J. D. (1988). *Certain aspects of the transmission of heartwater, the occurrence of the organism in ticks and in vitro culture* (DVSc Thesis). University of Pretoria.

Bezuidenhout, J. D., & Prozesky, L. (1994). Heartwater. In Infectious diseases of Livestock with special reference to Southern Africa. Cape Town, South Africa: Oxford University Press.

Bezuidenhout, J. D. (1990). Recent research findings on cowdriosis. In C. Williams & I. Kakoma (Eds.), *Ehrlichiosis* (Vol. J, pp. 125–135). Kluwer Academic Press. doi:10.1007/978-94-009-1998-3_11

Bezuidenhout, J. D., & Bigalke, R. D. (1987). The control of heartwater by means of tick control. *The Onderstepoort Journal of Veterinary Research*, *54*(3), 525–528. PMID:3329336

Bezuidenhout, J. D., & Jacobsz, C. J. (1986). Proof of transovarial transmission of Cowdria ruminantium by Amblyomma hebraeum. *The Onderstepoort Journal of Veterinary Research*, *53*(1), 31–34. PMID:3960489

bin Tarif, A., Lasecka, L., Holzer, B., & Baron, M. D. (2012). Ganjam virus/Nairobi sheep disease virus induces a pro-inflammatory response in infected sheep. *Veterinary Research, 43*(1), 71. doi:10.1186/1297-9716-43-71

Bishop, J. P., & Adams, L. G. (1973). Combination thin and thick blood films for the detection of Babesia parasitemia. *American Journal of Veterinary Research*, *34*(9), 1213–1214.

Bishop, P., Rae, P. F., Philips, L. P., Boid, R., & Luckins, A. G. (1995). T. equiperdum: Detection of Trypanosomal antibodies and antigen by enzyme-linked immunosorbent assay. *The British Veterinary Journal*, *151*(6), 715–720. doi:10.1016/S0007-1935(95)80152-9 PMID:8605584

Bishop, R. P., Odongo, D., Ahmed, J., Mwamuye, M., Fry, L. M., Knowles, D. P., Nanteza, A., Lubega, G., Gwakisa, P., Clausen, P. H., & Obara, I. (2020). A review of recent research on *Theileria parva*: Implications for the infection and treatment vaccination method for control of East Coast fever. *Transboundary and Emerging Diseases*, *67*(S1), 56–67. doi:10.1111/tbed.13325 PMID:32174044

Biswas, S. (2003). Role of veterinarian in the care and management during harvest of skin in livestock species. *Proceedings of the National Seminar Leather Industry in Today's perspective.*

Blacklock, B., & Yorke, W. (1913). The trypanosomes causing dourine (Mal de Coitor Beschalseuche). *Proceedings of the Royal Society of London. Series B, Containing Papers of a Biological Character*, *87*(593), 89–96. doi:10.1098/rspb.1913.0063

Black, S., & Mansfield, J. (2016). Prospects for vaccination against pathogenic African trypanosomes. Parasite Immunology. *Parasite Immunology*, *38*(12), 735–743. Advance online publication. doi:10.1111/pim.12387 PMID:27636100

Blagburn, B. L., & Dryden, M. W. (2009). Biology, treatment and control of flea and tick infestations. *Veterinary Clinics: Small Animal Practice*, *39*(6), 1173–1200. doi:10.1016/j.cvsm.2009.07.001 PMID:19932369

Bock, R., Jackson, L., de Vos, A., & Jorgensen, W. (2004). Babesiosis of cattle. *Parasitology, 129*, S247-269. doi:10.10170031182004005190

Bock, R., Jackson, L., De Vos, A., & Jorgensen, W. (2004). Babesiosis of cattle. *Parasitology*, *129*(S1), S247–S269. doi:10.1017/S0031182004005190 PMID:15938514

Bock, R., & Vos, A. (2001). Immunity following use of Australian tick fever vaccine: A review of the evidence. *Australian Veterinary Journal*, *79*(12), 832–839. https://doi.org/10.1111/j.1751-0813.2001.tb10931.x

Bonomi, A., Bassetti, F., Gabrieli, P., Beadell, J., Falchetto, M., Scolari, F., Gomulski, L. M., Regazzini, E., Ouma, J. O., Caccone, A., Okedi, L. M., Attardo, G. M., Guglielmino, C. R., Aksoy, S., & Malacrida, A. R. (2011). Polyandry is a common event in wild populations of the Tsetse fly Glossina fuscipes fuscipes and may impact population reduction measures. *PLoS Neglected Tropical Diseases*, *5*(6), e1190–e1190. doi:10.1371/journal.pntd.0001190 PMID:21666797

Bosch-Camós, L., Lopez, E., & Rodriguez, F. (2020). African swine fever vaccines: A promising work still in progress. *Porcine Health Management*, *6*(1), 14. doi:10.118640813-020-00154-2 PMID:32626597

Bose, R., Jorgensen, W. K., Dalgliesh, R. J., Friedhoff, K. T., & de Vos, A. J. (1995). Current state and future trends in the diagnosis of babesiosis. *Veterinary Parasitology, 57*(1–3), 61–74. (94)03111-9 doi:10.1016/0304-4017

Bossche, V. D. P., Shumba, W., & Makhambera, P. (2000). The distribution and epidemiology of bovine trypasonomosis in Malawi. *Veterinary Parasitology*, *88*(3-4), 163–176. doi:10.1016/S0304-4017(99)00222-8 PMID:10714455

Botros, B. A. M., Elmolla, M. S., Salib, A. W., Calamaio, C. A., Dasch, G. A., & Arthur, R. R. (1995). Canine ehrlichiosis in Egypt: Sere-epidemiological survey. *The Onderstepoort Journal of Veterinary Research*, *62*, 41–43. PMID:8539033

Boucader, D., Modibo, D., Oumar, D., & Boubacar, B. (2019). A national atlas of tsetse and African animal trypanosomosis in Mali. *Parasites & Vectors*, *12*(1), 466. doi:10.118613071-019-3721-3 PMID:31597558

Bouchard, C., Dibernardo, A., Koffi, J., Wood, H., Leighton, P. A., & Lindsay, L. R. (2019). N Increased risk of tick-borne diseases with climate and environmental changes. *Canada communicable disease report = Releve des maladies transmissibles au Canada, 45*(4), 83–89. doi:10.14745/ccdr.v45i04a02

Boulange, A., Pillay, D., Chevtzoff, C., Biteau, N., Come de Graca, V., Rempeters, L., Theodoridis, D., & Baltz, T. (2017). Development of a rapid antibody test for point-of-care diagnosis of animal African trypanosomosis. *Veterinary Parasitology*, *233*, 32–38. doi:10.1016/j.vetpar.2016.11.017 PMID:28043385

Bourn, D., Reid, R., Rogers, D., Snow, B., & Wint, W. (2001). Environmental change and the autonomous 476 control of tsetse and trypanosomiasis in sub-Saharan Africa: case histories from Ethiopia, The 477 Gambia, Kenya, Nigeria, and Zimbabwe. Oxford, UK: Environmental Research Group.

Bourn, D. A. V. I. D., Grant, I., Shaw, A., Torr, S., & Protection, C. E. (2005). Cheap and safe tsetse control for livestock production and mixed farming in Africa. *Aspects of Applied Biology*, *75*, 81.

Bouyer, J., Bouyer, F., Donadeu, M., Rowan, T., & Napier, G. (2013). Community-and farmer-based management of animal African trypanosomosis in cattle. *Trends in Parasitology*, *29*(11), 519–522. doi:10.1016/j.pt.2013.08.003 PMID:24012356

Bowie, M. V., de la Fuente, J., Kocan, K. M., Blouin, E. F., & Barbet, A. F. (2002). Conservation of major surface protein 1 genes of *Anaplasma marginale* during cyclic transmission between ticks and cattle. *Gene*, *282*(1-2), 95–102. doi:10.1016/S0378-1119(01)00845-9 PMID:11814681

Bradley, D. (1995). *Genetic characterisation of cattle in West and Central Africa*. Department of Genetics, Trinity College.

Bradway, D. S., Torioni de Eschaide, S., Knowles, D. P., Hennager, S. G., & McElwain, T. F. (2001). Sensitivity and specificity of the complement fixation test for detection of cattle persistently infected with *Anaplasma marginale*. Journal of. Veterinary. Diagnostic atic. *Investigation.*, *13*, 79–81. PMID:11243370

Bram, R. (1983). Bram, R. A. (1983). Tick-borne livestock diseases and their vectors. *FAO Animal Production and Health Paper*, 36.

Branger, S., Rolain, J. M., & Raoult, D. (2004). Evaluation of antibiotic susceptibilities of Ehrlichia canis, Ehrlichia chaffeensis, and Anaplasma phagocytophilum by real-time PCR. *Antimicrobial Agents and Chemotherapy*, *48*(12), 4822–4828. doi:10.1128/AAC.48.12.4822-4828.2004 PMID:15561862

Bremer, W. G., Schaefer, J. J., Wagner, E. R., Ewing, S. A., Rikihisa, Y., Needham, G. R., Jittapalapong, S., Moore, D. L., & Stich, R. W. (2005). Transstadial and intrastadial experimental transmission of Ehrlichia canis by male Rhipicephalus sanguineus. *Veterinary Parasitology*, *131*(1-2), 95–105. doi:10.1016/j.vetpar.2005.04.030 PMID:15941624

Brizuela, C. M., Ortellado, C. A., Sanabria, E., Torres, O., & Ortigosa, D. (1998). The safety and efficacy of Australian tick-borne disease vaccine strains in cattle in Paraguay. *Veterinary Parasitology*, *76*(1-2), 27–41. doi:10.1016/S0304-4017(97)00047-2 PMID:9653988

Brouqui, P., Davoust, B., Haddad, S., Vidor, E., & Raoult, D. (1991). Serological evaluation of Ehrlichia canis infections in military dogs in Africa and Reunion Island. *Veterinary Microbiology*, *26*(1-2), 103–105. doi:10.1016/0378-1135(91)90046-I PMID:2024432

Brown, C. G. D. (1990). Theileriosis. In M. M. H. Sewell (Ed.), Handbook on animal Diseases in the Tropics (4th ed.). Academic Press.

Brown, A. A., Penrith, M. L., Fasina, F. O., & Beltran-Alcrudo, D. (2018). The African swine fever epidemic in West Africa, 1996–2002. *Transboundary and Emerging Diseases*, *65*(1), 64–76. doi:10.1111/tbed.12673 PMID:28710823

Brown, W. C., Norimine, J., Knowles, D. P., & Goff, W. L. (2006). Immune control of Babesia bovis infection. *Veterinary Parasitology*, *138*(1–2), 75–87. https://doi.org/10.1016/j.vetpar.2006.01.041

Brun, R., & Lun, Z. R. (1994). Drug sensitivity of Chinese T. evansi and T. equiperdum isolates. *Veterinary Parasitology*, *52*, 37–46. doi:10.1016/0304-4017(94)90033-7 PMID:8030186

Burridge, M. J., & Kimber, C. D. (1972). The indirect fluorescent antibody test for experimental East Coast fever (Theileria parva infection of cattle). Evaluation of a cell culture schizont antigen. *Research in Veterinary Science*, *13*(5), 451–455. doi:10.1016/S0034-5288(18)33992-4 PMID:4564334

Burridge, M. J., & Kimber, C. D. (1973). Duration of serological response to the indirect fluorescent antibody test of cattle recovered from Theileria parva infection. *Research in Veterinary Science*, *14*(2), 270–271. doi:10.1016/S0034-5288(18)33929-8 PMID:4217920

Burridge, M. J., Kimber, C. D., & McHardy, N. (1973). Detection of antibodies to Babesia bigemina in dried blood samples using the indirect fluorescent antibody test. *Annals of Tropical Medicine and Parasitology*, *67*(2), 191–195. https://doi.org/10.1080/00034983.1973.11686876

Burridge, M. J., Kimber, C. D., & Young, A. S. (1973). The use of indirect fluorescent antibody technique in serological studies of Theileria lawrencei infections in cattle. *American Journal of Veterinary Research*, *13*, 897. PMID:4198343

Burt, F. J., Spencer, D. C., Leman, P. A., Patterson, B., & Swanepoel, R. (1996). Investigation of tick-borne viruses as pathogens of humans in South Africa and evidence of Dugbe virus infection in a patient with prolonged thrombocytopenia. *Epidemiology and Infection*, *116*(3), 353–361. https://doi.org/10.1017/S0950268800052687

Buscher, G. (1988). The infection of various tick species with Babesia bigemina, its transmission and identification. *Parasitology Research*, *74*(4), 324–330. https://doi.org/10.1007/bf00539453

Buscher, P., Ngoyi, D. M., Kabore, J., Lejon, V., Robays, J., Jamonneau, V., Bebronne, N., Van der Veken, W., & Bieler, S. (2009). Improved models of mini Anion exchange centrifugation technique (MAECT) and modified single centrifugation (MSC) for sleeping sickness diagnosis and staging. *PLoS Neglected Tropical Diseases*, *3*(11), e471. doi:10.1371/journal.pntd.0000471 PMID:19936296

Buxton, P. A. (1955). *The natural history of tsetse flies: an account of the biology of the genus Glossina (Diptera)* (Vol. 10). London: H.K. Lewis.

Byaruhanga, C., Oosthuizen, M. C., Collins, N. E., & Knobel, D. (2015b). Using participatory epidemiology to investigate management options and relative importance of tick-borne diseases amongst transhumant zebu cattle in Karamoja Region, Uganda. *Preventive Veterinary Medicine*, *122*(3), 287–297. doi:10.1016/j.prevetmed.2015.10.011 PMID:26527312

CABI Data Mining. (2001). CAB Abstracts Data Mining. CABI, current year. *Fallopia japonica*. In *Invasive Species Compendium*. Wallingford, UK: CAB International. www.cabi.org/isc

CABI. (2020). *Theileria annulata infections. Invasive Species Compendium*. CAB International. https://www.cabi.org/isc/datasheet/62107#tooverview

Calisher, C. H., & Fenner, F. J. (2000). Macroecology and Microecology of Viruses of Terrestrial Mammals. In Viral Ecology (pp. 493–517). Elsevier. https://doi.org/10.1016/B978-012362675-2/50014-8.

Caljon, G., De Vooght, L., & Van Den Abbeele, J. (2014). The biology of tsetse-trypanosome interactions. In The Biology of Tsetse–Trypanosome Interactions. Springer.

Callow, L. L., Dalgliesh, R. J., & De Vos, A. J. (1997). Development of effective living vaccines against bovine babesiosis—The longest field trial? *International Journal for Parasitology*, *27*(7), 747–767. (97)00034-9 doi:10.1016/S0020-7519

Camus, E., Martinez, D., Beauperthuy, L., Benderdouche, A., Coisne, S., Corbette, C., Denormandie, N., Garrís, G., Harris, E., King, T., Lannoy, L., Louvet, A., Lunel, E., Montrose, M., Nisbett, B., Nyack, B., Robinson, I., Rouchosse, P., Swanston, G. B., ... Thye, G. D. (1993). Heartwaterin Guadeloupe and the Caribbean.InProceedings of the 2nd Biennial Meeting of the American Society of Tropical Veterinary Medicine.2-6 February. *Revue d'Élevage et de Médecine Vétérinaire des Pays Tropicaux*, *46*, 109–114. PMID:8134620

Carasco-Lacombe, C., Lefrançois, T., Meyer, D. F., Martinez, D., Boulouis, H. J., & Haddad, N. (2012). A new typing technique for the Rickettsiales Ehrlichia ruminantium: Multiple-locus variable number tandem repeat analysis. *Journal of Microbiological Methods*, *88*(2), 205–211. doi:10.1016/j.mimet.2011.11.011 PMID:22143037

Cardoso, L., Oliveira, A. C., Granada, S., Nachum-Biala, Y., Gilad, M., Lopes, A. P., Sousa, S. R., Vilhena, H., & Baneth, G. (2016). Molecular investigation of tick-borne pathogens in dogs from Luanda, Angola. *Parasites & Vectors*, *9*(1), 1–6. doi:10.118613071-016-1536-z PMID:27160839

Carelli, G., Decaro, N., Lorusso, A., Elia, G., Lorusso, E., Mari, V., Ceci, L., & Buonavoglia, C. (2007). Detection and quantification of *Anaplasma marginale*DNA in blood samples of cattle by real-time PCR. *Veterinary Microbiology*, *124*(1-2), 107–114. doi:10.1016/j.vetmic.2007.03.022 PMID:17466470

Carlton, W., McGavin, M. D., & Thomson, R. G. (Eds.). (1995). *Thomson's special veterinary pathology* (2nd ed.). Mosby.

Cattand, P. (2001). African Trypanosomiasis, human. In Encyclopedia of arthropod–transmitted infections of man and domestic animals (pp. 13–23). CABI.

CDC - Laboratory Identification of Parasitic Diseases of Public Health Concern. (2017). Centers for Disease Control and Prevention.

Cecchi, G., Mattioli, R. J. S., & De la Rocque, S. (2008). Land cover and tsetse fly distributions in sub-Saharan Africa. *Medical and Veterinary Entomology*, *22*(4), 364–373. Advance online publication. doi:10.1111/j.1365-2915.2008.00747.x PMID:18785934

Cecchi, G., Paone, M., Argiles, H. R., Vreysen, M., & Mattioli, R. (2015). Developing a continental atlas of the distribution and trypanosomal infection of tsetse flies (Glossina species). *Parasites & Vectors*, *8*(1), 284. doi:10.118613071-015-0898-y PMID:25994757

Cecchi, G., Paone, M., Feldmann, U., Vreysen, M. J., Diall, O., & Mattioli, R. C. (2014). Assembling a geospatial database of tsetse-transmitted animal trypanosomosis for Africa. *Parasites & Vectors*, *7*(1), 39. doi:10.1186/1756-3305-7-39 PMID:24447638

Chaisi, M. E., Sibeko, K. P., Collins, N. E., Potgieter, F. T., & Oosthuizen, M. C. (2011). Identification of *Theileria parva* and *Theileria* sp. (buffalo) 18S rRNA gene sequence variants in the African buffalo (*Syncerus caffer*) in southern Africa. *Veterinary Parasitology*, *182*(2-4), 150–162. doi:10.1016/j.vetpar.2011.05.041 PMID:21700394

Channumsin, M., Ciosi, M., Masiga, D., Mable, B., Michael, C., & Turner, R. (2018). Sodalis glossinidius presence in wild tsetse is only associated with presence of trypanosomes in complex interactions with other tsetse-specific factors. *BMC Microbiology*, *18*(S1), 163. doi:10.118612866-018-1285-6 PMID:30470184

Chartier, C., Itard, J., Morel, P., & Troncy, P. (2000). *Précis de parasitologie vétérinaire tropicale. Universités francophones*. Aupelf Uref.

Chauvin, A., Moreau, E., Bonnet, S., Plantard, O., & Malandrin, L. (2009). Babesia and its hosts: Adaptation to long-lasting interactions as a way to achieve efficient transmission. *Veterinary Research*, *40*(2), 37. https://doi.org/10.1051/vetres/2009020

Chema, S. (1984). Evaluation of new acaricides in Kenya and views on selection for acaricide resistance in past and future. In *Report of the Third FAO Expert Consultation on Research on Tick-borne Diseases and Their Vectors* (p. 29). FAO.

Chema, S., Chumo, R. S., Dolan, T. T., Gathuma, J. M., James, A. D., Irvin, A. D., & Young, A. S. (1987). Clinical trials of halofunone lactate for treatment of East Coast fever in Kenya. *The Veterinary Record*, *120*(24), 575–577. doi:10.1136/vr.120.24.575 PMID:3303642

Chema, S., Waghala, S., James, A. D., Dolan, T. T., Young, A. S., Masiga, W. N., Irvin, A. D., Mulela, G. H. M., & Wekesa, L. M. (1986). Clinical trials of parvaquone for the treatment of East Coast fever in Kenya. *The Veterinary Record*, *118*(21), 588–589. doi:10.1136/vr.118.21.588-a PMID:3526698

Chenais, E., Depner, K., Guberti, V., Dietze, K., Viltrop, A., & Stahl, K. (2019). Epidemiological considerations on African swine fever in Europe 2014-2018. *Porcine Health Management*, *5*(1), 6. doi:10.118640813-018-0109-2 PMID:30637117

Chenais, E., Sternberg-Lewerin, S., Boqvist, S., Emanuelson, U., Aliro, T., Tejler, E., & Ståhl, K. (2015). African swine fever in Uganda: Qualitative evaluation of three surveillance methods with implications for other resource-poor settings. *Frontiers in Veterinary Science*, *2*, 51. doi:10.3389/fvets.2015.00051 PMID:26664978

Cheng, Q., & Aksoy, S. (1999). Tissue tropism, transmission and expression of foreign genes *in vivo* in midgut symbionts of tsetse flies. *Insect Molecular Biology*, *8*(1), 125–132. doi:10.1046/j.1365-2583.1999.810125.x PMID:9927181

Cheng, Q., Ruel, T., Zhou, W., Moloo, S., Majiwa, P., O'Neill, S., & Aksoy, S. (2000). Tissue distribution and prevalence of Wolbachia infections in tsetse flies, Glossina spp. *Medical and Veterinary Entomology*, *14*(1), 44–50. doi:10.1046/j.1365-2915.2000.00202.x PMID:10759311

Chiduwa, G., Chimonyo, M., Halimani, T. E., Chisambara, S. R., & Dzama, K. (2008). Herd dynamics and contribution of indigenous pigs to the livelihoods of rural farmers in a semi-arid area of Zimbabwe. *Tropical Animal Health and Production*, *40*(2), 125–136. doi:10.100711250-007-9071-8 PMID:18422255

Chiejina, S. N., Behnke, J. M., Nnadi, P. A., Ngongeh, L. A., & Musongong, G. A. (2009). The responses of two ecotypes of Nigerian West African Dwarf goat to experimental infections with Trypanosoma brucei and Haemonchus contortus. *Small Ruminant Research*, *85*(2–3), 91–98. doi:10.1016/j.smallrumres.2009.07.004

Chikowore, G., Dicko, A., Chinwada, P., Zimba, M., Shereni, W., Roger, F., Bouyer, J., & Guerrini, L. (2017). A pilot study to delimit tsetse target populations in Zimbabwe. *PLoS Neglected Tropical Diseases*, *11*(5), e0005566. doi:10.1371/journal.pntd.0005566 PMID:28467409

Chira, R. M. (2003). *Changes in wildlife habitat and numbers in Embu and Mbeere districts*. Eastern Province.

Chitanga, S., Namangala, B., De Deken, R., & Marcotty, T. (2013). Shifting from wild to domestic hosts: The effect on the transmission of Trypanosoma congolense to tsetse flies. *Acta Tropica*, *125*(1), 32–36. doi:10.1016/j.actatropica.2012.08.019 PMID:23000543

Claes, F., Agbo, E. C., Radwanska, M., Te Pas, M. F. W., Baltz, T., De Waal, D. T., Goddeeris, B. M., Claassen, E., & Büscher, P. (2003). How does T. equiperdum fit into the Trypanozoon group? A cluster analysis by RAPD and multiplex-endonuclease genotyping approach. *Parasitology*, *126*(5), 425–431. doi:10.1017/S0031182003002968 PMID:12793646

Clausen, P. H., Gebreselassie, G., Abditcho, S., Mehlitz, D., & Staak, C. (1999). Detection of Trypanosoma DNA in serological positive but aparasitemic horses suspected of dourine in Ethiopia. *The Tokai Journal of Experimental and Clinical Medicine*, *23*, 303–308. PMID:10622626

Claxton, J., Faye, J., & Rawlings, P. (1991). Trypanosome infections in warthogs (Phacochoerus aethiopicus) in the Gambia. *Veterinary Parasitology*, *41*(3–4), 179–187. PMID:1502780

Cobo, F. (2014). Viral haemorrhagic fevers. In Imported Infectious Diseases (pp. 243–255). Elsevier. https://doi.org/10.1533/9781908818737.243.

Coles, T. B., & Dryden, M. W. (2014). Insecticide/acaricide resistance in fleas and ticks infesting dogs and cats. *Parasites & Vectors*, *7*(1), 1–10. doi:10.1186/1756-3305-7-8 PMID:24393426

Combrink, Waal, & Troskie. (1997). Evaluation of a 3 ml heartwater (cowdriosis) infective blood vaccine dose. *The Onderstepoort Journal of Veterinary Research*, *64*(4), 309–311. PMID:9551483

Conceiçao. (1949). *A heartwater (Rickettsia ruminantium) in Angola*. Academic Press.

Conner, R. J., & Mulkangi, D. J. A. (1986). . Concurrent outbreak of pseudo-lumpy skin disease and acute *T. vivax* infection in cattle. *Tropical Animal Health Production, 18*, 127 – 132.

Conrad, P. A., Iams, K., Brown, W. C., Sohanpal, B., & Ole-Moi Yoi, O. (1987b). DNA probes detect genomic diversity in *Theileria parva* stocks. *Molecular and Biochemical Parasitology*, *25*(3), 213–226. doi:10.1016/0166-6851(87)90085-5 PMID:2892129

Conrad, P. A., Ole-Moi Yoi, O., Baldwin, C. L., Dolan, T. T., O'Callaghan, C. J., Njamungeh, R. E. G., Grootenhuis, J. G., Stagg, D. A., Leitch, B. L., & Young, A. S. (1989). Characterisation of buffalo derived theilerial parasites with monoclonal antibodies and DNA probes. *Parasitology*, *98*(2), 179–183. doi:10.1017/S0031182000062089 PMID:2569709

Conrad, P. A., Stagg, D. A., Grootenhuis, J. G., Irvin, A. D., Newson, J., Njamungeh, R. E. G., Rossiter, P. B., & Young, A. S. (1987a). Isolation of *Theileria* parasites from the African buffalo (*Syncerus caffer*) and characterization with anti-schizont monoclonal antibodies. *Parasitology*, *94*(3), 413–423. doi:10.1017/S0031182000055761 PMID:3112700

Constable, P., Hinchcliff, K., Done, S., & Gruenberg, W. (2016). *Veterinary Medicine: A Textbook of the Diseases of Cattle*. Horses, Sheep, Pigs and Goats.

Cordon-Obras, C., Berzosa, P., Ndong-Mabale, N., Bobuakasi, L., Buatiche, J. N., Ndongo-Asumu, P., Benito, A., & Cano, J. (2009). Trypanosoma brucei gambiense in domestic livestock of Kogo and Mbini foci (Equatorial Guinea). *Tropical Medicine & International Health*, *14*(5), 535–541. doi:10.1111/j.1365-3156.2009.02271.x PMID:19320872

Cordoves, C. O., Fernandez, C., Garcia, A. J., & Gonzalez, B. R. (1982). Trypano- soma vivax, Ziemann, 1905. Lista de transmisores mecanicos en Cuba. *Revista Cubana de Ciencias Veterinarias*, *13*, 219–221.

Cowdry, E. V. (1925a). Studies on the etiology of heartwater I. Observation of a rickettsia, Rickettsia ruminantium (n. sp.), in the tissues of infected animals. *The Journal of Experimental Medicine*, *42*(2), 231–252. doi:10.1084/jem.42.2.231 PMID:19869049

Cowdry, E. V., & Danks, W. B. C. (1933). Studies on East Coast fever. II Behaviour of the parasite and the development of distinctive lesions in susceptible animals. *Parasitology*, *25*(1), 1–63. doi:10.1017/S003118200001920X

Cox, F. E. G. (1979). Pathogenesis of animal trypanosomiasis. *Nature*, *277*(5698), 603–604. doi:10.1038/277603a0 PMID:423960

Craighead, J. E. (2000). Hemorrhagic Fever Viruses. In Pathology and Pathogenesis of Human Viral Disease (pp. 277–296). Elsevier. https://doi.org/10.1016/B978-012195160-3/50020-0.

Cresswell, P., Ackerman, A. L., Peaper, D. R., Wearsch, P. A., Peaper, D. R., & Wearsch, P. A. (2005). Mechanisms of MHC class I-restricted antigen processing and cross- presentation. *Immunological Reviews*, *207*, 145–157.

Cumming, G. S. (2002). Comparing climate and vegetation as limiting factors for species ranges of African ticks. *Ecology*, *83*(1), 255–268. doi:10.1890/0012-9658(2002)083[0255:CCAVAL]2.0.CO;2

Cunningham, M. P. (1981). Biological control of ticks with particular reference to Rhipicephalus appendiculatus. In Advances in the Control of Theileriosis. Martinus Nijhoff Publishers.

Cunningham, M. P. (1966). Immunity in bovine trypanosomiasis. *East African Medical Journal*, *43*(9), 394–397. PMID:5965864

Cunningham, M. P. (1977). Immunisation of cattle against Theileria parva. In L. H. Miller & J. J. McKelvey Jr., (Eds.), *Immunity to Blood Parasites of Animal and Man* (pp. 1–132). Planum Press. doi:10.1007/978-1-4615-8855-9_11

Cunningham, M. P., Brown, C. D. G., Burridge, M. J., & Purnell, R. E. (1973b). Cryopreservation of infective particles of Theileria parva. *International Journal for Parasitology*, *3*(5), 583–587. doi:10.1016/0020-7519(73)90082-9 PMID:4200405

Cunningham, M. P., Brown, C. G. D., Burridge, M. J., Musoke, A. J., Purnell, R. E., Radley, D. E., & Sempebwa, C. (1974). East Coast fever: Titration in cattle of suspension of *Theileria parva* derived from ticks. *The British Veterinary Journal*, *130*(4), 336–345. doi:10.1016/S0007-1935(17)35836-0 PMID:4214585

Cunningham, M. P., Brown, C. G. D., Joyner, L. P., Purnell, R. E., & Bailey, K. P. (1973a). Infection of cattle with East Coast fever by inoculation of the infective stage of *Theileria parva* harvested from the tick vector *Rhipicephalus appendiculatus. Bulletin of Epizootic Diseases of Africa*, *21*, 235–238.

d'Ieteren, G. D., Authié, E., Wissocq, N., & Murray, M. (1998). Trypanotolerance, an option for sustainable livestock production in areas at risk from trypanosomosis. *Revue Scientifique et Technique (International Office of Epizootics)*, *17*(1), 154–175. doi:10.20506/rst.17.1.1088 PMID:9638808

DAD-IS. (2020). *Domestic Animal Diversity Information System (DAD-IS) | Food and Agriculture Organization of the United Nations*. Food and Agriculture Organization of the UN. http://www.fao.org/dad-is/en/

Daffa, J., Byamungu, M., Nsengwa, G., Mwambembe, E., & Mleche, W. (2013). Tsetse distribution in Tanzania: 2012 status. *Tanzania Veterinary Journal, 28*.

DAGRIS. (2020). *Domestic Animal Genetic Resources Information System (DAGRIS)*. International Livestock Research Institute. http://www.dagris.ilri.cgiar.org

Dahmani, M., Davoust, B., Sambou, M., Bassene, H., Scandola, P., Ameur, T., Raoult, D., Fenollar, F., & Mediannikov, O. (2019). Molecular investigation and phylogeny of species of the Anaplasmataceae infecting animals and ticks in Senegal. *Parasites & Vectors*, *12*(1), 1–5. doi:10.118613071-019-3742-y PMID:31640746

Dahmani, M., Loudahi, A., Mediannikov, O., Fenollar, F., Raoult, D., & Davoust, B. (2015). Molecular detection of Anaplasma platys and Ehrlichia canis in dogs from Kabylie, Algeria. *Ticks and Tick-Borne Diseases*, *6*(2), 198–203. doi:10.1016/j.ttbdis.2014.12.007 PMID:25583345

Dalgliesh, R. J., & Stewart, N. P. (1982). Some effects of time, temperature and feeding on infection rates with Babesia bovis and Babesia bigemina in Boophilus microplus larvae. *International Journal for Parasitology, 12*(4), 323–326. (82)90035-2 doi:10.1016/0020-7519

Dalgliesh, R. J., & Stewart, N. P. (1978). The extraction of infective Babesia bovis and Babesia bigemina from tick eggs and B. bigemina from unfed larval ticks. *Australian Veterinary Journal*, *54*(9), 453–454. https://doi.org/10.1111/j.1751-0813.1978.tb05582.x

Dandawate, C. N. (1977). Studies on Ganjam virus with reference to viremia in vertebrate hosts & development of antibodies. *Indian Journal of Experimental Biology*, *15*(11), 1058–1059.

Dandawate, C. N., Rajagopalan, P. K., Pavri, K. M., & Work, T. H. (1969). Virus isolations from mosquitoes collected in North Arcot district, Madras state, and Chittoor district, Andhra Pradesh between November 1955 and October 1957. *The Indian Journal of Medical Research*, *57*(8), 1420–1426. https://www.ncbi.nlm.nih.gov/pubmed/4395257

Dandawate, C. N., & Shah, K. V. (1969). Ganjam virus: A new arbovirus isolated from ticks Haemaphysalis intermedia Warburton and Nuttall, 1909 in Orissa, India. *The Indian Journal of Medical Research*, *57*(5), 799–804.

Dandawate, C. N., Work, T. H., Webb, J. K., & Shah, K. V. (1969). Isolation of Ganjam virus from a human case of febrile illness: A report of a laboratory infection and serological survey of human sera from three different states of India. *The Indian Journal of Medical Research*, *57*(6), 975–982.

Dantas, T. (2015). Climate change, biodiversity, ticks and tick-borne diseases: The butterfly effect. *International Journal for Parasitology. Parasites and Wildlife*, *4*(3), 452–461. https://doi.org/10.1016/j.ijppaw.2015.07.001

Dantas-Torres, F. (2008). The brown dog tick Rhipicephalus sanguineus (Latreille, 1806) (Acari:Ixodidae) from taxonomy to control. *Veterinary Parasitology*, *152*(3-4), 173–185. doi:10.1016/j.vetpar.2007.12.030 PMID:18280045

Dantas-Torres, F., Ferreira, D. R., de Melo, L. M., Lima, P. A. C., Siqueira, D. B., Rameh-de-Albuquerque, L. C., de Melo, A. V., & Ramos, J. A. C. (2010). Ticks on captive and free-living wild animals in northeastern Brazil. *Experimental & Applied Acarology*, *50*(2), 181–189. doi:10.100710493-009-9296-5 PMID:19693679

Dantas-Torres, F., Figueredo, L. A., & Faustino, M. A. G. (2004). Ectoparasitos de cães provenientes de alguns municípios da região metropolitana do Recife, Pernambuco, Brasil. *Revista Brasileira de Parasitologia Veterinária*, *13*(4), 151–154.

Dantas-Torres, F., Melo, M. F., Figueredo, L. A., & Brandão-Filho, S. P. (2009). Ectoparasite infestation on rural dogs in the municipality of São Vicente Férrer, Pernambuco, Northeastern Brazil. *Revista Brasileira de Parasitologia Veterinária*, *18*(3), 75–77. doi:10.4322/rbpv.01803014 PMID:19772781

Dantas-Torres, F., Oliveira-Filho, E. F., Soares, F. A. M., Souza, B. O. F., Valença, R. B. P., & Sá, F. B. (2008). Ticks infesting amphibians and reptiles in pernambuco. *Revista Brasileira de Parasitologia Veterinária*, *17*(4), 218–221. doi:10.1590/S1984-29612008000400009 PMID:19265581

Daubney, R. (1939). Heartwater (Rickettsia ruminantium). Department of Agriculture Kenya.

Daubney, R., & Hudson, J. R. (1931). Nairobi Sheep Disease. *Parasitology*, *23*(4), 507–524. https://doi.org/10.1017/S0031182000013895

Davidson, D. E., Dill, G. S., Tingpalapong, M., Premabutra, S., Nguen, P. L., Stephenson, E. H., & Ristic, M. (1978). Prophylactic and therapeutic use of tetracycline during an epizootic of ehrlichiosis among military dogs. *Journal of the American Veterinary Medical Association*, *172*(6), 697. PMID:640933

Davies, F. G. (1978a). Nairobi sheep disease in Kenya. The isolation of virus from sheep and goats, ticks and possible maintenance hosts. *The Journal of Hygiene*, *81*(2), 259–265. https://doi.org/10.1017/s0022172400025092

Davies, F. G. (1997). Nairobi sheep disease. *Parassitologia*, *39*(2), 95–98. https://doi.org/10.1201/9780429280276-10

Davies, F. G., Casals, J., Jesset, D. M., & Ochieng, P. (1978). The serological relationships of Nairobi sheep disease virus. *Journal of Comparative Pathology*, *88*(4), 519–523. https://doi.org/10.1016/0021-9975(78)90005-1

Davies, F. G., Mungai, J. N., & Taylor, M. (1977). The laboratory diagnosis of Nairobi sheep disease. *Tropical Animal Health and Production*, *9*(2), 75–80. https://doi.org/10.1007/bf02236383

Davies, F. G., & Mwakima, F. (1982). Qualitative studies of the transmission of Nairobi sheep disease virus by Rhipicephalus appendiculatus (ixodoidea, ixodidae). *Journal of Comparative Pathology*, *92*(1), 15–20. https://doi.org/10.1016/0021-9975(82)90039-1

Davilla, A., & Silva, R. (2000). *Animal trypanosomiasis in South America: Current status, partnership, and information technology*. Academic Press.

Davita, P., Julien, I., Regassa, F., Philipe, B., Hermogenes, M., Luis, N., Alain, B., Momar, T. S., Jeremy, B., Grant, B. N., Cyrille, C., Virginie, C., & Theo, B. (2013). Trypanosoma vivax GM6 Antigen: A Candidate Antigen for Diagnosis of African Animal Trypanosomosis in Cattle. *PLoS One*, *8*(10), e78565. Advance online publication. doi:10.1371/journal.pone.0078565 PMID:24205263

Davoust, B., Bourry, O., Gomez, J., Lafay, L., Casali, F., Leroy, E., & Parzy, D. (2006). Surveys on Seroprevalence of Canine Monocytic Ehrlichiosis among Dogs Living in the Ivory Coast and Gabon and Evaluation of a Quick Commercial Test Kit Dot-ELISA. *Annals of the New York Academy of Sciences*, *1078*(1), 464–469. doi:10.1196/annals.1374.086 PMID:17114755

De Beer, C. J., Venter, G. J., Kappmeier Green, K., Esterhuizen, J., De Klerk, D. G., Ntshangase, J., Vreysen, M. J. B., Pienaar, R., Motloang, M., Ntantiso, L., & Latif, A. A. (2016). An update of the tsetse fly (Diptera: Glossinidae) distribution and African animal trypanosomosis prevalence in north-eastern KwaZulu- Natal, South Africa. *The Onderstepoort Journal of Veterinary Research*, *1172*(1). Advance online publication. doi:10.4102/ojvr.v83i1.1172 PMID:27380653

De Beer, C., Venter, G., & Vreysen, M. (2015). Determination of the optimal mating age of colonised Glossina brevipalpis and Glossina austeni using walk-in field cages in South Africa. *Parasites & Vectors*, *8*(467), 467. Advance online publication. doi:10.118613071-015-1073-1 PMID:26380975

De Castro, J. J. (1991). Resistance to ixodid ticks in cattle with an assessment and its role in tick control in Africa. In J. B. Owen & R. F. E. Axford (Eds.), *Breeding for Disease Resistance in farm Animals* (pp. 244–262). C.A.B. International.

De Castro, J. J., & Newson, R. M. (1993). Host resistance in cattle tick control. *Parasitology Today (Personal Ed.)*, *9*(1), 13–17. doi:10.1016/0169-4758(93)90154-8 PMID:15463656

De Castro, J. J., Young, A. S., Dransfield, R. D., Cunnighm, M. P., & Dolan, T. T. (1985). Effects of tick infestation on Boran (Bos indicus) cattle immunized against theileriosis in an endemic area of Kenya. *Research in Veterinary Science*, *39*(3), 279–288. doi:10.1016/S0034-5288(18)31714-4 PMID:4081331

De Castro, M. B., Machado, R. Z., de Aquino, L. P., Alessi, A. C., & Costa, M. T. (2004). Experimental acute canine monocytic ehrlichiosis: Clinicopathological and immunopathological findings. *Veterinary Parasitology*, *119*(1), 73–86. doi:10.1016/j.vetpar.2003.10.012 PMID:15036578

De Deken, R., Martin, V., Saido, A., Madder, M., Brandt, J., & Geysen, D. (2007). An outbreak of East Coast Fever on the Comoros: A consequence of the import of immunized cattle from Tanzania? *Veterinary Parasitology*, *143*(3-4), 245–253. doi:10.1016/j.vetpar.2006.08.018 PMID:16996692

de la Fuente, J., Almazan, C., Canales, M., de la Lastra, J. M. P., Kocan, K. M., & Willadsen, P. (2007). A ten-year review of commercial vaccine performance for control of tick infestations on cattle. *Animal Health Research Reviews*, *8*(1), 23–28. doi:10.1017/S1466252307001193 PMID:17692140

De la Fuente, J., Lew, A., Lutz, H., Meli, M. L., Hofmann-Lehmann, R., Shkap, V., Molad, T., Mangold, A. J., Alamaza'n, C., Naranjo, V., Gorta'zar, C., Torina, A., Caracappa, S., Garcı'a-Pe'rez, A. L., Barral, M., Oporto, B., Ceci, L., Carelli, G., Blouin, E. F., & Kocan, K. M. (2005). Genetic diversity of Anaplasma species major surface proteins and implications for anaplasmosis sero-diagnosis and vaccine development. *Animal Health Research Reviews*, *6*(1), 75–89. doi:10.1079/AHR2005104 PMID:16164010

De La Rocque, S., Augusseau, X., & Guillobez, S. (2001). The changing distribution of two riverine tsetse flies over 15 years in an increasingly cultivated area of Burkina Faso. *Bulletin of Entomological Research*, *91*, 157–166. PMID:11415469

De Leeuw, P. N. (1990). *Interactive effects of environment, management and mortality on cattle productivity in livestock systems in sub-Saharan Africa*. Academic Press.

De Meneghi, D., Stachurski, F., & Adakal, H. (2016). Experiences in tick control by acaricide in the traditional cattle sector in Zambia and Burkina Faso: Possible environmental and public health implications. *Frontiers in Public Health*, *4*, 239. doi:10.3389/fpubh.2016.00239 PMID:27882313

De Tommasi, A. S., Otranto, D., Dantas-Torres, F., Capelli, G., Breitschwerdt, E. B., & Caprariis, D. (2013). Are vector-borne pathogen co-infections complicating the clinical presentation in dogs? *Parasites & Vectors*, *6*(1), 97. doi:10.1186/1756-3305-6-97 PMID:23587324

De Vos, A. J., De Waal, D. T., & Jackson, L. A. (2004). Bovine babesiosis. In J. A. W. Coetzer & R. C. Tustin (Eds.), *Infectious Diseases of Livestock* (pp. 406–424). Oxford University Press Southern Africa.

De Vos, S., Zeinstra, L., Taoufik, O., Willadsen, P., & Jongejan, F. (2001). Evidence for the utility of the Bm86 antigen from *Boophilus microplus* in vaccination against other tick species. *Experimental & Applied Acarology*, *25*(3), 245–261. doi:10.1023/A:1010609007009 PMID:11523920

De Waal, D. (2019). *Porcine Babesiosis*. Anipedia.

De Waal, D. T. (2000, December). Anaplasmosis control and diagnosis in South Africa. *Annals of the New York Academy of Sciences*, *916*(1), 474–483. doi:10.1111/j.1749-6632.2000.tb05327.x PMID:11193662

DeCastro, J. J. (1997). Suitable tick and tick-borne disease control in livestock improvement. *Veterinary Parasitology*, *71*, 77–97. doi:10.1016/S0304-4017(97)00033-2 PMID:9261972

Deem, S. L., Perry, B. D., Katende, J. M., McDermott, J. J., Mahan, S. M., Maloo, S. H., Morzaria, S. P., Musoke, A. J., & Rowlands, G. J. (1993). Variations in prevalence rates of tick-borne diseases in Zebu cattle by agro-ecological zone: Implications for the East Coast fever immunization. *Preventive Veterinary Medicine*, *16*(3), 171–187. doi:10.1016/0167-5877(93)90064-Z

Delafosse, A., Thébaud, E., Desquesnes, M., & Michaux, Y. (2006). Epidemiology of Trypanosoma vivax infection in cattle in the tse-tse free area of Lake Chad. *Preventive Veterinary Medicine*, *74*(2-3), 108–119. doi:10.1016/j.prevetmed.2005.10.006 PMID:16321453

Delgado, C., Rosegrant, M., Steinfeld, H., Ehui, S., & Courbois, C. (2001). Livestock to 2020: The Next Food Revolution. *Outlook on Agriculture*, *30*(1), 27–29. doi:10.5367/000000001101293427

Dennis, J., Durkin, S., Horsley, D., Hamill, L., Anderson, N., & MacLeod, E. (2014). Sodalis glossinidius prevalence and trypanosome presence in tsetse from Luambe National Park, Zambia. *Parasites & Vectors*, *7*(1), 378. doi:10.1186/1756-3305-7-378 PMID:25138709

Depner, D., Staubach, C., Probst, C., Globig, A., Blome, S., & Dietze, K., C., S.-L, Conraths, F. (2016). African Swine fever – epidemiological considerations and consequences for disease control. *TierärztlicheUmschau Heft*, *3/2016*, 72–78.

Desowitz, R. S. (1959). Studies on Immunity and Host-Parasite Relationships: I.—The Immunological Response of Resistant and Susceptible Breeds of Cattle to Trypanosomal Challenge. *Annals of Tropical Medicine and Parasitology*, *53*(3), 293–313. doi:10.1080/00034983.1959.11685927 PMID:13849587

Desquesnes, M. (2017). *Compendium of Diagnostic Protocols of the OIE Reference Laboratory for Animal Trypanosomoses of African Origin*. OIE. https://www.oie.int/nttat/Attached/A16-REC-OMPENDIUM_PROTOCOLES_TRYPANO-En.pdf

Desquesnes, M. (1997). International and regional standardization of immunoenzyme tests: Methods, concerns and limitations. *Scientific and Technical Review*, *16*, 809–823. PMID:9567307

Desquesnes, M. (2004). *Livestock trypanosomoses and their vectors in Latin America*. OIE World Organization for Animal Health.

Desquesnes, M., Dargantes, A., Lai, D.-H., Lun, Z.-R., Holzmuller, P., & Jittapalapong, S. (2013). *Trypanosoma evansi* and Surra: A Review and Perspectives on Transmission, Epidemiology and Control, Impact, and Zoonotic Aspects. *BioMed Research International*, *2013*, 1–20. Advance online publication. doi:10.1155/2013/321237 PMID:24151595

Desquesnes, M., De La Rocque, S., & Peregrine, A. S. (1995). French Guyana stock of Trypanosoma vivax resistant to diminazine aceturate but sensitive to isometamidium chloride. *Acta Tropica*, *60*, 133–136. doi:10.1016/0001-706X(95)00117-W PMID:8610541

Desquesnes, M., & Dia, M. L. (2003). Mechanical transmission of Trypanosoma congolense in cattle by the African tabanid Atylotus agrestis. *Experimental Parasitology*, *105*(3), 226–231. doi:10.1016/j.exppara.2003.12.014 PMID:14990316

Desquesnes, M., & Gardiner, P. R. (1993). Épidémiologie de la trypanosome bovine (Trypanosoma vivax) en Guyane Française. *Revue d'Élevage et de Médecine Vétérinaire des Pays Tropicaux*, *46*, 463–470. PMID:7910693

Desquesnes, M., & Holzmuller, P. (2013). Trypanosoma evansi and surra: A review and perspectives on origin, history, distribution, taxonomy, morphology, hosts, and pathogenic effects. *BioMed Research International*, 1–22.

Dhliwayo, S., Chihambakwe, B., Taonezvi, K., Chikerema, S. M., Tivapasi, M. T., & Pfukenyi, D. M. (2019). Seroprevalence of Canine Ehrlichiosis and Microscopic Screening for Canine Babesiosis in Dogs in Harare, Zimbabwe, 2016-2017. *Veterinary Medicine International*, *2019*, 4130210. doi:10.1155/2019/4130210 PMID:31885847

Dia, M., & Desquesnes, M. (2004). Evaluation of a New Trap for Tabanids and Stomoxyines. *Bulletin Société royale belge d'Entomologie Koninklijke Belgische Vereniging voor Entomologie, 140*, 72-81.

Diall, O., Cecchi, G., Wanda, G., Argiles-Herrero, R., Vreysen, M. J. B., Cattoli, G., Viljoen, G. J., Mattioli, R., & Bouyer, J. (2017). Developing a Progressive Control Pathway for African Animal Trypanosomosis. *Trends in Parasitology*, *33*(7), 499–509. doi:10.1016/j.pt.2017.02.005 PMID:28456474

Dicko, A. H., Percoma, L., Sow, A., Adam, Y., Mahama, C., Sidibe, I., Dayo, G.-K., Thévenon, S., Fonta, W., Sanfo, S., Djiteye, A., Salou, E., Djohan, V., Cecchi, G., & Bouyer, J. (2015). A Spatio-temporal Model of African Animal Trypanosomosis Risk. *PLoS Neglected Tropical Diseases*, *9*(7), e0003921. doi:10.1371/journal.pntd.0003921 PMID:26154506

Dirie, M., Wallbanks, K., Aden, A., Bornstein, S., & Ibrahim, M. (1989). Camel trypanosomiasis and its vectors in Somalia. *Veterinary Parasitology*, *32*(4), 285–291. doi:10.1016/0304-4017(89)90039-3 PMID:2781715

Dixit, A. K., Dixit, P., & Shukla, P. C. (2012). Canine Monocytic Ehrlichiosis and its therapeutic management in a dog. *Intas Polivet*, *13*(1), 140–141.

Dixon, L. K., Chapman, D. A. G., Netherton, C. L., & Upton, C. (2013). African swine fever virus replication and genomics. *Virus Research*, *173*(1), 3–14. doi:10.1016/j.virusres.2012.10.020 PMID:23142553

Dixon, L. K., Stahl, K., Jori, F., Vial, L., & Pfeiffer, D. U. (2020). African swine fever epidemiology and control. *Annual Review of Animal Biosciences*, *8*, 221–246.

Dobbelaere, D. A. E., Shapiro, S. Z., & Webster, P. (1985). Identification of a surface antigen on *Theileria parva* sporozoites by monoclonal antibodies. *Proceedings of the National Academy of Sciences of the United States of America*, *82*(6), 1771–1775. doi:10.1073/pnas.82.6.1771 PMID:3920654

Dolan, R. (1998). *The Orma Boran: A trypanotolerant east African breed: Fifteen years of research on Galana Ranch in Kenya*. Kenya Trypanosomiasis Research Institute.

Dolan, R. B. (1987). Genetics and trypanotolerance. *Parasitology Today (Personal Ed.)*, *3*(5), 137–143. doi:10.1016/0169-4758(87)90197-9 PMID:15462937

Dolan, T. T. (1981). Progress in the chemotherapy of theileriosis. In A. D. Irvin, M. P. Cunnigham, & A. S. Young (Eds.), *Advances in the control of theileriosis* (pp. 186–208). Martinus Nijhoff Publishers. doi:10.1007/978-94-009-8346-5_34

Dolan, T. T. (1989). Theileriasis: A comprehensive review. *Rev. Sci. Tech. Off. Int. Epiz*, *8*(1), 11–36. doi:10.20506/rst.8.1.398 PMID:32731664

Dolan, T. T., Linyonyi, A., McHardy, N., Bond, A. L., & Clampit, R. B. (1988). Chemotherapy of East Coast fever: Parvaquone treatment of *Theileria parva parva* at intervals after infection. *Research in Veterinary Science*, *44*(1), 15–20. doi:10.1016/0034-5288(88)90005-7 PMID:3375580

Dolan, T. T., & McHardy, N. (1978). The chemotherapy of experimental *T. parva* infection. In J. K. H. Wilde (Ed.), *Tick-borne Diseases and Their Vectors* (pp. 318–323). University of Edinburgh.

Dolan, T. T., Young, A. S., Losos, G. J., McMillan, T., Minder, Ch. E., & Soulsby, K. (1984). Dose-dependent responses of cattle to *Theileria parva* stabilate. *International Journal for Parasitology*, *14*(1), 89–95. doi:10.1016/0020-7519(84)90017-1 PMID:6423557

Domingos, A., Antunes, S., Borges, L., & Do Rosario, V. E. (2013). Approaches towards tick and tick-borne diseases control. *Revista da Sociedade Brasileira de Medicina Tropical*, *46*(3), 265–269.

Dominguez, M., Echaide, I., de Echaide, S. T., Wilkowsky, S., Zabal, O., Mosqueda, J. J., … Florin-Christensen, M. (2012). Validation and Field Evaluation of a Competitive Enzyme-Linked Immunosorbent Assay for Diagnosis of Babesia bovis Infections in Argentina. *Clinical and Vaccine Immunology, 19*(6), 924–928. doi:10.1128/CVI.00015-12

Donatien, A., & Lestoquard, F. (1937). State of the present knowledge concerning rickettsiosis of animals. *Archives. Institut Pasteur d'Algerie*, *15*, 142–187.

Doudoumis, V., Tsiamis, G., Wamwiri, F., Brelsfoard, C., Alam, U., Aksoy, E., Dalaperas, S., Abd-Alla, A., Ouma, J., Takac, P., Aksoy, S., & Bourtzis, K. (2012). Detection and characterization of Wolbachia infections in laboratory and natural populations of different species of tsetse flies (genus Glossina). *BMC Microbiology*, *12*(Suppl 1), S3. doi:10.1186/1471-2180-12-S1-S3 PMID:22376025

Drummond, R. (1970). *Current worldwide research on control of ticks involved in animal diseases*. Entomol Soc Amer Misc Publ.

du Plessis, J. L. (1982). *Mice infected with a Cowdria ruminantium-like agent as a model in the study of heartwater* (DVSc thesis). University of Pretoria.

du Plessis, J. L. (1970). Pathogenesis of heartwater: I. Cowdria ruminantium in the lymph nodes of domestic ruminants. *The Onderstepoort Journal of Veterinary Research*, *37*(2), 89–95. PMID:4115310

du Plessis, J. L. (1985). A method for determining the Cowdria ruminantium infection rate of Amblyomma hebraeum: Effects in mice injected with tick homogenates. *The Onderstepoort Journal of Veterinary Research*, *52*(2), 55–61. PMID:3900855

du Plessis, J. L. (1990). Increased pathogenicity of an Ehrlichia-like agent after passage through Amblyomma hebraeum: A preliminary report. *The Onderstepoort Journal of Veterinary Research*, *57*(4), 233–237. PMID:2293132

du Plessis, J. L., Jansen, B. C., & Prozesky, L. (1983). Heartwater in Angora goats. I. Immunity subsequent to artificial infection and treatment. *The Onderstepoort Journal of Veterinary Research*, *50*(2), 137–143. PMID:6634087

du Plessis, J. L., & Malan, L. (1987). The application of the indirect fluorescent antibody test in research on heartwater. *The Onderstepoort Journal of Veterinary Research*, *54*(3), 319–325. PMID:3329322

Dubie, T., Mohammed, Y., Terefe, G., Muktar Y., & Tesfaye, J. (2014). An insight review on canine ehrlichiosis with emphasis on its epidemiology and pathogenesity importance. *Global Journal of Veterinary Medicine and Research, 2*(4), 59–67.

Ducheyne, H., Mweempwa, C., De Pus, C., Vernieuwe, H., De Deken, R., Hendrickx, G., & Van den Bossche, P. (2009). The impact of habitat fragmentation on tsetse abundance on the plateau of eastern Zambia. *Preventive Veterinary Medicine*, *91*(1), 11–18. doi:10.1016/j.prevetmed.2009.05.009 PMID:19523702

Duguma, R., Tasew, S., Olani, A., Damena, D., Alemu, D., Mulatu, T., Alemayehu, Y., Yohannes, M., Bekana, M., Hoppenheit, A., Abatih, E., Habtewold, T., Delespaux, V., & Duchateau, L. (2015). Spatial distribution of Glossina sp. and Trypanosoma sp. in south-western Ethiopia. *Parasites & Vectors*, *8*(1), 430. doi:10.118613071-015-1041-9 PMID:26286484

Dumler, J. S., Barbet, A. F., Bekker, C. P. J., Dasch, G. A., Palmer, G. H., Ray, S., Rikihisa, C. Y., & Rurangirwa, F. R. (2001). Reorganization of the genera in the families Rickettsiaceae and Anaplasmataceae in the order Rickettsiales: Unification of some species of *Ehrlichia* with *Anaplasma,Cowdria* with *Ehrlichia* and *Ehrlichia* with *Neorickettsia*, descriptions of six new species combinations and designation of *Ehrlichia equi* and "HGE agent" as subjective synonyms of *Ehrlichia phagocytophila. International Journal of Systematic and Evolutionary Microbiology*, *51*(6), 2145–2165. doi:10.1099/00207713-51-6-2145 PMID:11760958

Dumler, J. S., Madigan, J. E., Pusterla, N., & Bakken, J. S. (2007). Ehrlichioses in humans: Epidemiology, clinical presentation, diagnosis, and treatment. *Clinical Infectious Diseases*, *45*(1, Supplement_1), 45–51. doi:10.1086/518146 PMID:17582569

Duzgun, A., Schunter, C. A., Wright, I. G., Leatch, G., & Waltisbuhl, D. J. (1988). A sensitive ELISA technique for the diagnosis of *Anaplasma marginale* infections. *Veterinary Parasitology*, *29*(1), 1–7. doi:10.1016/0304-4017(88)90002-7 PMID:3176298

Dwinger, R. H., Agyemang, K., Snow, W. F., Rawlings, P., Leperre, P., & Bah, M. L. (1994). Productivity of trypanotolerant cattle kept under traditional management conditions in the Gambia. *The Veterinary Quarterly*, *16*(2), 81–86. doi:10.1080/01652176.1994.9694424 PMID:7985361

Dzimianski, J. V., Beldon, B. S., Daczkowski, C. M., Goodwin, O. Y., Scholte, F. E. M., Bergeron, É., & Pegan, S. D. (2019). Probing the impact of nairovirus genomic diversity on viral ovarian tumor domain protease (vOTU) structure and deubiquitinase activity. *PLoS Pathogens* (Vol. 15). doi:10.1371/journal.ppat.1007515

Eddlestone, S. M., Diniz, P. P., Neer, T. M., Gaunt, S. D., Corstvet, R., Cho, D., Hosgood, G., Hegarty, B., & Breitschwerdt, E. B. (2007). Doxycycline clearance of experimentally induced chronic Ehrlichia canis infection in dogs. *Journal of Veterinary Internal Medicine*, *21*(6), 1237–1242. doi:10.1111/j.1939-1676.2007.tb01944.x PMID:18196732

Eddlestone, S. M., Neer, T. M., Gaunt, S. D., Corstvet, R., Gill, A., Hosgood, G., Hegarty, B., & Breitschwerdt, E. B. (2006). Failure of imidocarb dipropionate to clear experimentally induced Ehrlichia canis infection in dogs. *Journal of Veterinary Internal Medicine*, *20*(4), 840–844. doi:10.1111/j.1939-1676.2006.tb01795.x PMID:16955806

Edelsten, R. M. (1975). The distribution and prevalence of Nairobi Sheep disease and other tick-borne infections of sheep and goats in northern Somalia. *Tropical Animal Health and Production*, *7*(1), 29–34. https://doi.org/10.1007/BF02383239

Egeru, A., Opio, J., Siya, A., Barasa, B., Magaya, J., & Namaalwa, J. J. (2020). Tsetse Invasion as an Emerging Threat to Socioecological Resilience of Pastoral Communities in Karamoja, Uganda. *Sustainability*, *12*(4), 1599. doi:10.3390u12041599

Elliot, M., Janes, N.F., & Potter, C. (1978). The future of pyrethroids in insect control. *Annal rev Entomology, 23*, 443-469.

Emery, D. L., MacHugh, N. D., & Morrison, W. I. (1988). *Theileria parva* (Muguga) infects bovine T lymphocytes *in vivo* and induces coexpression of BoT4 and BoT8. *Parasite Immunology*, *10*(4), 379–391. doi:10.1111/j.1365-3024.1988.tb00228.x PMID:3262855

Emery, D. L., Morrison, W. I., Nelson, R. T., & Murray, M. (1981). The induction of cell mediated immunity in cattle inoculated with cell lines parasised with *Theileria parva*. In A. D. Irvin, M. P. Cunnigham, & A. S. Young (Eds.), *Advances in the control of theileriosis* (pp. 295–310). Martinus Nijhoff Publishers. doi:10.1007/978-94-009-8346-5_51

Enwezor & Sackey. (2005). *Camel trypanosomosis, Nigeria institute for Trypanosomosis Research (NITR).* Department of veterinary surgery medicine. Ahmadu Bello University (ABU) Zaria, Kaduna State.

Erlank, E., Koekemoer, L. L., & Coetzee, M. (2018). The importance of morphological identification of African anopheline mosquitoes (Diptera: Culicidae) for malaria control programmes. *Malaria Journal*, *17*(1), 43. Advance online publication. doi:10.118612936-018-2189-5 PMID:29357873

Esterhuizen, J., Rayaisse, J. B., Tirados, I., Mpiana, S., Solano, P., Vale, G. A., Lehane, M. J., & Torr, S. J. (2011). Improving the cost-effectiveness of visual devices for the control of riverine tsetse flies, the major vectors of human African trypanosomiasis. *PLoS Neglected Tropical Diseases*, *5*(8), e1257. doi:10.1371/journal.pntd.0001257 PMID:21829743

Estrada, A., Bouattour, A., Camicas, J.-L., Guglielmone, A., Horak, I., Jongejan, F., … Walker, A. R. (2006). The known distribution and ecological preferences of the tick subgenus Boophilus (Acari: Ixodidae) in Africa and Latin America. *Experimental & Applied Acarology, 38*(2–3), 219–235. doi:10.100710493-006-0003-5

Estrada-Peña, A., Mangold, A. J., Nava, S., Venzal, J. M., Labruna, M. B., & Guglielmone, A. A. (2010). A systematic review of the tick family argasidae (Ixodida). Acarologia, 50(3), 317–333.

Estrada-Peña, A., Venzal, J. M., González-Acuña, D., & Guglielmone, A. A. (2003). *Argas* (*Persicargas) keiransin.* sp. (Acari: Argasidae), a parasite of the Chimango, *Milvago c. chimango* (Aves: Falconiformes). *Journal of Medical Entomol*ogy, *40*(6), 766-769.

Estrada-Peña, A. (2003). Climate change decreases habitat suitability for some tick species (Acari: Ixodidae) in South Africa. *The Onderstepoort Journal of Veterinary Research*, *70*(2), 79–93. PMID:12967169

Estrada-Peña, A., Horak, I. G., & Petney, T. N. (2008). Changes in climate and their suitability for the ticks *Amblyomma hebraeum* and *Amblyomma variegatum (Ixodidae)* in Zimbabwe (1974–1999). *Veterinary Parasitology*, *151*(2-4), 256–267. doi:10.1016/j.vetpar.2007.11.014 PMID:18083309

Estrada-Peña, A., & Jongejan, F. (1999). Ticks feeding on humans: A review of records on human-biting Ixodoidea with special reference to pathogen transmission. *Experimental & Applied Acarology*, *23*(9), 685–715. doi:10.1023/A:1006241108739 PMID:10581710

Estrada-Peña, A., Venzal, J. M., Nava, S., Mangold, A., Guglielmone, A. A., Labruna, M. B., & de la Fuente, J. (2012). Reinstatement of Rhipicephalus (Boophilus) australis (Acari: Ixodidae) with redescription of the adult and larval stages. *Journal of Medical Entomology*, *49*(4), 794–802. https://doi.org/10.1603/me11223

Etter, E. M. C., Seck, I., Grosbois, V., Jori, F., Blanco, E., Vial, L., & Akakpo, A. J., Alhambedji, R., Kone, P., & Roger, F. L. (2011). Sero-prevalence of African swine fever in Senegal. *Emerging Infectious Diseases*, *17*(1), 49–54.

Ettinger, S., Feldman, F., & Edward, C. (1995). *Textbook of Veterinary Internal Medicine* (4th ed.). W.B. Saunders Company.

Evans, D. E. (1992). Tick infestsation of livestockand tick control methods in Brazil: A situation report. *Insect Science and Its Application*, *13*(4), 629–643.

Evans, G. (1880). Report on "Surra" disease in the Dera Ismail Khan District of India. *Punjab Government Miltary*, (493), 446.

Eygelaar, D., Jori, F., Mokopasetso, M., Sibeko, K. P., Collins, N. E., Vorster, I., Troskie, M., & Oosthuizen, M. C. (2015). Tick-borne haemoparasites in African buffalo (*Syncerus caffer*) from two wildlife areas in Northern Botswana. *Parasites & Vectors*, *8*(1), 26. doi:10.118613071-014-0627-y PMID:25589064

Faburay, B., Geysen, D., Munstermann, S., Taoufik, A., Postigo, M., & Jongejan, F. (2007). Molecular detection of Ehrlichia ruminantium infection in Amblyomma variegatum ticks in The Gambia. *Experimental & Applied Acarology*, *42*(1), 61–74. doi:10.100710493-007-9073-2 PMID:17476576

Fahrimal, Y., Goff, W., & Jasmer, D. (1992). Detection of Babesia bovis carrier cattle by using polymerase chain reaction amplification of parasite DNA. *Journal of Clinical Microbiology*, *30*(6), 1374–1379. https://europepmc.org/articles/PMC265295

Fakhar, M., Hajihasani, A., Maroufi, S., Alizadeh, H., Shirzad, H., Piri, F., & Pagheh, A. S. (2012). An epidemiological survey on bovine and ovine babesiosis in Kurdistan Province, western Iran. *Tropical Animal Health and Production*, *44*(2), 319–322. https://doi.org/10.1007/s11250-011-0023-y

FAO. (1977). *Review of recent progress in the control of ticks and tick-borne diseases in East*. Africa. FAO.

FAO. (1982). *Tsetse Biology, Systematics and Distribution: Techniques* (Vol. 1). Food and Agriculture Organisation of the United Nations.

FAO. (1984). *Ticks and TBDs control; in a practical field manual* (Vol. 1). FAO.

FAO. (1992). Training Manual for Tsetse control personnel (vol. 4). FAO.

FAO. (1992). *Use of attractive devices for tsetse survey and control* (Vol. 4). FAO.

FAO. (1992a). *Ecology and behaviour of tsetse*. FAO.

FAO. (1992b). *Hosts of glossina and transmission of disease*. FAO.

FAO. (1993). *Insecticides for tsetse and trypanosomiasis control using attractive bait techniques* (Vol. 5). FAO.

FAO. (2004). The State of Food Insecurity in the World (SOFI). Food and Agriculture Organization of the United Nations.

FAO. (2006). *Corporate Document Repository. A field guide for the diagnosis, treatment and prevention of African animal trypanosomosis*. Food and Agriculture Organization of the United Nations. www.fao.org/DOCREP/2006

FAO. (2011). *A value chain approach to animal diseases management. Technical foundations and practical framework for field application. FAO Animal Production and Health Guidelines No. 4*. Food and Agriculture Organization of the United Nations.

FAO. (2019). *R*. FAOSTAT.

FAO/OIE. (2015). *Global Strategy for the Control and Eradication of PPR*. FAO/OIE.

FAO/OIE/WB. (2010). *Good practices for biosecurity in the pig sector – Issues and options in developing and transition countries. FAO Animal Production and Health Paper No. 169*. Food and Agriculture Organization of the United Nations.

Faria, J. L. M., Dagnone, A. S., Munhoz, T. D., João, C. F., Pereira, W. A. B., Machado, R. Z., & Tinucci-Costa, M. (2010). Ehrlichia canis morulae and DNA detection in whole blood and spleen aspiration samples. *Revista Brasileira de Parasitologia Veterinária*, *19*(2), 98–102. doi:10.1590/S1984-29612010000200006 PMID:20624346

Fawcet, D. W., Young, A. J., & Leitch, B. L. (1985). Sporogony in Theileria (Ampicomplexa: Piroplasmida). A comparative ultrastructural study. *Journal of Submicroscopic Cytology*, *17*, 299–314.

Faye, D., Pereira, P. J., de Almeida, B., Osaer, S., Ndao, M., Berkvens, D., Speybroeck, N., Nieberding, F., & Geerts, S. (2001). Prevalence and incidence of trypanosomosis in horses and donkeys in The Gambia. *Veterinary Parasitology*, *101*(2), 101–114. doi:10.1016/S0304-4017(01)00503-9 PMID:11587839

Fenner, F. J. (2017). Fenner's Veterinary Virology (N. J. M. E. J. Dubov, Ed.; 4th ed.). Elsevier., https://doi.org/10.1016/C2013-0-06921-6.

Fenner, F., Peter, A., & Bachmann, D. O. W. (1987). *Veterinary Virology*. Elsevier. doi:10.1016/C2013-0-10647-2

Fevre, E., Coleman, P., Odiit, M., Magona, J. W., Welburn, S. C., & Woolhouse, M. E. (2001). Occurrence, Clinical Manifestation and the Epidemiological Implications of Naturally Occurring Canine Trypanosomosis in Western Kenya. *Lancet*, *358*(9282), 625–628. PMID:11530149

Figueroa, J. V., Chieves, L. P., Johnson, G. S., & Buening, G. M. (1993). Multiplex polymerase chain reaction based assay for the detection of *Babesia bigemina, Babesia bovis* and *Anaplasma marginale* DNA in bovine blood. *Veterinary Parasitology*, *50*(1-2), 69–81. doi:10.1016/0304-4017(93)90008-B PMID:8291198

Figueroa, J. V., Chieves, L. P., Johnson, G. S., Goff, W. L., & Buening, G. M. (1994). Polymerase chain reaction-based diagnostic assay to detect cattle chronically infected with Babesia bovis. *Revista Latinoamericana de Microbiologia*, *36*(1), 47–55.

Figueroa, J., Chieves, L., Johnson, G., & Buening, G. (1992). Detection of Babesia bigemina-infected carriers by polymerase chain reaction amplification. *Journal of Clinical Microbiology*, *30*(10), 2576–2582. https://europepmc.org/articles/PMC270481

Fikru, R., Gari, Hagos, A., Alemu, T., Goddeeris, B. M., & Claes, F. (2010). Comparative diagnosis of parasitological, serological, and molecular tests in dourine-suspected horses. *Tropical Animal Health and Production*, *42*, 1649–1654.

Filip, C., Philipps, B., Louis, T., & Bruno, M. (2005). Trypanosoma equiperdum: Master of disguise or historical mistake? *Trends in Parasitology*, *21*(7), 316–321. doi:10.1016/j.pt.2005.05.010 PMID:15923142

FITCA. (2004). *Farming in Tsetse Controlled Controlled Areas Kenya Project 1999-2004 Lessons Learned*. FAITCA.

Florin-Christensen, M., Suarez, C. E., Rodriguez, A. E., Flores, D. A., & Schnittger, L. (2014). Vaccines against bovine babesiosis: Where we are now and possible roads ahead. *Parasitology*, *141*(12), 1563–1592. doi:10.1017/S0031182014000961 PMID:25068315

Foil, L. (1989). Tabanids as vectors of disease agents. Parasitology Today, 5:88-96. *Parasitology Today (Personal Ed.)*, *5*, 88–96. doi:10.1016/0169-4758(89)90009-4 PMID:15463186

Food and Agriculture Organization of the United Nations and African Union-Inter-African Bureau for Animal Resources and International Livestock Research Institute. (2017). *Regional strategy for the control of African Swine Fever in Africa*. www.fao.org/publications

Food and Agriculture Organization of the United Nations. (Ed.). (2011). *FAO in the 21st Century: Ensuring Food Security in a Changing World*. FAO.

Forbes, B. A., Sahm, D. F., & Weissfeld, A. S. (2007). Obligate intracellular and nonculturable bacterial agents. In Bailey and Scott's diagnostic microbiology (12th ed.). Miss. Mosby.

Ford, J. (1971). *The role of the trypanosomiases in African ecology: A study of the tsetse fly problem*. Clarendon Press.

Ford, J., & Katondo, K. (1977). Maps of tsetse flies (Glossina) distribution in Africa, 1973 according to sub-generic groups on a scale of 1: 5 000 000. *Bulletin of Animal Health and Production in Africa*, *25*, 187–193.

Ford, J., & Katondo, K. M. (1977). Maps of tsetse flies (Glossina) distribution in Africa, 1973 according to sub-generic groups on scale of 1:5 000 000. *Bulletin of Animal Health and Production in Africa*, 187–193.

Fouche, G., Ramafuthula, M., Maselela, V., Mokoena, M., Senabe, J., Leboho, T., Sakong, B. M., Adenubi, O. T., Eloff, J. N., & Wellington, K. W. (2016). Acaricidal activity of the organic extracts of thirteen South African plants against *Rhipicephalus* (*Boophilus*) *decoloratus* (Acari: Ixodidae). *Veterinary Parasitology*, *224*, 39–43. doi:10.1016/j.vetpar.2016.05.011 PMID:27270388

Franco, J. R., Cecchi, G., Priotto, G., Paone, M., Diarra, A., Grout, L., Simarro, P. P., Zhao, W., & Argaw, D. (2018). Monitoring the elimination of human African trypanosomiasis: Update to 2016. *PLoS Neglected Tropical Diseases*, *12*(12), e0006890. doi:10.1371/journal.pntd.0006890 PMID:30521525

Franco, J. R., Simarro, P. P., Diarra, A., & Jannin, J. G. (2014). Epidemiology of human African trypanosomiasis. *Clinical Epidemiology*, *6*, 257–275. doi:10.2147/CLEP.S39728 PMID:25125985

Frezoulis, P. S., Angelidou, E., Karnezi, D., Oikonomidis, I. L., Kritsepi-Konstantinou, M., Kasabalis, D., & Mylonakis, M. E. (2017). Canine pancytopoenia in a Mediterranean region: A retrospective study of 119 cases (2005 to 2013). *The Journal of Small Animal Practice*, *58*(7), 395–402. doi:10.1111/jsap.12647 PMID:28247969

Friedhoff, K. T. (1988). Transmission of Babesia. In *Babesiosis of Domestic Animals and Man*. CRC Press, Inc.

Friedhoff, K. T. (1997). Tick-borne diseases of sheep and goats caused by Babesia, Theileria or Anaplasma spp. *Parassitologia*, *39*(2), 99–109.

Frolov, B. A. (1974). Chemicaland biological methods of controlling poultry ecto-parasites. *Aria, 12*, 66–8.

Gachohi, J. M., Bett, B., & Murill, G. A. (2009). Factors influencing prevalence of trypanosomosis in orma boran (trypanotolerant) and teso zebu (trypanosusceptible) crosses in Teso district, western Kenya. *Bulletin of Animal Health and Production in Africa*, *57*(4), 4.

Gachohi, J., Skilton, R., Hansen, F., Ngumi, P., & Kitala, P. (2012). Epidemiology of East Coast fever (*Theileria parva* infection) in Kenya: Past, present and the future. *Parasites & Vectors*, *5*(1), 194. doi:10.1186/1756-3305-5-194 PMID:22958352

Gago da Câmara, N. J. (1933). História da pestesuínaem Angola. *Pecuária: Anais Dos Serviços Pecuários Da Colónia De Angola*, *1932*(1), 25–40.

Gaithuma, A., Yamagishi, J., Martinelli, A., Hayashida, K., Kawai, N., Marsela, M., & Sugimoto, C. (2019). A single test approach for accurate and sensitive detection and taxonomic characterization of Trypanosomes by comprehensive analysis of internal transcribed spacer 1 amplicons. *PLoS Neglected Tropical Diseases*, *13*(2), e0006842. Advance online publication. doi:10.1371/journal.pntd.0006842 PMID:30802245

Gale, K. R., Dimmock, C. M., Gartside, M., & Leatch, G. (1996). *Anaplasma marginale*: Detection of carrier cattle by PCR-ELISA. *International Journal for Parasitology*, *26*(10), 1103–1109. doi:10.1016/S0020-7519(96)80009-9 PMID:8982791

Galiza, G., Garcia, H., Assis, A., Oliveira, D., Pimentel, L., Dantas, A. F., Simoes, S. V., Teixeira, M. M., & Riet-Correa, F. (2011). High mortality and lesions of the central nervous system in trypanosomosis by Trypanosoma vivax in Brazilian hair sheep. *Veterinary Parasitology*, *182*(2–4), 359–363. doi:10.1016/j.vetpar.2011.05.016 PMID:21664764

Gallardo, M. C., & de la Torre Reoyo, A. (2015). Fernández-Pinero. J., Iglesias I., Munoz M. J., Arias M. L. (2015). African swine fever: A global view of the current challenge. *PHM*, *1*, 21. doi:10.118640813-015-0013-y

García-García, J. C., Montero, C., Redondo, M., Vargas, M., Canales, M., & Boue, O. … de la Fuente, J. (2000). Control of ticks resistant to immunization with Bm86 in cattle vaccinated with the recombinant antigen Bm95 isolated from the cattle tick, Boophilus microplus. *Vaccine, 18*(21), 2275–2287. (99)00548-4 doi:10.1016/S0264-410X

Garrison, A. R., Alkhovsky Альховский Сергей Владимирович, S. V., Avšič-Županc, T., Bente, D. A., Bergeron, É., Burt, F., … Consortium, I. R. (2020). ICTV Virus Taxonomy Profile: Nairoviridae. *The Journal of General Virology, 101*(8), 798–799. doi:10.1099/jgv.0.001485

Gaunt, S. D., Beall, M. J., Stillman, B. A., Lorentzen, L., Diniz, P. P. V. P., Chandrashekar, R., & Breitschwerdt, E. B. (2010). Experimental infection and co-infection of dogs with Anaplasma platys and Ehrlichia canis: Hematologic, serologic and molecular findings. *Parasites & Vectors*, *3*(1), 33. doi:10.1186/1756-3305-3-33 PMID:20377870

Geerts, S., Holmes, P. H., Eisler, M. C., & Diall, O. (2001). African bovine trypanosomiasis: The problem of drug resistance. *Trends in Parasitology*, *17*(1), 25–28. doi:10.1016/S1471-4922(00)01827-4 PMID:11137737

Geerts, S., Osaer, S., Goossens, B., & Faye, D. (2009). Trypanotolerance in small ruminants of sub-Saharan Africa. *Trends in Parasitology*, *25*(3), 132–138. doi:10.1016/j.pt.2008.12.004 PMID:19200783

George, J. E., Pound, J. M., & Davey, R. B. (2004). Chemical control of ticks on cattle and the resistance of these parasites to acaricides. *Parasitology*, *129*(S1), S353–S366. doi:10.1017/S0031182003004682 PMID:15938518

Getahun, A., Yacob, H., & Basu, A. (2016). Overview of the Biology, Epidemiology and Control Methods against Hard Ticks. *Global Journal of Science Frontier Research, 16*(2).

Getahun, M., Cecchic, G., & Seyouma, E. (2014). Population studies of Glossina pallidipes in Ethiopia: Emphasis on cuticular hydrocarbons and wing morphometric analysis. *Acta Tropica*, *138*, S12–S21. doi:10.1016/j.actatropica.2014.04.015 PMID:24751419

Ghorbel, A., Ben Ayed, M., Diwan, F., Ghram, A., Landolsi, F., Messaadp, L., Zrellp, S., & Chabchoub, A. (2001). Incidence and seroprevalence of canine ehrlichiosis in the Medjez El Bab region (northwestern Tunisia) during 1994, 1995 and 1996. *Archives de l'Institut Pasteur de Tunis*, *78*(1-4), 41–47. PMID:14658238

Ghorbel, A., Clerc, B., & Djaiem, A. (1994). Canine ehrlichiosis in Tunisia. Seroepidemiological survey. *Revue d'elevage et de medecine veterinaire despays tropicaux. Revue d'Élevage et de Médecine Vétérinaire des Pays Tropicaux*, *47*(3), 271–275. PMID:7709025

Gianopoulos, A., Mylonakis, M. E., Theodorou, K., & Christopher, M. M. (2016). Quantitative and qualitative leukocyte abnormalities in dogs with experimental and naturally occurring acute canine monocytic ehrlichiosis. *Veterinary Clinical Pathology*, *45*(2), 281–290. doi:10.1111/vcp.12359 PMID:27142117

Gikonyo, N. K., Hassanali, A., Njagi, P. G., Gitu, P. M., & Midiwo, J. O. (2002). Odor composition of preferred (buffalo and ox) and nonpreferred (waterbuck) hosts of some Savanna tsetse flies. *Journal of Chemical Ecology*, *28*(5), 969–981. doi:10.1023/A:1015205716921 PMID:12049234

Gikonyo, N. K., Hassanali, A., Njagi, P. G., & Saini, R. K. (2003). Responses of Glossina morsitans morsitans to blends of electroantennographically active compounds in the odors of its preferred (Buffalo and ox) and nonpreferred (waterbuck) hosts. *Journal of Chemical Ecology*, *29*(10), 2331–2345. doi:10.1023/A:1026230615877 PMID:14682515

Gimonneau, Rayaisse, & Bouyer. (2018). Integrated control of trypanosomosis. *Pests and vector-borne diseases in the livestock industry*, 147 - 174. doi:10.3920/978-90-8686-863-6_6

Gomes, A. F., & Neves, L. (2018). *Rhipicephalus microplus* (Acarina, Ixodidae) in Angola: Evidence of its establishment and expansion. *Experimental & Applied Acarology*, *74*(1), 117–122. doi:10.100710493-017-0207-x PMID:29380168

Gong, S., He, B., Wang, Z., Shang, L., Wei, F., Liu, Q., & Tu, C. (2015). Nairobi Sheep Disease Virus RNA in Ixodid Ticks, China, 2013. *Emerging Infectious Diseases*, *21*(4), 718–720. https://doi.org/10.3201/eid2104.141602

Gonzatti, M. I., González-Baradat, B., Aso, P. M., & Reyna-Bello, A. (2014). *Trypanosoma (duttonella) vivax and typanosomosis in latin America: Secadera/Huequera/Cacho Hueco* (Vol. 9783709115565). Academic Press.

Götsch, S., Leschnik, M., Duscher, G., Burgstaller, J. P., Wille-Piazzai, W., & Joachim, A. (2009). Ticks and haemoparasites of dogs from Praia, Cape Verde. *Veterinary Parasitology*, *66*(1-2), 171–174. doi:10.1016/j.vetpar.2009.08.009 PMID:19729247

Gough, J. M., Jorgensen, W. K., & Kemp, D. H. (1998). Development of Tick Gut Forms of Babesia bigemina In Vitro. *The Journal of Eukaryotic Microbiology*, *45*(3), 298–306. https://doi.org/10.1111/j.1550-7408.1998.tb04540.x

Gouteux, J. (1987). Une nouvelle glossine du Congo: Glossina (Austenina) frezili sp. nov. (Diptera; Glossinidae). *Tropical Medicine and Parasitology*, *38*, 97–100. PMID:3629143

Graf, J.-F., Gogolewski, R., Leach-Bing, N., Sabatini, G. A., Molento, M. B., Bordin, E. L., & Arantes, G. J. (2004). Tick control: An industry point of view. *Parasitology*, *129*(S1), S427–S442. https://doi.org/10.1017/S0031182004006079

Grant, C., Anderson, N., & Machila, N. (2015). Stakeholder narratives on trypanosomiasis, their effect on policy, and the scope for One Health. *PLoS Neglected Tropical Diseases*, *9*(12), e0004241. doi:10.1371/journal.pntd.0004241 PMID:26658646

Grant, I. F. (2001). Insecticides for tsetse and trypanosomiasis control: Is the environmental risk acceptable? *Trends in Parasitology*, *17*(1), 10–14. doi:10.1016/S1471-4922(00)01848-1 PMID:11137734

Gray, A. R., & Luckins, A. G. (1978). An extravascular site of development of Trypanosoma congolense. *Nature*, *272*(5654), 613–614. doi:10.1038/272613a0 PMID:643055

Gray, J. S. (1999). The ecology of Lyme borreliosis vector ticks. In J. Bruin, L. P. S. van der Geest, & M. W. Sabelis (Eds.), *Ecology and Evolution of the Acari. In Entomologica* (Vol. 55, pp. 269–276). Springer.

Gray, J. S., Dautel, H., Estrada-Peña, A., Kahl, O., & Lindgren, E. (2009). Effects of Climate Change on Ticks and Tick-Borne Diseases in Europe. *Interdisciplinary Perspectives on Infectious Diseases*, *2009*, 1–12. https://doi.org/10.1155/2009/593232

Gray, J. S., Estrada-Peña, A., & Zintl, A. (2019). Vectors of Babesiosis. *Annual Review of Entomology*, *64*(1), 149–165. https://doi.org/10.1146/annurev-ento-011118-111932

Green, C. (1994). Bait methods for tsetse fly control. *Advances in Parasitology*, *34*, 229–291. doi:10.1016/S0065-308X(08)60140-2 PMID:7976751

Greene, C. E., & Harvey, J. W. (1984). Canine ehrlichiosis. In C. E. Greene (Ed.), *Clinical microbiology and infectious diseases of the dog and cat*. The W. B. Saunders Company.

Griffith, B., Weiss, B., Aksoy, E., Mireji, P., Auma, J., Wamwiri, F., Echodu, R., Murilla, G., & Aksoy, S. (2018). Analysis of the gut-specific microbiome from field-captured tsetse flies, and its potential relevance to host trypanosome vector competence. *BMC Microbiology*, *18*(S1), 146. doi:10.118612866-018-1284-7 PMID:30470178

Griffith, F. (1898). *The Petrie Papyri: Hieratic Papyri from Kahun and Gurob (Principally of the Middle Kingdom)*. Bernard Quaritch.

Gummery, L., Jallow, S., Raftery, A., Bennet, E., Rodgers, J., & Sutton, D. (2020). Comparison of loop-mediated isothermal amplification (LAMP) and PCR for the diagnosis of infection with Trypanosoma brucei ssp. In equids in The Gambia. *PLoS One*, *15*(8), e0237187. Advance online publication. doi:10.1371/journal.pone.0237187 PMID:32833981

Gupta, M. P., Kumar, H., & Singla, L. D. (2009). Trypanosomosis concurrent to tuberculosis in black bucks. *The Indian Veterinary Journal*, *86*, 727–728.

Gustafson, C. R., VanWormer, E., Kazwala, R., Makweta, A., Paul, G., Smith, W., & Mazet, J. A. K. (2015). Educating pastoralists and extension officers on diverse livestock diseases in a changing environment in Tanzania. *Pastoralism*, *5*(1), 1–12. doi:10.118613570-014-0022-5

Gutierrez, C. N., Martinez, M., Sanchez, E., De Vera, M., Rojas, M., Ruiz, J., & Triana-Alonso, F. J. (2008). Cultivation and molecular identification of Ehrlichia canis and Ehrlichia chaffeensis from a naturally co-infected dog in Venezuela. *Veterinary Clinical Pathology*, *37*(3), 258–265. doi:10.1111/j.1939-165X.2008.00046.x PMID:18761516

Gutiérrez, C., Montoya, J. A., Padron, M., Corbera, J. A., Juste, M. C., & Molina, J. M. (1998). Description of a case of trypanosomiasis due to Trypanosoma evansi in a dromedary in the Canary Islands. *Medicina Veterinária*, *15*(6), 356–357.

Habela, M., Van, H. A. I., Rol, J. A., Navarrete, I., & Jongejan, F. (1999). Transmisión experimental.de *Theileriaan-nulata. Acta Parasitologica PortuguesaActa.*, *1*(2), 8.

Habila, N., Inuwa, M. H., Aimola, I. A., Udeh, M. U., & Haruna, E. (2012). Pathogenic mechanisms of Trypanosoma evansi infections. *Research in Veterinary Science*, *93*(1), 13–17. doi:10.1016/j.rvsc.2011.08.011 PMID:21940025

Hagos, A., Abebe, G., Buscher, P., Goddeeris, B. M., & Claes, F. (2010). Serological and parasitological survey of dourine in the Arsi–Bale highlands of Ethiopia. *Tropical Animal Health and Production*, *42*(4), 769–776. doi:10.100711250-009-9485-6 PMID:19924557

Haig, D. A., & Lund, A. S. (1948). Transmission of the South African strain of dourine to laboratory animals. *Onderstepoort Journal of Veterinary Science*, *23*, 59–61. PMID:18863432

Handistatus, O. I. E. (2002). *World Animal Health Publication and Handistatus II (dataset for 2001)*. Office International des Epizooties.

Handistatus, O. I. E. (2003). *World Animal Health Publication and Handistatus II (dataset for 2002)*. Office International des Epizooties.

Handistatus, O. I. E. (2004). *World Animal Health Publication and Handistatus II (data set for 2003)*. Office International des Epizooties.

Handistatus, O. I. E. (2005). *World Animal Health Publication and Handistatus II (data set for 2004)*. Office International des Epizooties.

Hanotte, O. (2002). African Pastoralism: Genetic Imprints of Origins and Migrations. *Science*, *296*(5566), 336–339. doi:10.1126cience.1069878 PMID:11951043

Hanotte, O., Ronin, Y., Agaba, M., Nilsson, P., Gelhaus, A., Horstmann, R., Sugimoto, Y., Kemp, S., Gibson, J., Korol, A., Soller, M., & Teale, A. (2003). Mapping of quantitative trait loci controlling trypanotolerance in a cross of tolerant West African N'Dama and susceptible East African Boran cattle. *Proceedings of the National Academy of Sciences of the United States of America*, *100*(13), 7443–7448. doi:10.1073/pnas.1232392100 PMID:12805560

Hansen, T. H., & Bouvier, M. (2009). MHC class I antigen presentation : Learning from viral evasion strategies. *Nature Reviews. Immunology*, *9*(7), 503–513. https://doi.org/10.1038/nri2575

Happi, A. N., Toepp, A. J., Ugwu, C. A., Petersen, C. A., & Sykes, J. E. (2018). Detection and identification of blood-borne infections in dogs in Nigeria using light microscopy and the polymerase chain reaction. *Veterinary Parasitology. Regional Studies and Reports*, *11*, 55–60. doi:10.1016/j.vprsr.2017.12.002 PMID:31014619

Haresnape, J. M., Lungu, S. A. M., & Mamu, F. D. (1987). An updated survey of African swine fever in Malawi. *Epidemiology and Infection*, *99*, 723–732.

Hargrove, J. W., Omolo, S., Msalilwa, J. S. I., & Fox, B. (2000). Insecticide-treated cattle for tsetse control: The power and the problems. *Medical and Veterinary Entomology*, *14*(2), 123–130. doi:10.1046/j.1365-2915.2000.00226.x PMID:10872856

Harley, K. (1966). Studies on the survival of the non-parasitic stages of the cattle tick Boophilus microplus in three climatically dissimilar districts of North Queensland. *Australian Journal of Agricultural Research*, *17*(3), 387. https://doi.org/10.1071/AR9660387

Harrus, S., Alleman, A. R., Bark, H., Mahan, S. M., & Waner, T. (2002). Comparison of three enzyme-linked immunosorbant assays with the indirect immunofluorescent antibody test for the diagnosis of canine infection with Ehrlichia canis. *Veterinary Microbiology*, *86*(4), 361–368. doi:10.1016/S0378-1135(02)00022-6 PMID:11955786

Harrus, S., Kass, P. H., Klement, E., & Waner, T. (1997). Canine monocytic ehrlichiosis: A retrospective study of 100 cases, and an epidemiological investigation of prognostic indicators for the disease. *The Veterinary Record*, *141*(14), 360–363. doi:10.1136/vr.141.14.360 PMID:9351183

Harrus, S., Kenny, M., Miara, L., Aizenberg, I., Waner, T., & Shaw, S. (2004). Comparison of simultaneous splenic sample PCR with blood sample PCR for diagnosis and treatment of experimental Ehrlichia canis infection. *Antimicrobial Agents and Chemotherapy*, *48*(11), 4488–4490. doi:10.1128/AAC.48.11.4488-4490.2004 PMID:15504892

Harrus, S., & Waner, T. (2011). Diagnosis of canine monocytotropic ehrlichiosis (Ehrlichia canis): An overview. *Veterinary Journal (London, England)*, *187*(3), 292–296. doi:10.1016/j.tvjl.2010.02.001 PMID:20226700

Harrus, S., Waner, T., Aizenberg, I., & Bark, H. (1998b). Therapeutic effect of doxycycline in experimental subclinical canine monocytic ehrlichiosis: Evaluation of a 6-week course. *Journal of Clinical Microbiology*, *36*(7), 2140–2142. doi:10.1128/JCM.36.7.2140-2142.1998 PMID:9650986

Harrus, S., Waner, T., Aizenberg, I., Foley, J. E., Poland, A. M., & Bark, H. (1998a). Amplification of ehrlichial DNA from dogs 34 months after infection with Ehrlichia canis. *Journal of Clinical Microbiology*, *36*(1), 73–76. doi:10.1128/JCM.36.1.73-76.1998 PMID:9431923

Harrus, S., Waner, T., & Neer, T. M. (2012). *Ehrlichia canis infection. Infectious diseases of the dog and cat* (4th ed.). Elsevier Saunders.

Hassanain, M. A., El-Garhy, M. F., Abdul-Ghaffar, F. A., El-Sharaby, A., & Abdul Megeed, K. N. (1997). Biological control studies of soft and hard ticks. The effect of *Bacillus thuringiensis*varieties on soft and hard ticks (Ixodidae). *Parasitology Research*, *83*, 208–1.

Hay, S., Tucker, C., Rogers, D., & Packer, M. (1996). Remotely sensed surrogates of meteorological data for the study of the distribution and abundance of arthropod vectors of disease. *Annals of Tropical Medicine and Parasitology*, *90*(1), 1–19. doi:10.1080/00034983.1996.11813021 PMID:8729623

Hegarty, B. C., de Paiva Diniz, P. P., Bradley, J. M., Lorentzen, L., & Breitschwerdt, E. (2009). Clinical relevance of annual screening using a commercial enzyme-linked immunosorbent assay (SNAP 3Dx) for canine ehrlichiosis. *Journal of the American Animal Hospital Association*, *45*(3), 118–124. doi:10.5326/0450118 PMID:19411647

Hegarty, B. C., Levy, M. G., Gager, R. F., & Breitschwerdt, E. B. (1997). Immunoblot analysis of the immunoglobulin G response to *Ehrlichia canis* in dogs: An international survey. *Journal of Veterinary Diagnostic Investigation*, *9*(1), 32–38. doi:10.1177/104063879700900106 PMID:9087922

Hemmink, J. D., Sitt, T., Pelle, R., Klerk-Lorist, L. M., Shiels, B., Toye, P. G., Morrison, W. I., & Weir, W. (2018). Ancient diversity and geographical sub-structuring in African buffalo *Theileria parva* populations revealed through metagenetic analysis of antigen-encoding loci. *International Journal for Parasitology*, *48*(3-4), 287–296. doi:10.1016/j.ijpara.2017.10.006 PMID:29408266

Hendrickx, G., & Napala, A. (1997). *Le controle de la trypanosome a la carte: Une approche integree basee sur un Systeme d'Information Geographique*. Acadamie Royale des Sciences d'Oture Mer. Memoires Classe Sciences Naturelles and Medicales.

Henning, M. W. (1956). Heartwater. In *Animal Diseases in South Africa* (3rd ed., pp. 1155–1178). Central News Agency Ltd.

Hewetson, R. W. (1972). The inheritance of resistance by cattle to cattle ticks. *Australian Veterinary Journal*, *48*(5), 299–303. doi:10.1111/j.1751-0813.1972.tb05161.x PMID:5068812

Hill, P. D. S., & Campbell, J. A. (1973). The production of symbiont-free *Glossina morsitans* and an associated loss of female fertility. *Transactions of the Royal Society of Tropical Medicine and Hygiene*, *67*(5), 727–728. doi:10.1016/0035-9203(73)90051-5 PMID:4798390

Hoare, C. (1972). The Trypanosomes of Mammals, a Zoological Monograph. Blackwell Scientific Publications.

Hoare, C. A. (1965). Vampire Bats as Vectors and Hosts of Equine and Bovine Trypanosomes. *Acta Tropica*, *22*(3), 204–216. PMID:4379528

Holmes, P. (2015). On the road to elimination of rhodesiense human African trypanosomiasis: First WHO meeting of stakeholders. *PLoS Neglected Tropical Diseases*, *9*(4), e0003571. doi:10.1371/journal.pntd.0003571 PMID:25836650

Holzer, B., Bakshi, S., Bridgen, A., & Baron, M. D. (2011). Inhibition of interferon induction and action by the nairovirus Nairobi sheep disease virus/Ganjam virus. *PLoS One*, *6*(12), e28594. https://doi.org/10.1371/journal.pone.0028594

Homewood, K., Trench, P., Randall, S., Lynen, G., & Bishop, B. (2006). Livestock health and socio-economic impacts of a veterinary intervention in Maasai land: Infection-and-treatment vaccine against East Coast fever. *Agricultural Systems*, *89*(2-3), 248–271. doi:10.1016/j.agsy.2005.09.004

Hoof, L., Van Henrard, C. & Peel, E. (1948). Quelques observations sur les trypanosomes del grands mammifères au Congo Belge. *Acta Tropica, 5*, 327-344.

Horak, I. G., Camicas, J.-L., & Keirans, J. E. (2002). The Ixodidae,Amblyommidae and Nuttalliellidae (Acari: Ixodida): a world list of valid tick names. *Experimental & Applied Acarology*, *28*(1-4), 7–54. doi:10.1023/A:1025381712339

Hornok, S., Kontschán, J., Estrada-Peña, A., de Mera, I. G. F., Tomanović, S., & de la Fuente, J. (2015). Contributions to the morphology and phylogeny of the newly discovered bat tick species, *Ixodesariadnae* in comparison with *I. vespertilionis* and *I. simplex*. *Parasites & Vectors*, *8*(1), 47. doi:10.118613071-015-0665-0 PMID:25616683

Houre, C. A. (1976). Trypanosomes. *Veterinary Parasitology*, *42*, 9–29.

Hove, P., Chaisi, M. E., Brayton, K. A., Ganesan, H., Catanese, H. N., Mtshali, M. S., Mutshembele, A. M., Oosthuizen, M. C., & Collins, N. E. (2018). Co-infections with multiple genotypes of *Anaplasma marginale* in cattle indicate pathogen diversity. *Parasites & Vectors*, *11*(1), 5. doi:10.118613071-017-2595-5 PMID:29298712

Howell, J. M., Ueti, M. W., Palmer, G. H., Scoles, G. A., & Knowles, D. P. (2007). Transovarial Transmission Efficiency of Babesia bovis Tick Stages Acquired by Rhipicephalus (Boophilus) microplus during Acute Infection. *Journal of Clinical Microbiology*, *45*(2), 426–431. https://doi.org/10.1128/JCM.01757-06

Hunfeld, K.-P., Hildebrandt, A., & Gray, J. S. (2008). Babesiosis: Recent insights into an ancient disease. *International Journal for Parasitology*, *38*(11), 1219–1237. https://doi.org/10.1016/j.ijpara.2008.03.001

Hungerford, L. L., & Smith, R. D. (1997). Variations in seroprevalence and host factors for bovine anaplasmosis in Illinois. *Veterinary Research Communications*, *21*(1), 9–18. doi:10.1023/B:VERC.0000009696.68224.a8 PMID:9060138

Hurtado, O. J. B., & Giraldo-Ríos, C. (2018). Economic and Health Impact of the Ticks in Production Animals. In *Ticks and Tick-Borne Pathogens*. https://www.intechopen.com/books/ticks-and-tick-borne-pathogens/economic-and-health-impact-of-the-ticks-in-production-animals

Hutcheson, H. J., Dergousoff, S. J., & Lindsay, L. R. (2019). Haemaphysalis longicornis: A tick of considerable veterinary importance, now established in North America. *The Canadian Veterinary Journal =. The Canadian Veterinary Journal. La Revue Veterinaire Canadienne*, *60*(1), 27–28. https://www.ncbi.nlm.nih.gov/pubmed/30651646

Hutchings, C. L., Li, A., Fernandez, K. M., Fletcher, T., Jackson, L. A., Molloy, J. B., … Cooke, B. M. (2007). New insights into the altered adhesive and mechanical properties of red blood cells parasitized by Babesia bovis. *Molecular Microbiology, 65*(4), 1092–1105. doi:10.1111/j.1365-2958.2007.05850.x

Huxoll, D. L. (1976). Canine ehrlichiosis (tropical canine pancytopaenia): A review. *Veterinary Parasitology*, *2*(1), 49–60. doi:10.1016/0304-4017(76)90052-2

IDRC. (2018). Development and deployment of a novel vaccine against babesiosis. United States Department of Agriculture, Monash University, and Biosciences Eastern and Central Africa-International Livestock Research Institute (BecA-ILRI) Hub.

ILARD. (1990). *Problems of current control method.* Academic Press.

Ilemobade, A. A. (1976). *Study on heartwater and the causative agent Cowdria ruminantium (Cowdry, 1925) in Nigeria* (PhD thesis). Ahmadu Bello University.

ILRAD. (1989). *Annual report of the International Laboratory for Research on Animal Diseases*. ILRAD.

Inokuma, H., Oyamada, M., Davoust, B., Boni, M., Dereure, J., Bucheton, B., Hammad, A., Watanabe, M., Itamoto, K., Okuda, M., & Brouqui, P. (2006). Epidemiological survey of Ehrlichia canis and related species infection in dogs in eastern Sudan. *Annals of the New York Academy of Sciences*, *1078*(1), 461–463. doi:10.1196/annals.1374.085 PMID:17114754

Inokuma, H., Tamura, K., & Onishi, T. (1996). Seasonal occurrence of Rhipicephalus sanguineus in Okayama Prefecture, Japan and effect of temperature on development of the tick. *The Journal of Veterinary Medical Science*, *58*(3), 225–228. doi:10.1292/jvms.58.225 PMID:8777229

Iori, A., Lanfranchi, P., & Manilla, G. (1996). Contribution to the knowledge of Ixodidae ticks of wild mammals of Somalia. *Parassitologia*, *38*(3), 571–573. PMID:9257347

Irungu, P. (2000). Cattle Keeping Practices Of The Orma People A Household Survey In Tana River District. Academic Press.

Irvin, A. D. (1987). Characterisation of species and strains of *Theileria. Advances in Parasitology*, *26*, 145–197. doi:10.1016/S0065-308X(08)60296-1 PMID:3118658

Irvin, A. D., & Morrison, W. I. (1987). Immunopathology, immunology and immunoprophylaxis of Theileria infections. In E. J. L. Soulsby (Ed.), *Immune Responses in Parasite Infections: Immunopathology, Immunology and Immunoprophylaxis* (Vol. 3, pp. 223–274). CRC Press.

Isaac, C., Ohiolei, J., Ebhodaghe, F., Igbinosa, I., & Eze, A. (2017). Animal African trypanosomiasis in Nigeria: A long way from elimination/eradication. *Acta Tropica*, *176*, 323–331. doi:10.1016/j.actatropica.2017.08.032 PMID:28870536

Ismail, N., & McBride, J. W. (2017). Tick-borne emerging infections: Ehrlichiosis and anaplasmosis. *Clinics in Laboratory Medicine*, *37*(2), 317–340. doi:10.1016/j.cll.2017.01.006 PMID:28457353

Itty, P., Rowlands, G. J., Minengu, M., Ngamuna, S., Van Winkel, F., & d'Ieteren, G. D. M. (1995). The economics of recently introduced village cattle production in a tsetse affected area (I): Trypanotolerant N'Dama cattle in Zaire. *Agricultural Systems*, *47*(3), 347–366. doi:10.1016/0308-521X(95)90747-H

Jacob, F., Melachio, T., Njitchouang, G., Gimonneau, G., Njiokou, F., Abate, L., Christen, R., Reveillaud, J., & Geiger, A. (2017). Intestinal bacterial communities of trypanosome-infected and uninfected Glossina palpalis palpalis from three human African trypanosomiasis foci in Cameroon. *Frontiers in Microbiology*, *8*, 1464. doi:10.3389/fmicb.2017.01464 PMID:28824591

Jacobs, P. A. H., Fourie, L. J., Kok, D. J., & Horak, I. G. (2001). Diversity, seasonality and sites of attachment of adult ixodid ticks on dogs in the central region of the Free State Province. Academic Press.

James, M. A. (1989). Application of exoantigens of Babesia and Plasmodium in vaccine development. *Transactions of the Royal Society of Tropical Medicine and Hygiene, 83*, 67–72. (89)90607-X doi:10.1016/0035-9203

Jittapalapong, S., Stich, R. W., Gordon, J. C., Wittum, T. E., & Barriga, O. O. (2000). Performance of female Rhipicephalus sanguineus (Acari: Ixodidae) fed on dogs exposed to multiple infestations or immunization with tick salivary gland or midgut tissues. *Journal of Medical Entomology*, *37*(4), 601–611. doi:10.1603/0022-2585-37.4.601 PMID:10916303

Jones, T., & Akinbamijo, O. (1999). Further investigations on the role of a 33 KD antigen in the trypanotolerance of working N'dama bulls. 25th International Scientific Council for Trypanosomosis Research and Control, Mombasa, Kenya.

Jongejan, E. & Ulineberg, G. (1994). Ticks and control methods. *Revue Scientifiqueet Technique International Office of Epizootics, 13*(4), 1201-1226.

Jongejan, F., & Zandbergen, T. A. (1990). *The tick-borne rickettsia Cowdria ruminantium has a Chlamydia-like developmental cycle. In: Tick host interaction and disease transmission with special reference to Cowdria ruminantium (Rickettsiales)* (PhD thesis). Utrecht University.

Jongejan, F. (2007). Immunisation of sheep against heartwater in The Gambia using inactivated and attenuated Ehrlichia ruminantium vaccines. *Vaccine*, *25*(46), 7939–7947. doi:10.1016/j.vaccine.2007.09.002 PMID:17920167

Jongejan, F., & Uilenberg, G. (2004). The global importance of ticks. *Parasitology*, *129*(S1), S3–S14. doi:10.1017/S0031182004005967 PMID:15938502

Jongejan, F., Zivkovic, D., Pegram, R. G., Tatchell, R. J., Fison, T., Latif, A. A., & Paine, P. (1987). Ticks of the blue and White Nile ecosystems in the Sudan with particular reference to the *Rhipicephalus sanguineus* group. *Experimental & Applied Acarology*, *3*, 331–346. doi:10.1007/BF01193169 PMID:3331134

Jongejan, & Zandbergen, Wiel, Groot, & Uilenberg. (1991). The tick borne rickettsia Cowdria ruminantium has a Chlamydia-like developmental cycle. *The Onderstepoort Journal of Veterinary Research*, *58*(4), 227–237. PMID:1780122

Jonsson, N. N. (2006). The productivity effects of cattle tick (*Boophilus microplus*) infestation on cattle, with particular reference to *Bos indicus* cattle and their crosses. *Veterinary Parasitology*, *137*(1-2), 1–10. doi:10.1016/j.vetpar.2006.01.010 PMID:16472920

Jonsson, N. N., Bock, R. E., Jorgensen, W. K., Morton, J. M., & Stear, M. J. (2012). Is endemic stability of tick-borne disease in cattle a useful concept? *Trends in Parasitology*, *28*(3), 85–89. https://doi.org/10.1016/j.pt.2011.12.002

Jonsson, N. N., Mayer, D. G., Matschoss, A. L., Green, P. E., & Ansell, J. (1998). Production effects of cattle tick (*Boophilus microplus*) infestation of high yielding dairy cows. *Veterinary Parasitology*, *78*(1), 65–77. doi:10.1016/S0304-4017(98)00118-6 PMID:9703620

Jordan, A. M. (1972). The insemination potential of male Glossina austeni Newstead and Glossina morsitans morsitans Westw. (Dipt.: Glossinidae). *Bulletin of Entomological Research*, *62*(2), 319–325. doi:10.1017/S000748530004774X

Jori, F., & Bastos, A. D. S. (2009). Role of wild suids in the epidemiology of African swine fever. *EcoHealth*, *6*, 296–310. https://doi.org/10.1007/s10393-009-0248-7

Jori, F., Vial, L., Penrith, M. L., Pérez-Sánchez, R., Etter, E., Albina, E., & Roger, F. (2013). Review of the sylvatic cycle of African swine fever in sub-Saharan Africa and the Indian ocean. *Virus Research*, *173*, 212–227. https://doi.org/10.1016/j.virusres.2012.10.005

Joshi, M. V., Geevarghese, G., Joshi, G. D., Ghodke, Y. S., Mourya, D. T., & Mishra, A. C. (2005). Isolation of Ganjam Virus from Ticks Collected off Domestic Animals Around Pune, Maharashtra, India. *Journal of Medical Entomology*, *42*(2), 204–206. https://doi.org/10.1093/jmedent/42.2.204

Joshi, P. P., Shegokar, V. R., Powar, R. M., & Herder, S. (2005). Human trypanosomiasis caused by Trypanosoma evansi in India: The first case report. *The American Journal of Tropical Medicine and Hygiene*, *73*(3), 491–495. doi:10.4269/ajtmh.2005.73.491 PMID:16172469

Joyner, L. P., & Donnelly, J. (1979). The Epidemiology of Babesial Infections. In *Advances in Parasitology* (Vol. 17, pp. 115–140). (08)60550-3 doi:10.1016/S0065-308X

Kaewmongkol, G., Maneesaay, P., Suwanna, N., Tiraphut, B., Krajarngjang, T., Chouybumrung, A., Kaewmongkol, S., Sirinarumitr, T., Jittapalapong, S., & Fenwick, S. G. (2016). First detection of *Ehrlichia canis* in cerebrospinal fluid from a nonthrombocytopenic dog with meningoencephalitis by broad-range PCR. *Journal of Veterinary Internal Medicine*, *30*(1), 255–259. doi:10.1111/jvim.13788 PMID:26610407

Kaiser, M. N., Sutherst, R. W., Bourne, A. S., Gorissen, L., & Flyod, R. B. (1988). Population dynamics of ticks on Ankole cattle in five ecological zones in Burundi and strategies for their control. *Preventive Veterinary Medicine*, *6*(3), 199–122. doi:10.1016/0167-5877(88)90031-1

Kaitho, T., Ndeereh, D., & Ngoru, B. (2013). An outbreak of anthrax in endangered Rothschild's giraffes in Mwea National Reserve, Kenya. *Veterinary Medicine: Research and Reports*, *4*, 45–48.

Kalu, A. U., Oboegbulem, S., & Uzoukwu, M. (2001). Trypanosomosis in small ruminants maintained by low riverine tsetse population in central Nigeria. Small Ruminant Research. *The Journal of the International Goat Association*, *40*(10), 2015–2017.

Kamani, J., Apanaskevich, D. A., Gutiérrez, R., Nachum-Biala, Y., Baneth, G., & Harrus, S. (2017). Morphological and molecular identification of *Rhipicephalus* (*Boophilus*) *microplus* in Nigeria, West Africa: A threat to livestock health. *Experimental & Applied Acarology*, *73*(2), 283–296. doi:10.100710493-017-0177-z PMID:28887701

Kamani, J., Baneth, G., Mumcuoglu, K. Y., Waziri, N. E., Eyal, O., Guthmann, Y., & Harrus, S. (2013). Molecular detection and characterization of tick-borne pathogens in dogs and ticks from Nigeria. *PLoS Neglected Tropical Diseases*, *7*(3), e2108. doi:10.1371/journal.pntd.0002108 PMID:23505591

Kame-Ngasse, G., Njiokou, F., Melachio-Tanekou, T., Farikou, O., Simo, G., & Geiger, A. (2018). Prevalence of symbionts and trypanosome infections in tsetse flies of two villages of the "Faro and Déo" division of the Adamawa region of Cameroon. *BMC Microbiology*, *18*(S1), 159. doi:10.118612866-018-1286-5 PMID:30470177

Kanduma, E. G., Emery, D., Githaka, N. W., Nguu, E. K., Bishop, R. P., & Šlapeta, J. (2020). Molecular evidence confirms occurrence of *Rhipicephalus microplus* Clade A in Kenya and sub-Saharan Africa. *Parasites & Vectors*, *13*(1), 432. doi:10.118613071-020-04266-0 PMID:32854747

Kanté, T. S., Farikou, O., Njiokou, F., & Simo, G. (2018). Prevalence of Sodalis glossinidius and different trypanosome species in Glossina palpalis palpalis caught in the Fontem sleeping sickness focus of the southern Cameroon. *Parasite (Paris, France)*, *25*, 44. doi:10.1051/parasite/2018044 PMID:30117802

Karanja, S. M. (2006). *Epidemiology and importance of trypanosomosis, helminthosis and tick-borne diseases on the performance of cattle in Busia district* [PhD Thesis].

Karunakaran, S., Pillai, U., & Kuriakose, A. (2011). *Babesia bigemina infection in a 20 day old calf.* Academic Press.

Katende, J. M., Musoke, A. J., Nantulya, V. M., & Gooderis, B. M. (1987). A new method for fixation and preservation of trypanosomal antigens for use in the indirect immunofluorescence antibody test for diagnosis of bovine trypanosomosis. *Tropical Medicine and Parasitology*, *38*, 41–44. PMID:3299658

Kawai, T., & Akira, S. (2006). *Innate immune recognition of viral infection.* doi:10.1038/ni1303

Kawai, T., & Akira, S. (2011). Review Toll-like Receptors and Their Crosstalk with Other Innate Receptors in Infection and Immunity. *Immunity*, *34*(5), 637–650. https://doi.org/10.1016/j.immuni.2011.05.006

Kazimírová, M., Thangamani, S., Bartíková, P., Hermance, M., Holíková, V., Štibrániová, I., & Nuttall, P. A. (2017). Tick-Borne Viruses and Biological Processes at the Tick-Host-Virus Interface. *Frontiers in Cellular and Infection Microbiology, 7*. doi:10.3389/fcimb.2017.00339

Keating, M. I. (1983). Tick control by chemical ixodicides in Kenya: A review 1912-1981. *Tropical Animal Health and Production*, *15*(1), 1–6. doi:10.1007/BF02250752 PMID:6342241

Kelly, P. J. (2000). Canine ehrlichiosis: An update. *Journal of the South African Veterinary Association*, *71*(2), 77–86.

Kelly, P. J., Matthewman, L. A., Mahan, S. M., Semu, S., Peter, T., Mason, P. R., Brouqui, P., & Raoult, D. (1994). Serological evidence for antigenic relationships between Ehrlichia canis and Cowdria ruminantium. *Research in Veterinary Science*, *56*(2), 170–174. doi:10.1016/0034-5288(94)90100-7 PubMed

Kemp, D. H., Agbede, R. I. S., Johnson, L. A. Y., & Gough, J. M. (1986). Immunisation of cattle against Boophilus microplus using extracts derived from adult female ticks: Feeding and survival of the parasite on vaccinated cattle. *International Journal for Parasitology*, *16*(2), 115–120. doi:10.1016/0020-7519(86)90096-2 PMID:3721702

Kennedy, P. (2009). Cytokines in central nervous system trypanosomiasis: Cause, effect or both? *Transactions of the Royal Society of Tropical Medicine and Hygiene*, *103*(3), 213–214. doi:10.1016/j.trstmh.2008.08.013 PMID:18814893

KENTTEC. (2011). *Strategy for tsetse and trypanosomiasis eradication in Kenya 2011-2021*. Ministry of Livestock Development Dovs2011. 449 8.

KENTTEC. (2019). *Draft KENTTEC Strategic Plan*. Retrieved April 27th, 2020 from https://www.kenttec.go.ke/wp-content/uploads/2019/08/Draft-KENTTEC-Strategic-Plan.pdf

KENTTEC. (2019). https://www.kenttec.go.ke/wp-content/uploads/2019/08/Parasitological-and-Entomologuical-monitoring-report_Pate_Nov_Dec-2016.pdf

KENTTEC. (2020). Tsetse and Trypanosomiasis Problem in Kenya. *KENTTEC*. https://www.kenttec.go.ke/tsetse-and-trypanosomiasis-problem-in-kenya/

Kenya Law. (2012). *State Corporations Act (Kenya Tsetse and Trypanosomiasis Eradication Council) Order, 2012*. Retrieved April 27th, 2020 from, www. http://kenyalaw.org/kl/index.php?id=784

Kenya Tsetse and Trypanosomiasis Eradication Council (KENTTEC). (2020). *Achievements*. https://www.kenttec.go.ke/achievements/

KETRI. (1997). *Tsetse distribution map of Kenya* [Research Report]. Kenya Trypanosomiasis Research Institute.

Kettle, D. S. (1990). *Medical and Veterinary Entomology*. C.A.B. International.

KidambasiK.MasigaD.VillingerJ.CarringtonM.BargulJ. (2019). Detection of blood pathogens in camels and their associated ectoparasitic camel biting keds: the potential application of keds in xenodiagnosis of camel haemopathogens *Hippobosca camelina* AAS Open Research *164*. doi:10.12688/aasopenres.13021.1

Kim, J., Hanotte, O., Mwai, O. A., Dessie, T., Bashir, S., Diallo, B., Agaba, M., Kim, K., Kwak, W., Sung, S., Seo, M., Jeong, H., Kwon, T., Taye, M., Song, K.-D., Lim, D., Cho, S., Lee, H.-J., Yoon, D., ... Kim, H. (2017). The genome landscape of indigenous African cattle. *Genome Biology*, *18*(1), 34. doi:10.118613059-017-1153-y PMID:28219390

King, K., Chapman, D., Argilaguet, J. M., Fishbourne, E., Hutet, E., Cariolet, R., Hutchings, G., Oura, C. A., Netherton, C. L., Moffat, K., Taylor, G., Le Potier, M. F., Dixon, L. K., & Takamatsu, H. H. (2011). Protection of European domestic pigs from virulent African isolates of African swine fever virus by experimental immunization. *Vaccine*, *29*, 4593–4600.

Kirade, P., Shukla, S., Shrivastava, N., Audarya, S. D., & Jatav, G. P. (2018). Rickettsial and haemoparasitic infections in dogs of malwa region, Madhya Pradesh-An etiopathological study. *Intas Polivet*, *19*(1), 3–6.

Kitaa, J. M. A., Mulei, C. M., Mande, J. D., & Wabacha, J. (2014). A retrospective study of canine ehrlichiosis in Kenya. *International Journal of Veterinary Science*, *3*(3), 122–124.

Kitron, U., Otieno, L., Hungerford, L., Odulaja, A., Brigham, W. U., Okello, O. O., Joselyn, M., Mohamed-Ahmed, M. M., & Cook, E. (1996). Spatial analysis of the distribution of tsetse flies in the Lambwe valley, Kenya, using Landsat TM satellite imagery and GIS. *Journal of Animal Ecology*, *65*(3), 371–380. doi:10.2307/5883

Kivaria, F.M. (2006). Estimated direct economic costs associated with tick-borne diseases on cattle in Tanzania. *Tropical Animal Health and Production, 38*, 291-299.

Kivaria, F. M., Kapanga, A. M., Mbassa, G. K., Mtui, P. F., & Wani, R. J. (2012). Epidemiological perspectives of ticks and tick-borne diseases in South Sudan: Cross-sectional survey results. *The Onderstepoort Journal of Veterinary Research*, *79*(1), E1–E10. doi:10.4102/ojvr.v79i1.400 PMID:23327317

Knowles, G., Betschart, B., Kukla, B. A., Scott, J. R., & Majiwa, P. A. O. (1988). Genetically discrete populations of Trypanosoma congolense from livestock on the Kenyan coast. *Parasitology*, *96*(3), 461–474. doi:10.1017/S0031182000080100 PMID:3405633

Kocan, K. M., Halbur, T., Blouin, E. F., Onet, V., Fuente, J. L., Garcia-Garcia, J. C., & Saliki, J. T. (2001). Immunization of cattle with *Anaplasma marginale* derived from tick cell culture. *Veterinary Parasitoloby, 102*, 151-161.

Kocan, K. M., & Bezuidenhout, J. D. (1987). Morphology and development of Cowdria ruminantium in Amblyo mma ticks. *The Onderstepoort Journal of Veterinary Research*, *54*(3), 177–182. PMID:3448559

Kocan, K. M., Blouin, E. F., & Barbet, A. F. (2000). Anaplasmosis control. Past, present, and future. *Annals of the New York Academy of Sciences*, *916*(1), 501–509. doi:10.1111/j.1749-6632.2000.tb05329.x PMID:11193665

Kocan, K. M., de la Fuente, J., Blouin, E. F., Coetzee, J. F., & Ewing, S. A. (2010). The natural history of *Anaplasma marginale. Veterinary Parasitology*, *167*(2-4), 95–107. doi:10.1016/j.vetpar.2009.09.012 PMID:19811876

Kocan, K. M., de la Fuente, J., Blouin, E. F., & Garcia-Garcia, J. C. (2004). *Anaplasmamarginale* (Rickettsiales: Anaplasmataceae):recent advances in defining host-pathogen adaptations of a tick-borne rickettsia. *Parasitology*, *129*(S1), S285–S300. doi:10.1017/S0031182003004700 PMID:15938516

Kocan, K. M., de la Fuente, J., Guglielmone, A. A., & Meléndez, R. D. (2003). Antigens and Alternatives for Control of *Anaplasma marginale* Infection in Cattle. *Clinical Microbiology Reviews*, *16*(4), 698–712. doi:10.1128/CMR.16.4.698-712.2003 PMID:14557295

Koch, R. (1903). *Interim report on Rhodesia Redwater or 'African Coast fever'.* Argus Printing and Publishing Co. Ltd.

Koch, H. G. (1982). Seasonal incidence and attachment sites of ticks (Acari: Ixodidae) on domestic dogs in southeastern Oklahoma and northwestern Arkansas, USA. *Journal of Medical Entomology*, *19*(3), 293–298. doi:10.1093/jmedent/19.3.293 PMID:7120309

Koch, H. G., & Popham, T. W. (1987). Evaluation of acaricide-impregnated plastic collars for control of lone star ticks, Amblyomma americanum (L.)(Acari: Ixodidae) on domestic dogs in Oklahoma, U.S.A. *Preventive Veterinary Medicine*, *5*(1), 63–67. doi:10.1016/0167-5877(87)90007-9

Koch, H. G., & Tuck, M. D. (1986). Molting and survival of the brown dog tick (Acari: Ixodidae) under different temperatures and humidities. *Annals of the Entomological Society of America*, *79*(1), 11–14. doi:10.1093/aesa/79.1.11

Kordick, S. K., Breitschwerdt, E. B., Hegarty, B. C., Southwick, K. L., Colitz, C. M., Hancock, S. I., Bradley, J. M., Rumbough, R., Mcpherson, J. T., & MacCormack, J. N. (1999). Co-infection with multiple tick-borne pathogens in a walker hound kennel in North Carolina. *Journal of Clinical Microbiology*, *37*(8), 2631–2638. doi:10.1128/JCM.37.8.2631-2638.1999 PMID:10405413

Kosgey, I. S., & Okeyo, A. M. (2007). Genetic improvement of small ruminants in low-input, smallholder production systems: Technical and infrastructural issues. *Small Ruminant Research*, *70*(1), 76–88. doi:10.1016/j.smallrumres.2007.01.007

Koudstaal, D., Kemp, D. H., & Kerr, J. D. (1978). Boophilus microplus: Rejection of larvae from British breed cattle. *Parasitology*, *76*(3), 379–386. https://doi.org/10.1017/s0031182000048241

Kristjanson, P. B. M., Rowlands, G., Kruska, R., & deLeeuw, P. (1999). Measuring the costs of African animal trypanosomosis: The potential benefits of control and returns to research. *Agricultural Systems*, *59*(1), 79–98. doi:10.1016/S0308-521X(98)00086-9

Kruss, H. A., Weber, A., Appel, M., Enders, B., Isenberg, H. D., Schiefer, H. G., Slenczka, W., von Graevenitz, A., & Zahner, H. (2003). *Zoonoses: infectious diseases transmissible from animals to humans*. ASM Press. doi:10.1128/9781555817787

Kuhn, J. H., Wiley, M. R., Rodriguez, S. E., Bào, Y., Prieto, K., & Travassos da Rosa, A. P. A. … Palacios, G. (2016). Genomic Characterization of the Genus Nairovirus (Family Bunyaviridae). *Viruses, 8*(6). doi:10.3390/v8060164

Kulohoma, B., Wamwenje, S., Wangwe, I., Masila, N., Mirieri, C., & Wambua, L. (2020). Drug resistance in Shimba Hills, Kwale County, Kenya. *BMC Research Notes*, *13*(1), 234. doi:10.118613104-020-05077-3 PMID:32349785

Kumar, S., Kumar, R., & Sugimoto, C. (2009). A perspective on Theileria equi infections in donkeys. *The Japanese Journal of Veterinary Research*, *56*(4), 171–180.

Kuttler, K. L. (1980). Pharmacotherapeutics of drugs used in treatment of anaplasmosis and babesiosis. *Journal of the American Veterinary Medical Association, 176*(10), 1103–1108.

Kuttler, K. L. (2018). World-wide impact of babesiosis. In *Babesiosis of domestic animals and man* (pp. 1–22). CRC Press.

Kuzoe, F. A. S., & Schofield, C. J. (2004). Strategic review of traps and targets for tsetse and African trypanosomiasis control. WHO/ TDR.

Kuzoe, F. A. S., & Schofield, C. J. (2005). *Strategic review of traps and targets for tsetse and African trypanosomiasis control*. Academic Press.

Kuzoe, F., & Schofield, C. (2004). *Strategic review of traps and targets for tsetse and African trypanosomiasis control* [Research report]. WHO.

Laha, R., Das, M., Goswami, A., & Singh, P. (2012). *Losses of milk production due to Babesia bigemina infection in a cross bred cow: A case study*. Academic Press.

Laha, R., Das, M., & Sen, A. (2015). Morphology, epidemiology, and phylogeny of Babesia: An overview. *Tropical Parasitology*, *5*(2), 94. https://doi.org/10.4103/2229-5070.162490

Lanza-Perea, M., Zieger, U., Qurollo, B. A., Hegarty, B. C., Pultorak, E. L., Kumthekar, S., Bruhl-Day, R., & Breitschwerdt, E. B. (2014). Intraoperative bleeding in dogs from Grenada seroreactive to Anaplasma platys and Ehrlichia canis. *Journal of Veterinary Internal Medicine*, *28*(6), 1702–1707. doi:10.1111/jvim.12442 PMID:25274547

Lasecka, L., & Baron, M. D. (2014a). The molecular biology of nairoviruses, an emerging group of tick-borne arboviruses. *Archives of Virology*, *159*(6), 1249–1265. https://doi.org/10.1007/s00705-013-1940-z

Lasecka, L., & Baron, M. D. (2014b). The Nairovirus Nairobi Sheep Disease Virus/Ganjam Virus Induces the Translocation of Protein Disulphide Isomerase-Like Oxidoreductases from the Endoplasmic Reticulum to the Cell Surface and the Extracellular Space. *PLoS One*, *9*(4), e94656. https://doi.org/10.1371/journal.pone.0094656

Lasecka, L., Bin-Tarif, A., Bridgen, A., Juleff, N., Waters, R. A., & Baron, M. D. (2015). Antibodies to the core proteins of Nairobi sheep disease virus/Ganjam virus reveal details of the distribution of the proteins in infected cells and tissues. *PLoS One*, *10*(4), e0124966. https://doi.org/10.1371/journal.pone.0124966

Latif, A. A., & Walker, A. R. (2004). *An introduction to the biology and control of ticks in Africa*. www.alanrwalker.com/assets/PDF/tick-biology-africa.pdf

Latif, A. A., Rowlands, G. J., Punyua, D. K., Hassan, S. M., & Capstick, P. B. (1994). An epidemiological study of tick-borne diseases and their effects on productivity of zebu cattle under traditional management on Rusinga Island, western Kenya. *Preventive Veterinary Medicine*, *22*(3), 169–181. doi:10.1016/0167-5877(94)00408-B

Lauzi, S., Maia, J. P., Epis, S., Marcos, R., Pereira, C., Luzzago, C., Santos, M., Puente-Payo, P., Giordano, A., Pajoro, M., Sironi, G., & Faustino, A. (2016). Molecular detection of Anaplasma platys, Ehrlichia canis, Hepatozoon canis and Rickettsia monacensis in dogs from Maio Island of Cape Verde archipelago. *Ticks and Tick-Borne Diseases*, *7*(5), 964–969. doi:10.1016/j.ttbdis.2016.05.001 PMID:27177475

Lawrence, J. A., Perry, B. D., & Williamson, S. M. (1992). Infectious Diseases of Livestock. Oxford University Press.

Lawrence, J. A., Norval, R. A. I., & Uilenberg, G. (1983). *R. zambeziensis* as a vector of bovine *Theileria. Tropical Animal Health and Production*, *15*(1), 39–42. doi:10.1007/BF02250760 PMID:6405516

Lawrence, J. A., Perry, B. D., & Williamson, S. M. (2004a). Corridor disease. In J. A. W. Coetzer & R. C. Tustin (Eds.), *Infectious Diseases of Livestock* (2nd ed., pp. 468–471). Oxford University Press Southern Africa.

Lawrence, J. A., Perry, B. D., & Williamson, S. M. (2004b). East Coast fever. In J. A. W. Coetzer & R. C. Tustin (Eds.), *Infectious Diseases of Livestock* (2nd ed., pp. 448–467). Oxford University Press Southern Africa.

Lawrence, J. A., Perry, B. D., & Williamson, S. M. (2004c). Zimbabwe theileriosis. In J. A. W. Coetzer & R. C. Tustin (Eds.), *Infectious Diseases of Livestock* (2nd ed., pp. 472–474). Oxford University Press Southern Africa.

Leak, S. (1998). *Tsetse Biology and Ecology. Their Role in the Epidemiology and Control of Trypanosomosis.* CABI Publishing in association with ILRI.

Leak, S. (1999). *Tsetse biology and ecology: their role in the epidemiology and control of trypanosomosis*. ILRI.

Leak, S. G. A., Ejigu, D., & Vreysen, M. J. B. (2008). *Collection of entomological baseline data for tsetse area-wide integrated pest management programmes*. Joint FAO/IAEA Programme of Nuclear Techniques in Food and Agriculture.

Leak, S. G. A., Mulatu, W., Authié, E., d'Ieteren, G. D. M., Peregrine, A. S., Rowlands, G. J., & Trail, J. C. M. (1993). Epidemiology of bovine trypanosomiasis in the Ghibe valley, southwest Ethiopia 1. Tsetse challenge and its relationship to trypanosome prevalence in cattle. *Acta Tropica*, *53*(2), 121–134. doi:10.1016/0001-706X(93)90024-6 PMID:8098898

Leperre, P., & Claxton, J. R. (1994). Comparative study of trypanosomosis in zebu and N'Dama cattle in the Gambia. *Tropical Animal Health and Production*, *26*(3), 139–145. doi:10.1007/BF02241070 PMID:7809985

Lessard, P., Le Plattenier, R., Norval, R. A. I., Kundert, K., Dolan, T. T., Croze, H., Walker, J. B., Irvin, A. D., & Perry, B. D. (1990). Geographical information systems for studying the epidemiology of cattle diseases caused by *Theileria parva. The Veterinary Record*, *126*, 255–262. PMID:2327044

Leta, S., Habtamu, Y., Alemayehu, G., Ayele, B., Chanie, M., Tesfaye, S., & Masele, F. (2015). *Spatial analysis of the distribution of tsetse flies in Ethiopia using high resolution environmental datasets and Maxent modeling technique. National Symposium on Trends and Challenges in Adoption of Science, Technology and Innovation in Local Development Endeavours*, Ethiopia.

Levine, N. D., Corliss, J. O., Cox, F. E. G., Deroux, G., Grain, J., Honigberg, B. M., Leedale, G. F., Loeblich, A. R., Lom, I. J., Lynn, D., Merinfeld, E. G., Page, F. C., Poljansky, G., Sprague, V., Vavra, J., & Wallace, F. G. (1980). A Newly Revised Classification of the Protozoa. *The Journal of Protozoology*, *27*(1), 37–58. doi:10.1111/j.1550-7408.1980.tb04228.x PMID:6989987

Lewis, E. (1950). Tsetse Flies carried by Railway Trains in Kenya Colony. *Bulletin of Entomological Research*, *40*(4), 511–531. doi:10.1017/S0007485300024421

Lewis, E. A. (1946). Nairobi sheep disease: The survival of the virus in the tick Rhipicephalus appendiculatus. *Parasitology*, *37*(1–2), 55–59. https://doi.org/10.1017/S0031182000013159

Lichoti, J. K., Davies, J., Kitala, P. M., Githigia, S. M., Okoth, E., Maru, Y., & Bishop, R. P. (2016). Social network analysis provides insights into African swine fever epidemiology. *Preventive Veterinary Medicine*, *126*, 146210.

Lindh, J. M., Torr, S. J., Vale, G. A., & Lehane, M. J. (2009). Improving the cost-effectiveness of artificial visual baits for controlling the tsetse fly Glossina fuscipes fuscipes. *PLoS Neglected Tropical Diseases*, *3*(7), e474.

Lindh, J., & Lehane, M. (2011). The tsetse fly Glossina fuscipes fuscipes (Diptera: Glossina) harbours a surprising diversity of bacteria other than symbionts. *Antonie van Leeuwenhoek*, *711*(3), 20–29. doi:10.100710482-010-9546-x PMID:21203841

Lisbôa, R. S., Teixeira, R. C., Rangel, C. P., Santos, H. A., Massard, C. L., & Fonseca, A. H. (2017). *Avian spirochetosis in chickens following experimental transmission of Borrelia anserina by Argas (Persicaargas) Miniatus.* National Library.

Loo, Y., & Gale, M. (2011). Review Immune Signaling by RIG-I-like Receptors. *Immunity*, *34*(5), 680–692. https://doi.org/10.1016/j.immuni.2011.05.003

Louly, C. C. B., Fonseca, I. N., Oliveira, V. F. D., Linhares, G. F. C., Menezes, L. B. D., & Borges, L. M. F. (2007). Seasonal dynamics of Rhipicephalus sanguineus (Acari: Ixodidae) in dogs from a police unit in Goiânia, Goias, Brazil. *Ciência Rural*, *37*(2), 464–469. doi:10.1590/S0103-84782007000200026

Louly, C. C. B., Fonseca, I. N., Oliveira, V. F., & Borges, L. M. F. (2006). Occurrence of Rhipicephalus sanguineus in workers of veterinary clinics and kennels, in the city of Goiânia, GO. *Brazilian Animal Science*, *7*(1), 103–106.

Lubembe, D. M., Odongo, D. O., Salih, D. A., & Sibeko-Matjila, K. P. (2020). Microsatellite and minisatellite genotyping of *Theileria parva* population from southern Africa reveals possible discriminatory allele profiles with parasites from eastern Africa. *Ticks and Tick-Borne Diseases*, *11*(6), 101539. doi:10.1016/j.ttbdis.2020.101539 PMID:32993948

Luckins, A. G., & Dwinger, R. H. (2004). Non-tsetse-transmitted animal trypanosomiases. In The Trypanosomiases. Epidemiology and Diagnosis. Cabi Publishing.

Luckins, A. G. (1972). Studies on bovine trypanosomiasis. Serum immunoglobulin levels iin zebu callte exposed to natural infection in East Africa. *The British Veterinary Journal*, *128*(10), 523–528. doi:10.1016/S0007-1935(17)36737-4 PMID:4630575

Luckins, A. G. (1977). Detection of antibodies in trypanosome-infected cattle by means of a microplate enzyme-linked immunosorbent assay. *Tropical Animal Health and Production*, *9*(1), 53–62. doi:10.1007/BF02297393 PMID:333677

Luckins, A. G. (1988). Trypanosoma evansi in Asia. *Parasitology Today (Personal Ed.)*, *4*(5), 137–142. doi:10.1016/0169-4758(88)90188-3 PMID:15463067

Luckins, A. G. (1992). Methods for diagnosis of trypanosomosis in livestock. *World Animal Review*, *70*(71), 15–20.

Luckins, A. G. (1994). Equine trypanosomiasis. Exotic disease series. *Equine Veterinary Education*, *6*(5), 259–262. doi:10.1111/j.2042-3292.1994.tb01149.x

Luckins, A. G. (1998). Epidemiology of Surra: Unanswered Questions. *Journal. Protozoological Research*, *8*, 106–119.

Luther, N. J., Majiyagbe, K. A., Shamaki, D., Lombin, L. H., Antiagbong, J. A., Bitrus, Y., & Owolodun, O. (2007). Detection of African swine fever virus genomic DNA in a Nigerian Red River hog (*Potamochoerusporcus*). *The Veterinary Record*, *160*, 58–59.

Lynen, G., Zeman, P., Bakuname, C., Di Giulio, G., Mtui, P., Sanka, P., & Jongejan, F. (2008). Shifts in the distribution ranges of *Boophilus* ticks in Tanzania: Evidence that a parapatric boundary between *Boophilus microplus* and *B. decoloratus* follows climate gradients. *Experimental & Applied Acarology*, *44*(2), 147–164. doi:10.100710493-008-9134-1 PMID:18266058

M'ghirbi, Y., Ghorbel, A., Amouri, M., Nebaoui, A., Haddad, S., & Bouattour, A. (2009). Clinical, serological, and molecular evidence of ehrlichiosis and anaplasmosis in dogs in Tunisia. *Parasitology Research Journal*, *104*(4), 767–774. doi:10.100700436-008-1253-4 PMID:19018569

Machila, N., Emongor, R., Shaw, A., Welburn, S. C., McDermott, J., Maudlin, I., & Eisler, M. C. (2007). A community education intervention to improve bovine trypanosomiasis knowledge and appropriate use of trypanocidal drugs on smallholder farms in Kenya. *Agricultural Systems*, *94*(2), 261–272. doi:10.1016/j.agsy.2006.09.004

Mackenstedt, U., Gauer, M., Fuchs, P., Zapf, F., Schein, E., & Mehlhorn, H. (1995). DNA measurements reveal differences in the life cycles of Babesia bigemina and B. canis, two typical members of the genus Babesia. *Parasitology Research*, *81*(7), 595–604. https://doi.org/10.1007/bf00932027

Mackenstedt, U., Gauer, M., Mehlhorn, H., Schein, E., & Hauschild, S. (1990). Sexual cycle of Babesia divergens confirmed by DNA measurements. *Parasitology Research*, *76*(3), 199–206. https://doi.org/10.1007/bf00930815

Madder, M., Thys, E., Achi, L., Touré, A., & De Deken, R. (2011). Rhipicephalus (Boophilus) microplus: A most successful invasive tick species in West-Africa. *Experimental & Applied Acarology*, *53*(2), 139–145. https://doi.org/10.1007/s10493-010-9390-8

Maggi, R. G., & Krämer, F. (2019). A review on the occurrence of companion vector-borne diseases in pet animals in Latin America. *Parasites & Vectors*, *12*(1), 145. doi:10.118613071-019-3407-x PMID:30917860

Magona, J. W., Walubengo, J., & Odimin, J. T. (2008). Acute haemorrhagic syndrome of bovine trypanosomosis in Uganda. *Acta Tropica*, *107*(2), 186–191. doi:10.1016/j.actatropica.2008.05.019 PMID:18599006

Mahama, C., Koné, A., de la Rocque, S., De Deken, R., & Geert, S. (2005). Prediction of the distribution of Glossina tachinoides (Diptera: Glossinidae) in the Volta basin of northern Ghana. *Bulletin of Entomological Research*, *95*(1), 63–67. doi:10.1079/BER2004340 PMID:15705216

Mahan, Andrew, Tebele, Burridge, & Barbet. (1995). Immunisation of sheep against heartwater with inactivated Cowdria ruminantium. *Research in Veterinary Science, 58*(1), 46-49.

Mahan, S. M., Peter, T. F., Semu, S. M., Simbi, B. H., Norval, R. A. I., & Barbet, A. F. (1995). Laboratory reared *Amblyomma hebraeum* and *Amblyomma variegatum* ticks differ in their susceptibility to infection with *Cowdria ruminantium*. *Epidemiology and Infection, 115*(2), 345–353. doi:10.1017/S0950268800058465 PMID:7589273

Mahmoud, M. M., & Gray, A. R. (1980). Trypanosomosis due to Trypanosoma evansi (Steel, 1885) Balbiani, 1888. A review of recent research. *Tropical Animal Health and Production, 12*(1), 35–47. doi:10.1007/BF02242629 PMID:7370098

Mahoney, D. F. (1974). *The application of epizootiological principals in the control of babesiosis in cattle*. Academic Press.

Mahoney, D. F., & Ross, D. (n.d.). *Epizootiological factors in the control of bovine babesiosis*. Academic Press.

Mahoney, D. F., & Ross, D. R. (1972). Epizootiological factors in the control of bovine babesiosis. *Australian Veterinary Journal, 48*(5), 292–298. https://doi.org/10.1111/j.1751-0813.1972.tb05160.x

Maichomo, M. (2008). *Comparative productivity and performance of Orma/Zebu and Sahiwal/Zebu cross breed cattle under trypanosomosis, tick borne diseases, and anaplasmosis and helminthosis challange in Kajiado district, Kenya* [PhD Thesis]. University of Nairobi.

Maichomo, M., Kosura, W., Gathuma, J., Gitau, G., Ndungú, J., & Nyamwaro, S. (2009). Economic assessment of the performance of trypanotolerant Orma/Zebu cross cattle in a pastoral production system in Kenya. *Journal of South African Veterinary Association, 80*(3), 157–162.

Maichomo, M., Olubai, W., Mwendia, C. M. T., & Mapenay, I. M. (2001). Farmer responses on acceptability of the Orma Boran, a trypanotolerant breed of cattle introduced into Nguruman, kajiado District in Kenya. *OAU/STRC Publication, 120*, 293–295.

Maichomo, M.W., Ndung'u, J.M., Ngare, P.M., & Ole-Mapenay, I.M. (2005). The performance of OrmaBoran and Maasai Zebu crossbreeds in a trypanosomosis endemic area of Nguruman, southwestern Kenya. *Onderstepoort Journal of Veterinary Research, 72*, 87–93.

Maichomo, M., Gitau, G., Gathuma, J., Ndungú, J., Kosura, W., & Hanotte, O. (2008). Extent and implications of incorrect offspring-sire relationships in pastoral production system in Kajiado District, Kenya. *Livestock Research for Rural Development, 20*, •••. http://www.lrrd.org/lrrd20/5/maic20080.htm

Maichomo, M., Mwendia, C. M. T., & Ole-Mapenay, I. M. (1999). Preliminary performance of the Orma bull, a trypanotolerant breed at Nguruman. *The Kenya Veterinarian, 24*, 129–131.

Maichomo, M., Ndung'u, J. M., Ngare, P., & Ole-Mapenay, I. M. (2005). The performance of Orma Boran and Maasai Zebu crossbreeds in a trypanosomosis endemic area of Nguruman, South-western Kenya. *The Onderstepoort Journal of Veterinary Research, 72*(1), 87–93. doi:10.4102/ojvr.v72i1.225 PMID:15991707

Maichomo, M., Nginyi, J. M., Caruso-Varès, A., Mungube, E. O., Pagny, G., Chevtzoff, C., Njanja, J. C., Lumumba, P. A., Collin, J. F., & Isaka, N. (2019). Efficacy of VERYL® in the treatment of cattle naturally infected with gastrointestinal nematodes in Kenya. *Tropical Animal Health and Production, 52*(4), 1699–1705. doi:10.100711250-019-02179-4 PMID:31865537

Majekodunmi, A., Fajinmi, A., Dongkum, C., Picozzi, K., Thrusfield, M., & Welburn, S. (2013). A longitudinal survey of African animal trypanosomiasis in domestic cattle on the Jos Plateau, Nigeria: Prevalence, distribution and risk factors. *Parasites & Vectors, 6*(1), 239. doi:10.1186/1756-3305-6-239 PMID:23958205

Malele, I., Nyingilili, H., Lyaruu, E., Tauzin, M., Ollivier, B., Cayol, J., & al., e. (2018). Bacterial diversity obtained by culturable approaches in the gut of Glossina pallidipes population from a non-sleeping sickness focus in Tanzania: preliminary results. *BMC Microbiology, 18*, 164.

Malele, I. H., & Msangi, A. (2011). Factors defining the distribution limit of tsetse infestation and the implication for livestock sector in Tanzania. *African Journal of Agricultural Research*, *6*(10), 2341–2347.

Malele, I. I. (2011). Fifty years of tsetse control in Tanzania: Challenges and prospects for the future. *Tanzania Journal of Health Research*, *13*(5).

Malele, I. I., Magwisha, H. B., Nyingilili, H. S., Mamiro, K. A., Rukambile, E. J., Daffa, J. W., Lyaruu, E. A., Kapange, L. A., Kasilagila, G. K., Lwitiko, N. K., Msami, H. M., & Kimbita, E. N. (2011). Multiple Trypanosoma infections are common amongst Glossina species in the new farming areas of Rufiji district, Tanzania. *Parasites & Vectors*, *4*(1), 217–217. doi:10.1186/1756-3305-4-217 PMID:22093363

Malmquist, W. A., Nyindo, M. B. A., & Brown, C. G. D. (1970). East Coast fever: Cultivation in vitro of bovine spleen cell lines infected and transformed by *Theileria parva*. *Tropical Animal Health and Production*, *2*(3), 139–150. doi:10.1007/BF02359680

Mamabolo, M. V., Ntantiso, L. A. L., & Majiwa, P. A. O. (2009). Natural infection of cattle and tsetse flies in South Africa with two genotypic groups of Trypanosoma congolense. *Parasitology*, *136*(4), 425–431. doi:10.1017/S0031182009005587 PMID:19250578

Mangwiro, T. N. C., Torr, S. J., Cox, J. R., & Holloway, M. T. P. (1999). The efficacy of various pyrethroid insecticides for use on odour-baited targets to control tsetse. *Medical and Veterinary Entomology*, *13*, 315–323.

Manyarara, R., Tubbesing, U., Soni, M., & Noden, B. H. (2015). Serodetection of Ehrlichia canis amongst dogs in central Namibia. *Journal of the South African Veterinary Association*, *86*(1), 272. doi:10.4102/jsava.v86i1.1272 PMID:26244587

Manzano-Román, Díaz-Martín, de la Fuente, & Pérez-Sánchez. (2012). *Soft ticks as pathogen vectors: distribution, surveillance and control, parasitology*. In Tech. Available from: https://www.intechopen.com/books/parasitology/softticks-as-pathogen-vectors-distribution-surveillance-and-control

Marcelino, I., Sousa, M. F. Q., Veríssimo, C., Cunha, A. E., Carrondo, M. J. T., & Alves, P. M. (2006). Process development for the mass production of Ehrlichia ruminantium. *Vaccine*, *24*(10), 1716–1725. doi:10.1016/j.vaccine.2005.08.109 PMID:16257480

Marcelino, I., Vachiéry, N., Amaral, A. I., Roldão, A., Lefrançois, T., Carrondo, M. J. T., Alves, P. M., & Martinez, D. (2007). Effect of the purification process and the storage conditions on the efficacy of an inactivated vaccine against heartwater. *Vaccine*, *25*(26), 4903–4913. doi:10.1016/j.vaccine.2007.04.055 PMID:17531356

Marcellino, W. L. (2004). Current situation of East Coast fever in Bahr el Jebel State. *East Coast Fever in southern Sudan. Proceedings of a workshop*, 8-9.

Marczinke, B. I., & Nichol, S. T. (2002). Nairobi Sheep Disease Virus, an Important Tick-Borne Pathogen of Sheep and Goats in Africa, Is Also Present in Asia. *Virology*, *303*(1), 146–151. https://doi.org/10.1006/viro.2002.1514

Maritim, A. C., Kariuki, D. P., Young, A. S., & Mutugi, J. J. (1989b). The importance of carrier state of *Theileria parva* in the epidemiology of theileriosis and its control by immunization. In T. T. Dolan (Ed.), *Theileriosis in Eastern, Central and Southern Africa* (pp. 121–128). ILRAD Nairobi.

Maritim, A. C., Young, A. S., Lesan, A. C., Ndungu, S. G., Stagg, D. A., & Ngumi, P. N. (1992). Transformation of Theileria parva derived from the African buffalo *(Syncerus caffer)* by tick passage in cattle and its use in infection and treatment immunization. *Veterinary Parasitology*, *43*(1-2), 1–14. doi:10.1016/0304-4017(92)90043-9 PMID:1496792

Marques, R., Krüger, R. F., Peterson, A. T., de Melo, L. F., Vicenzi, N., & Jiménez-García, D. (2020). Climate change implications for the distribution of the babesiosis and anaplasmosis tick vector, *Rhipicephalus* (*Boophilus*) *microplus. Veterinary Research (Faisalabad)*, *51*(1), 81. doi:10.118613567-020-00802-z PMID:32546223

Marsh, T. L., Yoder, J., Deboch, T., McElwain, T. F., & Palmer, G. H. (2016). Livestock vaccinations translate into human capital and school attendance by girls. *Science Advances*, *2*(12), e1601410. doi:10.1126ciadv.1601410 PMID:27990491

Martinez, D., Lefrançois, T., Vachiery, N., & Frutos, R. (2009). MLST scheme of Ehrlichia ruminantium: Genomic stasis and recombination in strains from Burkina-Faso. *Infection, Genetics and Evolution*, *9*(6), 1320–1328. doi:10.1016/j.meegid.2009.08.003 PMID:19712754

Martinez, D., Maillard, J. C., Coisne, S., Sheikboudou, C., & Bensaid, A. (1994). Protection of goats against heartwater acquired by immunisation with inactivated elementary bodies of Cowdria ruminantium. *Veterinary Immunology and Immunopathology*, *41*(1/2), 153–163. doi:10.1016/0165-2427(94)90064-7 PMID:8066991

Martinez, D., Vachiéry, N., Stachurski, F., Kandassamy, Y., Raliniaina, M., Aprelon, R., & Gueye, A. (2004). Nested PCR for detection and genotyping of Ehrlichia ruminantium: Use in genetic diversity analysis. *Annals of the New York Academy of Sciences*, *1026*(1), 106–113. doi:10.1196/annals.1307.014 PMID:15604477

Martin, H. M., Barnett, S. F., & Vidler, B. O. (1964). Cyclic development and longevity of *Theileria parva* in the tick *Rhipicephalus appendiculatus. Experimental Parasitology*, *15*(6), 527–555. doi:10.1016/0014-4894(64)90045-1 PMID:14242265

Martin, V., Chevalier, V., Ceccato, P., Anyamba, A., De Simone, L., Lubroth, J., de La Rocque, S., & Domenech, J. (2008). The impact of climate change on the epidemiology and control of Rift Valley fever. *Revue Scientifique et Technique (International Office of Epizootics)*, *27*(2), 413–426. doi:10.20506/rst.27.2.1802 PMID:18819669

Marufu, M. C., Chimonyo, M., Mans, B. J., & Dzama, K. (2013). Cutaneous hypersensitivity responses to *Rhipicephalus* tick larval antigens in pre-sensitized cattle. *Ticks and Tick-Borne Diseases*, *4*(4), 311–316. doi:10.1016/j.ttbdis.2012.12.001 PMID:23453577

Mashatise, E., Hamudikuwanda, H., Dzama, K., Chimonyo, M., & Kanengoni, A. (2005). Socio-economic roles, traditional management systems and reproductive patterns of Mukota pigs semi-arid north-eastern Zimbabwe. *Bunda Journal of Agriculture. Environmental Science & Technology*, *3*, 97–105.

Masiga, D. K., Ndung'u, K., Tweedie, A., Tait, A., & Turner, C. M. R. (2006). Trypanosoma evansi: Genetic variability detected using amplified restrictionfragment length polymorphism (AFLP) and random amplified polymorphic DNA (RAPD) analysis of Kenyan isolates. *Experimental Parasitology*, *114*(3), 147–153. doi:10.1016/j.exppara.2006.03.002 PMID:16624308

Mason, C. A., & Norval, R. A. I. (1980). The ticks for Zimbabwe. 1. The genus Boophilus. *Zimbabwe Veterinary Journal*, *11*, 36–43.

Matete, G. O. (2003). Occurrence, clinical manifestation and the epidemiological implications of naturally occurring canine trypanosomosis in western Kenya. *The Onderstepoort Journal of Veterinary Research*, *70*(4), 317–323. doi:10.4102/ojvr.v70i4.296 PMID:14971734

Matheron, G., Barre, N., Camus, E., & Gogue, J. (1987). Genetic resistance of Guadeloupe native goats to heartwater. *The Onderstepoort Journal of Veterinary Research*, *54*(3), 337–340. PMID:3448569

Matjila, P. T., Leisewitz, A. L., Jongejan, F., & Penzhorn, B. L. (2008). Molecular detection of tick-borne protozoal and ehrlichial infections in domestic dogs in South Africa. *Veterinary Parasitology*, *155*(1-2), 152–157. doi:10.1016/j.vetpar.2008.04.012 PMID:18502588

Matthewman, L. A., Kelly, P. J., Mahan, S. M., Semu, D., Tagwira, M., Bobade, P. A., Brouqui, P., Mason, P. R., & Raoult, D. (1993). Western blot and indirect fluorescent antibody testing for antibodies reactive with Ehrlichia canis in sera from apparently healthy dogs in Zimbabwe. *Journal of the South African Veterinary Association*, *64*(3), 111. PMID:8176682

Matthysse, J. G., & Colbo, M. H. (1987). *The ixodid ticks of Uganda*. Entomological Society of America.

Mattioli, R. C., Pandey, V. S., Murray, M., & Fitzpatrick, J. L. (2000). Immunogenetic influences on tick resistance in African cattle with particular reference to trypano-tolerant N'Dama (*Bos taurus*) and trypano-susceptible Gobra zebu (*Bos indicus*) cattle. *Acta Tropica*, *75*(3), 263–277. doi:10.1016/S0001-706X(00)00063-2 PMID:10838210

Maxie, M. G., Dolan, T. T., Jura, W. G. Z., Tabel, H., & Flowers, M. J. (1982). A comparative study of the disease in cattle caused by *Theileria parva* or *Theileria lawrencei*: II haematology, clinical chemistry, coagulation studies and complement. *Veterinary Parasitology*, *10*(1), 1–19. doi:10.1016/0304-4017(82)90002-4 PMID:7201709

Maxie, M. G., Losos, G. J., & Tabel, H. (1979). Experimental bovine trypanosomiasis (*T. vivax and T. congolense*) 1. Symptomatology and clinical pathology. *Tropical Medicine and Parasitology*, *30*, 275–282.

McAlpine, J. (1989). Phylogeny and classification of the Muscomorpha. In J. McAlpine & D. Wood (Eds.), *Manual of Nearctic Diptera* (Vol. 3, pp. 1397–1518). Agriculture Canada.

McBride, J. W., Corstvet, R. E., Gaunt, S. D., Chinsangaram, J., Akita, G. Y., & Osburn, B. I. (1996). PCR detection of acute Ehrlichia canis infection in dogs. *Journal of Veterinary Diagnostic Investigation*, *8*(4), 441–447. doi:10.1177/104063879600800406 PMID:8953528

McClure, J. C., Crothers, M. L., Schaefer, J. J., Stanley, P. D., Needham, G. R., Ewing, S. A., & Stich, R. W. (2010). Efficacy of a doxycycline treatment regimen initiated during three different phases of experimental ehrlichiosis. *Antimicrobial Agents and Chemotherapy*, *54*(12), 5012–5020. doi:10.1128/AAC.01622-09 PMID:20921310

McHardy, N. (1978). *In vitro* studies on the action of menoctone and other compounds on *Theileria parva* and *Theileria annulata. Annals of Tropical Medicine and Parasitology*, *72*(6), 501–511. doi:10.1080/00034983.1978.11719354 PMID:104669

McHardy, N. (1984). Recent advances in the chemotherapy of theileriosis. In H. P. Riemann & M. J. Burridge (Eds.), *Impact of Diseases on Livestock Production in the Tropics* (pp. 179–192). Elsevier. doi:10.1016/0167-5877(84)90061-8

McHardy, N., Haigh, A. J. B., & Dolan, T. T. (1976). Chemotherapy of *Theileria parva* infection. *Nature*, *261*(5562), 698–699. doi:10.1038/261698a0 PMID:934316

McHardy, N., & Wekesa, L. S. (1985). Buparvaquone (BW 720), a new anti-theilerial naphthoquinone, its role in therapy and prophylaxis. In A. D. Irvin (Ed.), *Immunisation Against Theileriosis in Africa* (p. 88). ILRAD.

Medlock, J. M., Hansford, K. M., Bormane, A., Derdakova, M., Estrada-Peña, A., George, J.-C., … Van Bortel, W. (2013). Driving forces for changes in geographical distribution of Ixodes ricinus ticks in Europe. *Parasites & Vectors,* *6*(1), 1. doi:10.1186/1756-3305-6-1

Medzhitov, R., Schneider, D. S., & Soares, M. P. (2012). Disease Tolerance as a Defense Strategy. *Science*, *335*(6071), 936–941. doi:10.1126cience.1214935 PMID:22363001

Mehlhorn, H., Schein, E., & Voigt, W. P. (1980). Light and electron microscopic study on developmental stages of Babesia canis within the gut of the tick Dermacentor reticulatus. *The Journal of Parasitology*, *66*(2), 220–228.

Melhorn, H., & Schein, E. (1984). The piroplasms: Lifecycle and sexual stages. *Advances in Parasitology*, *23*, 37–103. doi:10.1016/S0065-308X(08)60285-7 PMID:6442536

Meltzer, M. I., Norval, R. A. I., & Donachie, P. L. (1995). Effects of tick infestation and tick-borne disease infections (heartwater, anaplasmosis and babesiosis) on the lactation and weight-gain of Mashona cattle in southeastern Zimbabwe. *Tropical Animal Health and Production*, *27*(3), 129–144. doi:10.1007/BF02248956 PMID:7502343

Mendes, A. M. (1994). A história da pestesuínaafricanaem Angola. *Revista Portuguesa de Ciências Veterinárias*, *89*, 110–120.

Mercy, M. (2016). Laboratory and field evaluation of entomopathogenic fungi for the control of amitraz-resistant and susceptible strains of *Rhipicephalus decoloratus*. *Veterinary Parasitology*, *225*, 12–18. doi:10.1016/j.vetpar.2016.05.026 PMID:27369570

Merino, O., Alberdi, P., Pérez de la Lastra, J. M., & de la Fuente, J. (2013). Tick vaccines and the control of tick-borne pathogens. *Frontiers in Cellular and Infection Microbiology*, *3*. doi:10.3389/fcimb.2013.00030

Metcalf, R. L. (2000). *Insect Control*. Ullmann's Encyclopedia of Industrial Chemistry. doi:10.1002/14356007.a14_263

Mettam, R. W. M., & Carmichael, J. (1936). Turning disease, a protozoan encephalitis of cattle in Uganda. Its relationship with East Coast fever. *Parasitology*, *28*(2), 254–283. doi:10.1017/S0031182000022447

Meyer, A., Holt, H., Selby, R., & Guitian, J. (2016). Past and Ongoing Tsetse and Animal Trypanosomiasis Control Operations in Five African Countries: A Systematic Review. *PLoS Neglected Tropical Diseases*, *10*(12), e0005247. Advance online publication. doi:10.1371/journal.pntd.0005247 PMID:28027299

Minami, T., Spooner, P. R., Irvin, A. D., Ocama, J. G. R., Dobbelaere, D. A. E., & Fujinaga, T. (1983). Characterisation of stocks of Theileria parva by monoclonal antibody profiles. *Research in Veterinary Science*, *35*(3), 334–340. doi:10.1016/S0034-5288(18)32029-0 PMID:6420853

Mining, S. K. (1992). *The investigation of maternally derived immunity in calves* (PhD Thesis). University of Liverpool.

Minjauw, B., & McLeode, A. (2003). Tick-borne diseases and poverty. The impact of ticks and tick-borne diseases on the livelihood of small scale and marginal livestock owners in India and eastern and southern Africa. Research Report, DFID Animal Health Programme, Centre for Tropical Veterinary Medicine, University of Edinburgh, UK.

Minjauw, B., & McLeod, A. (2003). *Tick-borne diseases and poverty. The impact of ticks and tick-borne diseases on the livelihood of small-scale and marginal livestock owners in India and eastern and southern Africa. Research report, DFID Animal Health Programme*. Centre for Tropical Veterinary Medicine, University of Edinburgh.

Minshull, J. I., & Norval, R. A. I. (1982). Factors influencing the spatial distribution of Rhipicephalus appendiculatus in Kyle Reacreational Park in south-eastern Zimbabwe. *South African Journal of Wildlife Research*, *12*, 118–123.

Misao, T., & Kobayashi, Y. (1954). Studies on infectious mononucleosis. I. Isolation of etiologic agent from blood, bone marrow, and lymph node of a patient with infectious mononucleosis by using mice. *Tokyo Iji Shinshi*, *71*, 683–686.

Misinzo, G., Jumapili, F., Ludosha, M., Mafie, E., Silialis, J., Mushi, R., & Van Doorsselaere, J. (2012). Genotyping of African swine fever virus from an outbreak in Tanzania in 2009. *Research Opinions in Animal & Veterinary Sciences*, *2*, 334–338.

Molia, S., Frebling, M., Vachiéry, N., Pinarello, V., Petitclerc, M., Rousteau, A., Martinez, D., & Lefrançois, T. (2008). Amblyomma variegatum in cattle in Marie Galante, French Antilles: Prevalence, control measures, and infection by Ehrlichia ruminantium. *Veterinary Parasitology*, *153*(3/4), 338–346. doi:10.1016/j.vetpar.2008.01.046 PMID:18406061

Moll, G., Lodhing, A., Young, A. S., & Leitch, B. L. (1986). Epidemiology of theileriosis in an endemic area in Kenya. *Veterinary Parasitology*, *19*(3-4), 255–273. doi:10.1016/0304-4017(86)90073-7 PMID:3085323

Moloo, S.K. (1993). The distribution of Glossina species in Africa and their natural hosts. *Insect Science 480 and its Application, 14*, 511-28.

Molyneux, D. H. (1975). Diagnostic methods in animal trypanosomiases. *Veterinary Parasitology, 1*(1), 5–17. doi:10.1016/0304-4017(75)90003-5

Molyneux, D. H., Hopkins, D. R., & Zagaria, N. (2004). Disease eradication, elimination and control: The need for accurate and consistent usage. *Trends in Parasitology, 20*(8), 347–351. doi:10.1016/j.pt.2004.06.004 PMID:15246314

Montenegro-James, S. (1992). Prevalence and control of Babesiosis in the Americas. *Memorias do Instituto Oswaldo Cruz, 87*(suppl 3), 27–36. https://doi.org/10.1590/S0074-02761992000700003

Montenegro-James, S., & Guillen, A.T., Ma, S.-J., Tapang, P., Abdel-Gawad, A., Tori, M., & Ristic, M. (1990). Use of the dot enzyme-linked immunosorbent assay with isolated *Anaplasma marginale* initial bodies for serodiagnosis of anaplasmosis in cattle. *American Journal of Veterinary Research, 51*, 1518–1521. PMID:2240769

Montenegro-James, S., James, M. A., & Ristic, M. (1985). Modified indirect fluorescent antibody test for the serodiagnosis of *Anaplasma marginale* infections in cattle. *American Journal of Veterinary Research, 46*, 2401–2403. PMID:3907438

Montenegro-James, S., Toro, M., Leon, E., & Guillen, A. T. (1992). Field evaluation of an exoantigen-containing Babesia vaccine in Venezuela. *Memorias do Instituto Oswaldo Cruz, 87*(suppl 3), 283–288. https://doi.org/10.1590/S0074-02761992000700048

Montgomery, R. E. (1921). On a form of swine fever occurring in British East Africa (Kenya Colony). *Journal of Comparative Pathology, 34*(159–191), 243– 262.

Montgomery, E. (1917). On a Tick-borne Gastroenteritis of Sheep and Goats occurring in British East Africa. *The Journal of Comparative Pathology and Therapeutics, 30*(1), 28–57.

Montgomery, R. M. (1917). On a tick-borne gastroenteritis of sheep and goats occurring in Kenya. *The Journal of Comparative Pathology and Therapeutics, 30*(1), 28–57.

Moran, N. A., & Baumann, P. (2000). Endosymbionts in animals. *Current Opinions in Microbiology*, 270–275.

Morand, S., & Frutos, R. (2010). Clonal origin of emerging populations of Ehrlichia ruminantium in Burkina Faso. *Infection, Genetics and Evolution, 10*(7), 903–912. doi:10.1016/j.meegid.2010.05.011 PMID:20553966

Morrison, W. I., Buscher, G., Emery, D. L., Nelson, R. I., & Murray, M. (1981). The kinetics of infection with Theileria parva in cattle and the relevance of the development of immunity. In A. D. Irvin, M. P. Cunningham, & A. S. Young (Eds.), *Advances in the Control of Theileriosis* (pp. 311–327). Martinus Nijhoff Publishers. doi:10.1007/978-94-009-8346-5_52

Morrison, W. I., Goddeeris, B. M., Brown, W. C., Baldwin, C. L., & Teale, A. J. (1989). *Theileria parva* in cattle: Characterisation of infected lymphocytes and immune responses they provoke. *Veterinary Immunology and Immunopathology, 20*(3), 213–237. doi:10.1016/0165-2427(89)90003-2 PMID:2497579

Morrison, W. I., Goddeeris, B. M., & Teale, A. J. (1987b). Bovine cytotoxic T cell clones which recognize lymphoblasts infected with two antigenically different stocks of the protozoan parasite *Theileria parva. European Journal of Immunology, 17*(12), 1703–1709. doi:10.1002/eji.1830171205 PMID:3121356

Morrison, W. I., Goddeeris, B. M., Teale, A. J., Groocock, C. M., Kemp, S. J., & Stagg, D. A. (1987a). Cytotoxic cells elicited in cattle challenged with *Theileria parva* (Muguga): Evidence for restriction by class 1 MHC determinants and parasite strain specificity. *Parasite Immunology, 9*(5), 563–578. doi:10.1111/j.1365-3024.1987.tb00530.x PMID:3120135

Morrison, W. I., Hemmink, J. D., & Toye, P. G. (2020). *Theileria parva*: A parasite of African buffalo, which adapted to infect and undergo transmission in cattle. *Parasitology*, *50*, 403–412. PMID:32032592

Mortelmans, J., & Kageruka, P. (1976). Trypanotolerant cattle breeds in Zaire. *World Animal Review*, *19*, 24.

Morzaria, S. P., Young, A. S., & Hudson, E. B. (1977). Babesia bigemina in Kenya: Experimental transmission by Boophilus decoloratus and the production of tick-derived stabilates. *Parasitology*, *74*(3), 291–298. https://doi.org/10.1017/s0031182000047910

Morzaria, S., Katende, J., Kairo, A., Nene, V., & Musoke, A. (1992). New methods for the diagnosis of Babesia bigemina infection. *Memorias do Instituto Oswaldo Cruz*, *87*(suppl 3), 201–205. https://doi.org/10.1590/S0074-02761992000700033

Mosi, R., & Mbuuza, F. (1996). Animal recording, characterization and improvement schemes in East Africa. Proceedings of Livestock Improvement Association, 75th Anniversarry, 10–16.

Mosqueda, J., Olvera-Ramirez, A., Aguilar-Tipacamu, G., & Canto, G. J. (2012). Current Advances in Detection and Treatment of Babesiosis. *Current Medicinal Chemistry*, *19*(10), 1504–1518. https://doi.org/10.2174/092986712799828355

Mossaad, E., Salim, B., Suganuma, K., Musinguzi, P., Hassan, M. A., Elamin, E. A., Mohammed, G. E., Bakhiet, A. O., Xuan, X., Satti, R. A., & Inoue, N. (2017). Trypanosoma vivax is the second leading cause of camel trypanosomosis in Sudan after Trypanosoma evansi. *Parasites & Vectors*, *10*(1), 176–186. doi:10.118613071-017-2117-5 PMID:28403897

Msangi, A., Mramba, F., Kiwia, N., Kitwika, W., Malele, I., Kasilagila G, Byamungu M, Chalo O, & Athuman J. (1999). Eradication of tsetse fly from Mafia Island after the Zanzibar success. *25th International Scientific Council for Trypanosomosis Research and Control.*

Mtshali, K., Nakao, R., Sugimoto, C., & Thekisoe, O. (2017). Occurrence of *Coxiella burnetii, Ehrlichia canis, Rickettsia* species and *Anaplasma phagocytophilum*-like bacterium in ticks collected from dogs and cats in South Africa. *Journal of the South African Veterinary Association*, *88*(0), a1390. doi:10.4102/jsava.v88i0.1390 PMID:28582983

Mudaliar, S. V., Achary, G. R., & Alwar, V. S. (1950). On a species of Babesia in an Indian wild cat (Felis catus). *The Indian Veterinary Journal*, *26*(5), 392–395.

Mugera, G. M., & Chema, S. (1967). Nairobi sheep disease: A study of its pathogenesis in sheep, goats and suckling mice. *Bulletin of Epizootic Diseases of Africa. Bulletin Des Epizooties En Afrique*, *15*(4), 337–354.

Mugunieri, L., Matete, G., Otieno, K., & Olubai, W. (2003). *Assessment of the impact of animal trypanosomosis on dairy cattle productivity in western Kenya* [Research Report].

Muhairwa, A. P., Mwakijungu, E. O., Msoffe, P. L. M., & Mtambo, M. M. A. (2012). Seroprevalence and factors affecting canine monocytic ehrlichiosis and canine brucellosis in Tanzania. *Research Opinions in Animal & Veterinary Sciences*, *2*(3), 181–188.

Muhammed, S.T., & Luerman, L.H. (1975). Effect of humoral antibodies on the course of *Theileria parva* (East Coast fever) of cattle). *American Journal of Veterinary Research*, *36*, 389–402. PMID:47727

Muhanguzi, D., Byaruhanga, J., Amanyire, W., Ndekezi, C., Ochwo, S., Nkamwesiga, J., Mwiine, F. N., Tweyongyere, R., Fourie, J., Madder, M., Schetters, T., Horak, I., Juleff, N., & Jongejan, F. (2020). Invasive cattle ticks in East Africa: Morphological and molecular confirmation of the presence of *Rhipicephalus microplus* in south-eastern Uganda. *Parasites & Vectors*, *13*(1), 165. doi:10.118613071-020-04043-z PMID:32245511

Muhanguzi, D., Mugenyi, A., Bigirwa, G., Kamusiime, M., Kitibwa, A., Akurut, G. G., Ochwo, S., Amanyire, W., Okech, S. G., Hattendorf, J., & Tweyongyere, R. (2017). African animal trypanosomiasis as a constraint to livestock health and production in Karamoja region: A detailed qualitative and quantitative assessment. *BMC Veterinary Research*, *13*(1), 355. doi:10.118612917-017-1285-z PMID:29178951

Mukhebi, A. W., Chamboko, T., O'collaghan, C. J., Peter, T. F., Kruska, R. L., Medley, G. F., Mahan, S. M., & Perry, B. D. (1999). An assessment of the economic impact of heartwater (*Cowdria ruminantium* infection) and its control in Zimbabwe. *Preventive Veterinary Medicine*, *39*(3), 173–189. doi:10.1016/S0167-5877(98)00143-3 PMID:10327437

Mukhebi, A. W., Perry, B., & Kruska, R. (1992). Estimated economics of theileriosis control in Africa. *Preventive Veterinary Medicine*, *12*(1-2), 73–85. doi:10.1016/0167-5877(92)90070-V

Mulumba-Mfumu, L. K., Saegerman, C., Dixon, L. K., Madimba Eric Kazadi, K. C., Mukalakata, N. T., Oura, C. A. L., Chenais, E., & Masembe Karl Ståhl, C. (2019, July). African swine fever: Update on Eastern, Central and Southern Africa. *Transboundary and Emerging Diseases*, *66*(4), 1462–1480. https://doi.org/10.1111/tbed.13187

Munsterman, S., Mbura, R., Maloo, S., & Lohr, C. (1992). Trypanosomiasis control in Boran cattle in Kenya: A comparison between chemoprophylaxis and a parasite detection and treatment. *Tropical Animal Health and Production*, *24*, 17–27. doi:10.1007/BF02357230 PMID:1306914

Munz, E., Göbel, E., Krolopp, C., Reimann, M., & Davies, F. G. (1981). Electron microscopic studies on the morphology of Nairobi sheep disease virus. *Zentralblatt Fur Veterinarmedizin. Reihe B. Journal of Veterinary Medicine. Series B*, *28*(7), 553–563. doi:10.1111/j.1439-0450.1981.tb01773.x

Munz, E., Reimann, M., & Meier, K. (1983). An enzyme-linked immunosorbent assay (ELISA) for the detection of antibodies to Nairobi sheep disease virus in comparison with an indirect immunofluorescent and hemagglutination test. I. Development and testing of the ELISA with sera of experimentally inf. *Zentralblatt Fur Veterinarmedizin. Reihe B. Journal of Veterinary Medicine. Series B*, *30*(7), 473–479. doi:10.1111/j.1439-0450.1983.tb01870.x

Muraguri, G. R., McLeod, A., McDermott, J. J., & Taylor, N. (2005). The incidence of calf morbidity and mortality due to vector-borne infections in smallholder dairy farms in Kwale District, Kenya. *Veterinary Parasitology*, *130*(3–4), 305–315. doi:10.1016/j.vetpar.2004.11.026 PMID:15885914

Murilla, G. A., Ndung'u, K., Thuita, J. K., Gitonga, P. K., Kahiga, D. T., Auma, J. E., Ouma, J. O., Rutto, J. J., & Ndung'u, J. M. (2014). Kenya Trypanosomiasis Research Institute cryobank for human and animal trypanosome isolates to support research: Opportunities and challenges. *PLoS Neglected Tropical Diseases*, *8*(5), e2747. https://doi.org/10.1371/journal.pntd.0002747

Murray, M., Clifford, D. J., & Mcintyre, W. I. M. (1979). Diagnosis of African trypanosomosis in the bovine. *Transactions of the Royal Society of Tropical Medicine and Hygiene*, *73*(1), 120–121. doi:10.1016/0035-9203(79)90148-2 PMID:442172

Murray, M., Trail, J. C. M., & D'Ieteren, G. D. M. (1990). Tripanotolerance in cattle and prospects for the control of trypanosomiasis by selective breeding. *Scientific and Technical Review*, *9*, 369–386. PMID:2132686

Murray, M., Trail, J., & D'Ieteren, G. (1990). Trypanotolerance in cattle and prospects for the control of trypanosomiasis by selective breeding. *Scientific and Technical Review Office International Des Epizooties*, *9*(2), 369–386. doi:10.20506/rst.9.2.506 PMID:2132686

Murrell, A., & Barker, S. C. (2003). Synonomy of *Boophilus*Curtice, 1981 with *Rhipicephalus* Koch, 1844 (Acari: Ixodidae). *Systematic Parasitology*, *56*(3), 169–172. doi:10.1023/B:SYPA.0000003802.36517.a0 PMID:14707501

Musoke, A. J., Nantulya, V. M., Buscher, G., Masake, R. A., & Otim, B. (1982). Bovine immune response to Theileria parva: Neutralizing antibodies to sporozoites. *Immunology*, *45*, 663–668. PMID:6802745

Musoke, R. A., Twenyongyere, R., Bizimenyera, E., Waiswa, C., Mugisha, A., Biryomumaisho, S., & McHardy, N. (2004). Treatment of East Coast fever of cattle with a combination of parvaquone and frusemide. *Tropical Animal Health and Production*, *36*(3), 233–245. doi:10.1023/B:TROP.0000016832.69136.68 PMID:15080540

Mutugi, J. J., Ndungu, S. G., Linyonyi, A., Maritim, A. C., Mining, S. K., Ngumi, P. N., Kariuki, D. P., Leitch, B. L., D'Souza, D., Maloo, S., & Lohr, K. F. (1991). Responses to vaccine trial for East Coast fever in two cattle herds at the Kenya Coast. *Preventive Veterinary Medicine*, *10*(3), 173–183. doi:10.1016/0167-5877(91)90002-J

Mutugi, J. J., Young, A. S., Martim, A. C., Linyonyi, A., Mbogo, S. K., & Leitch, B. L. (1988). Immunisation of cattle using varying doses of *Theileria parva lawrencei* sporozoites derived from an African buffalo (*Syncerus caffer*) and treatment with buparvaquone. *Parasitology*, *96*(2), 391–402. doi:10.1017/S0031182000058376 PMID:3131722

Muturi, C., Ouma, J., Malele, I., Ngure, R., Rutto, J., Mithöfer, K., Enyaru, J., & Masiga, D. K. (2011). Tracking the Feeding Patterns of Tsetse Flies (Glossina Genus) by Analysis of Bloodmeals Using Mitochondrial Cytochromes Genes. *PLoS One*, *6*(2), e17284. doi:10.1371/journal.pone.0017284 PMID:21386971

Mwacharo, J. M., & Drucker, A. G. (2005). Production objectives and management strategies of livestock keepers in south-east Kenya: Implications for a breeding programme. *Tropical Animal Health and Production*, *37*(8), 635–652. doi:10.100711250-005-4253-8 PMID:16619880

Mwai, O., Hanotte, O., Kwon, Y.-J., & Cho, S. (2015). African Indigenous Cattle: Unique Genetic Resources in a Rapidly Changing World. *Asian-Australasian Journal of Animal Sciences*, *28*(7), 911–921. doi:10.5713/ajas.15.0002R PMID:26104394

Mwangi, E., Stevenson, P., Gettinby, G., Reid, S., & Murray, M. (1998). Susceptibility to trypanosomosis of three Bos indicus cattle breeds in areas of differing tsetse fly challenge. *Veterinary Parasitology*, *79*(1), 1–17. doi:10.1016/S0304-4017(98)00147-2 PMID:9777722

Mwongela, G., Kovatch, R., & Fazil, M. (2005). Acute trypanosoma vivax infection in dairy cattle in coast province, Kenya. *Tropical Animal Health and Production*, *13*(1), 63–69. doi:10.1007/BF02237893 PMID:7233561

Mylonakis, M. E., Ceron, J. J., Leontides, L., Siarkou, V. I., Martinez, S., Tvarijonaviciute, A., Koutinas, A. F., & Harrus, S. (2011). Serum acute phase proteins as clinical phase indicators and outcome predictors in naturally occurring canine monocytic ehrlichiosis. *Journal of Veterinary Internal Medicine*, *25*(4), 811–817.

Mylonakis, M. E., Koutinas, A. F., Breitscwerdt, E. B., Hegarty, B. C., Billinis, C. D., Leontides, L. S., & Kontos, V. S. (2004). Chronic canine ehrlichiosis (Ehrlichia canis): A retrospective study of 19 natural cases. *Journal of the American Animal Hospital Association*, *40*(3), 174–184. doi:10.5326/0400174 PMID:15131097

Mylonakis, M. E., Kritsepi-Konstantinou, M., Dumler, J. S., Diniz, P. P., Day, M. J., Siarkou, V. I., Breitschwerdt, E. B., Psychas, V., Petanides, T., & Koutinas, A. F. (2010). Severe hepatitis associated with acute Ehrlichia canis infection in a dog. *Journal of Veterinary Internal Medicine*, *24*(3), 633–638. doi:10.1111/j.1939-1676.2010.0501.x PubMed

Mylonakis, M. E., Siarkou, V. I., & Koutinas, A. F. (2010). Myelosuppressive canine monocytic ehrlichiosis (Ehrlichia canis): An update on the pathogenesis, diagnosis and management. *Israel Journal of Veterinary Medicine*, *65*(4), 129–135.

Naessens, J. (2006). ovine trypanotolerance: A natural ability to prevent severe anaemia and haemophagocytic syndrome? *International Journal for Parasitology*, *36*(5), 521–528. doi:10.1016/j.ijpara.2006.02.012 PMID:16678182

Naessens, J., Teale, A. J., & Sileghem, M. (2002). Identification of mechanisms of natural resistance to African trypanosomiasis in cattle. *Veterinary Immunology and Immunopathology*, *87*(3–4), 187–194. doi:10.1016/S0165-2427(02)00070-3 PMID:12072233

Nagel, P. (1995). *Ecological side effect of tsetse control. Abstracts of VIII International Conference of Institutions of Tropical Veterinary Medicine*, Berlin, Germany.

Nah, K., Bede-Fazekas, A., Trájer, A. J., & Wu, J. (2020). The potential impact of climate change on the transmission risk of tick-borne encephalitis in Hungary. *BMC Infectious Diseases*, *20*(1), 34. doi:10.118612879-019-4734-4 PMID:31931734

Nakaghi, A. C. H., Machado, R. Z., Costa, M. T., André, M. R., & Baldani, C. D. (2008). Clinical, hematological, serological and molecular survey of canine ehrlichiosis. *Ciência Rural, Santa Maria*, *38*(3), 776–770. doi:10.1590/S0103-84782008000300027

Nakao, R., Morrison, L. J., Zhou, L., Magona, J. W., Jongejan, F., & Sugimoto, C. (2012). Development of multiple-locus variable-number tandem-repeat analysis for rapid genotyping of *Ehrlichia ruminantium* and its application to infected *Amblyomma variegatum* collected in heartwater endemic areas in Uganda. *Parasitology*, *139*(1), 69–82. doi:10.1017/S003118201100165X PMID:22008706

Namangala, B., & Odongo, S. (2014). *Animal African trypanosomosis in sub-saharan Africa and beyond African borders*. Academic Press.

Namangala, B. (2012). Contribution of Innate Immune Responses towards Resistance to African Trypanosome Infections: Innate Mechanisms of Resistance to Trypanosomiasis. *Scandinavian Journal of Immunology*, *75*(1), 5–15. doi:10.1111/j.1365-3083.2011.02619.x PMID:21916913

Namangala, B., Hachaambwa, L., Kajino, K., Mweene, A. S., Hayashida, K., Simuunza, M., Simukoko, H., Choongo, K., Chansa, P., Lakhi, S., Moonga, L., Chota, A., Ndebe, J., Nsakashalo-Senkwe, M., Chizema, E., Kasonka, L., & Sugimoto, C. (2012). The use of Loop-mediated Isothermal Amplification (LAMP) to detect the re-emerging Human African Trypanosomiasis (HAT) in the Luangwa and Zambezi valleys. *Parasites & Vectors*, *5*(1), 282. doi:10.1186/1756-3305-5-282 PMID:23211002

Nantulya, V. M. (1986). Immunological approaches to the control of ani- mal trypanosomiasis. *Parasitology Today (Personal Ed.)*, *2*(6), 168–173. doi:10.1016/0169-4758(86)90148-1 PMID:15462812

Nantulya, V. M., & Lindqvist, K. J. (1989). Antigen-detection enzyme immunoassays for the diagnosis of Trypanosoma vivax, T. congolense and T. brucei infections in cattle. *Tropical Medicine and Parasitology*, *40*, 267–272. PMID:2617031

Narahashi, T. (2013). *Toxicology of Halogenated Hydrocarbons Health and Ecological Effects*. doi:10.1016/B978-0-08-027530-7.50023-6

Naylor, D. C. (1971). The Haematology and Histopathology of Trypanosoma Congolense Infection in Cattle. Ii. Haematology (Including Symptoms). *Tropical Animal Health and Production*, *3*(3), 159–168. doi:10.1007/BF02360746 PMID:4945970

Nazar, M., Khan, M. A., Shah, A. A., Rahman, S. U., Khan, I., Ullah, A., & Khan, I. U. (2018). Occurrence and transplacental transmission of Anaplasma marginale in dairy cattle. *Slovenian Veterinary Research*, *55*(3), 183–191. https://www.cabi.org/isc/datasheet/91734. doi:10.26873/SVR-499-2018

Ndip, L. M., Ndip, R. N., Esemu, S. N., Dickmu, V. L., Fokam, E. B., Walker, D. H., & McBride, J. W. (2005). Ehrlichial infection in Cameroonian canines by Ehrlichia canis and Ehrlichia ewingii. *Veterinary Microbiology*, *111*(1-2), 59–66. doi:10.1016/j.vetmic.2005.08.010 PMID:16181750

Ndip, L. M., Ndip, R. N., Esemu, S. N., Walker, D. H., & McBride, J. W. (2010). Predominance of Ehrlichia chaffeensis in Rhipicephalus sanguineus ticks from kennel-confined dogs in Limbe. *Experimental & Applied Acarology*, *50*(2), 163–168. doi:10.100710493-009-9293-8 PMID:19693681

Ndungu, S. G., Brown, C. G. D., & Dolan, T. T. (2005). *In vivo* comparison of susceptibility and disease resistance between *Bos indicus* and *Bos taurus* cattle types to *Theileria parva* Muguga infection. *The Onderstepoort Journal of Veterinary Research*, *72*, 13–22. doi:10.4102/ojvr.v72i1.220 PMID:15991702

Neave, S. (1912). Notes on the blood-sucking insects of Eastern Tropical Africa. *Bulletin of Entomological Research*, *3*(3), 275–323. doi:10.1017/S0007485300001942

Neer, T. M., & Harrus, S. (2006). *Canine monocytotropic ehrlichiosis and neorickettsiosis. E. canis, E. chaffeensis, E. ruminantium, N. sennetsu*. Academic Press.

Neer, T. M., Eddlestone, S. M., Gaunt, S. D., & Corstvet, R. E. (1999). Efficacy of enrofloxacin for the treatment of experimentally induced *Ehrlichia canis* infection. *Journal of Veterinary Internal Medicine*, *13*(5), 501–504. doi:10.1111/j.1939-1676.1999.tb01470.x PMID:10499737

Neitz, W. 0. (1966). Nairobi Sheep Disease. In *Encyclopedic Reference of Parasitology* (pp. 362–362). Berlin: Springer Berlin Heidelberg. doi:10.1007/3-540-29835-5_2337

Neitz, W. O. (1950). The specific action of pamaquin on the haemotropic parasites of *Theileria parva*. *South African Journal of Science*, *46*, 218–219.

Neitz, W. O. (1953). Aureomycin in *Theileria parva* infection. *Nature*, *171*(4340), 34–35. doi:10.1038/171034a0 PMID:13025470

Neitz, W. O. (1968). Heartwater. *Bulletin - Office International des Epizooties*, *70*(1), 329–336. PMID:5761407

Neitz, W. O., & Alexander, R. A. (1941). The immunisation of calves against heartwater. *Journal of the South African Veterinary Medical Association*, *12*(4), 103–111.

Nene, V., Kiara, H., Lacasta, A., Pelle, R., Svitek, N., & Steina, L. (2016). The biology of Theileria parva and control of East Coast fever- current status and future trends. *Ticks and Tick-Borne Diseases*, *7*(4), 549–564. doi:10.1016/j.ttbdis.2016.02.001 PMID:26972687

Newson, R. M., & Chiera, J. W. (1989). Development of resistance in calves to nymphs of *Rhipicephalus appendiculatus* (Acarina:Ixodidae) during test feeds. *Experimental & Applied Acarology*, *6*(1), 19–27. doi:10.1007/BF01193230 PMID:2707106

Newson, R. M., Chiera, J. W., Young, A. S., Dolan, T. T., Cunningham, M. P., & Radley, D. E. (1984). Survival of Rhipicephalus appendiculatus (Acarina:Ixodidae) and persistence of Theileria parva (Ampicomplexa:Theileriidae) in the field. *International Journal for Parasitology*, *14*(5), 483–489. doi:10.1016/0020-7519(84)90029-8 PMID:6439668

Ngari, N. N., Gamba, D. O., Olet, P. A., Zhao, W., Paone, M., & Cecchi, G. (2020). Developing a national atlas to support the progressive control of tsetse-transmitted animal trypanosomosis in Kenya. *Parasites & Vectors*, *13*(1), 1–12.

Ngari, N., Gamba, D., Olet, P., Zhao, W., Paone, M., & Cecchi, G. (2020). Developing a national atlas to support the progressive control of tsetse-transmitted animal trypanosomosis in Kenya. *Vectors*, *13*(1), 286. doi:10.118613071-020-04156-5 PMID:32503681

Ngongolo, K., Estes, A. B., Hudson, P. J., & Gwakisa, P. S. (2019). Assessing Risk Factors for Trypanosome Infections in Cattle and Wildlife Interface Areas in Northern Tanzania. *J Infect Dis Epidemiol*, *5*(3). Advance online publication doi:10.23937/2474-3658/1510078

Ngoune, J. M. T., Reveillaud, J., Sempere, G., Njiokou, F., Melachio, T. T., Abate, L., ... Geiger, A. (2019). The composition and abundance of bacterial communities residing in the gut of *Glossina palpalis palpalis* captured in two sites of southern Cameroon. *Parasites & Vectors*, *12*(1), 151. Advance online publication. doi:10.118613071-019-3402-2 PMID:30940213

Nicholson, W. L., Allen, K., McQuiston, J. H., Breitschwerdt, E. B., & Little, S. E. (2010). The increasing recognition of rickettsial pathogens in dogs and people. *Trends in Parasitology*, *26*(4), 205–212. doi:10.1016/j.pt.2010.01.007 PMID:20207197

NijhofA. M.GuglielmoneA. A.HorakI. G. (2016). www.catalogueoflife.org/col.ISSN

Njagu, Z., Mihok, S., Kokwaro, E., & Verloo, D. (1999). Isolation of Trypanosoma brucei from the monitor lizard (Varanus niloticus) in an endemic focus of Rhodesian sleeping sickness in Kenya. *Acta Tropica*, *72*(2), 137–148. doi:10.1016/S0001-706X(98)00092-8 PMID:10206114

Njiru, Z. K., Makumi, J. N., Okoth, S., Ndungu, J. M., & Gibson, W. C. (2004). Identification of trypanosomes in Glossina pallidipes and G. longipennis in Kenya. *Infection, Genetics and Evolution*, *4*(1), 29–35.

Njiru, Z., Constance, C., Guya, S., Crowther, K., Kiragu, J., Thompson, R., & Davila, A. (2005). The use of ITS1 rDNA PCR in detecting pathogenic African trypanosomes. *Parasitology Research*, *95*, 186–192. doi:10.100700436-004-1267-5 PMID:15619129

Njogu, A. R., Dolan, R. B., Wilson, A. J., & Sayer, P. D. (1985). Trypanotolerance in East African Orma Boran cattle. *The Veterinary Record*, *117*(24), 632–636. doi:10.1136/vr.117.24.632 PMID:4090214

Nogge, G. (1981). Significance of symbionts for the maintenance of an optimal nutritional state for successful reproduction in haematophagous arthropods. *Parasitology Today (Personal Ed.)*, *82*, 101–104.

Normile, D. (2018). Arrival of deadly pig disease could spell disaster for China. *Science*, *361*, 741. https://science.sciencemag.org/

Norval, R. A. I., Perry, B. D., & Young, A. S. (1992). Development of Theileria. In The Epidemiology of Theileriosis in Africa. Academic Press.

Norval, R. A. I. (1977). Tick problems in relation to land utilization in Rhodesia. *Rhodesia Veterinary Journal*, *8*, 33–38.

Norval, R. A. I., Andrew, H. R., & Yunker, C. E. (1990). Infection rates with Cowdria ruminantium of nymphs and adults of the bont tick Amblyomma hebraeum collected in the field in Zimbabwe. *Veterinary Parasitology*, *36*(3-4), 277–283. doi:10.1016/0304-4017(90)90039-E PMID:2399648

Norval, R. A. I., Lawrence, J. A., Young, A. S., Perry, B. D., Dolan, T. T., & Scott, J. (1991). *Theileria parva*: Influence of vectors, parasite and host relationships on the epidemiology of theileriosis in southern Africa. *Parasitiology*, *102*(3), 347–356. doi:10.1017/S0031182000064295 PMID:1907728

Norval, R. A. I., Perry, B. D., & Young, A. S. (1992). *The epidemiology of theileriosis in Africa. Academic Press.*

Norval, R. A. I., Perry, B. D., Young, A. S., Lawrence, J. A., Mukhebi, A. W., Bishop, R., & McKeever, D. (1992). *The Epidemiology of Theileriosis in Africa*. Academic press.

Norval, R. A. I., Sutherst, R. W., Kurki, K., Gibson, J. D., & Kerr, J. D. (1988). The effect of the brown ear tick Rhipicephalus appendiculatus on the growth of Sanga and European breed of cattle. *Veterinary Parasitology*, *30*(2), 149–164. doi:10.1016/0304-4017(88)90162-8 PMID:3245108

Norval, R. A. I., Walker, J. B., & Colborne, J. (1982). The ecology of *Rhipicephalus zambeziensis* and *Rhipicephalus appendiculatus* (Acarina: Ixodidae) wiyh particular reference to Zimbabwe. *The Onderstepoort Journal of Veterinary Research*, *49*, 181–190. PMID:7185034

Norval, R. A., Fivaz, B. H., Lawrence, J. A., & Daillecourt, T. (1983). Epidemiology of tick-borne diseases of cattle in Zimbabwe. I. Babesiosis. *Tropical Animal Health and Production*, *15*(2), 87–94. doi:10.1007/BF02239802 PMID:6868134

Notomi, T., Mori, Y., Tomita, N., & Kanda, H. (2015). Loop-mediated isothermal amplification (LAMP): Principle, features, and future prospects. *Journal of Microbiology (Seoul, Korea)*, *53*(1), 1–5. https://doi.org/10.1007/s12275-015-4656-9

Nuttall, P. A. (2009). Molecular characterization of tick-virus interactions. *Frontiers in Bioscience,* (14), 2466. doi:10.2741/3390

Nuttall, P. A. (2014). Tick-borne viruses. In Biology of Ticks (vol. 2, pp. 180–210). New York, NY: Oxford University Press.

Nuttall, P. A., & Labuda, M. (2004). *Tick-host interactions: saliva-activated transmission*. doi:10.1017/S0031182004005633

Nyangiwe, N., Goni, S., Hervé-Claude, L. P., Ruddat, I., & Horak, I. G. (2011). Ticks on pastures and on two breeds of cattle in the Eastern Cape Province, South Africa. *The Onderstepoort Journal of Veterinary Research*, *78*(1), 1–9. doi:10.4102/ojvr.v78i1.320 PMID:23327215

Nyangiwe, N., Haarrison, A., & Horak, I. G. (2013). Displacement of *Rhipicephalus decoloratus* by *Rhipicephalus microplus* (Acari: Ixodidae) in the Eastern Cape Province, South Africa. *Experimental & Applied Acarology*, *61*(3), 371–382. doi:10.100710493-013-9705-7 PMID:23771510

Nyangiwe, N., Yawa, M., & Muchenje, V. (2018). Driving forces for changes in geographic range of cattle ticks (Acari: Ixodidae) in Africa: A review. *South African Journal of Animal Science*, *48*(5), 829. doi:10.4314ajas.v48i5.4

Obara, I., Githaka, N., Nijhof, A., Krücken, J., Nanteza, A., Odongo, D., Lubembe, D., Atimnedi, P., Mijele, D., Njeri, A., Mwaura, S., Owido, G., Ahmed, J., Clausen, P. H., & Bishop, R. P. (2020). The *Rhipicephalus appendiculatus* tick vector of *Theileria parva* is absent from Cape buffalo (*Syncerus caffer*) populations and associated ecosystems in northern Uganda. *Parasitology Research*, *119*(7), 2363–2367. doi:10.100700436-020-06728-x PMID:32500369

Obereb, P. T., & Benzuidenhout, J. D. (1987). Heartwater in hosts other than domestic ruminants. *The Onderstepoort Journal of Veterinary Research*, *54*, 271–275. PMID:3329317

Obiero, G. F., Mireji, P. O., Nyanjom, S. R. G., Christoffels, A., Robertson, H. M., & Masiga, D. K. (2014). Odorant and Gustatory Receptors in the Tsetse Fly Glossina morsitans morsitans. *PLoS Neglected Tropical Diseases*, *8*(4), e2663. doi:10.1371/journal.pntd.0002663 PMID:24763191

Ocaido, M., Muwazi, R. T., & Opuda, J. A. (2009). Economic impact of ticks and tick-borne diseases on cattle production systems around Lake Mburo National Park in South Western Uganda. *Tropical Animal Health and Production*, *41*(5), 731–739. doi:10.100711250-008-9245-z PMID:18958579

Ochanda, H., Young, A. S., Mutugi, J. J., Mumo, J. M., & Omwoyo, P. L. (1988). The effect of temperature on the rate of transmission of *Theileria parva* infection to cattle by its tick vector *Rhipicephalus appendiculatus. Parasitology*, *97*, 239–245. PMID:3143989

Odeniran, P., & Ademola, I. (2018). A meta-analysis of the prevalence of African animal trypanosomiasis in Nigeria from 1960 to 2017. *Parasites & Vectors*, *11*(1), 280. doi:10.118613071-018-2801-0 PMID:29720251

Odeniran, P., Macleod, E., Ademola, I., & Welburn, S. (2019). Molecular identifcation of bloodmeal sources and trypanosomes in *Glossina* spp., *Tabanus* spp. and *Stomoxys* spp. trapped on cattle farm settlements in southwest Nigeria. *Medical and Veterinary Entomology*, *33*(2), 269–281. doi:10.1111/mve.12358 PMID:30730048

Odongo, D., Kamau, L., Skilton, R., Mwaura, S., Nitsch, C., Musoke, A., Taracha, E., Daubenberger, C., & Bishop, R. (2007). Vaccination of cattle with TickGARD induces cross-reactive antibodies binding to conserved linear peptides of Bm86 homologues in *Boophilus decoloratus*. *Vaccine*, *25*(7), 1287–1296. doi:10.1016/j.vaccine.2006.09.085 PMID:17070625

Ogunkoya, A. B., Adeyanju, J. B., & Aliu, Y. O. (1981). Experiences with the use of Imizol in treating canine blood parasites in Nigeria. *The Journal of Small Animal Practice*, *22*(12), 775–777. doi:10.1111/j.1748-5827.1981.tb00585.x PMID:7339205

OIE (World Organization for Animal Health). (2008). Chapter 2.4.1 Bovine Anaplasmosis. In *Manual of Diagnostic Tests and Vaccines for Terrestrial Animals* (6th ed., pp. 599-610). World Organization for Animal Health.

OIE WAHID. (2014). *Office International Des Epizooties–World Animal Health Information Database (WAHID) Interface*. http://www.oie.int/wahis/public. php? page=home

OIE. (1997). *Manual of standards for diagnostic tests and vaccines* (3rd ed.). World Organization for Animal Health.

OIE. (2009). *World Animal Health Information Database - Version: 1.4. World Animal Health Information Database*. Paris, France: World Organisation for Animal Health. http://www.oie.int

OIE. (2012). *Manual of Diagnostic Tests and Vaccines for Terrestrial Animals*. OIE.

OIE. (2012). Trypanosoma evansi infection (Surra). In Manual of Diagnostic Tests and Vaccines for Terrestrial Animals. Paris, France: OIE.

OIE. (2013). *Trypanosoma evansi infections (including surra)*. OIE.

OIE. (2014). *Manual of Diagnostic Tests and Vaccines for Terrestrial Animals*. OIE.

OIE. (2018). *OIE Terrestrial Manual. Animal trypanosomosis*. https://www.oie.int/fileadmin/Home/eng/Health_standards/tahm/3.04.16_TRYPANOSOMOSIS.pdf

Okech, G., Watson, E., Luckins, A., & Makawiti, D. (1996). The effect of experimental infection of Boran cattle in early and mid-pregnancy with Trypanosoma vivax. *The British Veterinary Journal*, *152*(4), 441–451. doi:10.1016/S0007-1935(96)80038-8 PMID:8791852

Okello-Onen, J., Tukahirwa, E. M., Perry, B. D., Rowlands, G. J., Nagda, S. M., Musisi, G., Bode, E., Heinonen, R., Mwayi, W., & Opuda-Asibo, J. (1999). Population dynamics of ticks on indigenous cattle in a pastoral dry to semi-arid rangeland zone of Uganda. *Experimental & Applied Acarology*, *23*(1), 79–88. doi:10.1023/A:1006058317111 PMID:10353189

Okoth, S. (2007). *Transmission dynamics and epidemiology of Rhodesian sleeping sickness in allopatric populations of Glossina pallidipes of Kenya* [PhD Thesis]. Kenyatta University.

Okoth, E., Gallardo, C., Macharia, J. M., Omore, A., Pelayo, V., Bulimo, D. W., & Bishop, R. P. (2013). Comparison of African swine fever virus prevalence and risk in two contrasting pig farming systems in South west and Central Kenya. *Preventive Veterinary Medicine*, *110*, 198205. https://doi.org/10.1016/j.prevetmed.2012.11.012

Olaho-Mikani, W., Nyang'ao, J. M. N., & Ouma, J. (1995). Elimination of drug-resistant Trypansoma evansi infection in a group of camels by sequential treatment with Melarsomine. *Journal of Camel Practice and Research*, *2*(2), 83–85.

Olaho-Mukani, W. (1981). *Studies on the epidemiology of Trypanosoma evansi in selected areas of Kenya* (Thesis MSc). University of Nairobi.

Olaho-Mukani, W., Kakaire, D., Matovu, E., & Enyaru, J. (1998). Prevalence of surra in dromedary camels in Uganda. *The Journal of Protozoology Research*, *8*(3), 120–125.

Olaho-Mukani, W., Munyua, W. K., Mutugi, M. W., & Njogu, A. R. (1993). Comparison of antibodyand antigen detection enzyme immunoassays for the diagnosis of Trypanosoma evansi infection in camels. *Veterinary Parasitology*, *45*(3-4), 231–240. doi:10.1016/0304-4017(93)90078-2 PMID:8447066

Olaho-Mukani, W., Nyanga'o, J. M. N., & Ouma, J. O. (1996). Use of Suratex for field diagnosis of patent and non-patent Trypanosoma evansi infections in camels. *The British Veterinary Journal*, *152*(1), 109–111. doi:10.1016/S0007-1935(96)80091-1 PMID:8634859

Olet, P., & Othieno, J. (2015). How to do mass media publicity for a neglected disease. Lessons from Tsetse and Trypanosomiasis in Kenya. *Journal of Science Communication*, *14*(3), Y03.

Olwoch, J. M., Reyers, B., Engelbrecht, F. A., & Erasmus, B. F. N. (2008). Climate change and the tick-borne disease, theileriosis (East Coast fever) in sub-Saharan Africa. *Journal of Arid Environments*, *72*(2), 108–120. doi:10.1016/j.jaridenv.2007.04.003

Omolo, M., Hassanali, A., Mpiana, S., Esterhuizen, J., Lindh, J., Lehane, M., Solano, P., Rayaisse, J. B., Vale, G. A., Torr, S. J., & Tirados, I. (2009). Prospects for developing odour baits to control Glossina fuscipes spp., the major vector of human African trypanosomiasis. *PLoS Neglected Tropical Diseases*, *3*(5), e435. doi:10.1371/journal.pntd.0000435 PMID:19434232

Omondi, P., Bitok, E., & Kagiri, J. (2004). Managing human–elephant conflicts: The Kenyan experience. *Pachyderm*, *36*, 80–86.

Onono, J. O., Wieland, B., & Rushton, J. (2013). Constraints to cattle production in a semiarid pastoral system in Kenya. *Tropical Animal Health and Production*, *45*(6), 1415–1422. doi:10.100711250-013-0379-2 PMID:23417827

Onyiche, T. E., Suganuma, K., Igarashi, I., Yokoyama, N., Xuan, X., & Thekisoe, O. (2019). A Review on Equine Piroplasmosis: Epidemiology, Vector Ecology, Risk Factors, Host Immunity, Diagnosis and Control. *International Journal of Environmental Research and Public Health*, *16*(10). https://doi.org/10.3390/ijerph16101736

Onyilagha, C., & Uzonna, J. (2019). Host Immune Responses and Immune Evasion Strategies in African Trypanosomiasis. *Frontiers in Immunology*, *10*, 2738. doi:10.3389/fimmu.2019.02738 PMID:31824512

Opiyo, E. A., Njogu, A. R., & Omuse, J. K. (1990). Use of impregnated targets for control of Glossina pallidipes in Kenya. *International Journal of Tropical Insect Science*, *11*(03), 417–425. doi:10.1017/S1742758400012844

Orenge, C. O., Munga, L., Kimwele, C. N., Kemp, S., Korol, A., Gibson, J. P., Hanotte, O., & Soller, M. (2012). Trypanotolerance in N'Dama x Boran crosses under natural trypanosome challenge: Effect of test-year environment, gender, and breed composition. *BMC Genetics*, *13*(1), 87. doi:10.1186/1471-2156-13-87 PMID:23075408

Orenge, C., Munga, L., Kimwele, C., Kemp, S., Korol, A., Gibson, J., Hanotte, O., & Soller, M. (2011). Expression of trypanotolerance in N'Dama x Boran crosses under field challenge in relation to N'Dama genome content. *BMC Proceedings*, *5*(Suppl 4), S23. doi:10.1186/1753-6561-5-S4-S23 PMID:21645303

Orji, C. U., Onyeocha, I. O., Shaida, S., & Ella, E. (2015). Molecular identification of tsetse fly (Diptera: Glossinidae) species based on mitochondrial DNA (COII and CytB) sequences. *African Journal of Biotechnology*, *14*, 1605–1613. doi:10.5897/AJB2015.14411

Osmonov, Y. D. (2019). *Methods for Acaricidal Solute Disinfection and Technical Means for their Implementation.* Academic Press.

Otieno, A. (2010). 'Reordering and Counterordering': Forestry Preservation, Bush Clearing and the Sociophysical Mapping of Chepalungu, Kericho District, Kenya, 1930–1963. *Environment and History, 16*(4), 381-408. Retrieved April 22, 2020, from www.jstor.org/stable/25799095

Ouedraogo, G. M. S., Demirbas-Uzel, G., Rayaisse, J.-B., Gimonneau, G., Traore, A. C., Avgoustinos, A., Parker, A. G., Sidibe, I., Ouedraogo, A. G., Traore, A., Bayala, B., Vreysen, M. J. B., Bourtzis, K., & Abd-Alla, A. M. (2018). Prevalence of trypanosomes, salivary gland hypertrophy virus and Wolbachia in wild populations of tsetse flies from West Africa. *BMC Microbiology, 18*(1), 153. doi:10.118612866-018-1287-4 PMID:30470187

Oura, C. A. L., Tait, A., Asiimwe, B., Lubega, G. W., & Weir, W. (2011). Haemoparasite prevalence and *Theileria parva* strain diversity in Cape buffalo (*Syncerus caffer*) in Uganda. *Veterinary Parasitology, 175*(3-4), 212–219. doi:10.1016/j.vetpar.2010.10.032 PMID:21074945

Owen, I. (1975). Survival of the Non-Parasitic Stages of the Cattle Tick, Boophilus Microplus, Under Simulated Field Conditions in Papua New Guinea. *Australian Journal of Zoology, 23*(2), 219. https://doi.org/10.1071/ZO9750219

Oyieke, F. (1989). Studies on Occurrence and Transmission of Camel Trypanosomiasis in Northern Kenya. PhD, University of Nairobi, Kenya.

PAAT. (2008). *On Target Against Poverty - The Programme Against African Trypanosomiasis 1997-2007*. PAAT Information Service.

Paling, R., Leak, S., Katende, J., Kamunya, G., & Moloo, S. (1987). Epidemiology of animal trypanosomiasis on a cattle ranch in Kilifi, Kenya. *Acta Tropica, 44*, 67–82. PMID:2884841

Palmer, G. H., Rurangirwa, F. R., & McElwain, T. F. (2001). Strain composition of the ehrlichia *Anaplasma marginale* within persistently infected cattle, a mammalian reservoir for tick transmission. *Journal of Clinical Microbiology, 39*(2), 631–635. doi:10.1128/JCM.39.2.631-635.2001 PMID:11158120

Papadopoulos, B., Morel, P. C., & Aeschlimann, A. (1996). Ticks of domestic animals in the Macedonia region of Greece. *Veterinary Parasitology, 63*(1-2), 25–40. doi:10.1016/0304-4017(95)00877-2 PMID:8792578

Paris, J., Murray, M., & McOdimba, A. (1982). A comparative evaluation of the parasitological techniques currently available for the diagnosis of African trypanosomiasis in cattle. *Acta Tropica, 39*, 307–316. PMID:6131590

Paris, J., Murray, M., & McOdimba, F. (1982). A comparative evaluation of the parasitological techniques currently available for the diagnosis of African trypanosomosis in cattle. *Acta Tropica, 39*, 307–316. PMID:6131590

Parola, P., Cornet, J. P., Sanogo, Y. O., Miller, R. S., Van Thien, H., Gonzalez, J. P., Raoult, D., Telford, S. R. III, & Wongsrichanalai, C. (2003). Detection of Ehrlichia spp, Anaplasma spp, Rickettsia spp, and other eubacteria in ticks from the Thai-Myanmar border and Vietnam. *Journal of Clinical Microbiology, 41*(4), 1600–1608. doi:10.1128/JCM.41.4.1600-1608.2003 PMID:12682151

Parola, P., Socolovschi, C., Jeanjean, L., Bitam, I., Fournier, P. E., Sotto, A., Labauge, P., & Raoult, D. (2008). Warmer weather linked to tick attack and emergence of severe rickettsioses. *PLoS Neglected Tropical Diseases, 2*(11), e338. doi:10.1371/journal.pntd.0000338 PMID:19015724

Parzy, D., Davoust, B., Raphenon, G., & Vidor, E. (1991). Canine ehrlichiosis in Senegal: human and canine seroepidemiological survey in Dakar. *Medicina tropical: revue du Corps de sante colonial, 51*(1), 59–63.

Pascucci, I., Andrea, D., Cesare, C., Gabriella, D., Paolo, C., Manuela, T., Nicola, F., Massimo, S., & Vincenzo, C. (2013). Diagnosis of dourine in outbreaks in Italy. *Veterinary Parasitology, 193*(1-3), 30–38. doi:10.1016/j.vetpar.2012.12.006 PMID:23298562

Payne, S. (2012). Bunyaviridae. In Virus Taxonomy (pp. 725–741). Elsevier. https://doi.org/10.1016/B978-0-12-384684-6.00059-8.

Pegram, R. G., & Banda, D. S. (1990). Ecology and phenology of cattle ticks in Zambia: Development and survival of free-living stages. *Experimental & Applied Acarology*, *8*(4), 291–301. doi:10.1007/BF01202139 PMID:2350995

Pegram, R. G., Keirans, J. E., Clifford, C. M., & Walker, J. B. (1987). Acari, Ixodoidae. *Systematic Parasitology*, *10*, 3–44. doi:10.1007/BF00009099

Pegram, R. G., Keirans, J. E., Clifford, C. M., & Walker, J. B. (1987). Clarification of the Rhipicephalus sanguineus group (Acari, Ixodoidea, Ixodidae). II. R. sanguineus (Latreille, 1806) and related species. *Systematic Parasitology*, *10*(1), 27–44. doi:10.1007/BF00009100

Pegram, R. G., Perry, B. D., Musisi, E. L., & Mwanaumo, B. (1986). Ecology and phenology of ticks in Zambia: Seasonal dynamics on cattle. *Experimental & Applied Acarology*, *2*(1), 25–45. doi:10.1007/BF01193353 PMID:3451861

Peixoto, C. C., Marcelino, I., Vachiéry, N., Bensaid, A., Martinez, D., Carrondo, M. J. T., & Alves, P. M. (2005). Quantification of Ehrlichia ruminantium by real time PCR. *Veterinary Microbiology*, *107*(3/4), 273–278. doi:10.1016/j.vetmic.2005.02.001 PMID:15863287

Penrith, M. L., Lopes Pereira, C., Lopes Da Silva, M. M. R., Quembo, C., Nhamusso, A., & Banze, J. (2007). African swine fever in Mozambique: Review, risk factors and considerations for control. *The Onderstepoort Journal of Veterinary Research*, *74*, 149–160.

Penrith, M. L., & Thomson, G. R. (2004). Special factors affecting the control of livestock diseases in sub-Saharan Africa. In J. A. W. Coetzer & R. C. Tustin (Eds.), *Infectious diseases of livestock* (2nd ed.). Oxford University Press.

Penrith, M. L., Vosloo, W., Jori, F., & Bastos, A. D. S. (2013). African swine fever virus eradication in Africa. *Virus Research*, *173*, 228–246.

Penzhorn, B. L., & Verwoerd, D. W. (2005). Infectious diseases of livestock, 2nd edition: Book review. *Journal of the South African Veterinary Association, 76*(4), 232. doi:10.4102/jsava.v76i4.435

Penzhorn, B. L., Netherlands, E. C., Cook, C. A., Smit, N. J., Vorster, I., Harrison-White, R. F., & Oosthuizen, M. C. (2018). Occurrence of Hepatozoon canis (Adeleorina: Hepatozoidae) and Anaplasma spp. (Rickettsiales: Anaplasmataceae) in black-backed jackals (Canis mesomelas) in South Africa. *Parasites & Vectors*, *11*(1), 1–7. doi:10.118613071-018-2714-y PMID:29554933

Percoma, L. A., & Pagabeleguem, S. (2018). Impact of an integrated control campaign on tsetse populations in Burkina Faso. *Parasites & Vectors*, *11*(1), 270. doi:10.118613071-017-2609-3 PMID:29703229

Peregrine, A. (1994). Chemotherapy and delivery systems: Haemoparasites. *Veterinary Parasitology*, *54*(1-3), 223–248. doi:10.1016/0304-4017(94)90092-2 PMID:7846852

Perry, B., Chamboko, T., Mahan, S., & Medley, G. (1998). *The economics of integrated tick and tick-borne disease control on commercial farms in Zimbabwe*. Academic Press.

Perry, B.D. (2016). The control of East Coast fever of cattle by live parasite vaccination. *One Health*, *2*, 103-114.

Perry, B. D. (2016). The control of East Coast fever of cattle by live parasite vaccination: A science-to-impact narrative. *One Health*, *2*, 103–114. doi:10.1016/j.onehlt.2016.07.002 PMID:28616483

Perry, B. D., Chamboko, T., Mahan, S. M., Medley, G. F., Minjauw, B., O'callaghan, C. J., & Peter, T. F. (1998). The economics of integrated tick and tick-borne disease control on commercial farms in Zimbabwe. *Zimbabwe Veterinary Journal*, *29*, 21–29.

Perry, B. D., Lessard, P., Norval, R. A. I., Kundert, K., & Kruska, R. (1990). Climate, vegetation and distribution of *Rhipicephalus appendiculatus* in Africa. *Parasitology Today (Personal Ed.)*, *6*(4), 100–110. doi:10.1016/0169-4758(90)90224-R PMID:15463309

Persing, D. H., Mathiesen, D., Marshall, W. F., Telford, S. R., Spielman, A., Thomford, J. W., & Conrad, P. A. (1992). Detection of Babesia microti by polymerase chain reaction. *Journal of Clinical Microbiology*, *30*(8), 2097–2103.

Petney, Horak, & Rechav. (1987). The ecology of the African vectors of heartwater, with particular reference to Amblyomma hebraeum and Amblyomma variegatum. *Onderstepoort Journal of Veterinary Research, 54*(3), 381-395.

Petney, T. N., & Horak, I. G. (1988). Comparative host usage by Amblyomma hebraeum and Amblyomma marmoreum, (Acari, Ixodidae) the South African vectors of the disease heartwater. *Journal of Applied Entomology*, *105*(5), 490–495. doi:10.1111/j.1439-0418.1988.tb00215.x

Pienaar, J. G. (1970). Electron microscopy of Cowdria (Rickettsia) ruminantium (Cowdry, 1926) in the endothelial cells of the vertebrate host. *The Onderstepoort Journal of Veterinary Research*, *37*(1), 67–78. PMID:4115395

Pinchbeck, G., Morrison, L. J., Tait, A., Langford, J., Meehan, L., Jallow, S., Jallow, J., Jallow, A., & Christley, R. M. (2008). Trypanosomosis in The Gambia: Prevalence in working horses and donkeys detected by whole genome amplification and PCR, and evidence for interactions between trypanosome species. *BMC Veterinary Research*, *4*(1), 7–13. doi:10.1186/1746-6148-4-7 PMID:18289378

Pipano, E., & Hadani, A. (1984). Control of bovine babesiosis. In M. Ristic, P. Ambroise-Thomas, & J. Kreier (Eds.), Malaria and Babesiosis (pp. 263–303). https://doi.org/10.1007/978-94-009-6042-8_15.

Plessis & Malan. (1987). Problems with the interpretation of epidemiological data in heartwater: A study on 23 farms. *The Onderstepoort Journal of Veterinary Research*, *54*(3), 427–433. PMID:3448573

Plessis, Camus, Oberem, & Malan. (1987). Heartwater serology: Some problems with the interpretation of results. *The Onderstepoort Journal of Veterinary Research*, *54*(3), 327–329. PMID:3329323

Plessis, Gas, Olivier, & Bezuidenhout. (1989). The heterogenicity of Cowdria ruminantium stocks: Cross-immunity and serology in sheep and pathogenicity to mice. *The Onderstepoort Journal of Veterinary Research*, *56*(3), 195–201. PMID:2812704

Plowright, W., Parker, J. A., & Peirce, M. A. (1969a). African swine fever virus in ticks (*Ornithodorosmoubata* Murray) collected from animal burrows in Tanzania. *Nature*, *221*, 1071–1073. https://doi.org/10.1038/2211071a0

Plowright, W., Parker, J., & Peirce, M. A. (1969b). The epizootiology of African swine fever in Africa. *The Veterinary Record*, *85*, 668–674.

Plowright, W., Perry, C. T., & Greig, A. (1974). Sexual transmission of African swine fever virus in the tick, *Ornithodorosmoubataporcinus* Walton. *Research in Veterinary Science*, *17*, 106–113.

Plowright, W., Thomson, G. R., & Neser, J. A. (1994). African swine fever. In J. A. W. Coetzer, G. R. Thomson, & R. C. Tustin (Eds.), *Infectious diseases of Livestock, with special reference to Southern Africa, 1* (pp. 568–599). Oxford University Press South Africa.

Pollock, J. N. (1982). Training manual for tsetse control personnel – volume 1: tsetse biology, systematics and distribution; techniques. FAO.

Pollock, J. N. (1992). *Tsetse biology, systematics and distribution techniques*. FAO.

Postigo, M., Bell-Sakyi, L., Paxton, E., & Sumption, K. (2002). Kinetics of experimental infection of sheep with Ehrlichia ruminantium cultivated in tick and mammalian cell lines. Experimental and Applied Acarology. *Ticks and Tick-Borne Pathogens. Proceedings of the 4th International Conference on Ticks and Tick-Borne Pathogens*, *28*(1/4), 187-193. 10.1023/A:1025390215007

Potgieter, F. T., & Els, H. J. (1977a). Light and electron microscopic observations on the development of Babesia bigemina in larvae, nymphae and non-replete females of Boophilus decoloratus. *The Onderstepoort Journal of Veterinary Research*, *44*(4), 213–231.

Potgieter, F. T., & Els, H. J. (1977b). The fine structure of intra-erythrocytic stages of Babesia bigemina. *The Onderstepoort Journal of Veterinary Research*, *44*(3), 157–168.

Potgieter, F. T., & Els, H. J. (1979). An electron microscopic study of intra-erythrocytic stages of Babesia bovis in the brain capillaries of infected splenectomized calves. *The Onderstepoort Journal of Veterinary Research*, *46*(1), 41–49.

Potgieter, F. T., & Stoltsz, W. H. (2004). Bovine anaplasmosis. In J. A. W. Coetzer & R. C. Tustin (Eds.), *Infectious Diseases of Livestock, 3* (2nd ed., pp. 594–616). Oxford University Press, Southern Africa.

Potgieter, F. T., & Van Rensburg, L. (1987). The persistence of colostral *Anaplasma* antibodies in incidence of in utero-transmission of *Anaplasma* infections in calves under laboratory conditions. *The Onderstepoort Journal of Veterinary Research*, *54*, 557–560. PMID:3444609

Preston, P. M. (1981). The role of macrophages in protective immunity and immunosuppression in bovine theileriosis. In A. D. Irvin, M. P. Cunnigham, & A. S. Young (Eds.), *Advances in the Control of Theileriosis* (pp. 354–356). Martinus Nijhoff Publishers. doi:10.1007/978-94-009-8346-5_57

Preston, P. M., & Brown, C. G. D. (1988). Macrophage-mediated cytostasis and lymphocyte cytotoxicity in cattle immunized with *Theileria annulata* sporozoites or macroschizont-infected cell lines. *Parasite Immunology*, *10*(6), 631–647. doi:10.1111/j.1365-3024.1988.tb00250.x PMID:3146048

Pretorius, A. M., & Kelly, P. J. (1998). Serological survey for antibodies reactive with Ehrlichia canis and E. chaffeensis in dogs from the Bloemfontein area of South Africa. *Journal of the South African Veterinary Association*, *64*(4), 126–128. doi:10.4102/jsava.v69i4.840 PMID:10192085

Price, J. E., & Karstad, L. H. (1980). Free-living jackals (*Canis mesomelas*) - potential reservoir hosts for *Ehrlichia canis* in Kenya. *Journal of Wildlife Diseases*, *16*(4), 469–473. doi:10.7589/0090-3558-16.4.469 PMID:7463598

Proboste, T., Kalema-Zikusoka, G., Altet, L., Solano-Gallego, L., Fernández de Mera, I. G., Chirife, A. D., Muro, J., Bach, E., Piazza, A., Cevidanes, A., Blanda, V., Mugisha, L., de la Fuente, J., Caracappa, S., & Millán, J. (2015). Infection and exposure to vector-borne pathogens in rural dogs and their ticks, Uganda. *Parasites & Vectors*, *8*(1), 306. doi:10.118613071-015-0919-x PMID:26043771

Punyua, D. K., Newson, R. M., & Mutinga, M. J. (1985). Diurnal and seasonal activity of unfed adult *Rhipicephalus appendiculatus* (Acarina: Ixodidae) in relation to some intrinsic factors. I. Factors regulating activity. *Insect Science and Its Application*, *6*, 63–70.

Purchase, H. S. (1945). A simple and rapid method for demonstrating Rickettsia ruminantium (Cowdry, 1925) in heartwater brains. *The Veterinary Record*, *57*(36), 413–414.

Purnell, R. E., Brown, C. G. D., Cunningham, M. P., Burridge, M. J., Kirimi, T. M., & Ledger, M. A. (1973). East Coast fever: Correlation between the morphology and infectivity of *Theileria parva* developing in its tick vector. *Veterinary Parasitology*, *66*(3), 539–544. doi:10.1017/S0031182000046084 PMID:4198658

Pypekamp & Prozesky. (1987). Heartwater. An overview of the clinical signs, susceptibility and differential diagnosis of the disease in domestic ruminants. *The Onderstepoort Journal of Veterinary Research*, *54*(3), 263–266. PMID:3329315

Qiu, Y., Kaneko, C., Kajihara, M., Ngonda, S., Simulundu, E., Muleya, W., Thu, M. J., Hang'ombe, M. B., Katakura, K., Takada, A., Sawa, H., Simuunza, M., & Nakao, R. (2018). Tick-borne haemoparasites and Anaplasmataceae in domestic dogs in Zambia. *Ticks and Tick-Borne Diseases*, *9*(1), 988–995. doi:10.1016/j.ttbdis.2018.03.025 PMID:29622515

Quembo, C., Jori, F., Heath, L., Pérez Sánchez, R., & Vosloo, W. (2016). Investigation into the epidemiology of African swine fever virus at the wildlife domestic interface of the Gorongosa National Park, Central Mozambique. *Transboundary and Emerging Diseases*, *63*, 443–451. https://doi.org/10.1111/tbed.12289

Qurollo, B. A., Davenport, A. C., Sherbert, B. M., Grindem, C. B., Birkenheuer, A. J., & Breitschwerdt, E. B. (2013). Infection with Panola Mountain E. hrlichia sp. in a Dog with Atypical Lymphocytes and Clonal T-Cell Expansion. *Journal of Veterinary Internal Medicine*, *27*(5), 1251–1255. doi:10.1111/jvim.12148 PMID:23875820

Radcliffe, R. W., & Khairani, K. O. (2019). Health of the Forest Rhinoceros of Southeast Asia. In *Fowler's Zoo and Wild Animal Medicine Current Therapy* (Vol. 9, pp. 707–715). Elsevier., doi:10.1016/B978-0-323-55228-8.00100-4

Radley, D. E. (1978). *Immunisation against East Coast fever by chemoprophylaxis*. FAO Technical Report 1, AG:DP, RAF/67/077, FAO, Rome.

Radley, D. E. (1981). Infection and treatment method against theileriosis. In A. D. Irvin, M. P. Cunnigham, & A. S. Young (Eds.), *Advances in the Control of Theileriosis* (pp. 227–237). Martinus Nijhoff Publishers. doi:10.1007/978-94-009-8346-5_41

Radley, D. E., Brown, C. G. D., Burridge, M. J., Cunnigham, M. P., Pierce, M. A., & Purnell, R. E. (1974). East Coast fever: Quantitative studies of *Theileria parva* in cattle. *Experimental Parasitology*, *36*(2), 278–287. doi:10.1016/0014-4894(74)90067-8 PMID:4213046

Radley, D. E., Brown, C. G. D., Burridge, M. J., Cunningham, M. P., Kirimi, I. M., Purnell, R. E., & Young, A. S. (1975). East Coast fever, I. chemophylactic immunization of cattle against *Theileria parva* (Muguga) and five *Theileria* strains. *Veterinary Parasitology*, *1*(1), 35–41. doi:10.1016/0304-4017(75)90005-9

Radostits, O., Blood, D., Gay, C., & Hinchcliff, K. (2000). *Veterinary Medicine: A textbook of the diseases of cattle, sheep, pigs, goats and horses*. Baillière Tindall.

Radostits, O., Gay, C., Hinchcliff, K., & Constable, P. (2006). *Veterinary Medicine:A textbook of the diseases of cattle, horses, sheep, pigs and goats* (10th ed.). Saunders Elsevier.

Radwanska, M., Vereecke, N., Deleeuw, V., Pinto, J., & Magez, S. (2018). Salivarian Trypanosomosis: A Review of parasites involved, their global distribution and their interaction with the innate and adaptive mammalian host immune system. *Frontiers in Immunology*, *9*(2253), 1–14. doi:10.3389/fimmu.2018.02253 PMID:30333827

Raidimi, E. N., & Kabiti, H. M. (2019). A review of the role of agricultural extension and training in achieving sustainable food security: A case of South Africa. *South African Journal of Agricultural Extension*, *47*(3). Advance online publication. doi:10.17159/2413-3221/2019/v47n3a520

Rajagopalan, P. K., Sreenivasan, M. A., & Paul, S. D. (1970). Isolation of Ganjam virus from the bird tick Haemaphysalis wellingtoni Nuttall and Warburton 1907. *The Indian Journal of Medical Research*, *58*(9), 1195–1196. https://www.ncbi.nlm.nih.gov/pubmed/5534070

Rajput, Z. I., Hu, S., Chen, W., Arijo, A. G., & Xiao, C. (2006). Importance of ticks and their chemical and immunological control in livestock. *J Zhejiang Univ Science B, 7*(11), 912-921. www.springerlink.com

Rajput, Z. I., Hu, S., Chen, W., Arijo, A. G., & Xiao, C. (2006). Importance of ticks and their chemical and immunological control in livestock. *Journal of Zhejiang University. Science. B.*, *7*(11), 912–921. https://doi.org/10.1631/jzus.2006.B0912

Ramanatha, K., Keshavamurthy, B., & Gopal, T. (1990). Acute trypanosomiasis in zebu cattle (Bos Indicus) in epidemic form-a report. *Indian Journal of Comparative Microbiology. Immunology & Infectious Diseases*, *11*, 158–159.

Rand, K. N., Moore, T., Sriskantha, A., Spring, K., Tellam, R., Willadsen, P., & Cobon, G. S. (1989). Cloning and expression of a protective antigen from the cattle tick *Rhipicephalus (Boophilus) microplus. Proceedings of the National Academy of Sciences of the United States of America*, *86*, 9657 – 9661.

Raskin, R. E., & Messick, J. B. (2012). Bone marrow cytologic and histologic biopsies: Indications, techniques, and evaluation. *Veterinary Clinics: Small Animal Practice*, *42*(1), 23–42. doi:10.1016/j.cvsm.2011.10.001 PMID:22285155

Rautenbach, G. H., Boomker, J., & De Villiers, I. L. (1991). A descriptive study of the canine population in a rural town in southern Africa. *Journal of the South African Veterinary Association*, *62*(4), 158–162. doi:10.4102/jsava.v62i4.1778 PMID:1770490

Rautenbach, R., Schoeman, J., & Goddard, A. (2018). Prevalence of canine Babesia and Ehrlichia co-infection and the predictive value of haematology. *The Onderstepoort Journal of Veterinary Research*, *85*(1), 1–5. doi:10.4102/ojvr.v85i1.1626 PMID:30326715

Rayaisse, J. B., Tirados, I., Kaba, D., Dewhirst, S. Y., Logan, J. G., Diarrassouba, A., & Esterhuizen, J. (2010). Article. *PLoS Neglected Tropical Diseases*, *4*, 632–638.

Rayaisse, J., Salou, E., & Courtin, F. (2015). Baited-boats: An innovative way to control riverine tsetse, vectors of sleeping sickness in West Africa. *Parasites & Vectors*, *8*(1), 236.

Rayaisse, J., Salou, E., Courtin, F., Yoni, W., Barry, I., Dofini, F., Kagbadouno, M., Camara, M., Torr, S. J., & Solano, P. (2015). Baited-boats: An innovative way to control riverine tsetse, vectors of sleeping sickness in West Africa. *Parasites & Vectors*, *8*(1), 236. doi:10.118613071-015-0851-0 PMID:25928366

Rebeski, D. E. (1999). Improved Methods for the Diagnosis of African Trypanosomosis. *Mem. Inst. Oswaldo Cruz, 94*(2), 249-253.

Rege, J. E. O., & Tawah, C. L. (1999). The state of African cattle genetic resources II. Geographical distribution, characteristics and uses of present-day breeds and strains. *Animal Genetic Resources/Resources Génétiques Animales/ Recursos Genéticos Animales, 26*, 1–25.

Rege, J. E. O., Kahi, A., Okomo-Adhiambo, M., Mwacharo, J. M., & Hanotte, O. H. (2001). *Zebu cattle of Kenya: Uses, performance, farmer preferences, measures of genetic diversity and options for improved use*. International Livestock Research Institute. https://cgspace.cgiar.org/handle/10568/4196

Regitano, L., & Prayaga, K. (2011). Ticks and tick-borne diseases in cattle. In S. Bishop, R. Axford, F. Nicholas, & J. Owen (Eds.), *Breeding for Disease Resistance in Farm Animals* (3rd ed., pp. 295–314). CAB International.

Reichard, R. E. (2002). Area-wide biological control of disease vectors and agents affecting wildlife. *Rev. sci. tech. Off. int. Epiz.*, *21*(1), 179–185. PMID:11974628

Reid, R. S., Kruska, R. L., Deichmann, U., Thornton, P. K., & Leak, S. G. A. (2000). Human population growth and the extinction of the tsetse fly. *Agriculture, Ecosystems & Environment*, *77*(3), 227–236. doi:10.1016/S0167-8809(99)00103-6

Reinbold, J. B., Coetzee, J. F., Hollis, L. C., Nickell, J. S., Riegel, C., Christopher, J. A., & Ganta, R. R. (2010). Comparison ofiatrogenic *Anaplasma marginale* transmission in Holstein steers by needle and needle-free injection techniques. *American Journal of Veterinary Research*, *71*(10), 1178–1188. doi:10.2460/ajvr.71.10.1178 PMID:20919904

Richey, E. J. (1981). Bovine anaplasmosis. In R. S. Howard (Ed.), *Current Veterinary Therapy Food Animal Practice* (pp. 767–772). W. B. Saunders.

Richey, E. J. (1991). Bovine anaplasmosis. *Proceedings of the 24th Annual Conference of the American Association of Bovine Practitioners*, 3-11.

Riek, R. (1964). The life cycle of Babesia bigemina (Smith and Kilborne, 1893) in the tick vector Boophilus microplus (Canestrini). *Australian Journal of Agricultural Research*, *15*(5), 802. https://doi.org/10.1071/AR9640802

Rikihisa, Y. (1991). The tribe Ehrlichieae and ehrlichial diseases. *Clinical Microbiology Reviews*, *4*(3), 286–308. doi:10.1128/CMR.4.3.286 PMID:1889044

Rikihisa, Y., Stills, H., & Zimmerman, G. (1991). Isolation and continuous culture of Neorickettsia helminthoeca in a macrophage cell line. *Journal of Clinical Microbiology*, *29*(9), 1928–1933. doi:10.1128/JCM.29.9.1928-1933.1991 PMID:1774317

Rio, R., Wu, Y., Filardo, G., & Aksoy, S. (2006). Dynamics of multiple symbiont density regulation during host development: Tsetse fly and its microbial flora. *Proceedings. Biological Sciences*, *273*(1588), 805–814. doi:10.1098/rspb.2005.3399 PMID:16618673

Ristic, M. (1981). Babesiosis. In M. Ristic & W. I. M. McIntyre (Eds.), Diseases of Cattle in the Tropics (pp. 443–468). https://doi.org/10.1007/978-94-011-9034-3_35.

Ristic, M., Huxsoll, D. L., Weisiger, R. M., Hildebrandt, P. K., & Nyindo, M. B. A. (1972). Serological diagnosis of tropical canine pancytopenia by indirect immunofluorescence. *Infection and Immunity*, *6*(3), 226–231. doi:10.1128/IAI.6.3.226-231.1972 PMID:4118045

Robays, J., Ebeja Kadima, A., Lutumba, P., Miaka mia Bilenge, C., Kande Betu Ku Mesu, V., De Deken, R., Makabuza, J., Deguerry, M., Van der Stuyft, P., & Boelaert, M. (2004). Human African trypanosomiasis amongst urban residents in Kinshasa: A case-control study. T. *Tropical Medicine & International Health*, *9*(8), 869–875. doi:10.1111/j.1365-3156.2004.01266.x PMID:15303991

Robbertse, L., Baron, S., Van der Merwe, N. A., Madder, M., Stoltsz, W. H., & Maritz-Olivier, C. (2016). Genetic diversity, acaricide resistance status and evolutionary potential of a *Rhipicephalus microplus* population from a disease-controlled cattle farming area in South Africa. *Ticks and Tick-Borne Diseases*, *7*(4), 595–603. doi:10.1016/j.ttbdis.2016.02.018 PMID:26965787

Roberts, J. A. (1968). Resistance of cattle to the tick boophilus microplus (canestrini). II. Stages of the life cycle of the parasite against which resistance is manifest. *The Journal of Parasitology*, *54*(4), 667–673.

Roberts, L., Wellde, B., Reardon, M., & Onyango, F. (1989). Mechanical transmission of Trypanosoma brucei rhodesiense by Glossina morsitans morsitans (Diptera: Glossinidae). *Annals of Tropical Medicine and Parasitology*, *83*(sup1), 127–131. doi:10.1080/00034983.1989.11812417 PMID:2619386

Roderick, S. (1995). *Pastoralist cattle productivity in a tsetse infested area of south west Kenya* [PhD Thesis]. University of Reading.

Rodgers, S. J., Welsh, R. D., & Stebbins, M. E. (1994). Seroprevalence of bovine anaplasmosis in Oklahoma from 1977 to 1991. *Journal of Veterinary Diagnostic Investigation*, *6*(2), 200–206. doi:10.1177/104063879400600211 PMID:8068752

Rodriguez, M., Rubiera, R., & Montesinos, R. (1994). High level expression of the *B. microplus*Bm86 antigen in the yeast *P. pastoris* forming highly immunogenic particles for cattle. *Journal of Biotechnology*, *33*(2), 135–146. doi:10.1016/0168-1656(94)90106-6 PMID:7764729

Rodriguez-Vivas, R. I., Jonsson, N. N., & Bhushan, C. (2018). Strategies for the control of Rhipicephalus microplus ticks in a world of conventional acaricide and macrocyclic lactone resistance. *Parasitology Research*, *117*(1), 3–29. https://doi.org/10.1007/s00436-017-5677-6

Rodriquez, M., Massard, C., Henrique da Fonseca, A., Fonseca, R. N., Machado, H., Labarta, V., & De La Fuente, J. (1995). Effect of vaccination with a recombinant Bm86 antigen preparation on natural infestations of Boophilus microplus in grazing dairy and cross-bred cattle in Brazil. *Vaccine*, *13*(18), 1804–1808. doi:10.1016/0264-410X(95)00119-L PMID:8701597

Roger, F., Ratovonjato, J., Vola, P., & Uilenberg, G. (2001). *Ornithodorosporcinus* ticks, bush pigs, and African swine fever in Madagascar. *Experimental & Applied Acarology*, *25*, 263–269.

Rogers, D. (1979). Tsetse Population Dynamics and Distribution: A New Analytical Approach. *Journal of Animal Ecology*, *48*(3), 825–849. doi:10.2307/4197

Rogers, D. J., Hay, S. I., & Packer, M. J. (1996). Predicting the distribution of tsetse flies in West Africa using temporal Fourier processed meteorological satellite data. *Annals of Tropical Medicine and Parasitology*, *90*(3), 225–241. doi:10.1080/00034983.1996.11813049 PMID:8758138

Rogers, D., & Randolph, S. E. (1985). Population Ecology of Tsetse. *Annual Review of Entomology*, *30*(1), 197–216. doi:10.1146/annurev.en.30.010185.001213 PMID:3882049

Rogoff, W. M. (1961). Chemical control of insect pests of domestic animals. *Advances in Pest Control Research*, *4*, 153.

Rohlf, F., & Marcus, L. (1993). A revolution in morphometrics. *Trends in Ecology & Evolution*, *8*(4), 129–132. doi:10.1016/0169-5347(93)90024-J PMID:21236128

Roopali, B., Vivek, R., Kasaralikar, N. A., Patil, R. B. G., Sandeep, H., & Dilipkumar, D. (2018). Clinico, haemato-biochemical changes and therapeutic management of canine ehrlichiosis. *Pharma Innovation Journal*, *7*(9), 1–6.

Roqueplo, C., Cheminel, V., Bourry, O., Gomez, J., Pre'vosto, J. M., Parzy, D., & Davoust, B. (2009). Canine ehrlichiosis in the Ivory Coast and Gabon: Alteration of biochemical blood parameters based on Ehrlichia canis serology. *Clinical Microbiology and Infection*, *15*(2), 41–42. doi:10.1111/j.1469-0691.2008.02168.x PMID:19793128

Rosenberg, R., Lindsey, N. P., Fischer, M., Gregory, C. J., Hinckley, A. F., Mead, P. S., Paz-Bailey, G., Waterman, S. H., Drexler, N. A., Kersh, G. J., Hooks, H., Partridge, S. K., Visser, S. N., Beard, C. B., & Petersen, L. R. (2018). Vital Signs: Trends in Reported Vector borne Disease Cases - United States and Territories, 2004-2016.Morbidity and Mortality Weekly Report. *Rep*, *67*, 496. PMID:29723166

Rosenstein, M. (1976). Paralysis in chickens caused by larvae of the poultry tick, ArgasPersicus. *Avian Diseases*, *20*(2), 407–409.

Rotondano, T. E. D. F., de Almeida, A. M. P., Lustosa, E. M. C., Cordeiro, A. A., Camboim, E. K. A., de Azevedo, S. S., de Andrade, P. P., & de Melo, M. A. (2012). An assessment of whole blood and fractions by nested PCR as a DNA source for diagnosing canine ehrlichiosis and anaplasmosis. *TheScientificWorldJournal*, *2012*, 2012. doi:10.1100/2012/605743 PMID:22973174

Rottcher, D., Schillinger, D., & Sweygarth, E. (1987). Trypanosomosis in the camel. *Rev. Sci. Tech. Off. Int. Epiz.*, *6*, 463–470.

Rotureau, B., & Van Den Abbeele, J. (2013). Through the dark continent: African trypanosome devel- opment in the tsetse ñy. *Frontiers in Cellular and Infection Microbiology*, *3*, 53. doi:10.3389/fcimb.2013.00053 PMID:24066283

Rousset, F., Bouchon, D., Pintureau, B., Juchault, P., & Solignac, M. (1992). *Wolbachia* endosymbionts responsible for various alterations of sexuality in arthropods. *Proceedings. Biological Sciences*, *250*(1328), 91–98. doi:10.1098/rspb.1992.0135 PMID:1361987

Rowlands, G.J., Kyule, M.N., & Perry, B.D. (1994). *Assessing impacts of animal health interventions on productivity*. Academic Press.

Rubaire-Akiiki, C., Okello-Onen, J., Nasinyama, G. W., Vaarst, M., Kabagambe, E. K., Mwayi, W., Musunga, D., & Wandukwa, W. (2004). The prevalence of serum antibodies to tick-borne infections in Mbale District, Uganda: The effect of agro-ecological zone, grazing management and age of cattle. *Journal of Insect Science*, *4*(1), 1–8. doi:10.1673/031.004.0801 PMID:15861224

Rudoler, N., Baneth, G., Eyal, O., van Straten, M., & Harrus, S. (2012). Evaluation of an attenuated strain of Ehrlichia canis as a vaccine for canine monocytic ehrlichiosis. *Vaccine*, *31*(1), 226–233. doi:10.1016/j.vaccine.2012.10.003 PMID:23072894

Rudzinska, M. A., Spielman, A., Lewengrub, S., Trager, W., & Piesman, J. (1983). Sexuality in piroplasms as revealed by electron microscopy in Babesia microti. *Proceedings of the National Academy of Sciences of the United States of America*, *80*(10), 2966–2970. https://doi.org/10.1073/pnas.80.10.2966

Ruprah, N. (1985). *Textbook of Clinical Protozoology*. Oxonian Press Pvt. Ltd.

Rutto, J., Osano, O., Thuranira, E. G., Kurgat, R. K., & Odenyo, V. A. O. (2013). Socio-economic and cultural determinants of human African trypanosomiasis at the Kenya—Uganda transboundary. PLoS Neglected Tropical Diseases, 7(4).

Sabu, L., Sreekrishnan, R., Devada, K., Rejitha, T., & Lakshmanan, B. (2013). *Babesia species in a cat-A case report*. Academic Press.

Saiki, R. K., Gelford, D. H., & Stoffel, S. (1988). Primer directed enzymatic amplification of DNA with thermostable DNA polymerase. *Science*, *239*(4839), 487–491. doi:10.1126cience.239.4839.487 PMID:2448875

Saimo, M., & Odongo, D. O. (2011). Recombinant *Rhipicephalus appendiculatus* gut (Ra86) and salivary gland cement (Trp64) proteins as candidate antigens for inclusion in tick vaccines: Protective effects of Ra86 on infestation with adult *R. appendiculatus. Vaccine: Development and Therapy*, *1*, 15–23. doi:10.2147/VDT.S20827

Saini, R. K., Orindi, B. O., Mbahin, N., Andoke, J. A., Muasa, P. N., Mbuvi, D. M., Muya, C. M., Pickett, J. A., & Borgemeister, C. W. (2017). Protecting cows in small holder farms in East Africa from tsetse flies by mimicking the odor profile of a non-host bovid. *PLoS Neglected Tropical Diseases*, *11*(10), e0005977–e0005977. doi:10.1371/journal.pntd.0005977 PMID:29040267

Sainz, A., Roura, X., Miró, G., Estrada-Peña, A., Kohn, B., Harrus, S., & Solano-Gallego, L. (2015). Guideline for veterinary practitioners on canine ehrlichiosis and anaplasmosis in Europe. *Parasites & Vectors*, *8*(1), 75. doi:10.118613071-015-0649-0 PMID:25649069

Sainz, A., Tesouro, M. A., Amusategui, I., Rodríguez, F., Mazzucchelli, F., & Rodríguez, M. (2000). Prospective comparative study of 3 treatment protocols using doxycycline or imidocarb dipropionate in dogs with naturally occurring ehrlichiosis. *Journal of Veterinary Internal Medicine*, *14*(2), 134–139. doi:10.1111/j.1939-1676.2000.tb02226.x PMID:10772483

Salih, A. D., Marcellino, W. L., & Elhussein, J. A. M. (2011). Economic impact of East Coast fever in central equatorial state of South Sudan. *International Research Journal of Agricultural Science and Soil Science*, *1*(6), 218–220.

Salih, D. A., Mwacharo, J. M., Pelle, R., Njahira, M. N., Odongo, D. O., Mbole-Kariuki, M. N., Marcellino, W. L., Malak, A. K., Kiara, H., El Hussein, A. R. M., Bishop, R. P., & Skilton, R. A. (2018). Genetic diversity and population structure of Theileria parva in South Sudan. *Ticks and Tick-Borne Diseases*, *9*(4), 806–813. doi:10.1016/j.ttbdis.2018.03.002 PMID:29534988

Sallau, A. B., Ibrahim, M. A., Salihu, A., & Yusuf, I. A. (2008). Bloodstream form of Trypanosoma evansi contains galactosidase. *Middle East Journal of Scientific Research*, *3*, 49–52.

Samuel, B. (1996). Bunyaviruses. In S. Baron (Ed.), *Medical Microbiology* (4th ed.). University of Texas Medical Branch at Galveston. Retrieved from https://www.ncbi.nlm.nih.gov/books/NBK8004/

Sánchez Botija, C. (1963). Reservoirs of African swine fever virus/ a study of the African swine fevere virus in arthropods by means of haemadsorption. *Bulletin - Office International des Epizooties*, *60*, 895–899.

Sanchez-Cordon, P. J., Reis, A. L., & Dixon, L. K. (2018). African swine fever: A re-emerging viral disease threatening the global pig industry. *Veterinary Journal (London, England)*, *233*. Advance online publication. doi:10.1016/j.tvjl.2017.12.025

Sanchez-Vizcaino, J. M., Mur, L., & Martinez-Lopez, B. (2013). African swine fever (ASF): Five years around Europe. *Veterinary Microbiology*, *165*, 45–50.

Sandra, B., Kati, F., & Martin, B. (2020). Review of African swine fever – A review of current knowledge. *Virus Research*, *287*, 198099. www.elsevier.com/locate/virusre

Sang, H., Miller, G., Lokhandwala, S., Sangewar, N., Waghela, S. D., Bishop, R. P., & Mwangi, W. (2020). Progress Toward Development of Effective and Safe African Swine Fever Virus Vaccines. *Frontiers in Veterinary Science*, *7*, 84.

Sarih, M., M'Ghirbi, Y., Bouattour, A., Gern, L., Baranton, G., & Postic, D. (2005). Detection and Identification of *Ehrlichia* spp. in Ticks Collected in Tunisia and Morocco. *Journal of Clinical Microbiology*, *43*(3), 1127–1132. doi:10.1128/JCM.43.3.1127-1132.2005 PMID:15750072

Sasaki, H., Kang'ethe, E., & Kaburia, H. (1995). Blood meal sources of Glossina pallidipes and G. longipennis (Diptera: Glossinidae) in Nguruman, southwest Kenya. *Journal of Medical Entomology*, *32*(3), 390–393. doi:10.1093/jmedent/32.3.390 PMID:7616534

Schaefer, J. J., Kahn, J., Needham, G. R., Rikihisa, Y., Ewing, S. A., & Stich, R. W. (2008). Antibiotic clearance of Ehrlichia canis from dogs infected by intravenous inoculation of carrier blood. *Annals of the New York Academy of Sciences*, *1149*(1), 263–269. doi:10.1196/annals.1428.087 PMID:19120226

Schaefer, J. J., Needham, G. R., Bremer, W. G., Rikihisa, Y., Ewing, S. A., & Stich, R. W. (2007). Tick acquisition of Ehrlichia canis from dogs treated with doxycycline hyclate. *Antimicrobial Agents and Chemotherapy*, *51*(9), 3394–3396. doi:10.1128/AAC.00358-07 PMID:17606682

Schein, E., & Voigt, W. P. (1979). Chemotherapy of bovine theileriosis with halofuginone. *Acta Tropica*, *36*, 391–394. PMID:44104

Schofield, C. J., & Kabayo, J. P. (2008). Trypanosomiasis vector control in Africa and Latin America. *Parasites & Vectors*, *1*(24), 24. doi:10.1186/1756-3305-1-24 PMID:18673535

Schofield, S., & Torr, S. (2002). A comparison of the feeding behaviour of tsetse and stable flies. *Medical and Veterinary Entomology*, *16*(2), 177–185. doi:10.1046/j.1365-2915.2002.00361.x PMID:12109712

Scoone, I., Dzingira, V., Anderson, N., MacLeod, E., Mangwanya, L., Matawa, F., ... Welburn, S. C. (2017). People, Patches, and Parasites: The Case of Trypanosomiasis in Zimbabwe. *Human Ecology*, *45*(5), 643–654. doi:10.100710745-017-9929-y PMID:29170590

Seidl, A., Dávila, A. M. R., & Silva, R. A. M. S. (1999). Estimated financial impact of Trypanosoma vivax on the Brazilian Pantanal and Bolivian lowlands. *Memorias do Instituto Oswaldo Cruz*, *94*(2), 269–272. doi:10.1590/S0074-02761999000200027 PMID:10224541

Seyoum, Z., Terefe, G., & Ashenafi, H. (2013). Farmers' perception of impacts of bovine trypanosomosis and tsetse fly in selected districts in Baro-Akobo and Gojeb river basins, Southwestern Ethiopia. *BMC Veterinary Research*, *9*(1), 214. doi:10.1186/1746-6148-9-214 PMID:24139090

Shaw, A. P. (2004). Economics of African trypanosomosis. In The Trypanosomiases. CABI Publishing.

Shaw, A. P. M., Cecchi, G., Wint, G. R. W., Mattioli, R. C., & Robinson, T. P. (2014). Mapping the economic benefits to livestock keepers from intervening against bovine trypanosomosis in Eastern Africa. *Preventive Veterinary Medicine*, *113*(2), 197–210.

Shaw, A. P. M., Wint, G. R. W., Cecchi, G., Torr, S. J., Mattioli, R. C., & Robinson, T. P. (2015). Mapping the benefit-cost ratios of interventions against bovine trypanosomosis in Eastern Africa. *Preventive Veterinary Medicine*, *122*(4), 406–416.

Shaw, M. K., Tilney, L. G., & Musoke, A. J. (1991). The entry of *Theileria parva* sporozoites in bovine lymphocytes. Evidence of MHC class 1 involvement. *The Journal of Cell Biology*, *113*(1), 87–101. doi:10.1083/jcb.113.1.87 PMID:1901066

Shereni, W., Anderson, N. E., Nyakupinda, L., & Cecchi, G. (2016). Spatial distribution and trypanosome infection of tsetse flies in the sleeping sickness focus of Zimbabwe in Hurungwe District. *Parasites & Vectors*, *9*(1), 605. doi:10.118613071-016-1879-5 PMID:27884172

Shipov, A., Klement, E., Reuveni-Tager, L., Waner, T., & Harrus, S. (2008). Prognostic indicators for canine monocytic ehrlichiosis. *Veterinary Parasitology*, *153*(1-2), 131–138. doi:10.1016/j.vetpar.2008.01.009 PMID:18289794

Shivairo, R. S., Muleke, C. I., & Lubembe, D. M. (2018). Text Book of Veterinary Parasitology and Parasitic Diseases. Academic Press.

Shivairo, R. S., Muleke, C. I., & Lubembe, D. M. (2018). *Students Guide Book* (1st ed.). Veterinary Parasitology and Parasitic Diseases.

Shkap, V., de Vos, A. J., Zweygarth, E., & Jongejan, F. (2007). Attenuated vaccines for tropical theileriosis, babesiosis and heartwater: The continuing necessity. *Trends in Parasitology*, *23*(9), 420–426. https://doi.org/10.1016/j.pt.2007.07.003

Short, N. J., Floyd, R. B., Norval, R. A. I., & Sutherst, R. W. (1989). Development rates, fecundity and survival of development stages of the tick *Rhipicephalus appendiculatus, Boophilus decoloratus* and *Boophilus microplus* under field conditions in Zimbabwe. *Experimental & Applied Acarology*, *6*(2), 123–141. doi:10.1007/BF01201643 PMID:2707110

Siddig, A., El Malik, K., & Hameida, T. A. (2003). Estimation of the financial losses during an outbreak of tropical Theilerosis in dairy farm. *Sudan J. Vet. Sci. Anim. Husb.*, *42*, 346.

Sidney, R., Andrew, McG., James, C. & Richard, N. (2013). Dourine – an emerging venereal threat to European horses. AHT / BEVA/DEFrA. *Equine Quarterly Disease Surveillance report, 6*, 7.

Silatsa, B. A., Kuiate, J. R., Njiokou, F., Simo, G., Feussom, J. M. K., Tunrayo, A., Amzati, G. S., Bett, B., Bishop, R., Githaka, N., Opiyo, S. O., Djikeng, A., & Pelle, R. (2019). A countrywide molecular survey leads to a seminal identification of the invasive cattle tick *Rhipicephalus* (*Boophilus*) *microplus* in Cameroon, a decade after it was reported in Cote d'Ivoire. *Ticks and Tick-Borne Diseases*, *10*(3), 585–593. doi:10.1016/j.ttbdis.2019.02.002 PMID:30765191

Silatsa, B. A., Simo, G., Githaka, N., Kamga, R., Oumarou, F., Tiambo, C. K., Machuka, E., Domelevo, J.-B., Odongo, D., Bishop, R., Kuiate, J.-R., Njiokou, F., Djikeng, A., & Pelle, R. (2020). First detection of *Theileria parva* in cattle from Cameroon in the absence of the main tick vector *Rhipicephalus appendiculatus*. *Transboundary and Emerging Diseases*, *67*(S1), 68–78. doi:10.1111/tbed.13425 PMID:32174039

Silveira, J. A., Passos, L. M., & Ribeiro, M. F. (2009). Population dynamics of Rhipicephalus sanguineus (Latrielle, 1806) in Belo Horizonte, Minas Gerais state, Brazil. *Veterinary Parasitology*, *161*(3-4), 270–275. doi:10.1016/j.vetpar.2009.01.028 PMID:19339119

Silvia, B., Rutili, D., & Guberti, V. (2016). Preventive measures aimedat minimizing the risk of African swine fever virus spread in pig farming systems. *Acta Veterinaria Scandinavica*, *58*(1), 82. doi:10.118613028-016-0264-x PMID:27899125

Simarro, P., Cecchi, G., Franco, J., Paone, M., Diarra, A., Ruiz-Postigo, J. A., & Jannin, J. G. (Eds.). (2012). *Estimating and Mapping the Population at Risk of Sleeping Sickness*. doi:10.1371/journal.pntd.0001859

Simarro, P., Jannin, J., & Cattand, P. (2008). Eliminating human African trypanosomiasis: Where do we stand and what comes next? *PLoS Medicine*, *5*(2), 174–180. doi:10.1371/journal.pmed.0050055 PMID:18303943

Simo, G., Asonganyi, T., Nkinin, S. W., Njiokou, F., & Herder, S. (2006). High prevalence of Trypanosoma brucei gambiense group 1 in pigs from the Fontem sleeping sickness focus in Cameroon. *Veterinary Parasitology*, *139*(1), 57–66. doi:10.1016/j.vetpar.2006.02.026 PMID:16567049

Simukoko, H., Marcotty, T., Phiri, I., Vercruysse, J., & Van den Bossche, P. (2007). Heterogeneity in the trypanosomosis incidence in Zebu cattle of different ages and sex on the plateau of eastern Zambia. *Acta Tropica*, *103*(2), 98–101. doi:10.1016/j.actatropica.2007.05.011 PMID:17618594

Simulundu, E., Chambaro, H. M., Sinkala, Y., Kajihara, M., Ogawa, H., Mori, A., & Mweene, A. S. (2018). Co-circulation of multiple genotypes of African swine fever viruses among domestic pigs in Zambia (2013–2015). *Transboundary and Emerging Diseases*, *65*, 114–122. https://doi.org/10.1111/tbed.12635

Simulundu, E., Lubaba, C., van Heerden, J., Kajihara, M., Mataa, L., Chambaro, H., & Mweene, A. (2017). The epidemiology of African swine fever in "non-endemic" regions of Zambia (1989–2015): Implications for disease prevention and control. *Viruses*, *9*, 9.

Singla, L. D., Juyal, P. D., & Sharma, N. S. (2010). Immune responses to haemorrhagic septicaemia (HS) vaccination in Trypanosoma evansi infected buffalo-calves. *Tropical Animal Health and Production*, *42*(4), 589–595. doi:10.100711250-009-9461-1 PMID:19784863

Sinshaw, A., Abebe, G., Desquesnes, M., & Yon, W. (2006). Biting flies and Trypanosoma vivax infection in three highland districts bordering Lake Tana, Ethiopia. *Veterinary Parasitology*, *142*(1-2), 35–46. doi:10.1016/j.vetpar.2006.06.032 PMID:16890359

Sitt, T., Henson, S., Morrison, W. I., & Toye, P. (2019). Similar levels of diversity in the gene encoding the p67 sporozoite surface protein of *Theileria parva* are observed in blood samples from buffalo and cattle naturally infected from buffalo. *Veterinary Parasitology*, *269*, 21–27. doi:10.1016/j.vetpar.2019.04.006 PMID:31079824

Sitt, T., Poole, E. J., Ndambuki, G., Mwaura, S., Njoroge, T., Omondi, G. P., Mutinda, M., Mathenge, J., Prettejohn, G., Morrison, W. I., & Toye, P. (2015). Exposure of vaccinated and naive cattle to natural challenge from buffalo-derived Theileria parva. *International Journal for Parasitology. Parasites and Wildlife*, *4*(2), 244–251. doi:10.1016/j.ijppaw.2015.04.006 PMID:26005635

Sivajothi, S., Rayulu, V., Malakondaiah, P., & Sreenivasulu, D. (2016). Diagnosis of Trypanosoma evansi in bovines by indirect ELISA. *Journal of Parasitic Diseases: Official Organ of the Indian Society for Parasitology*, *40*(1), 141–144. doi:10.100712639-014-0465-z PMID:27065614

Smith, R. D., Levy, M. G., Kuhlenschmidt, M. S., Adams, J. H., Rzechula, D. L., Hardt, T. A., & Kocan, K. M. (1986). Isolate of Anaplasma marginale not transmitted by ticks. *American Journal of Veterinary Research*, *47*, 127–129. PMID:3946891

Smith, R. D., Osorno, B. M., Brener, J., De La Rosa, R., & Ristic, M. (1978). Bovine babesiosis: Severity and reproducibility of Babesia bovis infections induced by Boophilus microplus under laboratory conditions. *Research in Veterinary Science*, *24*(3), 287–292.

SNV. (2013a). *Management of Livestock Drug Systems. Dairy Platform*. Policy Brief 03. SNV Netherlands Development Organization. http://www.snvworld.org/download/publications/ecf_brief_3_management_pdf

Snyder, A. K., Adkins, K. Z., & Rio, R. V. M. (2011). Use of the Internal Transcribed Spacer (ITS) Regions to Examine Symbiont Divergence and as a Diagnostic Tool for *Sodalis*-Related Bacteria. *Insects*, *2*(4), 515–531. doi:10.3390/insects2040515 PMID:26467831

Solano, P., De La Rocque, S., Cuisance, D., Geoffroy, B., De Meeus, T., Cuny, G., & Du Vallet, G. (1999). Intraspecific variability in natural populations of Glossina palpalis gambiensis from West Africa, revealed by genetic morphometric analysis. *Medical and Veterinary Entomology*, *13*(4), 401–407. doi:10.1046/j.1365-2915.1999.00189.x PMID:10608229

Solano, P., Ravel, S., & de Meeûs, T. (2010). How can tsetse population genetics contribute to African trypanosomiasis control? *Trends in Parasitology*, *26*(5), 255–263. doi:10.1016/j.pt.2010.02.006 PMID:20202905

Soulsby, E. (1986). *Helminths, Arthropods and Protozoa of Domesticated Animals* (7th ed.). Bailliere Tindall.

Soulsby, E. J. I. (1986). *Helminths, Arthropods and Protozoa of Domesticated Animals by E. J. I. Soulsby* (7th ed.). Lea and Febiger.

Spickler, A. (2018). *African Animal Trypanosomosis*. https://www.cfsph.iastate.edu/DiseaseInfo/factsheets.php

Spooner, P. B. (1990). The effect of oxytetracycline on *Theileria parva* in vitro. *Parasitology*, *100*(1), 11–17. doi:10.1017/S0031182000060066 PMID:2107502

Srvan & Oberem. (1987). The treatment of heartwater. *The Onderstepoort Journal of Veterinary Research*, *54*(3), 475–479. PMID:3329331

Stachurski, F. (2000). Invasion of West African cattle by the tick *Amblyomma variegatum*. *Medical and Veterinary Entomology*, *14*(4), 391–399. doi:10.1046/j.1365-2915.2000.00246.x PMID:11129703

Stagg, D. A., Dolan, T. T., Leitch, B. L., & Young, A. S. (1981). The initial stages of infection of cattle cells with *Theileria parva* sporozoites in vitro. *Parasitology*, *86*, 243–254. doi:10.1017/S0031182000050411 PMID:6791118

Stahl, K., Sternberg-Lewerin, S., Blome, S., Viltrop, A., Penrith, M. L., & Chenais, E. (2019). Lack of evidence for long term carriers of African swine fever virus - a systematic review. *Virus Research*, *272*, 197725.

Stanton, M. C., Esterhuizen, J., Tirados, I., Betts, H., & Torr, S. J. (2018). The development of high resolution maps of tsetse abundance to guide interventions against human African trypanosomiasis in northern Uganda. *Parasites & Vectors*, *11*(1), 340. doi:10.118613071-018-2922-5 PMID:29884213

Steck, W. (1928). *Pathological studies on heartwater*. 13th and 14th Report of the Director of Veterinary Education and Research.

Stendel, W. (1985). Experimental studies on the tickcidal effect of BayticolR pour-on. *Veterinary Medical Review*, 132–145.

Stephen, L. E. (1986). *Trypanosomiasis: a veterinary perspective.* Pergamon/A. Wheaton & Co.

Stewart, C. G. (1987). Specific immunity in farm animals to heartwater. *The Onderstepoort Journal of Veterinary Research, 54*, 341–342. PMID:3448570

Stewart, C. G. (1989). Cowdria ruminantium stocks: How significant are they? *Zimbabwe Veterinary Journal, 20*, 149–154.

Steyn, D. G. (1928). *Preliminary report on a South African virus disease amongst pigs. 13th and 14th Reports of the Director of Veterinary Education and Research, Union of South Africa*. Google Scholar.

Steyn, H. C., Pretorius, A., McCrindle, C. M. E., Steinmann, C. M. L., & Van Kleef, M. (2008). A quantitative real-time PCR assay for Ehrlichia ruminantium using pCS20. *Veterinary Microbiology, 131*(3/4), 258–265. doi:10.1016/j.vetmic.2008.04.002 PMID:18511221

Straten van der, G. (2008). *A survey of ticks, Ehrlichia canis and current control methods on dogs on the island of Aruba.* Research report, Utrecht Centre of Tick-borne diseases.

Strauss, J. H., & Strauss, E. G. (2008). Minus-Strand RNA Viruses. In *Viruses and Human Disease* (pp. 137–191). Elsevier. doi:10.1016/B978-0-12-373741-0.50007-6

Stuben, J. (2004). *Investigations on the prevalence of ehrlichiosis and babesiosis in dogs in Namibia with special consideration of the housing conditions* [Unpublished Master's Dissertation]. Free University.

Suarez, C. E., & Noh, S. (2011). Emerging perspectives in the research of bovine babesiosis and anaplasmosis. *Veterinary Parasitology, 180*(1–2), 109–125. https://doi.org/10.1016/j.vetpar.2011.05.032

Suarez, M., Rubi, J., Pérez, D., Cordova, V., Salazar, Y., Vielma, A., Barrios, F., Gil, C. A., Segura, N., Carrillo, Y., Cartaya, R., Palacios, M., Rubio, E., Escalona, C., Ramirez, R. C., Baker, R. B., Machado, H., Sordo, Y., Bermudes, J., ... Estrada, M. P. (2016). High impact and effectiveness of Gavac™ vaccine in the national program for control of bovine ticks *Rhipicephalus microplus* in Venezuela. *Livestock Science, 187*, 48–52. doi:10.1016/j.livsci.2016.02.005

Sudeep, A. B., Jadi, R. S., & Mishra, A. C. (2009). Ganjam virus. *The Indian Journal of Medical Research, 130*(5), 514–519. https://www.ncbi.nlm.nih.gov/pubmed/20090098

Suksawat, J., Hegarty, B. C., & Breitschwerdt, E. B. (2000). Seroprevalence of Ehrlichia canis, Ehrlichia equi, and Ehrlichia risticii in sick dogs from North Carolina and Virginia. *Journal of Veterinary Internal Medicine, 14*(1), 50–55. doi:10.1111/j.1939-1676.2000.tb01499.x PMID:10668817

Sumba, A., Mihok, S., & Oyieke, F. A. (1998). Mechanical transmission of Trypanosoma evansi and T. congolense by Stomoxys niger and S. taeniatus in a laboratory mouse model. *Medical and Veterinary Entomology, 12*(4), 417–422. doi:10.1046/j.1365-2915.1998.00131.x PMID:9824826

Sumption, K., Domenech, J., & Ferrari, G. (2012). Progressive control of FMD on a global scale. *The Veterinary Record, 170*(25), 637.

Sutherst, R., & Utech, K. (1981). *Controlling livestock parasites with host resistance. CRC Handbook of pest management in agriculture*. CRC Press Inc.

Swallow, B. (2000). *Impacts of trypanosomiasis on African agriculture.* Academic Press.

Swallow, B. M. (1999). *Impacts of trypanosomiasis on African agriculture. Report from the Programme Against African Trypanosomiasis*. Food and Agriculture Organization of the United Nations.

Synge, B. A. (1978). Brain biopsy for the diagnosis of heartwater. *Tropical Animal Health and Production*, *10*(1), 45–48. doi:10.1007/BF02235303 PMID:625799

Tabor, A. E., Ali, A., Rehman, G., Rocha Garcia, G., Zangirolamo, A. F., Malardo, T., & Jonsson, N. N. (2017). Cattle Tick Rhipicephalus microplus-Host Interface: A Review of Resistant and Susceptible Host Responses. *Frontiers in Cellular and Infection Microbiology*, *7*, 506. https://doi.org/10.3389/fcimb.2017.00506

Tagorti, G. (2019). Disease prevalence among young dogs in Grand Tunis, Tunisia: A retrospective study. *Veterinary World*, *12*(4), 489–495. doi:10.14202/vetworld.2019.489-495 PMID:31190703

Takken, W., & Weiss, M. (1990). The Sterile Insect Technique for Tsetse Flies in Africa. *International Atomic Energy Agency Bulletin*, *20*(3).

Tamarit, A., Gutierrez, C., Arroyo, R., Jimenez, V., Zagalá, G., Bosch, I., Sirvent, J., Alberola, J., Alonso, I., & Caballero, C. (2010). Trypanosoma evansi infection in mainland Spain. *Veterinary Parasitology*, *167*(1), 74–76. doi:10.1016/j.vetpar.2009.09.050 PMID:19864069

Tatchell, R. J. (1992). Ecology in relation to integrated tick management. *International Journal of Tropical Insect Science*, *13*(04), 551–561. https://doi.org/10.1017/S1742758400016131

Tavassoli, M., Tabatabaei, M., Mohammadi, M., Esmaeilnejad, B., & Mohamadpour, H. (2013). PCR-based Detection of Babesia spp. Infection in Collected Ticks from Cattle in West and North-West of Iran. *Journal of Arthropod-Borne Diseases*, *7*(2), 132–138.

Taylor, M. A., Coop, R. L., & Wall, R. L. (2013). *Veterinary Parasitology*. Retrieved from http://qut.eblib.com.au/patron/FullRecord.aspx?p=1158636

Taylor, M. A., Coop, R. L., & Wall, R. (2016). *Veterinary parasitology* (4th ed.). John Wiley and Sons, Inc.

Taylor, M. A., Coop, R. L., & Wall, R. L. (2015). Veterinary Entomology. In M. Taylor, R. Coop, & R. Wall (Eds.), *Veterinary Parasitology* (pp. 161–258). doi:10.1002/9781119073680.ch3

Taylor, M., Coop, R., & Wall, R. (2016). Veterinary entomology. In M. Taylor, R. Coop, & R. Wall (Eds.), *Veterinary Parasitology* (4th ed., pp. 161–258). Wiley-Blackwell.

Terpstra, C. (1990). Nairobi Sheep Disease Virus. In Virus Infections of Ruminants (pp. 495–500). Elsevier. https://doi.org/10.1016/B978-0-444-87312-5.50065-5.

Texas AgriLife Extension Service. (1997). *Integrated pest management of flies in Texas dairies*. animalscience.tamu.edu/images/pdf/dairy/dairy-integrated-pestmgmt

Thompson, R., Adams, H., Odoi, A., & Kennedy, M. (2020). Seroprevalence of viral and vector-borne bacterial pathogens in domestic dogs (Canis familiaris) in northern Botswana. *PLoS One*, *15*(1), e0220593. doi:10.1371/journal.pone.0220593 PMID:31914123

Thomson, G. R. (1985). The epidemiology of African swine fever: The role of free-living hosts in Africa. *The Onderstepoort Journal of Veterinary Research*, *52*, 201–209.

Tilki, T., Dagnachew, H. M., Yohanis, M., Biniyam, M., & Shawit, K. (2016). Epidemiology of trypanosomal infections of caprine in Nono district, south west showa zone Ethiopia. *Livestock Research for Rural Development*, *28*(10). http://www.lrrd.org/lrrd28/10/mulu28182.html

Tobe, S., & Langley, P. (1978). Reproductive physiology of *Glossina*. *Annual Review of Entomology*, *23*(1), 283–307. doi:10.1146/annurev.en.23.010178.001435 PMID:343707

Todorovic, R. A., Lopez, L. A., Lopez, A. G., & Gonzalez, E. F. (1975). Bovine babesiosis and anaplasmosis: Control by premunition and chemoprophylaxis. *Experimental Parasitology, 37*(1), 92–104. (75)90056-9 doi:10.1016/0014-4894

Tønnesen, M. H., Penzhorn, B. L., Bryson, N. R., Stoltsz, W. H., & Masibigiri, T. (2004). Displacement of *Boophilus decoloratus* by *Boophilus microplus* in the Soutpansberg region, Limpopo Province, South Africa. *Experimental & Applied Acarology*, *32*(3), 199–208. doi:10.1023/B:APPA.0000021789.44411.b5 PMID:15139085

Torioni De Echaide, S., Knowles, D. P., Mcguire, T. C., Palmer, G. H., Suarez, C. E., & Mcelwain, T. F. (1998). Detection of cattle naturally infected with *Anaplasma marginale* in a region of endemicity by nested PCR and a competitive enzyme-linked immunosorbent assay using recombinant major surface protein 5. *Journal of Clinical Microbiology*, *36*(3), 777–782. doi:10.1128/JCM.36.3.777-782.1998 PMID:9508311

Torr, S. J., & Hargrove, J. W. (1999). Behaviour of tsetse (Diptera: Glossinidae) during the hot season in Zimbabwe: the interaction of micro-climate and reproductive status. *Bulletin of Entomological Research*, *89*(4), 365–379. doi:10.1017/S0007485399000504

Torr, S. J., Holloway, M. T. P., & Vale, G. A. (1992). Improved persistence of insecticide deposits on targets for controlling Glossinapallidipes (Diptera: Glossinidae). *Bulletin of Entomological Research*, *82*, 525–533.

Trail, J., D'Ieteren, G., Murray, M., Ordner, G., Yangari, G., Collardelle, C., Sauveroche, B., Maille, J., & Viviani, P. (1993). Measurement of trypanotolerance criteria and their effect on reproductive performance of N'dama cattle. *Veterinary Parasitology*, *45*(3-4), 241–255. doi:10.1016/0304-4017(93)90079-3 PMID:8447067

Troy, G. C., & Forrester S. D. (1990). Canine ehrlichiosis. *Infectious diseases of the dog and cat*, 404–418.

Turnbull, D. O. (1932). *Nyasaland. Annual Report of the Veterinary Department, 1931. Swine fever*. Google Scholar.

Uilenberg, G. (1998). A field guide for the diagnosis, treatment and prevention of African animal. FAO.

Uilenberg, G. (1972). Diagnosis of Babesia Argentina Infection in Cattle Using Brain Smears. *Australian Veterinary Journal*, *48*(9), 534–534. https://doi.org/10.1111/j.1751-0813.1972.tb02327.x

Uilenberg, G. (1981). Heartwater disease. In M. Ristic & I. McIntyre (Eds.), *Diseases of Cattle in the Tropics* (pp. 345–360). Martinus Nijhoff Publishers.

Uilenberg, G. (1981). *Theileria* species of domestic livestock. In A. D. Irvin, M. P. Cunningham, & A. S. Young (Eds.), *Advances in the Control of Theileriosis* (pp. 4–37). Martinus Nijhoff Publishers. doi:10.1007/978-94-009-8346-5_2

Uilenberg, G. (1983). Heartwater (Cowdria ruminantium infection): Current status. *Advances in Veterinary Science and Comparative Medicine*, *27*, 427–480. PMID:6359836

Uilenberg, G. (1998). *A field guide for the diagnosis treatment and prevention of African animal trypanosomosis*. FAO.

Uilenberg, G. (2006). Babesia—A historical overview. *Veterinary Parasitology*, *138*(1–2), 3–10. https://doi.org/10.1016/j.vetpar.2006.01.035

Uilenberg, G. 1982. Disease problems associated with the introduction of European cattle in the tropics. *Proceedings of the 12th World Congress on Diseases of Cattle*, 1025-1032.

Uilenberg, G., & Camus, E. (1993). Heartwater (cowdriosis) In W. M. Ristic (Ed.), *Rickettsial and chlamydial diseases of domestic animals* (Vol. Z, pp. 293–332). Pergamon Press.

Uilenberg, G., Živković, D., Dwinger, R. H., Ter Huurne, A. A. H. M., & Perié, N. M. (1983). Cross immunity between strains of Cowdria ruminantium. *Research in Veterinary Science*, *35*(2), 200–205. doi:10.1016/S0034-5288(18)32179-9 PMID:6195703

University of Michigan. (2020). *Animal Diversity Web*. Retrieved from https://animaldiversity.org/accounts/Haemaphysalis_intermedia/classification/#Haemaphysalis_intermedia

Urquart, I. G. M., Armour, J., Dunncan, J. L., Dunn, A. M., & Jennings, F. W. (1995). Veterinary Parasitology. The University of Glasgow.

Urquhart, G. M., Armour, J., Ducan, J. L., Dunn, A. M., & Jennings, F. W. (1996). *Veterinary Parasitology (2nd ed.)*. Blackwell Science.

Valecha, N., Sharma, V. P., & Devi, C. U. (1998). A Rapid Immunochromatographic Test (ICT) for Diagnosis of Plasmodium falciparum. *Diagnostic Microbiology and Infectious Disease, 30*(4), 257–260. (98)00003-0 doi:10.1016/S0732-8893

Vale, G. A. (1977). Feeding responses of tsetse flies (Diptera: Glossinidae) to stationary hosts. *Bulletin of Entomological Research, 67*, 635–649.

Vale, G. A., Hall, D. R., & Gough, A. J. E. (1988). The olfactory responses of tsetse flies Glossina spp. (Diptera:Glossinidae) to phenols and urine in the field. *Bulletin of Entomological Research, 78*, 2.

Vale, G., Lovemore, D., Flint, S., & Cockbill, G. F. (1988). Odour-baited targets to control tsetse flies, Glossina spp. (Diptera: Glossinidae) in Zimbabwe. *Bulletin of Entomological Research, 78*(1), 31–49. doi:10.1017/S0007485300016059

Van den Bossche, P., & Delespaux, V. (2011). Options for the control of tsetse-transmitted livestock trypanosomosis. An epidemiological perspective. *Veterinary Parasitology, 181*(1), 37–42. doi:10.1016/j.vetpar.2011.04.021 PMID:21561717

Van Den Bossche, P., Rocque, S., Hendrickx, G., & Bouyer, J. (2010). A changing environment and the epidemiology of tsetse-transmitted livestock trypanosomiasis. *Trends in Parasitology, 26*(5), 236–243. doi:10.1016/j.pt.2010.02.010 PMID:20304707

Van den Bossche, P., & Rowlands, G. J. (2001). The relationship between the parasitological prevalence of trypanosomal infections in cattle and herd average packed cell volume. *Acta Tropica, 78*(2), 163–170. doi:10.1016/S0001-706X(00)00182-0 PMID:11230826

Van den Ingh, T. S., & Neijs-Bakker, M. H. (1979). Pancarditis in Trypano- soma vivax infection in cattle. *Tropenmedizin und Parasitologie, 30*, 239–243. PMID:483387

Van Heerden, J. (1979). The transmission of canine ehrlichiosis to the wild dog lycaon pictus (temminck) and blackbacked jackal Canis mesomelas Schreber. *Journal of the South African Veterinary Association, 50*(4), 245–248. PMID:553960

Van Heerden, J. (1982). A retrospective study on 120 natural cases of canine ehrlichiosis. *Journal of the South African Veterinary Association, 53*(1), 17–22. PMID:7097704

Van Heerden, J., Mills, M. G. L., Van Vuuren, M. J., Kelly, P. J., & Dreyer, M. J. (1995). An investigation into the health status and diseases of wild dogs (Lycaon *pictus)* in the Kruger National Park. *Journal of the South African Veterinary Association, 66*(1), 18–27. PMID:7629782

Van Heerden, J., & Immelman, A. (1979). The use of doxycycline in the treatment of canine ehrlichiosis. *Journal of the South African Veterinary Association, 50*(4), 241–244. PMID:553959

Venu, R., Sailaja, N., Srinivasa Rao, K., Jayasree, N., & Vara Prasad, W. L. N. V. (2015). Babesia bigemina infection in a 14-day old Jersey crossbred calf: A case report. *Journal of Parasitic Diseases: Official Organ of the Indian Society for Parasitology, 39*(2), 264–265. https://doi.org/10.1007/s12639-013-0338-x

Vercammen, F., De Deken, R., & Maes, L. (1995). Clinical and serological observations on experimental infections with Babesia canis and its diagnosis using the IFAT. *Parasite (Paris, France), 2*(4), 407–410.

Vergne, T. C., Fu, C., Li, S., Cappelle, J., Edwards, J., Martin, V., & Roger, F. L. (2017). Pig empire under infectious threat: Risk of African swine fever introduction into the People's Republic of China. *Veterinary Record*, *181*, 117. doi:10.1136/vr.103950

Verloo, D., Holland, W., My, L. N., Thanh, N. G., Tam, P. T., Goddeeris, B., Vercruysse, J., & Büscher, P. (2000). Comparison of serological tests for Trypanosoma evansi natural infections in water buffaloes from north Vietnam. *Veterinary Parasitology*, *92*(2), 87–96. doi:10.1016/S0304-4017(00)00284-3 PMID:10946132

Vickerman, K. (1985). Developmental cycles and biology of pathogenic trypanosomes. *British Medical Bulletin*, *41*(2), 105–114. doi:10.1093/oxfordjournals.bmb.a072036 PMID:3928017

Vincendeau, P., & Bouteille, B. (2006). Immunology and immunopathology of African trypanosomiasis. *Anais da Academia Brasileira de Ciências*, *78*(6), 645–665. doi:10.1590/S0001-37652006000400004 PMID:17143404

Vos, A. J., & Bock, R. E. (2006). Vaccination against Bovine Babesiosis. *Annals of the New York Academy of Sciences*, *916*(1), 540–545. https://doi.org/10.1111/j.1749-6632.2000.tb05333.x

Vreysen, M. J. B., Robinson, A., Hendrichs, J., & Kenmore, P. (2007). Area-wide integrated pest management (AW-IPM): Principles, practice and prospects. In Area-wide control of insect pests (pp. 3–33). Springer.

Vreysen, M.J. (2001). *Principles of Area Wide Integrated Tsetse Fly Control Using the Sterile Insect Technique*. Academic Press.

Vreysen, M. J. B., Robinson, A. S., Hendrichs, J., & Kenmore, P. (2007). Area-wide integrated pest management (AW-IPM): principles, practice and prospects. In *Area-wide control of insect pests* (pp. 3–33). Springer.

Vreysen, M. J., Seck, M., Sall, B., & Bouyer, J. (2013). Tsetse flies: Their biology and control using area-wide integrated pest management approaches. *Journal of Invertebrate Pathology*, *112*, 15–25. doi:10.1016/j.jip.2012.07.026 PMID:22878217

Vudriko, P., Okwee-Acai, J., Byaruhanga, J., Tayebwa, D. S., Okech, S. G., Tweyongyere, R., Wampande, E. M., Okurut, A. R. A., Mugabi, K., Muhindo, J. B., Nakavuma, J. L., Umemiya-Shirafuji, R., Xuan, X., & Suzuki, H. (2018a). Chemical tick control practices in southwestern and northwestern Uganda. *Ticks and Tick-Borne Diseases*, *9*(4), 945–955. doi:10.1016/j.ttbdis.2018.03.009 PMID:29606621

Vudriko, P., Okwee-Acai, J., Byaruhanga, J., Tayebwa, D. S., Omara, R., Muhindo, J. B., Lagu, C., Umemiya-Shirafuji, R., Xuan, X., & Suzuki, H. (2018b). Evidence-based tick acaricide resistance intervention strategy in Uganda: Concept and feedback of farmers and stakeholders. *Ticks and Tick-Borne Diseases*, *9*(2), 254–265. doi:10.1016/j.ttbdis.2017.09.011 PMID:28978456

Vudriko, P., Okwee-Acai, J., Tayebwa, D. S., Byaruhanga, J., Kakooza, S., Wampande, E., Omara, R., Muhindo, J. B., Tweyongyere, R., Owiny, D. O., Hatta, T., Tsuji, N., Umemiya-Shirafuji, R., Xuan, X., Kanameda, M., Fujisaki, K., & Suzuki, H. (2016). Emergence of multi-acaricide resistant *Rhipicephalus* ticks and its implication on chemical control in Uganda. *Parasites & Vectors*, *9*(1), 4. doi:10.118613071-015-1278-3 PMID:26727991

Wachira, B. M., Mireji, P. O., Okoth, S., Ng'ang'a, M. M., William, J. M., Murilla, G. A., & Hassanali, A. (2016). Responses of Glossinapallidipes and Glossinamorsitansmorsitans tsetse flies to analogues of δ-octalactone and selected blends. *Acta Tropica*, *160*, 53–57.

Walker & Bouattour. (2003). Ticks of domestic animals in Africa: A guide to identification of species. *Bioscience Reports*.

Walker, A.R., Young, A.S. & Leitch, B.L. (1981). Assessment of *Theileria* infections of *R. appendiculatus* ticks collected in the field. *Z. Parasitenkunde*, *65*, 63-69.

Walker, A. R. (1990). Parasitic adaptions in the transmission of Theileria by ticks. A review. *Tropical Animal Health and Production*, *22*(1), 23–33. doi:10.1007/BF02243494 PMID:2108514

Walker, A. R., Latif, A. A., Hans, N., Nijhof, A., & Jongejan, F. (2014b). *Electronic Tick Guide for Africa. Developed by the International Consortium on Ticks and Tick-borne Diseases*. ICTTD.

Walker, J. B. (1974). *The Ixodid ticks of Kenya. A review of Present Knowledge of Their Hosts and Distribution*. Commonwealth Institute of Entomology.

Walker, J. B. (1991). A review of the Ixodid ticks (Acari: Ixodidae) occurring in Southern Africa. *The Onderstepoort Journal of Veterinary Research*, *58*, 81–105. PMID:1881661

Walker, J. B., & Olwage, A. (1987). The tick vectors of *Cowdriaruminantium*(Ixodoidea, Ixodidae, genus *Amblyomma*) and their distribution. *The Onderstepoort Journal of Veterinary Research*, *54*, 353–379. PMID:3329325

Walker, J. G., Klein, E. Y., & Levin, S. A. (2014a). Disease at the wildlife-livestock interface: Acaricide use on domestic cattle does not prevent transmission of a tick-borne pathogen with multiple hosts. *Veterinary Parasitology*, *199*(3-4), 206–214. doi:10.1016/j.vetpar.2013.11.008 PMID:24315187

Wall, R., & Shearer, D. (2001). Veterinary Ectoparasites: Biology, Pathology and Control (2nd ed.). Blackwell Science.

Wall, R., & Shearer, D. (1997). *Veterinary Entomology*. Chapman and Hall Publishers. doi:10.1007/978-94-011-5852-7

Waltisbuhl, D. J., Goodger, B. V., Wright, I. G., Commins, M. A., & Mahoney, D. F. (1987). An enzyme linked immunosorbent assay to diagnose Babesia bovis infection in cattle. *Parasitology Research*, *73*(2), 126–131. https://doi.org/10.1007/BF00536468

Wamwiri, F., Alam, U., Thande, P., Aksoy, E., Ngure, R. M., Aksoy, S., Ouma, J. O., & Murilla, G. A. (2013). *Wolbachia, Sodalis* and trypanosome co-infections in natural populations of *Glossina austeni* and *Glossina pallidipes. Parasites & Vectors*, *6*(1), 232. doi:10.1186/1756-3305-6-232 PMID:23924682

Wamwiri, F., & Changasi, R. (2016). Tsetse Flies (Glossina) as Vectors of Human African Trypanosomiasis: A Review. *BioMed Research International*, *2016*, 1–8. Advance online publication. doi:10.1155/2016/6201350 PMID:27034944

Waner, T., & Harrus, S. (2013). Canine monocytic ehrlichiosis: From pathology to clinical manifestations. *Israel Journal of Veterinary Medicine*, *68*(1), 12–18.

Wang, Z. L. (1988). The similarities and differences of the characteristics between T. equiperdum and T. evansi. *Bulletin of Veterinary College*, *8*, 300–303.

Wanzala, W., Takken, W., Mukabana, W. R., Pala, A. O., & Hassanali, A. (2012). Ethnoknowledge of Bukusu community on livestock tick prevention and control in Bungoma district, western Kenya. *Journal of Ethnopharmacology*, *140*(2), 298–324. doi:10.1016/j.jep.2012.01.021 PMID:22306288

Warinda, E. (2001). *A socio-economic survey and economic analysis of land use options in Shompole and Olkiramatian group ranches–Magadi division. Report*. African Conservation Centre.

Warnes, M. L., Van den Bossche, P., Chihiya, J., Mudenge, D., Robinson, T. P., Shereni, W., & Chadenga, V. (1999). Evaluation of insecticide-treated cattle as a barrier to re-invasion of tsetse to cleared areas in northeastern Zimbabwe. *Medical and Veterinary Entomology*, *13*(2), 177–184.

Watson, D. W., Waldron, J. K., & Rutz, D. A. (1994). *Integrated management of flies in and around dairy and livestock barns*. Academic Press.

Weerathunga, D., Amarasinghe, A., Iddawel, D., & Wickramasinghe, S. (2019). Prevalence of canine tick-borne haemoparasites in three Divisional Secretariat Divisions (Rambewa, Tirappane, and Galenbidunuwewa) in the Anuradhapura district, Sri Lanka. *Sri Lankan Journal of Infectious Diseases*, *9*(2), 111–119. doi:10.4038ljid.v9i2.8254

Weinbren, M. P., Gourlay, R. N., Lumsden, W. H., & Weinbren, B. M. (1958). An epizootic of Nairobi sheep disease in Uganda. *Journal of Comparative Pathology*, *68*(2), 174–187. https://doi.org/10.1016/s0368-1742(58)80018-1

Weisss, K. E., Haig, D. A., & Alexander, R. A. (1952). Aureomycin in the treatment of heartwater. *The Onderstepoort Journal of Veterinary Research*, *25*, 41–50.

Weitz, B. (1963). The feeding habits of *Glossina. Bulletin of the World Health Organization*, *28*, 711–729. PMID:13999790

Welburn, S. C., & Maudlin, I. (1992). The nature of the teneral state in Glossina and its role in the acquisition of trypanosome infection in tsetse. . .. *Annals of Tropical Medicine and Parasitology*, *86*(5), 529–536. doi:10.1080/00034983.1992.11812703

Wellde, Chumo, Waema, Reardon, & Smith. (1989). A history of sleeping sickness in Kenya. *Annals of Tropical Medicine & Parasitology, 83*(S1), 1-11. DOI: doi:10.1080/00034983.1989.11812405

Wen, X., He, X., Zhang, X., Zhang, X., Liu, L., Guan, Y., Zhang, Y., & Bu, Z. (2019). Genome sequences derived from pig and dried blood pig feed samples provide important insights into the transmission of African swine fever virus in China in 2018. *Emerging Microbes & Infections*, *8*, 303–306.

Wernery, U., & Kaaden, O. R. (2002). *Infectious diseases in camelids* (2nd ed.). Wiley-Blackwel. doi:10.1002/9780470699058

Wharton, R. H. (1976). Tick-borne livestock diseases and their vectors. Acaricide resistance and alternative methods of tick control. *World Animal Review*, *2*, 8–15.

Whitehouse, C. A., Kuhn, J. H., Wada, J., & Ergunay, K. (2015). Family Bunyaviridae. In Global Virology I - Identifying and Investigating Viral Diseases (pp. 199–246). Springer New York. https://doi.org/10.1007/978-1-4939-2410-3_10.

Whitelaw, D. D., Bell, I. R., & Holmes, P. H. (1989). Immunological clearance of Se-labeled Trypanosoma brucei in goats. *Acta Tropica*, *46*(2), 101–106. doi:10.1016/0001-706X(89)90003-X PMID:2565070

WHO. (2012). *Research Priorities for Chagas Disease, Human African Trypanosomiasis and Leishmaniasis, WHO Technical Report Series*. WHO.

WHO. (2020). *World Health Organization (WHO) fact sheet*. https://www.who.int/news-room/fact-sheets/detail/trypanosomiasis-human-african-(sleeping-sickness)

Wickwire, K. B., Kocan, K. M., Barron, S. J., Ewing, S. A., & Smith, R. D. (1987). Infectivity of three *Anaplasma marginale* isolates for *Dermacentor andersoni. American Journal of Veterinary Research*, *48*, 96–99. PMID:3826850

Wilde, J. K. H., Brown, C. G. D., Hulliger, L., Gall, D., & MacLeod, W. G. (1968). Experiments with tissues of infected ticks. *The British Veterinary Journal*, *124*(5), 196–208. doi:10.1016/S0007-1935(17)39402-2 PMID:5649029

Willadsen, P. (1990). Perspectives for subunit vaccines for the control of ticks. InProc. FAO expert consultation on revision of strategies for the control of ticks and tick-borne diseases. *Parassitologia*, *32*, 195–200. PMID:2284131

Willadsen, P., & Jongejan, F. (1999). Immunology of tick-host interaction and the control of ticks and tick-borne diseases. *Parasitology Today (Personal Ed.)*, *15*(7), 258–262. doi:10.1016/S0169-4758(99)01472-6 PMID:10377526

Willadsen, P., Mckenna, R. V., & Riding, G. A. (1988). Isolation from the cattle tick, *Rhipicephalus (Boophilus) microplus*), of antigenic material capable of eliciting a protective immunological response in the bovine host. *International Journal for Parasitology*, *18*(2), 183–189. doi:10.1016/0020-7519(88)90059-8 PMID:3372125

William, B. L., & Steven, H. S. (2007). Infectious diseases of breeding stallion in. *Current Therapy in Large Animal Veterinary Research*, *55*, 117–119.

Wilson, S. G. (1953). The control of Glossina palpalis fuscipes Newstead in Kenya colony. *Bulletin of Entomological Research*, *44*, 711–728.

Wodaje, A., Adudna, B., & Hamid, M. (2019). *A review on Bovine Basiosis. International Journal of Advanced Research in Biological Sciences*. IJARBS.

Woo, P. T. K. (1970). The haematocrit centrifuge technique for the diagnosis of African Trypanosomosis. *Acta Tropica*, *27*, 384–386. PMID:4396363

Woo, T. (1970). The haematocrit centrifuge technique for the diagnosis of African trypanosomiasis. *Acta Tropica*, *27*, 384–386. PMID:4396363

Wosu, L. O., & Nwanta, J. A. (1989). Incidence of trypanosomes in settled Fulani cattle in the Anambra State, southern Nigeria. *Archives Roumaines de Pathologie Experimentales et de Microbiologie*, *48*(4), 367–372. PMID:2520674

Yadav, P. D., Vincent, M. J., Khristova, M., Kale, C., Nichol, S. T., Mishra, A. C., & Mourya, D. T. (2011). Genomic analysis reveals Nairobi sheep disease virus to be highly diverse and present in both Africa, and in India in the form of the Ganjam virus variant. *Infection, Genetics and Evolution*, *11*(5), 1111–1120. https://doi.org/10.1016/j.meegid.2011.04.001

Yaro, M., Munyard, K., Stear, M. J., & Groth, D. M. (2016). Combatting African Animal Trypanosomiasis (AAT) in livestock: The potential role of trypanotolerance. *Veterinary Parasitology*, *225*, 43–52. doi:10.1016/j.vetpar.2016.05.003 PMID:27369574

Yeoman, G. H., & Walker, J. B. (1967). *The ixodid ticks of Tanzania. A study of the zoogeography of the ixodidae of an East African country*. Commonwealth Institute of Entomology.

Yessinou, R. E., Adoligbe, C., Akpo, Y., Adinci, J., Youssao Abdou Karim, I., & Farougou, S. (2018). Sensitivity of Different Cattle Breeds to the Infestation of Cattle Ticks Amblyomma variegatum, Rhipicephalus microplus, and Hyalomma spp. On the Natural Pastures of Opkara Farm, Benin. *Journal of Parasitology Research, 2018*, 2570940. doi:10.1155/2018/2570940

Yonas, G. (2015). *Clinical signs, serum biochemical profiles and repro- ductive pathological lesions in horses naturally infected with dourine in Western Arsi zone, Ethiopia* (Unpublished MSc thesis). Addis Ababa University.

Young, A. S., De Castro, J. J., & Kiza-aura, P. P. (1985). Control of African ticks by application of acaricide impregnated ear tags to cattle. *Bulletin of Entomological Research*, *75*, 609–619. doi:10.1017/S0007485300015868

Young, A. S., Groocock, C. M., & Kariuki, D. P. (1988). Integrated control of ticks and tick-borne diseases of cattle in Africa. *Parasitology*, *96*(2), 403–432. doi:10.1017/S0031182000058388 PMID:3287285

Young, A. S., & Grotenhuis, J. G. (1985). Influence of wildlife on immunization of cattle against theileriosis in East Africa. In A. D. Irvin (Ed.), *Immunisation against Theileriosis in Africa* (pp. 104–109). ILRAD.

Young, A. S., & Leitch, B. L. (1981). Epidemiology of East Coast fever: Some effects of temperature on the development of *Theileria parva* in the tick vector, *Rhipicephalus appendiculatus. Parasitology*, *83*(1), 199–211. doi:10.1017/S0031182000050162 PMID:6791119

Young, A. S., Leitch, B. L., Morzaria, S. P., Irvin, A. D., Omwoyo, P. L., & De Castro, J. J. (1987). Development and survival of *Theileria parva parva* in *Rhipicephalus appendiculatus* exposed in the Trans Mara, Kenya. *Parasitology*, *94*(3), 433–441. doi:10.1017/S0031182000055785 PMID:3112702

Young, A. S., Leitch, B. L., & Mutugi, J. J. (1984). Some factors controlling the stimulation of sporogony of *Theileria parva* in its vector *Rhipicephalus appendiculatus* infecting cattle. *Parasitology*, *90*, 391–399. doi:10.1017/S0031182000051088

Young, A. S., Leitch, B. L., Newson, R. M., & Cunningham, M. P. (1986). Maintenance of *Theileria parva* infection in an endemic area of Kenya. *Parasitology*, *93*(1), 9–16. doi:10.1017/S0031182000049787 PMID:3092172

Young, A. S., Leitch, B. L., & Omwoyo, P. L. (1979). The induction of infective stages of Theileria parva by exposure of host tick to high temperature. *The Veterinary Record*, *105*, 531–533. PMID:119344

Yusuf, J. (2017). *Review on Bovine Babesiosis and its Economic Importance*. Academic Press.

Zable, T.A. & Agusto, F.B. (2018). *Transmission dynamics of Bovine Anaplasmosis in a cattle herd interdisciplinary perspectives in infectious diseases*. Academic Press.

Zhou, L., Evan, Y. W. Yu., Wang, S., & Changpo, S. (2019). African swine fever epidemic in China, Veterinary Record. *Journal of the British Veterinary Association*. doi:10.1136/vr.l4026

Zwart, D. (1989). Aspects of comparative pathology and pathogenesis of Trypanosomal infections in Africa. *Annales de la Société Belge de Medecine Tropicale*, *69*, 105–112.

Zweygarth, E., & Josemans, A. I. (2005). An attenuated Ehrlichia ruminantium (Welgevonden stock) vaccine protects small ruminants against virulent heartwater challenge. *Vaccine, 23*(14), 1695-1702. https://www.sciencedirect.com/science/journal/0264410X

About the Contributors

Caleb Orenge is a trained Veterinarian with a PhD in Veterinary Physiology. He is a seasoned research scientist in the field of Veterinary Sciences and Molecular genetics. He has special interest in genetics of disease resistance and stress in food animals and the impact this may have on the consumers, the human beings.

* * *

Popoola Moshood Abiola is a Nigerian, an early career scientist who has a strong background in several research fields. Over the last ten years, he has gained research experience in the area of genetics and animal breeding. He had all his degrees in Obafemi Awolowo University, Ile-Ife, Osun state, Nigeria. His Doctoral programme was completed in record time (2017) which was done through a trilateral engagements and collaborations with multiple global agencies and organizations including (a) EU-funded iLINOVA (www.ilinova.org), (b) Australian government-funded ABCF Fellowship and the (c) International Livestock Research Institute (ILRI), Nairobi, Kenya. Ever since the completion of his PhD programme, he has maintained links with these agencies and he is a regular visiting scientist at ILRI, Nairobi Kenya, where he is well known as a very promising, talented, dynamic and focused scientist. In 2016, he won the African Biosciences Challenge Fund (ABCF) Fellowship funded by the Australian Department for Foreign Affairs and Trade (DFAT) through the BecA-CSIRO partnership; the Syngenta Foundation for Sustainable Agriculture (SFSA); the Bill & Melinda Gates Foundation (BMGF); the UK Department for International Development (DFID) and the Swedish International Development Cooperation Agency (Sida). He also a member of research team for Projects on Management of Indigenous Livestock to Foster Agricultural Innovation in Eastern, Southern and Western Africa (iLINOVA); co-financed in the ACP-EU Cooperation Programme in Science and Technology II (S&T II). A programme of the ACP Group of States, with the financial assistance of the European Union Dr Popoola very rich Curriculum Vitae that include numerous publications in Animal Breeding and Genetics, livestock extension, travel grants and collaborations with the afore-mentioned agencies. He has attended several workshops and undergone trainings on Proposal Writing Skills, Scientific Paper Writing, Scientific Communication, Genomics Selection for Livestock, Climate Change and Livestock, Scientific Information Resources both within Nigeria and outside Nigeria. He is currently a lecturer at Federal College of Animal Health and Production Technology, Moor Plantation Ibadan. He is also a member of professional bodies like World Rabbit Science Association Sub-Saharan Africa Branch (SSA-WRSA), Genetics Society of Nigeria (GSN), Nigerian Society of Animal Production (NSAP), Animal Science Association of Nigeria (ASAN). His teaching and research experiences have span intellectually broad topics covering multiple disciplines. He has enjoyed the opportunity to design experiments, analyze results, present findings and publish papers both in the local and international gatherings and conferences. He has authored and co-authored over thirty scientific publications in different reputable journals both local and international journals. He has supervised over eighty students for projects. He has mentored and he is still mentoring lots of students and youths.

Daniel Ochieng Gamba is a graduate of the University of Nairobi in Zoology. His research interest is insect chemical ecology and has over 32 years' experience in tsetse and trypanosomiasis control. Dr. Gamba oversaw implementation of the PATTEC project in Kenya as the Entomologist in charge and currently Deputy Director Zoological services in the Directorates of Veterinary services of Kenya and the Head of technical services in KENTTEC.I have coordinated adaptive, operational research, and implementation of control programs. He is currently Co-PI in the field assessment of relative magnitude of attraction or repellence of *G. Pallidipes* in Kenya by the novel odor compounds 2) modeling concurrent deployment of the novel repellent (push)– attractant (pull) compounds to determine optimal performance and cost effectiveness of the push – pull technology in disease suppression in Kenya and 3) integration of the optimized model in the tsetse and trypanosomiasis eradication campaign in Kenya.

Charles Muleke Inyagwa, Prof. Muleke has Bachelor of Veterinary Medicine, (University of Nairobi -1989), and M.Sc. Tropical Veterinary Medicine, (Edinburgh University, 1995), and Ph.D. Preventive Veterinary Medicine, (Nanjing Agricultural University, 2006). He is an Associate Professor of Molecular Parasitology and Immunology at Egerton University, Kenya. He has a strong research profile, with over 45 refereed technical publications in diseases of public health and food safety importance in Africa, one patent, 2-book publications, and has attracted substantive research funds. He has vast professional expertise in guiding reforms at animal health training institutions in Kenya. Prof. Muleke has over 25 years experience in training and mentoring Animal Health Professionals in Kenya including curriculum development /reviews. He has extensive knowledge in livestock / animal health programme design, management, and evaluation leading to strengthened capacity for service delivering personnel. Prof. Muleke has been the Dean – Faculty of Veterinary Medicine and Surgery and Director for the Agro Based Science Park initiative a vision - 2030 project at Egerton University.

Abdul A. S. Katakweba is a Senior Research Fellow at Pest Management Centre (SPMC) in Sokoine University of Agriculture. He holds a PhD from Sokoine University of Agriculture (SUA), Tanzania (https://www.sua.ac.tz) specialized in Antimicrobial Resistance. He also holds MSc. Tropical Anima Production and Bachelor of Veterinary Medicine (BVM), both from Sokoine University of Agriculture. His main research interests are in public health aspects including zoonotic diseases (Leptospirosis, Toxoplasmosis and Plague) and antimicrobial resistance in humans and animals sharing the same environment. In the last 13 years he has conducted intensive research on rodent zoonotic diseases and antimicrobial resistance in humans and their livestock, pets, poultry and wild ungulates (Zebra, Wildebeest and Buffalo). He has published 56 articles in international peer reviewed journals. (https://scholar.google.com/citations?hl=en&view_op=search_authors&mauthors=sua.ac.tz) Dr. Katakweba has been involved in 9 projects at SUA Pest Management Centre. He is also a co-author of a book "Situation Analysis and Recommendations: Antibiotic Use and Resistance in Tanzania". Before joining SUA in 2003, Dr. Katakweba has served for 15 years (1987-2002) in the Tanzania Dairy Farm Company limited as a Veterinarian. He was involved in administration, control of tick-borne diseases and worms in Cattle by using different products of Acaricides and anthelmintics. Dr. Katakweba is a member of several committees which include; National Steering Committee of Public Health, Pesticides Management, Global Antibiotic Resistant Partnership (Tanzania, India & USA),SUA Integrity Committee and National Antimicrobial Resistance (AMR) Surveillance Technical Working Group (TWG).

Leonard Khaluhi is a trained and licensed veterinarian in Kenya and currently works as a Research Scientist at KALRO – VSRI. He has immense experience in animal Health. On the research front, he is a member of the Protozoology section in his institute and has worked extensively on Ticks and tick borne diseases on epidemiology, vaccinology and drug therapies. He also has had some experience with Trypanosomosis research. He is a Masters of Veterinary Medicine graduate of the Institute of Tropical Medicine Antwerp Belgium, specializing in Disease control

Esther Gwae Kimaro (PhD-2018) is a Veterinary Epidemiologist and Researcher at the Tropical Pesticides Research Institute, in the division of livestock disease vectors control. Her area of specialty is on veterinary epidemiology of livestock diseases. Currently, her research focus has been on the epidemiology of cattle vector-borne diseases particularly tick-borne diseases and tsetse borne diseases. She holds a Bachelor in Veterinary Medicine from Sokoine University of Agriculture, Tanzania; Masters in Veterinary Public Health from Massey University, Palmerstone North, New Zealand; and a Doctorate in Veterinary Science at the University of Sydney, Australia. Esther is a Senior Research Officer at the Tropical Pesticides Research Institute (TPRI) in the Division of Livestock and Human Diseases Vector Control. Esther's primary research interest is on the epidemiology of livestock vector-borne diseases including bovine theileriosis, animal and human African trypanosomiasis and zoonosis. Esther is currently building her research interest on the potential influence of climate change on livestock diseases. Specifically, on livestock vector, parasite and host relationships. Esther resides with her family in Arusha town. In her free time, she loves evening walking, attends sports club, reading academic and non-academic books also window shopping.

Erick Vitus Komba is an Associate Professor Department of Veterinary Medicine and Public Health of Sokoine University of Agriculture. At the University, since the year 2006, he has been involved in teaching veterinary epidemiology, veterinary medicine, animal welfare and research methods; to undergraduate and postgraduate students. He graduated from a veterinary school with a Bachelor of Veterinary Medicine, in the year 2005. He then undertook a 2 year Master Degree and then a 4 year PhD. During his graduate studies he conducted research in aspects of veterinary Epidemiology. After acquiring a PhD, Prof. Komba has been involved in research on livestock diseases particularly parasitic and bacterial zoonoses, with focus on diagnostics and molecular epidemiology. He recently participated in a project on disease surveillance using digital technologies. Currently, Prof. Komba works with the Southern African Centre for Infectious Disease Surveillance, being hosted is the community of practice on bacterial Zoonoses and Antimicrobial resistance. Prof. Komba has over 25 publications which involve journal articles, conference proceedings and a book chapter.

Sylvia Korir is principal zoologist KENTTEC. Zoologist in charge of Narok county.

Donald Lubembe is a veterinary surgeon, with a PhD in Veterinary Tropical Diseases from the University of Pretoria, South Africa. Interested in research aimed at control of vector-borne parasitic diseases of livestock and parasitic zoonoses.

Athumani Msalale Lupindu (BVM, MSc, PhD) is a veterinary epidemiologist at Department of Veterinary Medicine and Public Health of Sokoine University of Agriculture. He has more than 20 years of veterinary practice in tropical animal and zoonotic diseases. Dr. Lupindu has been teaching quantitative epidemiology, companion animal diseases, applied statistics and research methods to undergraduate and postgraduate students at the College of Veterinary Medicine and Biomedical Sciences for more than 10 years. His area of research interests are enteric bacteria and parasites of economic and public health importance. The most recent research by Dr. Lupindu is on assessment of pathogenic E. coli O157:H7 infection risk and transmission pathways in cattle keeping urban areas.

Monicah Maichomo is a veterinarian research scientist with more than 20 years of research experience. A graduate of the University of Nairobi, she has vast experience on livestock diseases including brucellosis, animal and human African trypanosomosis, helminthosis and tick-borne diseases among other infectious diseases, and accompanying publications in scientific journals. A great deal of her work also focuses on the trypanotolerant Orma boran breed that is undergoing multiplication in KALRO-Beef Research Institute. Other ongoing research interests include one-health, antimicrobial resistance (AMR) and vaccine development. She is currently serving as the Director of Veterinary Research Institute in the Kenya Agriculture and Livestock Research Organization.

Adrian Evance Materu (BVM) is a veterinary parasitologist at Department of Veterinary Microbiology, Parasitology and Biotechnology of Sokoine University of Agriculture. Dr. Materu has been teaching basic and applied parasitology to undergraduate and non-degree students at the College of Veterinary Medicine and Biomedical Sciences. His area of research interests are veterinary entomology, helminthology, and protozoology.

Samuel M. Mbuku is a Senior Research Scientist at the Kenya Agricultural and Livestock Research Organization. Samuel is a quantitative geneticist with over 13 years of experience in design and implementation of livestock genetic improvement programs - with special focus on indigenous breeds and population. He has minors in animal nutrition, physiology, livestock production systems and statistics. Dr. Mbuku has in the past led various genetic improvement initiatives for livestock in Kenya covering a wide range of thematic areas like characterization, sustainable use and conservation. Samuel has been heavily involved in undertaking strategic planning, priority setting, science policy, resource mobilization and research agenda for livestock genetics. These involves working in partnership and collaboratively with farmers, policy makers, academic institutions, private sector, international partners, Regional Economic Communities and non-profit institutions involved in development and upscaling of livestock technologies and innovations. Dr. Mbuku has and holds several positions with public interest including Chairman of the Animal Production Society of Kenya (APSK), Vice Chairman of the National Advisory Committee on Management of Animal Genetic Resources, Member of the National Taskforce for Livestock Standards, Policy and Regulations, Executive Committee Member of the All African Society of Animal Production (AASAP). He has (co)published widely and won several grants for research in animal production. Dr. Samuel Mbuku holds MSc Animal Production (Breeding and Genetics) and a PhD in Animal Sciences (Animal Breeding and Genetics) from Egerton University.

Ladslaus L. Mnyone holds a PhD in Medical Entomology from Wageningen University, the Netherlands. He also holds a bachelor degree in Veterinary Medicine (BVM) and MSc degree in Preventive Veterinary Medicine, both from Sokoine University of Agriculture. Dr. Mnyone is currently serving as the Director of Pest Management Centre. Main research interests include ecology, behavior, surveillance and control of insect disease vectors; insecticide resistance in insect disease vectors, and vector-host-pathogen interactions. His research accomplishments for the last 10 years focused largely on mosquito and flea vectors. He has been involved in more than five collaborative research projects as Principal Investigator and Co-Principal Investigator. He has published over 25 scientific articles in peer review Journals.

Eric Mungube is a senior research scientist and deputy institute director at the Veterinary Science Research Institute (VSRI), Muguga North. Mungube is a veterinarian by profession is holder of doctorate degree in Veterinary Parasitology. Mungube as a scientist has widely published in renowned international journals on a number of themes including but not limited to poultry disease, poultry parasites, mastitis, trypanosomiasis, tsetse flies and helminths over his over 18 years working experience. His research interests include disease diagnostics and epidemiology, zoonoses, antimicrobial resistance, reproduction and production. Currently, Mungube is collaborating on a project to validate climate smart vaccines and surveillance tools for the control of Newcastle and infectious bursal disease (Gumboro), both economically important diseases for indigenous chickens.

Michael Muthamia is a biomedical researcher interested in analyzing host-pathogen interactions influencing the outcome of infections.

Sammy Ndungu is a senior research scientist at the Veterinary Science Research Institute (VSRI), Muguga North. He is a veterinarian by profession (BVM, University of Nairobi) and holder of a Masters degree in Applied Immunology from Brunel University and an MPhil degree in Veterinary Parasitology from the University of Edinburgh. He has worked at VSRI, Protozoology Division for the last 40 years where he has been involved in various programs seeking a cure and vaccine for East Coast fever.

Jahashi Saidi Nzalawahe is a lecturer in Veterinary Parasitology, under Department of Microbiology, Parasitology and Biotechnology, Sokoine University of Agriculture (SUA), Morogoro, Tanzania. He holds Bachelor of Veterinary Medicine (2005), Master degree of Veterinary Medicine (2012) and PhD in Veterinary Parasitology from Sokoine University of Agriculture (2016). His research interest has been on epidemiology and control of parasitic diseases of veterinary and medical importance particularly trematode infections in cattle, zoonotic tapeworms of dogs and ticks and tick borne diseases in cattle, dogs and cats. Managed to publish 12 peer-review publications.

Fredrick Ojiambo Obonyo gained his Bachelor of Veterinary Medicine(BVM) degree from University of Nairobi in 2006. He then gained his MSc. degree in Applied Veterinary Parasitology from University of Nairobi in 2011. He worked as a research assistant at the University of Nairobi Institute of Development studies from 2011-2014 before joining Chuka University as a part-time lecturer in the Department of Animal Science. He is currently a Tutorial Fellow at Meru University of Science and

Technology in the Department of Animal Science and pursuing his PhD in Applied Veterinary Parasitology at the University of Nairobi [Biography pulled from the Profile]

William Olaho-Mukani is a Ugandan citizen and holds BVM, from Makerere University (1978), an MSc and a PhD in Veterinary Pathology and Microbiology from the University Nairobi. He worked as Graduate Assistant at the Faculty of Veterinary Medicine, Makerere University (1978-79) and did his Post doc studies at Free University of Berlin (1989) and was a Research Fellow at International Livestock Research Institute (ILRI), Nairobi (1981 & 1984-1985). He is an experienced researcher and worked at the Kenya Trypanosomiasis Research Institute, Muguga for 15 years (1980-1996) where he rose to the post of Acting Deputy Director of Research, before moving to Uganda to be Director of Research at the Livestock Research Institute (LIRI) Tororo, for five years (1996-2001) and got the National Agricultural Research Award of Outstanding Director of Research in 2000. From LIRI, he was appointed Director of Animal Resources in Ministry of Agriculture, Animal Industry and Fisheries in Uganda where he worked for 10 years in that position (2001-2011). Later served as AU-IBAR Programme Leader for Transboundary Animal Diseases and Zoonoses Programme and also coordinated the Participation of African Nations in Standards Setting Orgaisations (PANSPSO) project for the East African Communityin 2011. He was the Regional Coordinator for the EU-Funded AU-IBAR programme for Reinforcing Veterinary Governance in Africa (VETGOV) Programme for the EAC Region (211-2017). He was a member Boards and Task Forces of various national, regional and international Organisations and served as the Secretary General, Vice-President and President of the World Organisation for Animal Heath (OIE) Regional Commission for Africa from 2002-2010. He was the FAO/WHO Team Leader & Consultant for The Development of an Integrated National Action Plan (INAP) for Avian Influenza for Namibia in 2009 and recently, the National FAO consultant for Assessment of Animal Health Challenges caused by Ticks and Tick- Borne Diseases in Uganda; the national FAO Consultant for drafting the National Action Plan for Management of Ticks and Tick-Borne Diseases (2019-2024) and the East African Community (EAC) Consultant for the Drafting the EAC Strategy for Transboundary Animal Diseases- 2020-2025. He is a member of the Editorial Boards of a number of journals, namely: the African Journal of Animal and Biomedical Sciences (AJABS); African Journal of Food, Agriculture and Nutrition Development (AJFAND) & the Bulletin of Animal Health and Production in Africa and was Founding Member of the editorial team of the Journal of Camel Practice and Research and has reviewed articles on trypanosomosis for the Veterinary Journal (The British Veterinary Journal). He has lectured to undergraduate and postgraduate students and supervised undergraduate, MSc and PhD students at the University of Nairobi, Moi University and Makerere University. He has authored and co-authored more than 150 articles in scientific journals and conference proceedings. Currently he is one of the Directors of the ARD Consult Kenya (Ltd) and Vet Biotec, Uganda Ltd and a part-time lecturer at the College of Veterinary Medicine Animal Resources and Biosecurity (COVAB)-Makerere University.

Pamela Olet is chief executive officer of KENTTEC.

Moses Olum is a trained and licensed veterinarian in Kenya and currently works as a Research Scientist at KALRO – VSRI. He has immense experience in mixed animal practice. On the research front, he is a member of the Protozoology section in his institute and has worked extensively on Ticks and tick-borne diseases on epidemiology, vaccinology, and drug therapies. He is a Masters of Veterinary Medicine graduate of the University of Nairobi.

Florence Njeri Wamwiri received her PhD in Biochemistry from Egerton University in 2004. She is currently working at the Kenya Agricultural and Livestock Research Organisation (KALRO) Biotechnology Research Institute, where her areas of expertise focus on vector biology and control. She has wide ranging experience in the entomological aspects of tsetse and trypanosomiasis control.

James Wanjohi is a Senior Research Scientist at KALRO,VSRI-Muguga. He has worked extensively in the field of ticks and tick-borne diseases especially on East Coast fever culminating in the development on an ECF vaccine. He has wide experience on parasite isolation, Diagnostics and epidemiology of ticks and tick-borne diseases. He has been involved in the field delivery of the ECF vaccine. Currently he is involved in the development and validation of an ECF and Heartwater vaccines. Dr Wanjohi holds an MSc in Tropical Veterinary Medicine from the University of Edinburgh, Scotland, UK.

Sebastian Waruri is a trained veterinarian in the University of Nairobi Kenya and currently works as a Research Scientist at KALRO – VSRI. He has immense experience in mixed animal practice. He is a member of the Protozoology section in his institute and has worked extensively on Ticks and tick borne diseases on epidemiology, immunisation and general herd health management. He is a holder of Bachelor of Veterinary Medicine graduate of the University of Nairobi and currently pursuing Master's degree in Veterinary epidemiology and economics from the same institution. Dr. Waruri have also worked as Company Veterinarian for Farmer's Choice Limited being in charge of the daily and herd health for over 20,000 pigs for over 10 years.

Index

A

B

C

D

E

G

H

I

K

L

M

N

O

P

R

S

T

V

www.ingramcontent.com/pod-product-compliance
Ingram Content Group UK Ltd.
Pitfield, Milton Keynes, MK11 3LW, UK
UKHW050952290926
15470UKWH00005B/1782

9 781799 864332